Advances in Industrial Control

For further volumes:
www.springer.com/series/1412

Pål Johan From · Jan Tommy Gravdahl ·
Kristin Ytterstad Pettersen

Vehicle-Manipulator Systems

Modeling for Simulation, Analysis, and Control

Springer

Pål Johan From
Department of Mathematical Sciences and Technology
Norwegian University of Life Sciences
Ås, Norway

Jan Tommy Gravdahl
Department of Engineering Cybernetics
Norwegian Univ. of Science & Technology
Trondheim, Norway

Kristin Ytterstad Pettersen
Department of Engineering Cybernetics
Norwegian Univ. of Science & Technology
Trondheim, Norway

ISSN 1430-9491 ISSN 2193-1577 (electronic)
Advances in Industrial Control
ISBN 978-1-4471-5462-4 ISBN 978-1-4471-5463-1 (eBook)
DOI 10.1007/978-1-4471-5463-1
Springer London Heidelberg New York Dordrecht

Library of Congress Control Number: 2013947365

Printed on acid-free paper

Springer is part of Springer Science+Business Media (www.springer.com)

To our supportive families

Series Editors' Foreword

The series *Advances in Industrial Control* aims to report and encourage technology transfer in control engineering. The rapid development of control technology has an impact on all areas of the control discipline. New theory, new controllers, actuators, sensors, new industrial processes, computer methods, new applications, new philosophies..., new challenges. Much of this development work resides in industrial reports, feasibility study papers and the reports of advanced collaborative projects. The series offers an opportunity for researchers to present an extended exposition of such new work in all aspects of industrial control for wider and rapid dissemination.

In the original proposal for this monograph the authors Pål J. From, Jan T. Gravdahl and Kristin Y. Pettersen set themselves the ambitious objective of presenting a comprehensive modelling framework for the kinematics and dynamics of a general vehicle-manipulator system. This modelling framework was then to be used with analysis and simulation tools to study the control of specific system applications. The *Advances in Industrial Control* Series Editors are very pleased to welcome their monograph *Vehicle-Manipulator Systems: Modeling for Simulation, Analysis and Control* that reports the results of their successful research to the series.

A vehicle-manipulator system is defined as a robot comprising a base, able to move freely in a given environment, and one or more manipulator arms that are mounted on the base. Practical examples of these systems include: underwater remotely-operated vehicles (ROVs) with manipulator arms; field robots, which is a very large class of robots ranging from rover-type robots through to (domestic) service robots; and space robots.

To achieve the objective of creating a general model framework, the authors followed a thorough "top-down" approach that begins with theory and concludes with applications. The content of the monograph follows the structure of their methodology:

(i) a brief introduction to robot types—Chap. 1;
(ii) setting out of the mathematical concepts needed—Chap. 2;
(iii) development of the kinematical descriptions of rigid bodies, then a manipulator system and finally the vehicle-manipulator system—Chaps. 3–5;

(iv) progression to dynamical modelling of a rigid body, followed by that of a manipulator system and concluding with the vehicle-manipulator system—Chaps. 6–8, and investigation of some properties of the matrix versions of the dynamical modelling framework—Chap. 9; and
(v) specialisation and use of the modelling framework in specific real-world applications—Chaps. 10–13.

At the time of proposing the monograph, the target of the research seemed very ambitious, but the resulting monograph seems to have met all the authors' objectives. The "top-down" idea for a general vehicle-manipulator model was an exciting challenge. This methodology should be contrasted with a "bottom-up" approach that starts from a specific application or configuration and immediately builds in the particular physical characteristics and system constraints as the model is constructed. The *Advances in Industrial Control* monograph series has many works based on this approach but the top down approach has the advantage of generality, and the potential to cover many more application configurations through model specialisation.

The *Advances in Industrial Control* series has not published a purely robotics monograph since *Modelling and Identification in Robotics* by Krzysztof R. Kozłowski (ISBN 978-3-540-76240-9, 1998) and before that *Intelligent Seam Tracking for Robotic Welding* by Nitin R. Nayak and Asok Ray (ISBN 978-3-540-19826-0, 1993). It is therefore invaluable to have some profile in the robotics field once more, firstly with the publication of *Snake Robots* by Pål Liljebäck, Kristin Y. Pettersen, Øyvind Stavdahl and J. Tommy Gravdahl (ISBN 978-1-4471-2995-0, 2012) and then with this monograph on vehicle-manipulator systems.

In conclusion, this comprehensive treatment of robotic vehicle-manipulator systems enhances the *Advances in Industrial Control* series and should be of interest to a wide range of readers from both the robotics and the control communities.

Industrial Control Centre
Glasgow
Scotland, UK

M.J. Grimble
M.A. Johnson

Preface

During the last half century robotics has grown to become a mature technology with several and widespread applications. Due to the diversity in robotic applications, together with increasingly robust and sophisticated solutions, robotic technology has lead to both social and economic improvements that are apparent in our everyday life. In addition, several multi-billion industries are totally dependent on robotic solutions to keep production at competitive levels of price, quality, and efficiency.

The benefits on society that has arisen due to the increased use of robotic solutions are mainly found in manufacturing and assembly tasks on the factory floor using robotic manipulator arms. Examples include assembly of cars, pick-and-place tasks, packaging, spray paint and welding, assembly of electronic boards, and so on—all tasks that can be performed using a fixed-base manipulator arm bolted to the factory floor.

There is, however, a large potential in manufacturing also in locations outside the traditional factory floor. Distant locations and places that are hard or dangerous to reach for humans can also benefit from robotic solutions. In order to solve such tasks, the robots need to be both mobile—so that they can locomote themselves to the remote location—and they need to maintain their manipulation capabilities that have made robots so successful in factory manipulation tasks. In this book we will study this kind of robotic systems in great detail.

There are two main motivational factors that have encouraged the writing of this book. The first is that we believe the class of robotic systems that we will refer to as *vehicle-manipulator systems* deserves to be treated in detail as one topic. Even though the applications of these systems are many and diverse they can be described using the same framework, which makes this group of robots well suited to be treated in the form of a single book. Even though the applications of these systems are different in character, the modeling and control have several similarities. Particularly, the kinematic and dynamic coupling between the manipulator and the vehicle are present in all systems.

The second motivation for writing this book is the great impact that we believe these systems will have on the operation of remotely located fields or fields with difficult access that require some form of manipulation. In the same way that fixed-base robotic arms have revolutionized factory automation, it is our belief that mobile manipulators will change the way distant fields are operated in the future. The social and economic benefits of utilizing robotic technology in these fields will probably be as significant as the benefits that we have experienced from fixed-base manipulation over the last decades.

In addition to distant locations, vehicle-manipulator systems can also be used on or outside the factory floor in routine and time-consuming manipulation tasks. One example with great economic benefits is introducing agricultural robots in orchards and other fields, which will also relieve human workers from an ergonomically challenging working environment. Similarly, transportation and surveillance tasks can be performed more efficiently using robotic solutions in traditional factory production and manufacturing.

A final motivation for writing this book is to clarify some basic properties of vehicle-manipulator systems that have not been interpreted correctly in literature. When the dynamic equations are written in matrix form, we can identify several properties, especially of the inertia and Coriolis matrices, that greatly simplify the design of control laws. For example, the boundedness property of the inertia matrix is important both in simulation and in stability proofs of control laws. This property is always true for standard robotic arms, but not necessarily for vehicle-manipulator systems. The same is the case for the skew-symmetric property of the Coriolis matrix which is easily satisfied for single rigid bodies and standard robotic manipulators, but is not satisfied for the most commonly used representations of vehicle-manipulator dynamics. In this book we thus put great effort in clarifying for which representations these properties are in fact true. We also systematically derive the dynamic equations for which both these properties are satisfied.

After a brief introduction and a presentation of the most important concepts to be treated in the book in Chaps. 1 and 2, the first part of the book (Chaps. 3–5) treats the kinematics of rigid bodies, robotic manipulators, and vehicle-manipulator systems. We discuss several different representations of the state space and particularly representations that allow for robust formulations of the system kinematics and dynamics. Chapters 6–8 then discuss the dynamic equations of the same systems based on the formulations of the state space that we found in the previous chapters. Finally Chaps. 9–13 present a detailed treatment of different types of vehicle-manipulator systems aimed at different application areas such as underwater and space robots. In the introductory chapters on dynamics (Chaps. 6–8) only the rigid body properties of vehicle-manipulator systems are discussed. In the chapters that are dedicated to the different types of vehicle-manipulator systems we give a more specific presentation of these systems and include other considerations that apply only to that specific type of vehicle-manipulator system: for underwater systems we include buoyancy and gravitational forces as well as drag and added mass that come as a result of the mechanism being totally or partially submerged; for space manipulators we discuss how to choose the non-inertial reference frame and how to deal with a free-floating spacecraft in a free fall environment; for field robots non-holonomic constraints are discussed; and for a robotic manipulator mounted on a forced non-inertial base we show how the non-inertial motion of the base propagate to the manipulator dynamics.

There are several ways to read this book. The book can be used to give an introductory course in robotics by leaving out some of the final sections of each chapter, or as a more thorough treatment including several tools from differential geometry and Lie groups in order to obtain a more rigid formulation of the kinematics and dynamics.

Some sections that appear at the end of each chapter and large parts of Chap. 2 are not used in the subsequent chapters and can be left out. These sections discuss a more geometric approach to robot modeling using Lie groups and Lie algebras. The geometric formulations are important in order to derive a globally valid representation of the dynamics. The most common representations of vehicle-manipulator systems are not globally valid and we thus devote a large part of the book to show how to derive the dynamics without the presence of singularities. When working with vehicle-manipulator systems, this is a very important topic because the configuration space of the vehicle contains singularities when the standard vector representation is used to represent the position and velocity variables. There are many ways to solve this for a single rigid body, i.e., a vehicle with no manipulator attached, but for vehicle-manipulator systems the same approaches cannot be adopted directly. The geometric approach presented throughout the book can be used to derive the dynamics of single rigid bodies, robotic manipulators, and combinations of these without the presence of singularities.

The book can be read as an introductory course to robotics at bachelor level, or as a more advanced course including a geometric formulation on master/Ph.D. level:

Bachelor level

- For an overview of modeling of vehicle-manipulator systems and robotic systems in general, the reader can leave out the last sections of each chapter. The book can be used as an introductory course on robotics by including Chaps. 1–9 but leaving out the following sections:

  - Preliminaries: Sects. 2.5–2.10,
  - Kinematics: Sects. 3.3.6, 4.5.1, and 5.3.1,
  - Dynamics: Sects. 6.4.2–6.4.4, 7.6, and 8.3–8.3.3.

  The course will then start with the standard treatment of single rigid body modeling and modeling of robotic manipulators and multibody systems in general. The last part of the course on vehicle-manipulator systems will enhance the students understanding on state space representations, kinematic and dynamic coupling, and other aspects normally not found in standard textbooks on robotics.

Master/Ph.D. level

- For a more geometric treatment, the reader should include also the final sections of each chapter, starting with Chap. 2 (if not already familiar with differential geometry). In addition to the topics above, this will give an introduction to differential geometry and give a deeper understanding of how to derive well-defined and robust formulations of the kinematics and dynamics of general multibody mechanical systems.

Finally, the authors would like to express their gratitude to all those who have contributed to this book. First of all we would like to thank Professors Rolf Johansson, Olav Egeland, and Anton Shiriaev for their thorough feed-back on our early papers and the Ph.D. thesis of Pål Johan From. We would also like to show our gratitude to our colleagues at UC Berkeley, Vincent Duindam (now at Intuitive Surgical), Shankar Sastry, and Pieter Abbeel, who all have contributed by co-authoring several papers which first presented many of the results included in this book. They have also contributed with valuable discussions on several of the topics presented. We would also like thank Professor Stefano Stramigioli at University of Twente for discussions on the geometric approach used throughout this book. Finally we would also like to thank our students for proof-reading and valuable feedback.

Pål Johan From has received support from The Norwegian Nonfiction Writers And Translators Association (NFF).

Ås/Trondheim, Norway
December, 2012

Pål Johan From
Jan Tommy Gravdahl
Kristin Ytterstad Pettersen

Contents

Notation and Acronyms[1]

AUV	Autonomous underwater vehicle
BH equations	Boltzmann-Hamel equations
DEM	Dynamically equivalent manipulator
DoF	Degree(s) of freedom
SM system	Spacecraft-manipulator system
UAV	Unmanned aerial vehicle
URV	Underwater robotic vehicle
VM system	Vehicle-manipulator system
$\mathbb{R}$	The set of real numbers
$\mathbb{S}^1$	A circle
$\mathscr{H}$	The set of unit quaternions
H	An element of $\mathscr{H}$
$\mathscr{F}_0$	The inertial reference frame
$\mathscr{F}_b$	The reference frame of a single rigid body or the base of a manipulator
$\mathscr{F}_i$	The reference frame of link i or rigid body i
$\mathscr{O}_0$	The origin of $\mathscr{F}_0$
$\mathscr{O}_b$	The origin of $\mathscr{F}_b$
$\mathscr{O}_i$	The origin of $\mathscr{F}_i$
$\{e_x, e_y, e_z\}$	The Cartesian basis of an inertial reference frame
$\{e'_x, e'_y, e'_z\}$	The Cartesian basis of a non-inertial reference frame

[1]This book combines several different areas of mechanics, including applications as diverse as robotics, marine vessels, spacecraft and field robots, and combinations of these. The diversity in topics covered is challenging when it comes to notation, as different notation is used for the same quantities in the different areas. We have chosen the notation we find most appropriate for the different quantities, although this notation may differ from what the reader is used to. In general subscripts denote the frame that we want to describe while superscripts refer to the frame that this motion is observed from.

$\{n_x, n_y, n_z\}$	Unit vectors fixed in the same reference frame as the basis $\{e_x, e_y, e_z\}$ and with n_i pointing in the same direction as e_i for $i = x, y, z$
$\{E_x, E_y, E_z\}$	Basis elements in matrix form, given by the 3×3 matrix $\hat{E}_x$ where $E_x = [1\ 0\ 0]^\mathsf{T}$, etc.
$\{\hat{\upsilon}_x, \hat{\upsilon}_y, \hat{\upsilon}_z\}$	Basis elements in matrix form, given by the 4×4 matrix $\hat{\upsilon}_x$ where $\upsilon_x = [1\ 0\ 0\ 0\ 0\ 0]^\mathsf{T}$, etc.
$\{\hat{\omega}_x, \hat{\omega}_y, \hat{\omega}_z\}$	Basis elements in matrix form, given by the 4×4 matrix $\hat{\omega}_x$ where $\omega_x = [0\ 0\ 0\ 1\ 0\ 0]^\mathsf{T}$, etc.
$\mathscr{L}$	The Lagrangian
$\mathscr{K}$	Kinetic energy
$\mathscr{U}$	Potential energy
$\mathscr{M}, \mathscr{N}$	Topological space/manifold
O, U	Open sets
$\emptyset$	The empty set
$B(x, \rho)$	The open ball around the point x with radius ρ in $\mathbb{R}^n$
Φ	Homeomorphic mapping from $\mathbb{R}^n$ to $\mathscr{N}$
Ψ	Homeomorphic mapping from $\mathscr{N}$ to $\mathbb{R}^n$
(U, Ψ)	Coordinate chart
A, B	Matrices
$I, I_{n\times n}$	The $(n \times n)$ identity matrix
$\circ$	The group operator for group elements or the composite map for functions
$SE(3)$	The special Euclidean group
$SE(2)$	The special Euclidean group (planar)
$SO(3)$	The special orthogonal group
$SO(2)$	The special orthogonal group (planar)
$se(3)$	The Lie Algebra of $SE(3)$
$se(2)$	The Lie Algebra of $SE(2)$
$so(3)$	The Lie Algebra of $SO(3)$
$so(2)$	The Lie Algebra of $SO(2)$
$\mathscr{X}$	The Schönflies group
$\varkappa$	The Lie algebra of $\mathscr{X}$
C	The cylindrical group
c	The Lie algebra of C
$H(1)$	The helical group
$h(1)$	The Lie algebra of $H(1)$
G	A Lie group
g	An element of the Lie group G or a homogeneous transformation matrix
$\mathfrak{g}$	A Lie algebra
X, Y, Z	Elements of the Lie algebra $\mathfrak{g}$

g_{ab}	Homogeneous transformation matrix representing the position and orientation of $\mathscr{F}_b$ with respect to $\mathscr{F}_a$
$g^\vee$	The vector representation of the homogeneous transformation matrix g
X_{ab}^c	The twist representing the velocity of $\mathscr{F}_b$ with respect to $\mathscr{F}_a$ as viewed from $\mathscr{F}_c$
p_i	The direction of the linear motion for a prismatic joint i or the rotation axis for a revolute joint i
X_i^i	The body joint twist of joint i represented in $\mathscr{F}_i$
X_i	The spatial joint twist of joint i with respect to the base frame $\mathscr{F}_b$
X_i^e	The end-effector joint twist of joint i with respect to the end-effector frame $\mathscr{F}_e$
φ	Local position variable
$\dot{\varphi}$	The time derivative of the local position variable
$\mathscr{V}$	A vector space
$v_1, v_2 \in \mathscr{V}$	Elements of $\mathscr{V}$
F	A vector field
$x = [x_1\ x_2\ \cdots\ x_n]^\mathsf{T} \in \mathbb{R}^n$	A vector in $\mathbb{R}^n$
$y = [y_1\ y_2\ \cdots\ y_n]^\mathsf{T} \in \mathbb{R}^n$	A vector in $\mathbb{R}^n$
z	A complex number
$\dot{\gamma}$	Quasi-velocity
γ	Quasi-coordinate
$\bar{x}^a = [(x^a)^\mathsf{T}\ 1]^\mathsf{T} \in \mathbb{R}^4$	Homogeneous coordinates of a point $x \in \mathbb{R}^3$ fixed in the reference frame $\mathscr{F}_a$
$\bar{x}^a = [(x^a)^\mathsf{T}\ 0]^\mathsf{T} \in \mathbb{R}^4$	Homogeneous coordinates of a vector $x \in \mathbb{R}^3$ fixed in the reference frame $\mathscr{F}_a$
$p_{ab} = [x_{ab}\ y_{ab}\ z_{ab}]^\mathsf{T} \in \mathbb{R}^3$	The position of $\mathscr{O}_b$ with respect to $\mathscr{F}_a$
$s\phi, c\phi$	Short hand notation for $\sin\phi$ and $\cos\phi$
$\eta_1 = p_{0b} = [x_{0b}\ y_{0b}\ z_{0b}]^\mathsf{T} \in \mathbb{R}^3$	The position coordinates of a rigid body represented in the inertial frame
$\eta_2 = [\phi\ \theta\ \psi]^\mathsf{T} \in \mathbb{R}^3$	The Euler angle coordinates describing the orientation of a body in the inertial frame
$\eta = [\eta_1^\mathsf{T}\ \eta_2^\mathsf{T}]^\mathsf{T} \in \mathbb{R}^6$	The position and orientation variables of a rigid body in the inertial frame
$\dot{\eta} = [\dot{\eta}_1^\mathsf{T}\ \dot{\eta}_2^\mathsf{T}]^\mathsf{T} \in \mathbb{R}^6$	The linear and angular velocity variables of a rigid body in the inertial frame
$R_{ab} \in \mathbb{R}^{3\times 3}$	The rotation matrix describing the orientation of $\mathscr{F}_b$ with respect to $\mathscr{F}_a$
$\dot{\eta}_1 = [\dot{x}_{0b}\ \dot{y}_{0b}\ \dot{z}_{0b}]^\mathsf{T} \in \mathbb{R}^3$	The linear velocity of frame $\mathscr{F}_b$ with respect to $\mathscr{F}_0$ as viewed from $\mathscr{F}_0$

$v_{0b}^S = [u^0\ v^0\ w^0] \in \mathbb{R}^3$	The spatial linear velocity
$v_{0b}^B = [u\ v\ w] \in \mathbb{R}^3$	The body linear velocity
$\omega_{ab}^c = [\omega_x\ \omega_y\ \omega_z]^\mathsf{T} \in \mathbb{R}^3$	Vector of the angular velocity of frame $\mathscr{F}_b$ with respect to $\mathscr{F}_a$ as viewed from $\mathscr{F}_c$
$\omega_{0b}^S = [p^0\ q^0\ r^0]^\mathsf{T} \in \mathbb{R}^3$	Spacial angular velocity
$\omega_{0b}^B = [p\ q\ r]^\mathsf{T} \in \mathbb{R}^3$	Body angular velocity
$V_{ab}^c = [v_{ab}^c\ \omega_{ab}^c]^\mathsf{T} \in \mathbb{R}^6$	Twist variable describing the linear and angular velocities of frame $\mathscr{F}_b$ with respect to $\mathscr{F}_a$ as viewed from $\mathscr{F}_c$
$V_{0b}^S = [v_{0b}^S\ \omega_{0b}^S]^\mathsf{T} \in \mathbb{R}^6$	Spatial velocity twist
$V_{0b}^B = [v_{0b}^B\ \omega_{0b}^B]^\mathsf{T} \in \mathbb{R}^6$	Body velocity twist
$\tilde{V}_{0b}^B \in \mathbb{R}^m$	Body frame velocity variables for a rigid body with a configuration space with dimension m
$V_{0e}^S = [v_{0e}^S\ \omega_{0e}^S]^\mathsf{T} \in \mathbb{R}^6$	Spatial end-effector velocity twist
$V_{0e}^B = [v_{0e}^B\ \omega_{0e}^B]^\mathsf{T} \in \mathbb{R}^6$	Body end-effector velocity twist
$V_{0e}^0 = [v_{0e}^0\ \omega_{0e}^0]^\mathsf{T} \in \mathbb{R}^6$	The velocity of the end-effector frame $\mathscr{F}_e$ with respect to the inertial frame $\mathscr{F}_0$ as seen from $\mathscr{F}_0$
$V_{0\bar{i}}^S \in \mathbb{R}^6$	The spatial velocity twist of joint i
$V_{0i}^S \in \mathbb{R}^6$	The spatial velocity of the center of gravity of joint i
$V_{0\bar{i}}^B \in \mathbb{R}^6$	The body velocity twist of joint i
$V_{0i}^B \in \mathbb{R}^6$	The body velocity of the center of gravity of joint i
n	Number of robotic joints in a mechanism
m	Dimensionality of the vehicle's configuration space
r	Dimensionality of the configuration space of the end effector workspace
r_s	Controllable degrees of freedom of the vehicle
$q = [q_1\ q_2\ \dots\ q_n]^\mathsf{T} \in \mathbb{R}^n$	Generalized coordinates of a robot (Euclidean joints)
$\dot{q} = [\dot{q}_1\ \dot{q}_2\ \dots\ \dot{q}_n]^\mathsf{T} \in \mathbb{R}^n$	Generalized velocities of a robot (Euclidean joints)
$\ddot{q} = [\ddot{q}_1\ \ddot{q}_2\ \dots\ \ddot{q}_n]^\mathsf{T} \in \mathbb{R}^n$	Generalized accelerations of a robot (Euclidean joints)
$Q = \{Q_1, Q_2, \dots, Q_n\} \in \mathbb{R}^{4\times 4\times n}$	Robot joint positions (non-Euclidean joints)
$\zeta = [(V_{0b}^B)^\mathsf{T}\ \dot{q}_1\ \dot{q}_2\ \dots\ \dot{q}_n]^\mathsf{T} \in \mathbb{R}^{m+n}$	Joint velocities of a vehicle-manipulator system in terms of the body twist for vehicle configuration space $SE(3)$

$\dot{\zeta} = [(\dot{V}_{0b}^{B})^{\mathsf{T}}\ \ddot{q}_1\ \ddot{q}_2\ \ldots\ \ddot{q}_n]^{\mathsf{T}} \in \mathbb{R}^{m+n}$	Joint accelerations of a vehicle-manipulator system in terms of the body twist for vehicle configuration space $SE(3)$
$\tilde{\zeta} = [(\tilde{V}_{0b}^{B})^{\mathsf{T}}\ \dot{q}_1\ \dot{q}_2\ \ldots\ \dot{q}_n]^{\mathsf{T}} \in \mathbb{R}^{m+n}$	Joint velocities of a vehicle-manipulator system in terms of the body twist for vehicle configuration spaces that are subgroups of $SE(3)$
$\dot{\tilde{\zeta}} = [(\dot{\tilde{V}}_{0b}^{B})^{\mathsf{T}}\ \ddot{q}_1\ \ddot{q}_2\ \ldots\ \ddot{q}_n]^{\mathsf{T}} \in \mathbb{R}^{m+n}$	Joint accelerations of a vehicle-manipulator system in terms of the body twist for vehicle configuration spaces that are subgroups of $SE(3)$
$\xi = [\eta^{\mathsf{T}}\ q^{\mathsf{T}}]^{\mathsf{T}} \in \mathbb{R}^{6+n}$	Position vector for vehicle-manipulator systems in the inertial frame
$\dot{\xi} = [\dot{\eta}^{\mathsf{T}}\ \dot{q}^{\mathsf{T}}]^{\mathsf{T}} \in \mathbb{R}^{6+n}$	Velocity vector for vehicle-manipulator systems in the inertial frame
$\hat{\omega} \in \mathbb{R}^{3\times 3}$	The $\omega\times$ operator
$J_{b,p}(\eta_2) = R_{0b}(\eta_2) \in \mathbb{R}^{3\times 3}$	The Jacobian matrix relating the body linear velocities and the time derivatives of the position variables of a rigid body or vehicle
$J_{b,o}(\eta_2) = T_{0b}(\eta_2) \in \mathbb{R}^{3\times 3}$	The Jacobian matrix relating the body linear velocities and the time derivatives of the position variables of a rigid body or vehicle
$J_b(\eta_2) = \mathrm{diag}(J_{b,p}, J_{b,o}) \in \mathbb{R}^{6\times 6}$	The Jacobian matrix relating the body frame velocities and the time derivatives of the position and orientation variables of a rigid body or vehicle
$J_{b,p}^{S}(\eta_2) = R_{0b}(\eta_2) \in \mathbb{R}^{3\times 3}$	The Jacobian matrix relating the spatial linear velocities and the time derivatives of the position variables of a rigid body or vehicle
$J_{b,o}^{S}(\eta_2) = T_{0b}^{S}(\eta_2) \in \mathbb{R}^{3\times 3}$	The Jacobian matrix relating the spatial angular velocities and the time derivatives of the position variables of a rigid body or vehicle
$J_b^{S}(\eta_2) = \mathrm{diag}(J_{b,p}^{S}, J_{b,o}^{S}) \in \mathbb{R}^{6\times 6}$	The Jacobian matrix relating the spatial frame velocities and the time derivatives of the position and orientation variables of a rigid body or vehicle
$J_{m,g}(q) \in \mathbb{R}^{6\times n}$	The body geometric Jacobian of the manipulator
$J_{m,g}^{S}(q) \in \mathbb{R}^{6\times n}$	The spatial geometric Jacobian of the manipulator
$J_{m,a}(q) \in \mathbb{R}^{6\times n}$	The analytical Jacobian of the manipulator
$J_i(q) \in \mathbb{R}^{6\times(6+n)}$	The link Jacobian of link i of the manipulator
$\bar{J}_i(q) \in \mathbb{R}^{6\times(6+n)}$	The link Jacobian of the center of gravity of link i of the manipulator

$J_g^S(\eta_2) \in \mathbb{R}^{6\times(6+n)}$	The spatial geometric Jacobian of the vehicle-manipulator system
$J_a(\xi) \in \mathbb{R}^{6\times(6+n)}$	The analytical Jacobian of the vehicle-manipulator system
$J_w(\xi) \in \mathbb{R}^{6\times(6+n)}$	The workspace Jacobian of the vehicle-manipulator system
$J_{gi}^S(\xi) \in \mathbb{R}^{6\times(6+n)}$	The spatial geometric Jacobian of the center of gravity of link i of the vehicle-manipulator system
$J_{gi}^B(q) \in \mathbb{R}^{6\times(6+n)}$	The body geometric Jacobian of the center of gravity of link i of the vehicle-manipulator system
$\bar{J}_{gi}^S(\xi) \in \mathbb{R}^{6\times(6+n)}$	The spatial geometric Jacobian of link i of the vehicle-manipulator system
$\bar{J}_{gi}^B(q) \in \mathbb{R}^{6\times(6+n)}$	The body geometric Jacobian of link i of the vehicle-manipulator system
$J_{ge}^S(\xi) \in \mathbb{R}^{6\times(6+n)}$	The spatial geometric Jacobian of the end effector of the vehicle-manipulator system
$J_{ge}^B(q) \in \mathbb{R}^{6\times(6+n)}$	The body geometric Jacobian of the end effector of the vehicle-manipulator system
$J_{gi}(q) \in \mathbb{R}^{6\times(6+n)}$	The link geometric Jacobian of link i of the vehicle-manipulator system
$S(Q,\varphi)$	Velocity transformation matrix in local coordinates
$I_i \in \mathbb{R}^{6\times 6}$	Inertia matrix of link i in $\mathscr{F}_i$
$\mathrm{d}_i \in \mathbb{R}^{3\times 3}$	Moment of inertia tensor of link i in $\mathscr{F}_i$
m_i	Mass of link i
$m_t = \sum_i m_i$	Total mass of the mechanism
M, $M(x)$	Inertia matrix of the mechanism
$C(v)$, $C(x,v)$	Coriolis matrix of the mechanism
N, $N(x)$	Potential terms
M', $M'(x)$	Inertia matrix of the DEM
$C'(v)$, $C'(x,v)$	Coriolis matrix of the DEM
r_i	The vector connecting the center of mass of link i with joint $i+1$ of the space manipulator
l_i	The vector connecting joint i with the center of mass of link i of the space manipulator
W_i	The vector connecting i with joint $i+1$ of the DEM
l_{ci}	The vector connecting joint i with the center of mass of link i in the DEM
o	Higher order terms

Chapter 1
Introduction

Automation may be referred to as the technology of substituting humans with machines in manufacturing processes. This includes the physical handling of objects as well as information processing, supervising the process and decision making. There is a wide range of applications for which robots are used in the manufacturing process in factory automation, and over the last decades this has developed into a mature and robust technology. Some examples are pick-and-place tasks, packaging, spray paint and welding applications in the automotive industry, and assembly of electronic boards.

There is a strong desire to utilize robotic manipulators to automate processes other than the ones that can be found on the factory floor. Introducing robotic solutions to these new application areas is believed to bring a new era into robotics and automation. The 20th century was extremely important for industrial robots with the way they revolutionized factory automation is so many ways. In the same way, we will experience growth and advances in the use of mobile robots in the 21th century, most likely in the same way as we did with industrial robots in the previous century. Adding mobility to robots while maintaining the most important property of robotic manipulators, namely the ability to manipulate objects, will allow us to utilize automation in areas that are far more unstructured, more remotely located, and also in areas that are geographically larger than a factory floor. This will in turn lead to widespread use of robots in areas as diverse as operation of remotely located plants, sample taking in dangerous and remote areas, exploration, and surveillance. Furthermore, robots will also play an increasingly important role in household tasks, elderly and health care services, cleaning, security, and so on.

It is important to note, however, that automating processes in unstructured and sometimes harsh and distant environments brings out several challenging topics that are not present in factory automation: (1) First of all we need to have access to the process: if the process is sub-sea we need some kind of device to bring the robot to the ocean floor; if the task is to autonomously explore outer space we need to launch the robot into space and thereafter the robot needs to move around on its own; and for a personal assistant robot the robot needs to be able to locomote around the house. Mobile robots have been studied extensively for several different applica-

P.J. From et al., *Vehicle-Manipulator Systems*, Advances in Industrial Control,
DOI 10.1007/978-1-4471-5463-1_1, © Springer-Verlag London 2014

tions, see for example Siegwart and Nourbakhsh (2004), Dudek and Jenkin (2000), Tatnall et al. (2011), and Antonelli (2006). (2) The robot needs to carry the necessary equipment to perform the required tasks: for a robot exploring unknown areas this can be a camera or other sensing equipment, while for a robot performing maintenance operations on a remote site, this can be a manipulator arm with the necessary tools. (3) The robot needs be able to make intelligent decisions. As opposed to factory automation, the tasks are not always routine operations that are performed over and over again at pre-determined intervals. In an unstructured environment where unexpected events can occur, decisions need to be made to resolve a wide variety of problems. To complicate things further, there might not be a human supervisor present in the control loop to help the robot make the right decisions (Bishop 2006; Bertsekas 2001).

The applications of mobile robots are already numerous and there is no doubt that there will be a surge in the number of applications in the future. As more intelligent solutions to motion planning and decision making become available, the applications of these robots will become even more diverse and a widespread use of mobile robotics also in areas that we cannot possibly imagine today is likely. Furthermore, when these solutions become more robust and reliable they can be utilized to substitute humans in areas where this is not possible today. In fact, one of the most promising outcomes related to utilizing these systems is that they can prevent humans from being exposed to potentially hazardous working environments. This, together with the fact that these robots can explore and interact with sites where humans will never be able to go, illustrates the extreme potential of these systems.

From the discussion above it is clear that there is a need for mobile robots to be able to utilize robotic technology also in more remotely located plants and areas. This book is concerned with maybe the most promising way of bringing robots to distant areas, namely vehicle-manipulator systems. This is a class of mobile robots which is characterized by their ability to carry a robotic arm and therefore also manipulate objects. Bringing automation to distant fields and locations will be of great importance in operating existing and future fields and will lead to a whole new application area of mobile robots with robotic arms. In the same way, the exploration and research missions using mobile robots with robotic arms will take also these missions to a new level. The combination of mobility and manipulability is therefore what brings out the real potential of these systems, and this is also the main topic of this book.

1.1 Mobile Robotics

Mobile robots are characterized by their ability to move around freely in their environment, and unlike conventional industrial robots they can locomote within a given environment and possibly also between different environments. Mobile robots are normally classified by how they locomote, the environment in which they operate, the task they are set to perform, and so on. Today mobile robots are used for a wide variety of applications such as domestic cleaning, surveillance, exploration, transportation, and bomb disposal to name a few.

There is always a limitation in the area in which a mobile robot can locomote: an underwater vehicle will not operate on land and the operational area may be restricted by the umbilicals or communication ranges; a household robot will typically only operate in-door in a specific house; and space robots only operate in a free-floating environment. There are many types of mobile robots, with or without manipulator arms, but common for all of these is the property that allows them to move freely over a possibly large geographical area.

1.1.1 Autonomous Robots

An important aspect of mobile robots is how the robot navigates in its environment and solves its tasks. The level of involvement from the operator required to perform the tasks determines the degree of autonomy that the robot possesses. A robot can be totally controlled by an operator, or it can be completely autonomous. We will see that different mobile robots have different degrees of autonomy when it comes to navigation and decision making (Siegwart and Nourbakhsh 2004; Bekey 2005).

1.1.1.1 Manually Operated Robots

A manually operated robot is continuously operated by a human through an umbilical or wireless communication. The performance depends greatly on the communication delays between the controller and the robot, and stability is an issue with all teleoperated systems with large time delays. The operator will also get feedback from the robot. The most important feedback, and also the most common, is visual feedback. Most remotely operated robots are therefore equipped with cameras. Other important feedback mechanisms that the operator is supplied with are audio and haptic feedback.

1.1.1.2 Guarded Teleoperated Robots

It is also possible to implement certain algorithms to help the operator in teleoperation tasks. One example is obstacle avoidance where the robot continuously runs an obstacle avoidance algorithm without the operator having to actively avoid the obstacles. This is a great advantage in cluttered environments and where visual and haptic feedback is poor, and in many cases necessary to guarantee that the robot does not harm itself or its environment. The great advantage of these systems is that the operator can concentrate on the actual tasks that he is to perform while the control scheme deals with the secondary tasks such as collision avoidance, controlling the camera (visual feedback), positioning the base, and so on.

1.1.1.3 Partially Autonomous Robots

Some robots have a certain degree of autonomy and are able to perform specific tasks without being manually operated. What task it is to perform and when it is to be performed is specified by the operator. One example is a robot that is able to operate a part of a plant—for example opening and closing a valve—but not able to operate the entire plant without human intervention.

1.1.1.4 Autonomous Robots

An autonomous robot is a robot that can operate on its own without intervention from the operator. An autonomous mobile robot is typically able to

- collect information about its surroundings and environment and use this information to navigate and reach desired targets;
- work for an extended period of time without human intervention and guidance;
- move around safely in its environment;
- perform its mission or task without posing threat to human life or itself.

Numerous research groups work on making robots more autonomous and to remove the operator from the control loop. Increased autonomy will allow for robots to be utilized in more distant areas where traditional teleoperated systems will not work due to poor and slow communication. As the robots become more autonomous they can also perform routine tasks and substitute human workers in dangerous and repetitive tasks.

There are several reasons for not having a human in the control loop. First of all there might be considerable communication delays in the transmission of information between the robot and the operator which makes the performance of human control poor, or in the worst case unstable. Secondly, humans also respond slowly and often make wrong decisions. Furthermore, some environments, such as outer space or the dark side of the moon and planets, do not allow for continuous communication between the robot and the operator. Developing intelligent robotic solutions is thus necessary and therefore a very active research area that has undergone considerable advances over the last decades and will continue to do so also in the future.

1.1.1.5 Cognitive Robots

A cognitive robot is a fully autonomous robot with the additional capacity of learning and which itself—without the help of humans or operators—determines its goals and objectives. It is typically able to

- operate autonomously;
- solve problems that it has not encountered previously;
- learn and gain new experience based on previously encountered similar situations.

Cognitive robotics thus merges areas such as artificial intelligence and machine learning with robotics.

1.1.2 Other Classifications of Mobile Robots

There are many ways to classify mobile robots, for example by the environment in which the robots operate or by the way they locomote. We can also classify the robots by the tasks they are to perform. Mobile robotics is hence a very large field of research and includes a large variety of robots. As we will see, the robots are also very diverse in both the tasks they are to perform and the environment in which they operate. Some common classifications are the following:

- Classification based on the environment in which the robots locomote:
  - underwater vehicles (AUV),
  - unmanned aerial vehicles (UAV),
  - terrestrial robots,
  - space robots,
  - domestic robots.
- Classification based on how the robots locomote:
  - swimming/propelling robot,
  - flying robot,
  - legged robot,
  - wheeled robot,
  - robot on track or rail.
- Classification based on the task the robot is to perform:
  - surveillance robot,
  - maintenance robot,
  - sample-taking robot,
  - personal assistant robot,
  - bomb disposal robot,
  - domestic robot.

We will look at the different types of mobile robots in more detail in the next section and throughout the book.

1.2 Vehicle-Manipulator Systems

Robotic manipulators are, as the name suggests, designed to handle and manipulate objects with dexterity and accuracy. Unlike mobile robots, a standard industrial manipulator cannot, however, move around and can therefore only handle objects in its vicinity. In this sense, industrial robots are not very versatile. On the other hand, a vehicle that can move around freely—either by walking, rolling, flying, driving, or floating—possesses the mobility that industrial manipulators lack. A robotic manipulator mounted on a moving base possesses both the dexterity of robotic manipulators and the mobility of the vehicle and is thus a versatile solution which offers great potentials in a wide variety of applications (From et al. 2010).

1.2.1 Definition

A vehicle-manipulator system (VM system) is a robot that is intended to operate with dexterity in a workspace larger than that of a fixed-base manipulator, and that consists of two main parts:

1. A base, normally with actuation, which can move freely in a given environment.
2. One or more manipulator arms mounted on the base.

The base is what gives the VM system its ability to locomote over what is often a large geographical area. This base can be a vehicle in the normal sense, such as a car, satellite, ship, or underwater vehicle, or it can be another robotic device that allows for motion in a more restricted area such as a rail, gantry crane, or a Stewart platform. We will refer to all these systems as vehicle-manipulator systems.

We note that the geographical area is in many cases very restricted and will in general depend on the type of vehicle used. AUV-manipulator systems can for example only operate under water, and spacecraft-manipulator systems can only operate in space, which in both cases represent quite large but specific geographical areas and environments. In other cases, this area can actually be quite small, which is the case for a personal assistant robot that will only operate in one house, or maybe even only in one single room. Also, for a robot arm mounted on a gantry crane or a rail the workspace is limited to a very specific area. We thus adopt the rather wide interpretation of vehicle as any base on which we can mount a robotic arm. Common for all these systems is, however, that they can move freely in a restricted area and that the workspace of the VM system is considerably larger than for fixed-base manipulators.

We can imagine several different types of manipulator arms that we can attach to the base. Some examples are standard industrial manipulators and snake-like robotic arms. In general it will be smaller and lighter than industrial manipulators and ideally it should fold into the vehicle when it is not in use if friction, drag, or aerodynamics are present in the system. The most important property of the manipulator arm is that it has the necessary degrees of freedom to perform the tasks it is set to perform.

There are several different types of manipulator arms designed to be mounted on vehicles operating in different environments. The environment in which the robot operates determines the design and characteristics of the manipulator arm. We will see several examples of different arms in the last chapters where we discuss the different applications of VM systems in detail.

1.2.2 Classifications of Vehicle-Manipulator Systems

There are several different types of vehicle-manipulator systems. We will discuss some of these in more detail toward the end of the book where we devote one chapter to each of the systems that in our opinion constitute the most important present and future VM systems. In the following we present a short overview of some of the most important classes of vehicle-manipulator systems.

1.2.2.1 Underwater Robots

Autonomous underwater vehicles (AUVs) with robotic arms, or underwater robotic vehicles (URVs), represent one of the most promising applications of VM systems (Antonelli 2006). Underwater robots are used for a wide variety of applications ranging from inspection and surveillance of subsea fields and exploration and research missions, to interaction tasks such as installation, operation, and repair of underwater installations.

The underwater environment is not very accessible to humans. Therefore, as underwater robots can be used for operations both in deep waters and over large areas much focus has been given to these systems over the last decades. In other words, these robots can be utilized in places where humans cannot or do not want to operate. This includes operation at deep waters which pose extreme health dangers to human divers.

The petroleum industry has been driving the development of both intervention AUVs (I-AUVs) and underwater vehicles in general. Several underwater robots have been developed to perform inspection of underwater installations and also to explore the ocean floor. Most of these are remotely controlled by a human operator, often through an umbilical from a ship, but more autonomous systems are also being developed. Underwater vehicle-manipulator systems are discussed in detail in Chap. 10.

1.2.2.2 Free-Floating Space Robots

Another very important class of VM systems is spacecraft-manipulator systems (SM systems), see for example Ellery (2000). Introducing robotic arms to space applications has the potential of drastically reducing the time that astronauts are exposed to hostile environments. The operations performed by SM systems include assembling, repair, refueling, maintenance, and the daily operation of satellites and space stations. Due to the enormous risks and costs involved with launching humans into space, robotic solutions evolve as the most cost-efficient and reliable solution. In addition to maintenance and operation of space stations, spacecraft-manipulator systems can also be used for on-orbit maintenance and repair of satellites, connecting space stations with space shuttles, and so on.

Space manipulation does, however, involve quite a few challenges. Due to the enormous costs involved with transporting equipment to space, low weight is crucial in space robotics. As a result the links of a space robot are normally rather flexible, which complicates both the mathematical modeling and the control of these systems. This also drastically reduces the bandwidth of space robots.

One manipulator arm that is already operating is space is the Canadarm2 which is mounted on the International Space Station (ISS). The Canadarm2 is a 17.6 meter long robotic arm which is controlled by an astronaut on board the space station through visual feedback. The arm is mounted on the space station, which itself is a free-floating system, and can also move on a rail along the space station.

Space robotics is discussed in detail in Chap. 11.

1.2.2.3 Field Robots

The use of robots in remote areas such as the moon, distant planets, and the Arctic and Antarctic is only in its very infancy. Mobile robots with one or more manipulator arms have great potential when it comes to applications that range from maintenance and operation of remotely located sites to exploration and scientific research. Common for all these robots is that a manipulator arm is mounted on a wheeled or legged base. The robotic arm is typically used for sampling or manipulation.

Over the last decades we have seen the launching of wheeled robots to Mars with great success. Due to poor communication between the robot and the operator, these robots need to operate partially or completely on their own. First of all, this means that the robot must be able to perform the given task without continuous guidance from the operator. This may include tasks such as sample taking or monitoring a plant. Also, the robots need to make sure that they survive for a long enough period so that they can finish the task. For example, the Mars Exploration Rover needs to avoid getting stuck in sandy areas or between rocks, and it also needs to protect itself from sand storms and extreme temperature variations.

Another important application of field robotics is pruning and harvesting. An increasing demand for higher efficiency and quality in agricultural applications has led many research groups to investigate the use of robotic solutions. These are typically robots consisting of a wheeled platform with the possibility of installing a robotic arm for pruning the trees and plants and for harvesting.

A different type of field robots is a bomb disposal robot. Bomb disposal robots are either programmed or can be controlled to dispose of bombs and make them harmless. The bomb disposal robot consists of a moving base, which brings it to the bomb, and a manipulator arm, which is used to cut wires, etc. Normally the robot is equipped with a camera and is remotely operated through visual feedback.

The different types of field robots are discussed in more detail in Chap. 12.

1.2.2.4 Forced Base Motion

All the cases discussed above treat the case in which we have a vehicle that can be controlled in order to obtain our overall objective. There is, however, another class of vehicle-manipulator systems that consists of a base with a forced motion, i.e., we cannot control the motion of the base. We will denote a base motion where the base moves as a result of external forces, and not the actuators of the base, a *forced base motion*. Take for example a robotic manipulator mounted on a ship in high sea. In this case we can control the robot, but not the high-frequency motion of the ship. If the motion of this base is large this will add non-inertial terms to the dynamic equations of the manipulator. If these non-inertial forces are dominant, conventional motion planning and control approaches may lead to instability or poor robustness and performance. These non-inertial forces thus need to be included in the control and motion planning of the robot to guarantee robustness whenever they are significant.

Over the next decades we will probably see a dramatic change in the way ships are operated: the ships are becoming increasingly more autonomous, and the first completely autonomous ships are already on the drawing board. This will offer great savings to the shipping companies, will remove human workers from a potentially dangerous work environment and is also believed to be important in order combat the increasing problems with piracy around the world. Robust operation of these ships is extremely critical as they are loaded with large quantities of petrochemicals such as crude oil and gas in addition to the main freight. The use of robotic manipulators for surveillance, maintenance, and operation is thus subject to great challenges, especially when it comes to robustness. At times, ships sail at high sea where the inertial forces acting on the robot become very large. In Chap. 13 we thus discuss in detail the modeling, control and motion planning of robotic manipulators on a forced non-inertial base.

1.2.2.5 Domestic Robots

Domestic robots are normally programmed to perform domestic and household task on a regular basis. These robots are kinematically and dynamically equivalent to the field robots described above, but they differ in the way they operate and communicate with humans. While field robots to a certain extent are *controlled by* humans, the domestic robots *interact with* humans. This means that intelligent systems need to be developed to interpret the actions and commands of humans. The main objective of these robots is to perform various tasks to ease the everyday life of humans. This may be to perform repetitive household tasks or to help disabled or elderly in their everyday life.

For a domestic robot to be able to perform all the everyday tasks in a household environment, it is required to locomote around the house and manipulate a large variety of objects. This makes the domestic robot an advanced VM system which requires both mobility and extreme dexterity.

1.2.3 Challenges of Implementing Vehicle-Manipulator Systems

Vehicle-manipulator systems merge two areas of robotic research that are rather different in nature. The first system is the mobile base which is designed to be as mobile, versatile, and robust as possible. The second system is the robotic arm which is build with accuracy, repeatability, and speed as the main design objectives. Merging these systems into one vehicle-manipulator system thus leads to several open research questions.

First of all the manipulator arm needs to be built for environments far more challenging than what can be found on a factory floor and it must be given the same characteristics as the mobile base in terms of design, choice of materials, control, and intelligence. The robots are to operate in different types of environments and the design and choice of materials of the robotic arms therefore need to reflect the operating environment of the robots in different ways than for industrial robots.

Underwater manipulators for example need to be build with the same specifications and requirements as AUVs. This introduces aspects such as corrosion, wet environment, and high pressure to the design of the manipulator arm. Introducing robotic manipulators to an underwater environment therefore requires several advances compared to conventional industrial manipulators.

Another example is domestic and personal assistant robots. Mobile robots are often designed with safety as one of the most important design criteria. This is not the case with industrial robots which are extremely dangerous to humans and therefore only operate in separated cells where humans do not have access during operation. Utilizing robotic arms in a household environment, however, requires safe operation in cooperation with humans and we need to guarantee that the impact forces that the robot can impose on its environment do not do any harm. This has lead to the development of compliant robots that can interact with a dynamic environment safely.

Both the underwater and domestic robot arms are examples of areas that have undergone considerable advances over the last decades. There are, however, still several open research questions in the design and control of robotic arms that are to be used in environments other than the factory floor. One example is the level autonomy and intelligence implemented in the control of the robot arm. Much research has been done in this area in the setting of mobile robots and the same needs to be done with manipulator arms. Similarly, mobile robots are normally programmed to suppress environmental effects and disturbances. Also the manipulator arms need to be able to adapt to the environment in the same way.

In the same way that manipulator arms need to possess the characteristics normally associated with mobile robots, the mobile robots need to take on the characteristics of factory floor manipulators. The requirements for a mobile base that is to be used in manipulation tasks are thus far higher in terms of accuracy, repeatability, and speed.

Increased performance in terms of accuracy, speed, and repeatability requires good mathematical models of the vehicle-manipulator system. Accurate mathematical models are important both to obtain good simulation tools and for accurate control. Vehicle-manipulator systems are not naturally modeled using the standard framework multibody modeling and it is a great challenge to find a mathematical framework that is both robust, intuitive and computationally efficient. We will therefore put a lot of efforts into deriving the dynamics of vehicle-manipulator systems that possesses all these properties.

1.2.4 The Importance of Robust Solutions

Robustness is the property describing the ability to operate continuously without failure under a wide range of conditions (Soanes and Stevenson 2008). A mechanism, system or design is said to be robust if it is capable of coping well with variations in its operating environment with minimal or no damage, alteration, or loss of functionality. The design of robust robotic systems is one of the most challenging topics in robotics, and in unstructured environments with unpredictable and random variations this problem becomes even harder. This type of environments is what characterizes the environments in which most vehicle-manipulator systems operate.

It is impossible to design a system that is robust to all possible variations in the environment, so the main issue here is to design a robotic system that is robust enough so that utilizing it will lead to improvements over other existing systems. If operation is safer and more reliable using human workers nobody will spend a considerable amount of money robotizing the operation. Thus, robot manufacturers and developers need to search for areas where robots will improve both reliability and safety, and to do this in a cost efficient manner.

In this book we will see several different applications of mobile robots and common for all these is that robustness and safety are of vital importance. One class of robots for which robustness is important consists of robots operating in remotely located areas, such as a manipulator mounted on an autonomous underwater vehicle (AUV) operating subsea or a manipulator mounted on a spacecraft operating in space. The main strength of such systems is that they can operate in distant and harsh environments and far away from humans who may just observe the operations or remotely control them. Such systems thus need to be reliable as direct access may be impossible, very time consuming, or extremely costly. The vehicle-manipulator system may also be part of a larger system, such as a subsea oil installation or a space station. These systems will then depend on all the sub-systems—including the robotic systems—to work correctly to guarantee continuous operation and for assistance to the human operator both in routine operations and in emergency situations.

Another class of robots that will impose extreme safety requirements are personal assistant robots. These robots will operate close to or in direct contact with humans, performing tasks such as cleaning and tidying, but also lifting elderly or physically disabled people into and out of bed, as a few examples. Today, robotic arms are normally used in restricted access areas where humans do not have access to the operation. Considering the forces that these robots can apply to its surroundings, considerable advances are needed to guarantee the safe operation also in the presence of humans. Safe operation of robots in the presence of humans is currently a very active research area and over the last years we have seen considerable advance in this area, particularly in compliant behavior of robotic arms (Albu-Schäffer et al. 2007; Bischoff et al. 2010).

1.2.5 A Larger Perspective

It is our belief that the class of robots discussed in this book will play an important role when it comes to solving some of the most intriguing, challenging, and existential questions that the researchers of our time are facing. We believe this, not only based on the groundbreaking discoveries and achievements that have already come as a direct result of mobile robotics, but also due to the potential that lies in exploring areas that humans do not have access to without the help of mechanical devices. Improved vehicle-manipulator systems will therefore facilitate research in very diverse areas and be an important tool to researchers working in many and different fields of research.

Take into consideration that 71 % of the Earth lies under water and that only a fraction of this hidden world has been visited or explored by humans. This huge area has enormous potentials and we will gain much value from getting to know this area in more detail. To take full advantage of these resources we need robotic solutions that can take us to these extremely hostile places. In fact, robotic solutions have already been applied to these areas and has lead to the discoveries of several new and previously unknown animals and plants, new underwater formations and volcanoes, and even the chemosynthesis (Ballard and Hively 2002). We believe that improved robotic solutions will lead to innumerous new discoveries of existential importance to the planet Earth, its ecosystem and geology, and life on the planet.

Understanding the outer space is another area that has always been intriguing to scientists. Similarly to the underwater environment, outer space is extremely hostile and space traveling is very time consuming, so also here robotic solutions emerge as the most appropriate way to explore these distant places. Robots can travel over long distances and—if they are equipped with manipulator arms—they can be set to perform advanced research and exploration tasks. Advanced robotic solutions will therefore be important in getting to know and understand outer space, the most fundamental questions in physics, the existence of life in other parts of the universe, and how life arises and evolves over time.

References

Albu-Schäffer, A., Haddadin, S., Ott, C., Stemmer, A., Wimböck, T., & Hirzinger, G. (2007). The DLR lightweight robot: design and control concepts for robots in human environments. *Industrial Robot: An International Journal*, *34*(5), 376–385.

Antonelli, G. (2006). *Underwater robots. Motion and force control of vehicle-manipulator systems*. Berlin: Springer.

Ballard, R. D., & Hively, W. (2002). *The eternal darkness: a personal history of deep-sea exploration*. Princeton: Princeton University Press.

Bekey, G. (2005). *Intelligent robotics and autonomous agents. Autonomous robots: from biological inspiration to implementation and control*. Cambridge: MIT Press.

Bertsekas, D. P. (2001). *Athena scientific. Dynamic programming and optimal control, two volume set* (2nd ed.).

Bischoff, R., Kurth, J., Schreiber, G., Koeppe, R., Albu-Schaeffer, A., Beyer, A., Eiberger, O., Haddadin, S., Stemmer, A., Grunwald, G., & Hirzinger, G. (2010). The KUKA-DLR lightweight robot arm—a new reference platform for robotics research and manufacturing. In *Robotics (ISR), 2010 41st international symposium on and 2010 6th German conference on robotics (ROBOTIK)* (pp. 1–8).

Bishop, C. (2006). *Information science and statistics. Pattern recognition and machine learning*. Berlin: Springer.

Dudek, G., & Jenkin, M. (2000). *Computational principles of mobile robotics*. New York: Cambridge University Press.

Ellery, A. (2000). *Springer-praxis books in astronomy and space sciences. An introduction to space robotics*. Berlin: Springer.

From, P. J., Duindam, V., Pettersen, K. Y., Gravdahl, J. T., & Sastry, S. (2010). Singularity-free dynamic equations of vehicle-manipulator systems. *Simulation Modelling Practice and Theory, 18*(6), 712–731.

Siegwart, R., & Nourbakhsh, I. R. (2004). *Introduction to autonomous mobile robots*. Scituate: Bradford Company.

Soanes, C. & Stevenson, A. (Eds.) (2008). *Concise Oxford English dictionary*. Oxford: Oxford University Press.

Tatnall, A. R. L., Farrow, J. B., Bandecchi, M., & Francis, C. R. (2011). *Spacecraft system engineering* (4th ed.). Chichester: Wiley.

Chapter 2
Preliminary Mathematical Concepts

In this chapter we present several fundamental mathematical concepts that will be used to describe rigid body motion, which is the main topic of this book. A rigid body motion is a motion that preserves the distance between points in the rigid body, i.e., the body is not deformed and it can move freely in space or subject to a set of constraints on the admissible velocities and attainable positions. The main tool that we use to quantify the motion of a rigid body is that of *coordinate systems*. A coordinate system provides us with a measurement system which allows us to measure quantities such as position and orientation, angular and linear velocity, and angular and linear acceleration. We can also describe the forces and torques that act on the rigid body using coordinate systems.

To describe the motion of a rigid body with respect to some reference or observer, we use *reference frames*. A reference frame is a collection of points for which the distance between any two points is constant. It is thus natural to identify a reference frame with each rigid body in the system. In addition we choose a reference frame which we consider fixed in space. This inertial reference frame is often referred to as the world or Newtonian frame and is used as a reference for all other reference frames and thus also the rigid bodies in the system. The inertial reference frame is characterized by the requirement of zero acceleration and we will in most cases also assume zero velocity. On the other hand, a reference frame that accelerates with respect to the world frame is non-inertial, or non-Newtonian. Kinematics is the study of how these reference frames move with respect to the inertial frame and relative to one another. We will attach a non-inertial reference frame to every rigid body in the system. The framework of reference frames thus allows us to describe the positions, orientations, velocities, and accelerations of rigid bodies.

It is important to note that reference frames and coordinate systems are separate entities. While the reference frame determines *what* we want to measure, the coordinate systems determines *how* we want to measure the different quantities. Reference frames and coordinate systems are discussed in more detail in Sects. 2.1 and 2.2, respectively.

The kinematics of a rigid body is the study of the geometry of motion without regard to what causes the motion. Using the framework described above, rigid bodies

P.J. From et al., *Vehicle-Manipulator Systems*, Advances in Industrial Control,
DOI 10.1007/978-1-4471-5463-1_2, © Springer-Verlag London 2014

are naturally described using reference frames and we will study in great detail how one reference frame relates to another. In the setting of vehicle-manipulator systems, there is a fundamental difference between how the reference frame of a rigid body relates to the inertial frame on one hand, and on the other hand how the reference frames attached to each link in a robotic manipulator relate to each other. Both these transformations describe the location of one frame with respect to another frame, but the difference lies in what motions the transformations allow: For a general rigid object in the 3-dimensional Euclidean space, the transformation from the inertial frame is given by a free motion in six degrees of freedom, while for the links in a robotic manipulator the possible motions are restricted by kinematic constraints and the transformation from one link to another is given by a subgroup of $SE(3)$. For conventional robotic manipulators, the transformations allowed by the joints are the 1-parameter subgroups representing pure rotational or pure translational motion.

To get a clear understanding of this difference, we will distinguish between two types of transformations: Euclidean and non-Euclidean transformations. Whether a transformation is Euclidean or not depends on what type of coordinates that can be used to describe the transformation. More specifically, if the velocity variables can be written simply as the time derivative of the position variables, we will say that the transformation is Euclidean. For this type of transformations we can write the position as a vector $x \in \mathbb{R}^n$ and the velocity as a vector $v \in \mathbb{R}^n$ where $v = \dot{x} = \frac{\mathrm{d}x}{\mathrm{d}t}$. If the coordinates can be written in this way, we will see that this greatly simplifies the kinematic relations between two reference frames and the kinematics takes a very simple form. The configuration space of a standard robotic manipulator with 1-DoF joints can always be written in this way.

Unfortunately, there exists a rather large group of transformations that cannot be written in this form. This includes transformations such as pure rotational motion, representing for example the attitude of a rigid body, and free transformations of a rigid body in $\mathbb{R}^3$, which is the combination of three degrees of freedom translational motion and three degrees of freedom rotational motion. Take for example the attitude of a rigid body. In this case we cannot find a vector $\Theta \in \mathbb{R}^3$ describing the orientation of the body and at the same time find a vector $\omega \in \mathbb{R}^3$ describing the angular velocities for which $\omega = \dot{\Theta}$. In particular, this is not possible using a minimal representation in three coordinates. In this case we need to find a velocity transformation matrix $J(\Theta)$ so that the angular velocity relates to the orientation through the linear relation $\dot{\Theta} = J(\Theta)\omega$.

Finding a suitable mathematical representation of the velocity transformation matrix $J(\Theta)$ is challenging and as a result this transformation is often singularity prone. This means that at isolated points in the configuration space some elements of $J(\Theta)$ go to infinity. This singularity is a *mathematical singularity*. This means that it arises due to a choice of coordinates and not as a result of the physical properties of the system. This kind of singularities should always be avoided because it leads to unstable behavior when implemented in a simulation software or as a control law.

The velocity transformation matrix gives the relation between the velocity of the rigid body expressed in the body or spatial frame and the time derivative of the position coordinates. We will show how to find this transformation in several different ways. First we will find the relation using the Euler angles, and then we use a more geometric approach for which it is possible to derive this expression without the presence of singularities. Finding these relations is fundamental when deriving the rigid body kinematics, which is treated in detail in Chap. 3.

The remainder of this chapter is organized as follows: Reference frames and coordinate systems are discussed in Sects. 2.1 and 2.2. Section 2.3 describes in detail the difference between Euclidean and non-Euclidean transformations and Sect. 2.4 gives a short introduction to quasi-coordinates which can be used to represent transformations such as the one given by the velocity transformation matrix. We will also use this concept throughout the book to represent this relation in a mathematically sound way. To do this, however, we need some background on manifolds and Lie groups. Lie groups present us with a powerful mathematical tool for representing rigid body transformations in a geometrically meaningful way. Using this formalism we can write the position and velocity variables in a form that allows us to obtain a singularity free formulation of the kinematics and dynamics of these systems. A short introduction to these topics is presented in Sects. 2.5–2.9.

2.1 Reference Frames

A reference frame is a collection of points for which the distance between any two points is constant at all times (Rao 2006). Even though, as we will see from the definition, any choice of three or more non-collinear points can be used as a reference frame, the easiest way to think of a reference frame is to think of a rigid body. It is also in this context that we will use reference frames: first to identify a reference frame with a rigid body and then to study the motion of this reference frame with respect to an inertial reference frame. Reference frames and rigid bodies are thus used interchangeably throughout the book.

2.1.1 Inertial Reference Frames

The first reference frame that we need to identify is the inertial reference frame, also referred to as the world frame. This is chosen such that its points can be used as a reference for all other reference frames. An inertial reference frame is one whose points do not accelerate. There are two types of inertial reference frames depending on whether the points are fixed in space or move with constant velocity.

Choosing a set of points that are absolutely fixed in space and time can of course be done in several different ways. Assume that we choose a reference frame that is fixed at the surface of the Earth. This can be considered fixed in space and time for all objects on the surface of the Earth and will thus serve as a reference for observing the motion of these objects. On the other hand, if we want to observe the motion of the planets in the solar system relative to the sun, the Earth fixed reference frame is not inertial. In this case the inertial reference frame should be chosen as a

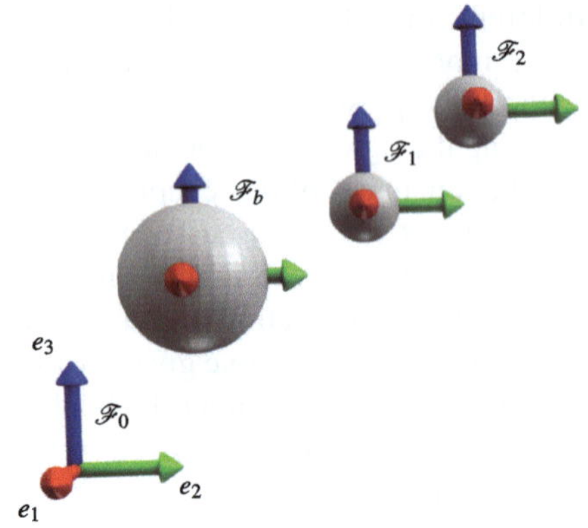

Fig. 2.1 The inertial reference frame $\mathscr{F}_0$, a principal non-inertial reference frame $\mathscr{F}_b$, and two other frames $\mathscr{F}_1$ and $\mathscr{F}_2$. The reference frame $\mathscr{F}_0$ has basis vectors e_1, e_2, and e_3

set of points on the sun, or alternatively at some distant star. How we choose the inertial reference frame thus depends on what we want to observe. For most of the applications that we will encounter, a reference frame fixed to the Earth will serve as an inertial frame.

We can also choose a reference frame attached to a system that moves relative to the Earth. Assume for example that we choose a set of points that orbit the Earth slowly at a constant velocity. This can be points on a space station or "imaginary points" that orbit the Earth with constant velocity which allow us to observe the motion of objects relative to a spacecraft or a satellite. From the perspective of the satellite and its nearby objects, such a reference frame will be fixed in space and time. This frame will thus also serve as an inertial reference frame.

2.1.2 Non-inertial Reference Frames

Once an inertial reference frame is chosen, we can attach a reference frame to each rigid body in the system. These reference frames will then accelerate with respect to the inertial frame and are thus non-inertial. We will use these reference frames to observe the motion of each rigid body *relative to the inertial frame*. In addition, by attaching a reference frame to each rigid body in the system it is also possible to describe the motion of a rigid body relative to the other rigid bodies in the system. Robot kinematics is the study of how these reference frames move with respect to the inertial frame and each other. This will be discussed in more detail in the subsequent chapters.

We will use $\mathscr{F}_0$ to denote an inertial reference frame. To observe the motion of a single rigid body, or a vehicle, we attach a reference frame $\mathscr{F}_b$ to this body. If more rigid bodies are present, for example the links of a robotic arm, we will attach one reference frame to each link, denoted $\mathscr{F}_i$ for $i = 1, \dots, n$ where n is the number of links in the robot. This is illustrated in Fig. 2.1.

2.2 Coordinate Systems

Reference frames provide us with a mean to observe motions of rigid bodies. However, we also need a way to describe this motion mathematically. One way to quantify the motion of a reference frame in mathematical terms is to choose a coordinate system that we attach to the reference frame. A coordinate system is determined by two choices:

1. We need to choose a point $\mathcal{O}$, called the *origin*, that is fixed in the reference frame.
2. We need to choose a basis in which we can represent vectors in $\mathbb{R}^3$. The basis must be a set of three linearly independent "directions" that are also fixed in the reference frame.

Choosing the origin is quite straight forward. However, if we have any information about the motion of the rigid body, we can greatly simplify the expressions if we make a well considered choice. For example, if there are points in the rigid body that do not move for the given motion, one of these points might be a smart choice. This is for example the case for a robotic link that rotates around a fixed axis and thus all the points on this axis will not move for a motion that satisfies the joint constraints. On the other hand, if the origin is chosen at the center of gravity and the basis vectors are chosen as the principal axes of inertia, this will simplify the dynamics because the inertia matrix can be written in a very simple form.

There are many ways to choose a basis $\{e_1, e_2, e_3\}$ for the coordinate system. In fact, any set of three linearly independent vectors can be chosen, but the most common representation is a *right-handed orthonormal basis*, i.e., a set of unit vectors that are mutually orthogonal and form a right-handed system. We will denote the scalar product (dot product) of two vectors x and y as $x \cdot y$ and the vector product as $x \times y$. Formally, a right-handed orthonormal basis is defined as a basis that satisfies the following three properties:

Property 2.1 (Unit vectors) The basis vectors e_1, e_2, and e_3 are unit vectors if

$$\begin{aligned} e_1 \cdot e_1 &= 1, \\ e_2 \cdot e_2 &= 1, \\ e_3 \cdot e_3 &= 1. \end{aligned} \tag{2.1}$$

Property 2.2 (Orthogonal vectors) The basis vectors e_1, e_2, and e_3 are mutually orthogonal if

$$\begin{aligned} e_1 \cdot e_2 &= e_2 \cdot e_1 = 0, \\ e_1 \cdot e_3 &= e_3 \cdot e_1 = 0, \\ e_2 \cdot e_3 &= e_3 \cdot e_2 = 0. \end{aligned} \tag{2.2}$$

Property 2.3 (Right-handed system) The basis vectors e_1, e_2, and e_3 form a right-handed coordinate system if

$$\begin{aligned} e_1 \times e_2 &= e_3, \\ e_2 \times e_3 &= e_1, \\ e_3 \times e_1 &= e_2. \end{aligned} \tag{2.3}$$

A right-handed orthogonal basis with basis vectors e_1, e_2, and e_3 is illustrated in Fig. 2.1.

2.2.1 Cartesian Basis

A *Cartesian basis* $\{e_x, e_y, e_z\}$ is a basis where the distance from a point $\mathscr{O}_0$, fixed in reference frame $\mathscr{F}_0$, to a point $\mathscr{O}_b$, fixed in reference frame $\mathscr{F}_b$, can be represented by a vector

$$p_{0b} = x_{0b} e_x + y_{0b} e_y + z_{0b} e_z \tag{2.4}$$

where x_{0b}, y_{0b}, and z_{0b} are the components of the position in the directions of the basis vectors e_x, e_y, and e_z, respectively, and the basis $\{e_x, e_y, e_z\}$ is fixed in $\mathscr{F}_0$. A Cartesian basis is defined such that each component that describes the position of the point $\mathscr{O}_b$ is the distance from the plane defined by the two other basis vectors in $\mathscr{F}_0$ to $\mathscr{O}_b$. For example, the distance from the point $\mathscr{O}_b$ to the plane that is spanned by e_x and e_y (the xy-plane), is given by the z-component, in our case z_{0b}. As a result, all points that have one common component lie in the same plane, and all points that have two equal components lie on the same line. The reference frame $\mathscr{F}_0$ in Fig. 2.1 has a Cartesian basis if we let $e_1 = e_x$, $e_2 = e_y$, and $e_3 = e_z$. The Cartesian basis is the most common choice of basis due to the convenient interpretation that each point can be projected into one of the three planes.

Let the position of the point $\mathscr{O}_b$ be given by (2.4) and thus described in reference frame $\mathscr{F}_0$. Then the velocity of $\mathscr{O}_b$ can be written using Cartesian coordinates as

$$v_{0b}^0 = \frac{\mathrm{d}p_{0b}}{\mathrm{d}t} = \dot{x}_{0b} e_x + \dot{y}_{0b} e_y + \dot{z}_{0b} e_z, \tag{2.5}$$

and the acceleration as

$$\dot{v}_{0b}^0 = \frac{\mathrm{d}v_{0b}^0}{\mathrm{d}t} = \ddot{x}_{0b} e_x + \ddot{y}_{0b} e_y + \ddot{z}_{0b} e_z, \tag{2.6}$$

where both the velocity and acceleration are observed from reference frame $\mathscr{F}_0$. It is important to notice that while a vector is independent of the reference frame in which it is observed, this is not the case with the rate of change of a vector. When writing the velocity and acceleration vectors we must thus always specify with respect to what reference frame the quantities are given. This is denoted by a

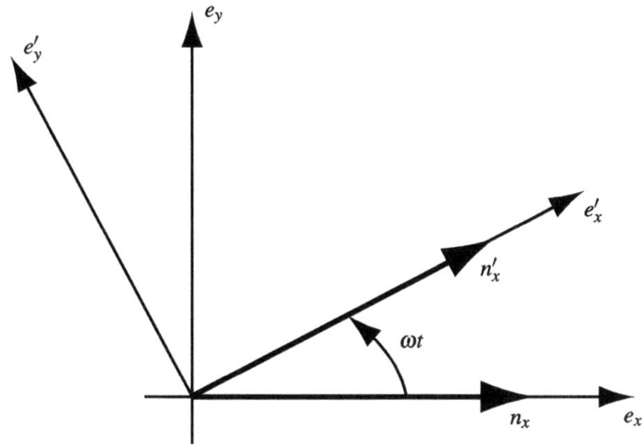

Fig. 2.2 The rotating reference frame in the plane as described in Example 2.1. The body frame $\{e'_x, e'_y\}$ rotates with constant velocity ω with respect to the inertial frame $\{e_x, e_y\}$

subscript so that v_0 is the velocity with respect to the inertial frame $\mathscr{F}_0$ and v_b is the velocity with respect to the body framez $\mathscr{F}_b$. This is most easily illustrated with an example:

Example 2.1 (Rao 2006) Consider the planar rotational motion illustrated in Fig. 2.2 where the body axes $\{e'_x, e'_y\}$ rotate with respect to a fixed set of inertial axes $\{e_x, e_y\}$ with a constant angular velocity ω.

We will now study how a unit vector n'_x, fixed in the set of axes $\{e'_x, e'_y\}$ and pointing in the direction of the e'_x, changes with respect to time. Of course, as this vector is fixed in the body frame $\{e'_x, e'_y\}$ the time derivative of n'_x is zero when written with respect to the body-fixed frame, i.e., $v_{bn'_x} = \frac{\mathrm{d}n'_x}{\mathrm{d}t} = 0$. However, if we represent n'_x in terms of the inertial basis $\{e_x, e_y\}$, then we have $n'_x = \cos(\omega t)e_x + \sin(\omega t)e_y$ which gives us $v_{0n'_x} = \frac{\mathrm{d}n'_x}{\mathrm{d}t} = (-\sin(\omega t)e_x + \cos(\omega t)e_y)\omega$. If we let n'_y be a vector of the same length as n'_x, but pointing in the direction e'_y, this can be written as $v_{0n'_x} = \frac{\mathrm{d}n'_x}{\mathrm{d}t} = n'_y\omega$ which is not identically equal to zero. In general, we thus obtain different rates of change of the same vector when it is written in different reference frames.

We see that the time derivative of the vector n'_x with respect to $\{e'_x, e'_y\}$ is zero. This is a general result and always true: The body velocity relative to the body frame is always zero. However, the body velocity relative to an inertial frame but *observed* from the body frame, is not identically equal to zero. On the contrary, the rate of change of a point moving with constant velocity on a circle is constant:

Example 2.2 The rate of change of a point n'_x parameterized by the $n'_x = \cos(\omega t)e_x + \sin(\omega t)e_y$ in the inertial frame $\{e_x, e_y\}$ is constant and given by

$$v_{bn'_x} = \begin{bmatrix} 0 \\ n_y \end{bmatrix} \omega \tag{2.7}$$

which represents a constant change in the direction of the y-axis of the body frame, i.e., $v_{bn'_x} = n'_y\omega$.

In this book we will normally assume that a Cartesian coordinate system is used to quantify the motion. Recall that reference frames and coordinate systems are separate entities, and we thus need to identify both a reference frame and a coordinate system with each rigid body in the system. We will often use the expression *coordinate frame* as short for a reference frame with a Cartesian coordinate system. When we say coordinate frame we thus refer to a reference frame and it is implicitly understood that a Cartesian coordinate system is used to quantify the motion.

2.3 Euclidean and Non-Euclidean Transformations

The framework presented in this book is especially suited for modeling robotic systems with joints or transformations that cannot be described by simple one degree of freedom motions. A vehicle-manipulator system is one such system due to the non-Euclidean configuration space of the vehicle. In this setting the difference between Euclidean and non-Euclidean transformations is very important so we start with the formal definitions. We first need to define the terms generalized coordinates and generalized velocities.[1]

Definition 2.1 (Generalized coordinates) A set of coordinates which uniquely describes the configuration of a body, or system of bodies, is called the generalized coordinates of the system.

The generalized coordinates thus uniquely determine the configuration of the system with respect to some initial configuration. This set is not unique in the sense that there are many different ways to choose these coordinates, but there is normally a set that will allow for an easier representation of the configuration space and a deeper physical insight. For robotic manipulators with 1-DoF joints the joint positions are normally chosen. The minimum number of independent generalized coordinates needed to describe the configuration of a system is known as the *degree of freedom* or *mobility* of the system.

The difference between Euclidean and non-Euclidean transformations lies in the way the velocity variables can be represented. We will start with generalized velocities:

[1] We note that there are many ways to define generalized velocity and generalized coordinates, depending on the application, and many different definitions can be found in literature. We will use the definitions presented in this section throughout the book. Note also that even though the correct usage of the term "configuration" refers to both position and orientation of the rigid body, we will sometimes use the term position to describe the configuration state of the system when there is no ambiguity.

Definition 2.2 (Generalized velocities) A generalized velocity is a velocity variable $\dot{x}_i$ associated with a generalized coordinate x_i defined as $\dot{x}_i = \frac{\mathrm{d}x_i}{\mathrm{d}t}$.

For each generalized coordinate x_i, there is a corresponding generalized velocity $\dot{x}_i$ that is simply the time derivative of the generalized coordinate. The generalized coordinates x together with the generalized velocities $\dot{x}$ represent a very convenient way to write the state space of a system. We note, however, that it is not always possible to find a set of generalized velocities to describe the velocity state of the system in this way, i.e., by the time derivative of the position variables. In this case it is thus necessary to describe the velocity state more generally without writing them as the time derivative of the generalized coordinates of the system. This is given in the following:

Definition 2.3 (Quasi-velocities) A quasi-velocity is a velocity variable $\dot{\gamma}$, which uniquely describes the velocity state of the system.

We note that we do not require $\dot{\gamma} = \frac{\mathrm{d}x}{\mathrm{d}t}$ to hold for a set of coordinates x describing the position of the system. Hence, the corresponding quasi-coordinates γ cannot be interpreted as the position variables of the system. We will study quasi-velocities in more detail in the next section.

Generalized velocities are easier to work with than quasi-velocities because they are in the form of Definition 2.2. Whenever we can find a set of coordinates for which Definition 2.2 is true, we chose to work with these coordinates. The main reason for this is that we in this case can apply Lagrange's equations directly to find the dynamics of the system. For a robotic manipulator the joint velocities are generalized velocities, and the state space is then given by $(q, \dot{q})$.

On the other hand, when we cannot find a set of coordinates that allows us to find generalized velocities in this way, we need to write the velocity state of the system in terms of quasi-velocities. In this case the velocities and positions do not simply relate by the time derivative, and we need to find a different relation between the two. This is the case for most vehicles that we will encounter, for which the state space is given by $(x, \dot{\gamma})$. This is discussed in detail later.

We now turn to the definition of Euclidean and non-Euclidean transformations. In short, a transformation gives a mapping between two different frames or the time evolution of a frame with respect its initial configuration. A Euclidean transformation is a transformation that allows us to write the velocity state in terms of generalized velocities, while a non-Euclidean transformation requires the velocity state to be written in terms of quasi-velocities. A Euclidean transformation is defined as follows:

Definition 2.4 (Euclidean transformation) A transformation is Euclidean if it can be parameterized in terms of generalized coordinates and a corresponding set of generalized velocities, i.e. the position variables are written as $x \in \mathbb{R}^n$ and the velocity variables as $v = \dot{x} \in \mathbb{R}^n$ where $\dot{x} = \frac{\mathrm{d}x}{\mathrm{d}t}$ (as defined in Definition 2.2).

Note that we only require that there exists (at least one) parameterization which allows for generalized velocities in the form of Definition 2.2. We will say that a joint is Euclidean if the transformation from one joint position to another is Euclidean. All 1-DoF joints are Euclidean and thus also the most commonly found robotic joints. Also joints with only translational motion are Euclidean. Similarly we will say that the state space is Euclidean if its elements can be written in vector form in this way, i.e., a vector $x \in \mathbb{R}^n$ representing the position variables and a corresponding vector $v = \dot{x} = \frac{\mathrm{d}x}{\mathrm{d}t}$ representing the velocity variables.

Definition 2.5 (Non-Euclidean transformation) A transformation is denoted non-Euclidean if it cannot be parameterized in terms of generalized coordinates and a corresponding set of generalized velocities, i.e. the position variables are written as $x \in \mathbb{R}^n$ and the velocity variables must be written as $\dot{\gamma} = S(x)\dot{x}$ for some transformation matrix $S(x) \neq I$ so that $\dot{\gamma} \neq \frac{\mathrm{d}x}{\mathrm{d}t}$.

A spherical joint which represents the attitude of a rigid body is thus a non-Euclidean joint. We see this if we represent the position variables as the Euler angles. In this case the state variables cannot be written in vector form where the velocity is the time derivative of the position. Because it is not possible to find a set of variables that describes the orientation of a rigid body and whose rate of change describes the instantaneous velocities, the transformation is non-Euclidean.

There is another way to distinguish Euclidean from non-Euclidean transformations. A Euclidean transformation is a transformation which can be written in terms of position variables x and velocity variables $\dot{x}$ for which the relation

$$x(t + \Delta t) = x(t) + \dot{x}(t)\Delta t \tag{2.8}$$

makes sense, i.e., $x(t + \Delta t)$ can be interpreted as a position variable in the same sense as $x(t)$.

For non-Euclidean transformations, on the other hand, this relations does not represent a new position of our system. One example is the Euler angles where

$$\begin{bmatrix} \phi(t+\Delta t) \\ \theta(t+\Delta t) \\ \psi(t+\Delta t) \end{bmatrix} = \begin{bmatrix} \phi(t) \\ \theta(t) \\ \psi(t) \end{bmatrix} + \begin{bmatrix} \dot{\phi}(t) \\ \dot{\theta}(t) \\ \dot{\psi}(t) \end{bmatrix} \Delta t \tag{2.9}$$

is meaningless unless Δt approaches zero. An element of $SO(3)$ therefore describes a non-Euclidean transformation.

We note that it is of course always possible to write the velocity variables as the time derivative of the position variables, but whenever this set of velocity variables does not represent a geometrically meaningful quantity, as in (2.9) we will say that the system is non-Euclidean and use quasi-velocities to describe the velocity of the system.

2.4 Quasi-coordinates and Quasi-velocities

As we have learned from the previous section, we sometimes need to use quasi-velocities to describe the velocity state of the system. One type of systems for which this is true is the class of rigid body motions in the 3-dimensional Euclidean space. From Definition 2.5 a transformation of a rigid body in this space is non-Euclidean because we need to include rotational motion in three degrees of freedom. We will encounter motions of this kind when dealing with single rigid bodies and the vehicle in VM systems. In the setting of this book, this kind of configuration space is thus very important and in this section we address in more detail how we can use quasi-velocities to describe the velocity state of these systems.

For systems with non-Euclidean configuration spaces, it is often easier to formulate the dynamic equations in terms of velocity variables that cannot be written simply as the time derivative of the position variables. For example, when dealing with angular motion, the angular velocity is not the rate of change at which a rotation angle changes, except for the planar case. There is thus no finite change in orientation that corresponds to the angular velocity. One way to deal with the fact that the integral of the velocity variable has no physical interpretation is to introduce quasi-velocities, often represented by $\dot{\gamma}$. When using quasi-velocities the corresponding quasi-coordinate γ does not have any useful physical interpretation itself whereas $\dot{\gamma}$ has the physical interpretation of $\mathrm{d}\gamma = \dot{\gamma}\mathrm{d}t$, i.e. it can be defined in terms of differential increments (Gingsberg 2007).

Remark 2.1 This observation presents us with an alternative definition of a Euclidean transformation, namely a transformation for which the integral of the velocity variables have the physical meaningful interpretation as the position variables of the system.

Because quasi-velocities are not restricted to be the time derivative of the position coordinates, we have more freedom in choosing the velocity variables. We will often denote the quasi-velocity by υ for a general velocity vector, ω for an angular velocity and by V for a twist. We will also use υ to describe pure translational velocity. We note that for ω and V the integral of the velocity variable does not have a geometric interpretation. To describe the configuration of the system we thus use generalized coordinates, and not quasi-coordinates. It is therefore necessary to find the relation between the quasi-velocities $\upsilon = \dot{\gamma}$ and the derivative of the configuration variables x (generalized coordinates). For all the configuration spaces that we will encounter in this book, υ is linear in $\dot{x}$ and thus of the form

$$\upsilon = S(x)\dot{x}. \tag{2.10}$$

In modeling of rigid bodies such as ships and satellites, relations of this kind will always arise.

Example 2.3 In robotics and modeling of the attitude of vehicles we often use the relation

$$\dot{\Theta} = J(\Theta)\omega \tag{2.11}$$

between the body angular velocity variables ω and the Euler angles Θ. This is a nice example of how quasi-coordinates are used in robotic applications. We see that if we write $x = \Theta$ and $\upsilon = \omega$ we get the relation

$$S(x) = J^{-1}(x) \tag{2.12}$$

and Eq. (2.11) is therefore in the form of Eq. (2.10). The matrix that defines the relation between the rate of change of the generalized coordinates and quasi-velocities is thus the inverse of the velocity transformation matrix that we find when we derive the system kinematics.

From this example we see that the quasi-coordinates present us with a well suited framework for describing non-Euclidean transformations. We note that this relation also exists for Euclidean transformations. In this case the mapping between the position and the velocity variables is found simply by taking the time derivative of the position variables. The velocity transformation matrix is then the identity matrix, i.e., $S(x) = I$.

We can also describe the dynamics in terms of local coordinates φ. It is often more convenient to derive the dynamics of systems with a non-Euclidean configuration space in terms of local coordinates because, as we will see in the next section, even though the configuration space is non-Euclidean globally, it is always Euclidean when looked at locally. The drawback of this approach is that the dynamic equations are only valid locally, i.e., in a neighborhood around a given configuration. We will often write the quasi-velocities in terms of the coordinates φ as

$$\upsilon = S(\varphi)\dot{\varphi}. \tag{2.13}$$

One advantage of writing the velocity transformation in terms of local coordinates is that we avoid singularities in the representation. As we will see, the relation in Eq. (2.11) is singular and as a result the dynamic equations are not valid globally. This can be avoided by writing the velocity transformation in the form of (2.13). To derive this relation we need some tools from topological spaces and manifolds.

2.5 Topological Spaces and Manifolds

We have already seen that when the state space is not Euclidean, the kinematics can pose challenges when it comes to finding a suitable mathematical representation. Thus, the study of evolutionary behavior of a system for which the state space is a curved surface, as opposed to a flat (Euclidean) surface, is important. These surfaces are called manifolds and allow us to describe motion on curved surfaces such as rigid body transformations in the 3-dimensional Euclidean space.

A manifold is a smooth and in general curved surface embedded in the Euclidean space. Locally, however, these surfaces look like the Euclidean space which means that, even though the global structure is curved, when looked at locally they are homeomorphic to the Euclidean space and can be described using vector algebra.

It is important to notice the difference between the Euclidean space and transformations in the Euclidean space. For the 3-dimensional Euclidean space any point (a point has no orientation) can be written as a vector in $\mathbb{R}^3$ with a corresponding velocity, also in $\mathbb{R}^3$, and this space is thus Euclidean. However, the transformation of a rigid body in the 3-dimensional Euclidean space is not Euclidean. As we have already seen, the transformation of a rigid body, for which both position and orientation need to be specified, is not Euclidean. The space that we use to describe rigid body motion is thus not Euclidean itself, but *embedded in the Euclidean space*. This is an important difference that we will discuss in more detail later.

The concepts of Lie groups and Lie algebras are important to describe the motion of rigid bodies where the configuration space is a manifold. Group theory dates back to the work of Cayley who introduced the abstract idea of groups in Cayley (1854). In the context of geometry, early contributions we made by Sophus Lie, see Lie (1888, 1890, 1893), which also gave the name to Lie groups. Other important contributions were made by, among others, Wilhelm Killing (1888), Eduard Study (1903), and Élie Cartan (Cartan and Adam 2000). Over the last decades, Lie theory has also become a very important tool in understanding the kinematics of rigid bodies and multibody systems, and this framework has also been adopted by many researchers in robotics. The reason why this framework has gained such popularity is to a large extent the way it allows us to describe rigid body motion in a mathematically rigorous manner. In the following we will show how to formulate the kinematics of a rigid body in terms of matrix Lie groups, which makes the equations valid globally and the singularities that appear for example in Eq. (2.11) in Sect. 2.4 are avoided. This will in turn be applied to multibody systems with a special focus on vehicle-manipulator systems.

The main objective of this section is to define the state space of rigid bodies, i.e., the set of possible transformations and admissible velocities that a rigid body can take. We first need to define topological spaces. All spaces that we will encounter in this book are topological spaces—in fact the definition of a topological space is very general (Bullo and Lewis 2000; Absil et al. 2008):

Definition 2.6 (Topological space) A set $\mathcal{M}$ is called a *topological space* if there exists a collection of open subsets of $\mathcal{M}$ for which the following axioms hold:

1. The union of a countable number of open sets is an open set.
2. The intersection of a finite number of open sets is an open set.
3. Both $\mathcal{M}$ and $\emptyset$ are open sets.

We see that the notion of open sets is very important in topology. The definition of an open set is a generalization of an open interval on $\mathbb{R}^n$. The open interval $(0, 1)$ on $\mathbb{R}$ includes all real numbers *between* 0 and 1, but not 0 and 1 themselves. On the other hand, the closed interval $[0, 1]$ includes all real numbers between 0 and 1, including 0 and 1. To choose an element of the open interval $(0, 1)$ we can thus choose a point arbitrarily close to the limits, but not the limits themselves.

Example 2.4 (The topology of $\mathbb{R}^n$) Any set $\mathcal{M} \in \mathbb{R}^n$ is a topological space. We can easily define open balls defined as $B(x, \rho) = \{x' \in \mathbb{R}^n \mid \|x' - x\| < \rho\}$ which define our open sets. We note that all open sets can be expressed as a countable union of open balls. This representation of $\mathbb{R}^n$ is not very practical and we often tend to other representations, but it is important to keep in mind that we can express open sets in $\mathbb{R}^n$ as open balls. We note that sets expressed in the form $\bar{h} = \{x \in \mathbb{R}^n \mid f(x) = 0\}$ or $\bar{H} = \{x \in \mathbb{R}^n \mid f(x) \leq 0\}$ define closed sets while sets in the form $H = \{x \in \mathbb{R}^n \mid f(x) < 0\}$ define open sets.

Another important set consists of all matrices for which $\det A \neq 0$. As this is the complement of the closed set defined by $\det A = 0$, this set is open. All the matrices that we will use to define the configuration spaces are subgroups of the general linear group that consists of all non-singular matrices i.e., all matrices for which $\det A \neq 0$. This set defines all matrices for which the inverse always exists which is the defining property of matrix Lie groups.

Example 2.5 (The Topology of $\mathbb{S}^1$) Another interesting topology is that of a circle. There are many ways to define a circle, but in topology we can use the notion of *identification*, i.e., to define two points as equivalent. For example, using polar coordinates, 0 and 2π represent the same point on the circle. Once 0 and 2π are identified as equivalent we can use the notion of open sets on $\mathbb{R}$ to describe the topology of a circle. For rotational joints, for example, we will describe the joint position as $q \in \mathbb{R}$, together with the equivalence that gives us the topology of a circle. We will see that for the special case of rotations in the plane (1-DoF rotational joints) this observation allows us to treat the transformation generated by the rotational joint as Euclidean, even though strictly speaking $\mathbb{S}^1$ is different from $\mathbb{R}$ in a topological sense. For the higher dimensional case, however, $\mathbb{S}^n$ is not a Euclidean transformation.

It is clear that the definition of topological spaces is general enough to include any configuration space that we will encounter in robotics. The definition of topology is in fact very general. It is thus useful to define some other properties that can be used to characterize the most important topological spaces in robotics. We will first define Hausdorff spaces.

Definition 2.7 (Hausdorff spaces) If, for any distinct $x_1, x_2 \in \mathcal{M}$, there exist open sets U_1 and U_2 such that $x_1 \in U_1, x_2 \in U_2$ and $U_1 \cap U_2 = \emptyset$, then the topological space $\mathcal{M}$ is a Hausdorff space.

Thus, for any two points in the set, it is possible to separate the points into two nonoverlapping open sets. The spaces that we are most interested in are manifolds, which are indeed Hausdorff spaces. For example, in $\mathbb{R}^n$ we can separate any two points simply by choosing open balls around each point that are sufficiently small.

Intuitively, a function is continuous if a small change in the input leads to a small change in the output. In topology, however, we adopt a rather different definition:

Definition 2.8 (Continuous function—topological definition) Let $f : \mathcal{M} \to \mathcal{N}$ denote a function from $\mathcal{M}$ to $\mathcal{N}$ where $\mathcal{M}$ and $\mathcal{N}$ are topological spaces. Then for any set $B \subseteq \mathcal{N}$, the preimage of B is defined by

$$f^{-1}(B) = \{x \in \mathcal{M} \mid f(x) \in B\}. \tag{2.14}$$

Then the function f is continuous if for every open set $U \subseteq \mathcal{N}$, $f^{-1}(U)$ is an open set.

We will make much use of mappings between topological spaces of this kind where it is important to know that the inverse mapping exists. Particularly, when we are to define mappings between the configuration manifold and the Euclidean space in the next section, it is important that these mappings—and their inverses—are well defined.

In topology, we will often need to check if two objects are equivalent. This equivalence relation is called a homeomorphism and allows us to recognize important properties that are true for a group of objects, but where the objects may appear different in most other aspects. The standard example is a donut and a coffee cup, which topologically speaking are the same (because they both have a single hole). On the other hand, a straight line and a circle are not the same in a topological sense.

Definition 2.9 (Homeomorphism) Given a one-to-one and onto (bijective) function $f : \mathcal{M} \to \mathcal{N}$ where $\mathcal{M}$ and $\mathcal{N}$ are topological spaces. Then if both f and f^{-1} are continuous, then f is a homeomorphism. If such a homeomorphism exists, the two spaces are homeomorphic.

Example 2.6 As we have seen, a circle and a straight line are not topologically the same. The reason is that we cannot find a continuous mapping from a circle to the real line. However, if we take a circle and cut away a small part, then the remaining of the circle is topologically the same as the real line $\mathbb{R}$, and not the circle $\mathbb{S}^1$. We can find a continuous mapping from this open circle to the real line by simply stretching out the circle into a line.

In group theory the corresponding equivalence relation is denoted an isomorphism. An isomorphism defines a one-to-one relation between the elements in two isomorphic groups. Isomorphic groups have the same properties and there is no need to distinguish between them: two isomorphic groups may differ in notation but they are identical for all practical purposes.

An important property of homeomorphisms is that, in addition to map points in a one-to-one manner, they also map open sets in a one-to-one manner. Thus, if $f : \mathscr{M} \to \mathscr{N}$ is a homeomorphism, then $\mathscr{M}$ and $\mathscr{N}$ are topologically the same. This is an important result because we will define such maps that allow us to describe the neighborhood of a point p on a manifold $\mathscr{M}$ using local Euclidean coordinates $\mathbb{R}^n$. Such a homeomorphism exists because a differentiable manifold of dimension n is topologically the same as $\mathbb{R}^n$ when looked at locally. The mapping f together with an open set U is called a chart and provides us with a set of local coordinates. We will discuss charts in more detail below.

The configuration spaces that we will encounter can all be defined on smooth manifolds. A manifold is an abstract mathematical space in which every point has a neighborhood which resembles the Euclidean space. In other words, around every point there is a neighborhood that is topologically the same as an open ball in $\mathbb{R}^n$. Globally, however, manifolds allow more complicated structures.

We will use the global properties of manifolds to correctly describe spaces with more complicated topology than those that can be described using vectors in $\mathbb{R}^n$, and we will use the local properties to perform operations on these spaces using calculus. Locally we can write both the position and velocity variables as vectors in $\mathbb{R}^n$ and consequently these are vector spaces.

We will see that in the context of robotics, the configuration space of the most important manifolds can be written as a matrix with certain constraints. If these matrices are smooth manifolds and also satisfy the group property, we will denote them matrix Lie groups. Matrix Lie groups allow us to represent the configuration of a rigid body so that the mathematical representation is consistent with the actual configuration of the rigid body. We will say that for such a representation the configuration does not leave the manifold. Recall that for certain transformations the integral of the velocity variable does not have a physically meaningful interpretation. Thus, if we integrate the velocity variables, the corresponding position variables do not describe a configuration that is attainable for the rigid body. In this case we have "left the manifold" and the actual displacement of the rigid body does not correspond to our mathematical representation. This illustrates the importance of finding a mathematically sound representation of the configuration space. Our main tool in obtaining this is to restrict the configuration space to a manifold. Formally, a manifold is defined in the following way:

Definition 2.10 (Manifold) A topological space is a *manifold* if for every $x \in \mathcal{M}$, there exists an open set $U \subset \mathcal{M}$ such that

1. $x \in U$,
2. U is homeomorphic to $\mathbb{R}^n$,
3. n is fixed for all $x \in \mathcal{M}$.

The most important property of a manifold is that locally it is homeomorphic to $\mathbb{R}^n$, i.e., it is a topological space with the nice property that locally it behaves like the Euclidean space. This allows us to do calculus on elements of this space, at least locally. This is a very important property that we will use frequently: we first map a point on the manifold to the tangent space on which we can perform an operation on the vector space, before we map back to manifold.

Every manifold has an underlying topology. For example, both a 1-DoF prismatic and a 1-DoF revolute robotic joint can be described by one parameter $q \in \mathbb{R}$, but the topology of the prismatic joint is a line while the topology of a revolute joint is a circle. The topology of a space thus gives us information about what this space looks like globally. The second condition in Definition 2.10 tells us that even though revolute and prismatic joints as spaces are different in a global sense, they both look like a line when looked at locally.

2.5.1 Coordinate Charts

A coordinate chart on a topological manifold is an invertible map between a subset of the manifold and the Euclidean space. More specifically, a chart, denoted (Ψ, U), is a bijection (one-to-one mapping) $\Psi : \mathcal{M} \to \mathbb{R}^n$ of a subset $U \subset \mathcal{M}$ onto an open subset of the Euclidean space $\mathbb{R}^n$. The inverse map maps points on $\mathbb{R}^n$ back to the manifold.

Recall that a homeomorphism not only maps isolated points, but also open sets in a one-to-one manner so that we can define an open set $U \subset \mathcal{M}$ for which U and $\Psi(U)$ are homeomorphic. A *chart* (Ψ, U) is thus defined by a mapping Ψ and an open set U for which Ψ maps U homeomorphically to $\Psi(U)$.

An important property of manifolds is that we can represent any manifold using a finite number of coordinate charts. The fact that each chart is homeomorphic to an open set in $\mathbb{R}^n$ means that we can always find a set of local coordinates, $(\varphi_1, \varphi_2, \dots, \varphi_n)$ that denotes a point in $\Psi(U)$. We will see that by introducing these local objects we are able to perform differentiation in the neighborhood of each point on the manifold. A chart that provides us with a set of local coordinates in this way is called a local coordinate system.

When two charts overlap, i.e., when a single point on the manifold can be described using two or more charts, we can define an *overlap map* which maps an open ball in $\mathbb{R}^n$ to the manifold and then back to another or the same open ball in $\mathbb{R}^n$. Overlap maps allow us to describe a single point using different coordinates, and are therefore often referred to as change of coordinates or coordinate transformations. For two mappings $\Psi_1 : U_1 \rightarrow \mathbb{R}^n$ and $\Psi_2 : U_2 \rightarrow \mathbb{R}^n$, the overlap map is defined as

$$\Psi_2 \circ \Psi_1^{-1} : \mathbb{R}^n \rightarrow \mathbb{R}^n. \tag{2.15}$$

We will now look at some of the most important coordinate charts and the corresponding overlap maps.

2.5.1.1 The Real Line

A straight or curved line is described using $\mathbb{R}$. This can thus be mapped to the Euclidean space using the trivial map $a = \Psi(a)$ for any point $a \in \mathbb{R}$. This space is itself Euclidean. The same is the case for the Euclidean space of higher dimension $\mathbb{R}^n$.

2.5.1.2 The Circle

Consider the circle defined by

$$x^2 + y^2 = 1. \tag{2.16}$$

The simplest description of a circle is obtained by identifying a point on $\mathbb{S}^1$ by the angular coordinate θ. However, this mapping does not take into account the topology of a circle, so in addition we need to define two points as identical if they differ with a multiple of 2π. This identification is what characterizes the topology of a circle.

We can also use the mappings shown in Fig. 2.3. The top half of the circle can be mapped into $\mathbb{R}$ by projecting onto the x-coordinate using the continuous and invertible chart

$$\Psi_{up}(x, y) = x. \tag{2.17}$$

This will map a point on the manifold to a point in $(-1, 1)$. A similar chart can be found for the lower half. We can also define similar charts mapping points to the right and the left, as shown in Fig. 2.3(a). The inverse mapping $\Psi_{up}^{-1}(x) = (x, y)$ takes us from a point on $\mathbb{R}$ back to the manifold. This mapping is simply given by $x = x$ and $y = \sqrt{1 - x^2}$ for Ψ_{up}. We can also define a similar map Ψ_{down} which maps the lower part of the circle ($y < 0$) to the real line. For Ψ_{down} the inverse map is given by $x = x$ and $y = -\sqrt{1 - x^2}$.

We see from Fig. 2.3(a) that the maps overlap, i.e., a point can be written in terms of two different charts. Take a point (a, b) in the upper right part of the circle. This can be mapped to $(-1, 1)$ through the right map $\Psi_{right}(a, b) = b$. Now, we can perform a manipulation on this point on $\mathbb{R}$ instead of on $\mathbb{S}^1$. We denote this by the function $b' = f_{right}(b)$. Finally we can map b' back to the circle using the inverse map $\Psi_{right}^{-1}(b') = (a', b')$. This is illustrated in Fig. 2.3(b). We can also perform the

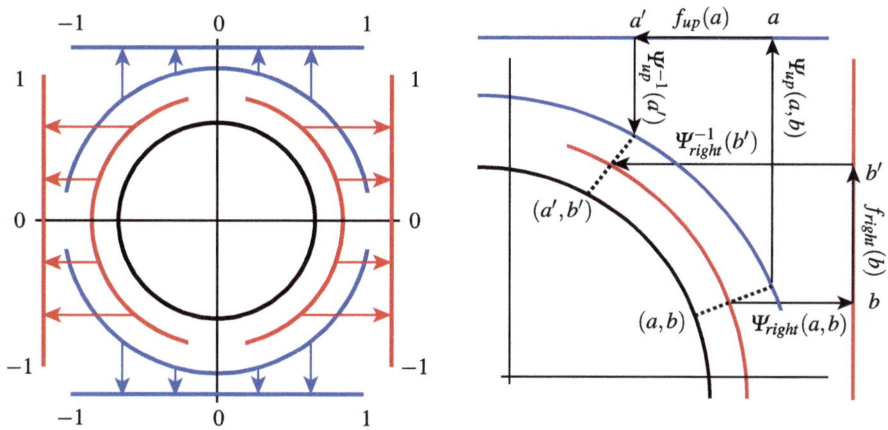

(a) Four different charts used to map $\mathbb{S}^1$ to $\mathbb{R}$. We note that the maps overlap and that the points can be represented in different charts.

(b) The map $\Psi_{right}(a,b) = b$ takes us from the manifold $\mathbb{S}^1$ to $\mathbb{R}$ on the right, while the map $\Psi_{up}(a,b) = a$ takes us from the manifold $\mathbb{S}^1$ to $\mathbb{R}$ upwards. The inverse mappings $\Psi_{right}^{-1}(b') = (a',b')$ and $\Psi_{up}^{-1}(a') = (a',b')$ take us from $\mathbb{R}$ back to $\mathbb{S}^1$. We see that manipulating the point (a,b) on the right chart and the upper chart both take us to the same point (a',b').

Fig. 2.3 The manifold of a circle $\mathbb{S}^1$ and one choice of coordinate charts that can be used to cover $\mathbb{S}^1$

same operation by mapping to the upper chart. Note that also in this case we get back to the same point (a', b').

We also observe that even though a point on the manifold is given by the two coordinates (a, b), $\mathbb{S}^1$ is a 1-dimensional manifold and locally it is homeomorphic to $\mathbb{R}$. $\mathbb{S}^1$ is a 1-dimensional manifold as it is described by a point in $\mathbb{R}^2$, but the constraint given in (2.16) reduces the dimension by one.

The overlap map from $\Psi_{up}(x, y)$ to $\Psi_{right}(x, y)$, i.e., a map that takes a point from (0, 1) above the circle to the interval (0, 1) on the right is given by

$$T(a) = \Psi_{right}\big(\Psi_{up}^{-1}(a)\big) \tag{2.18}$$

$$= \Psi_{right}\big(a, \sqrt{1-a^2}\big) \tag{2.19}$$

$$= \sqrt{1-a^2}. \tag{2.20}$$

This means that a point a on the interval above the circle and a point $\sqrt{1-a^2}$ on the interval to the right correspond to the same point on the circle. On Fig. 2.3(b) this means that $b = \sqrt{1-a^2}$ which we have already seen.

2.5.1.3 The n-Dimensional Sphere

It is not necessary to use four charts to describe the circle. In fact, for any n-dimensional sphere $\mathbb{S}^n$ it suffices to define two charts $U_1 = \mathbb{S}^n - \{(0, \dots, 0, 1)\}$ and $U_2 = \mathbb{S}^n - \{(0, \dots, 0, -1)\}$, i.e., removing the north and the south pole, respectively. This is the stereographic projection and is defined by the two maps

$$\Psi_1 : U_1 \to \mathbb{R}^n: \quad (x_1, \dots, x_{n+1}) \to \left(\frac{x_1}{1 - x_{n+1}}, \dots, \frac{x_n}{1 - x_{n+1}}\right), \tag{2.21}$$

$$\Psi_2 : U_2 \to \mathbb{R}^n: \quad (x_1, \dots, x_{n+1}) \to \left(\frac{x_1}{1 + x_{n+1}}, \dots, \frac{x_n}{1 + x_{n+1}}\right). \tag{2.22}$$

The overlap map is given by

$$\Psi_2 \circ \Psi_1^{-1} = \frac{x}{\|x\|^2}, \qquad \forall x \in \mathbb{R}^n - \{0\}. \tag{2.23}$$

For $\mathbb{S}^1$, $\Psi_1(x, y)$ maps all the points on the circle, except the north pole, to the x-axis while $\Psi_2(x, y)$ maps the all the points on the circle, except the south pole, to the x-axis. The stereographic projection for $\mathbb{S}^1$ is given by

$$\Psi_1(x, y) = \frac{x}{1 - y}, \tag{2.24}$$

$$\Psi_2(x, y) = \frac{x}{1 + y}. \tag{2.25}$$

Ψ_1 is illustrated in Fig. 2.4(a). The overlap map is given by

$$\Psi_2 \circ \Psi_1^{-1} = \frac{x}{x^2} = \frac{1}{x}, \qquad \forall x \in \mathbb{R} - \{0\}. \tag{2.26}$$

Similarly for $\mathbb{S}^2$, $\Psi_1(x, y, z)$ maps all the points on the sphere in $\mathbb{R}^3$, except the north pole, to the xy-plane at $z = 0$ while $\Psi_2(x, y, z)$ maps the all the points on the sphere in $\mathbb{R}^3$, except the south pole, to the xy-plane at the origin. The stereographic projection for $\mathbb{S}^2$ is given by

$$\Psi_1(x, y, z) = \left(\frac{x}{1 - z}, \frac{y}{1 - z}\right), \tag{2.27}$$

$$\Psi_2(x, y, z) = \left(\frac{x}{1 + z}, \frac{y}{1 + z}\right). \tag{2.28}$$

$\Psi_1(x, y, z)$ is illustrated in Fig. 2.4(b). The overlap map is given by

$$\Psi_2 \circ \Psi_1^{-1} = \left(\frac{x}{x^2 + y^2}, \frac{y}{x^2 + y^2}\right), \qquad \forall x, y \in \mathbb{R}^2 - \{0\}. \tag{2.29}$$

Fig. 2.4 The stereographic projection maps a point on the $\mathbb{S}^n$-sphere to a point on $\mathbb{R}^n$

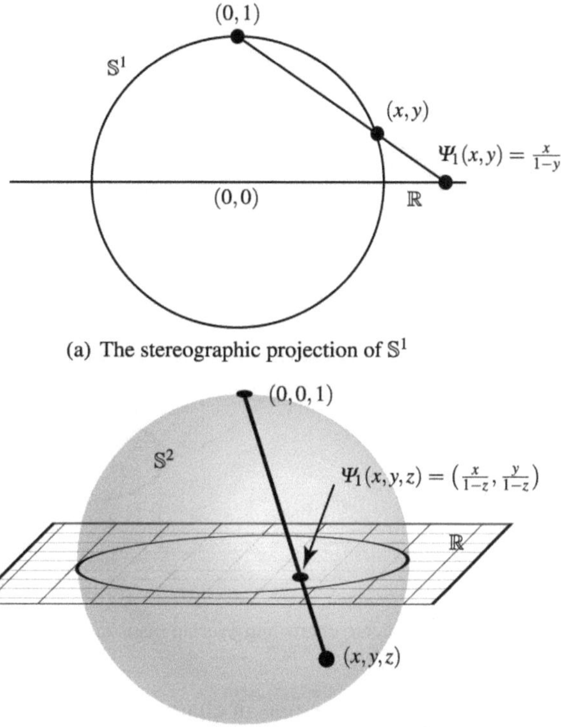

(a) The stereographic projection of $\mathbb{S}^1$

(b) The stereographic projection of $\mathbb{S}^2$

2.5.2 *Manifolds Again*

Based on the notion of charts and overlap functions we can give a manifold $\mathcal{M}$ a topological structure. Recall that the overlap function is in the form

$$f = \Psi_j \circ \Psi_i^{-1}. \tag{2.30}$$

If this mapping is C^∞, i.e., derivatives of all orders exist, we denote the underlying manifold a smooth manifold. Note that $f : \mathbb{R}^n \to \mathbb{R}^n$ is a mapping from one open set in $\mathbb{R}^n$ to another open set in $\mathbb{R}^n$ through an open set $U \subset \mathcal{M}$. Figure 2.5 illustrates the overlap functions on a manifold $\mathcal{M}$. We can now define manifolds as in Kwatny and Blankenship (2000):

Definition 2.11 An n-dimensional manifold is a set $\mathcal{M}$ together with a countable collection of subsets $U_i \subset \mathcal{M}$ and one-to-one mappings $\Psi_i : U_i \to V_i$ onto open subsets V_i of $\mathbb{R}^n$, where each pair (U_i, Ψ_i) is called a coordinate chart, with the following properties:

1. the coordinate charts cover $\mathcal{M}$, i.e.,

$$\bigcup_i U_i = \mathcal{M}, \tag{2.31}$$

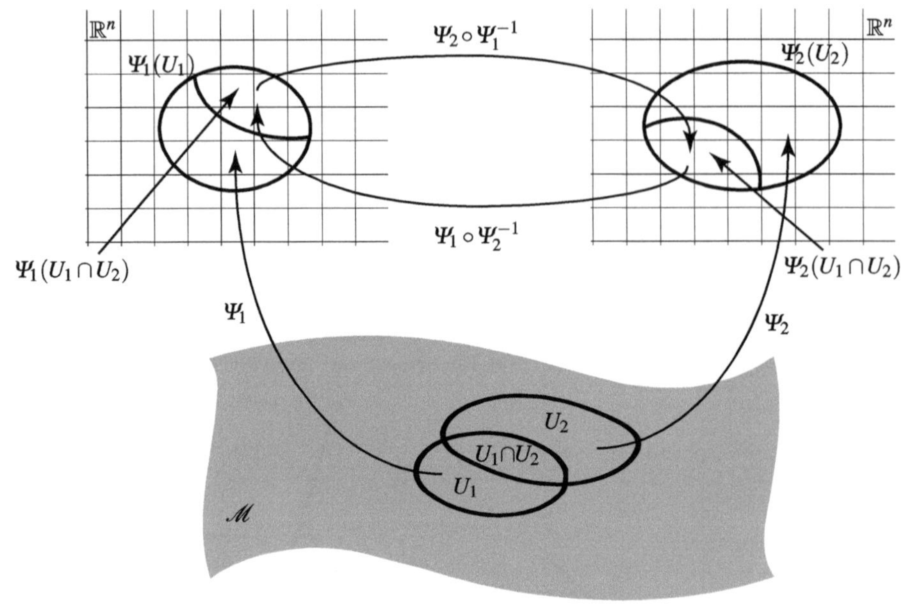

Fig. 2.5 The coordinate charts and overlap maps on a manifold $\mathcal{M}$

2. on the overlap of any pair of charts the composite map

$$f = \Psi_j \circ \Psi_i^{-1} : \Psi_i(U_i \cap U_j) \to \Psi_j(U_i \cap U_j) \tag{2.32}$$

   is a smooth function,
3. (Hausdorff topological spaces) if $p_i \in U_i$, $p_j \in U_j$ are distinct points of $\mathcal{M}$, then there are neighborhoods, O_i of $\Psi_i(p_i)$ in V_i and O_j of $\Psi_j(p_j)$ in V_j such that

$$\Psi_i^{-1}(O_i) \cap \Psi_j^{-1}(O_j) = \emptyset. \tag{2.33}$$

We see that as long as we can define a chart, i.e., a description of the topology of the set, in this way, the set is a manifold. The definition is rather general but again we see that the important aspect here is that there exists a mapping between an open set on the manifold and an open set in $\mathbb{R}^m$.

2.6 Lie Groups

The kinematics of a mechanical system, such as a rigid body or a robotic manipulator, can be derived globally in terms of Lie group and Lie algebra structures. More specifically, an element of the Lie group corresponds to a configuration of the mechanism while the velocity can be expressed as an element of the Lie algebra. We can thus write the state space in terms of the Lie group and algebra.

A group is a set of elements whose action on a space leaves some aspect of the space invariant. Also, two elements of a group can be combined to produce an element in the same group. Multiplication of two elements of the group is defined by the group operator. It is common to denote this action "multiplication" but this does not necessarily mean that we multiply in the normal sense. The group operator can represent operations such as addition, multiplication, matrix multiplication, complex multiplication, and the quaternion product. For the group of real numbers the group operator is addition while for matrix groups the group operation is matrix multiplication.

A fundamental property of groups is that multiplying two elements of a group will produce another element in the same group. For example, two consecutive rotations that can be represented by two elements of the group of right-handed orthonormal matrices on $\mathbb{R}^{n\times n}$, can also be written as a single rotation matrix which is also a right-handed orthonormal matrix on $\mathbb{R}^{n\times n}$. Furthermore, the inverse of an element of the group is also an element of the same group. For matrix groups the inverse is given by the matrix inverse and is used to represent the opposite operation. All these properties make the concept of groups a very powerful tool for describing rigid body motion.

Formally, we can define a group G by identifying four important properties on the elements $g_1, g_2, g_3 \in G$ given a group operation $\circ$:

Property 2.4 (Closure) A set G is closed under the group operation $\circ$ if for all $g_1, g_2 \in G$, then $g_1 \circ g_2 \in G$.

Property 2.5 (Identity) A set G has an identity element if it is possible to find an element $e \in G$ such that $g \circ e = e \circ g = g$ for every $g \in G$.

Property 2.6 (Inverse) A set G is invertible if for each $g \in G$, there exists a unique inverse $g^{-1} \in G$ such that $g \circ g^{-1} = g^{-1} \circ g = e$.

Property 2.7 (Associativity) A set G is called associative if for all $g_1, g_2, g_3 \in G$, then $(g_1 \circ g_2) \circ g_3 = g_1 \circ (g_2 \circ g_3)$.

Based on these properties we can formally define a group in the following way:

Definition 2.12 (Group) A set G with elements $g_1, g_2, g_3 \in G$ together with a binary operation $\circ$, is called a group if it satisfies Properties 2.4–2.7 above.

We see that elements of a group can be multiplied using the group operator and that the resulting element is a member of the same group. This property is very important as it allows us to perform two consecutive operations on a space for which the resulting operation has a meaningful geometrical interpretation. We will see several examples of this in the following. Property 2.6 is also important as it allows us to define "opposite" transformations. Opposite or inverse transformations and the fact that these are unique will also be used frequently in the following. The inverse of a rotation matrix can for example be interpreted geometrically as the opposite rotation.

Recall that the first reference frame that we need to specify is the inertial frame. This frame is used as a reference for all other frames and is usually chosen as the identity element of the group. All other group elements are thus represented with respect to the identity element and they relate to the identity in the same way that the non-inertial reference frames relate to the inertial world frame. In other words, each group element describes the position of a reference frame with respect to the inertial frame.

In robotics we are mainly interested in sets that are manifolds. Further we want these manifolds to be differentiable, i.e., manifolds that can be represented by several coordinate charts and can be patched together in a smooth manner. This takes us to the important definition of Lie groups:

Definition 2.13 (Lie Group) A Lie Group is a group G which is also a smooth manifold and for which the group operation and the inverse are smooth mappings.

The important thing to notice here is that because Lie groups are manifolds, we can always represent an element of a Lie group of dimension n in terms of local coordinates in $\mathbb{R}^n$. This is given by the second condition in Definition 2.10. We will use this property of Lie groups frequently in the subsequent chapters.

In addition to Properties 2.4–2.7 that are always true for Lie groups there are some additional properties that are only true for certain groups. One important property is the one of commutativity. This states whether the sequence in which we perform operations on a space is of importance or not:

Property 2.8 (Commutativity) A group G is commutative, or Abelian, if $g_1 \circ g_2 = g_2 \circ g_1$ for all $g_1, g_2 \in G$.

We note that whether a group is commutative or not depends on whether or not the group operator is commutative. Matrix multiplication, for example, is not commutative so in general matrix groups are not commutative. On the other hand, addition of real numbers is commutative, so the group of real numbers with addition as the group operator is commutative.

2.6.1 Some Important Lie Groups

There are several examples of Lie groups, many of which are widely used in robotics. The most important Lie group in our setting is the one of free rigid body motion which describes the position and orientation of a rigid body in the

3-dimensional Euclidean space. However, there are several other Lie groups that will be important in the subsequent chapters so we start off by introducing the most important groups briefly.

2.6.1.1 The Euclidean Space

The Euclidean space $\mathbb{R}^n$ with addition as the group operator is a group. Given two elements $x = [x_1\ x_2\ \cdots\ x_n]^\mathsf{T} \in \mathbb{R}^n$ and $y = [y_1\ y_2\ \cdots\ y_n]^\mathsf{T} \in \mathbb{R}^n$ the group operation is given by

$$x \circ y = \begin{bmatrix} x_1 + y_1 \\ x_2 + y_2 \\ \vdots \\ x_n + y_n \end{bmatrix} \in \mathbb{R}^n \tag{2.34}$$

and the inverse of an element is given by

$$x^{-1} = \begin{bmatrix} -x_1 \\ -x_2 \\ \vdots \\ -x_n \end{bmatrix} \in \mathbb{R}^n. \tag{2.35}$$

We see that the identity element is the vector $e = [0\,0\ \cdots\ 0]^\mathsf{T} \in \mathbb{R}^n$ and we note that Properties 2.4–2.7 are satisfied.

Property 2.9 The Euclidean space $\mathbb{R}^n$ with addition as group operator is a commutative group.

The manifold of this group is the vector space $\mathbb{R}^n$. This is the group of linear transformations in $\mathbb{R}^n$. If we choose $n = 1$ we get the motion of a prismatic robotic joint; if we choose $n = 2$ we get the group of linear transformations in the plane; and if we choose $n = 3$ we get the group of linear transformations in the 3-dimensional Euclidean space.

2.6.1.2 Complex Numbers of Unit Length

If we restrict ourselves to complex numbers written in the form $z = \cos\theta + i\sin\theta$ and let the group operation be complex multiplication we obtain another important Lie group. We first note that

$$\begin{aligned} z_1 z_2 &= (\cos\theta_1 + i\sin\theta_1)(\cos\theta_2 + i\sin\theta_2) \\ &= \cos(\theta_1 + \theta_2) + i\sin(\theta_1 + \theta_2) \end{aligned} \tag{2.36}$$

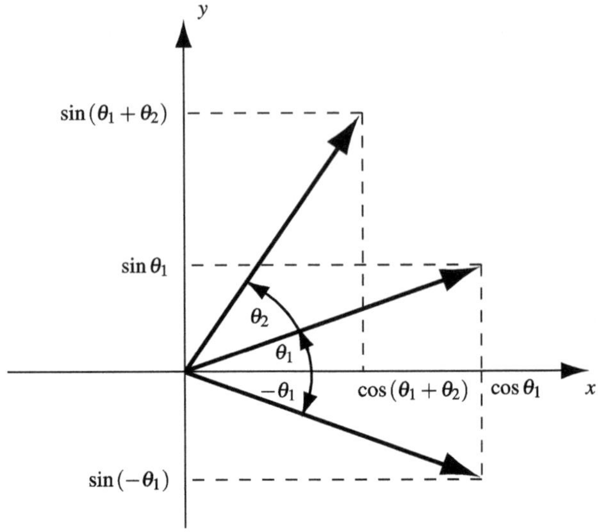

Fig. 2.6 The geometric interpretation of the group of complex numbers as rotations in the plane

which shows that the set is closed. We also note that this operation is continuous since addition of real numbers is continuous.

The topology of this group can be interpreted as in Fig. 2.6. We see that we obtain the same group element for $\theta = 0$ and for $\theta = 2\pi$. The manifold thus has the topology of a circle with unit radius, since $\|z\| = 1$. The transformation represented by complex numbers in this form is therefore pure rotational motion in the plane, i.e., anticlockwise rotation of the complex plane around the origin of the circle. The inverse element is given by the conjugate

$$\begin{aligned} z^{-1} &= \cos\theta - i\sin\theta \\ &= \cos(-\theta) + i\sin(-\theta). \end{aligned} \tag{2.37}$$

We see from Fig. 2.6 that geometrically this is the opposite rotation, i.e., the rotation of $-\theta$.

Property 2.10 Complex numbers in the form $z = \cos\theta + i\sin\theta$ is a commutative group.

This is easily seen from Eq. (2.36) if we use that addition of real numbers is commutative.

2.6.1.3 The Unit Quaternion

Similarly, the unit quaternion, i.e., one real and three imaginary numbers, is a group. A quaternion is written in terms of the basis 1, i, j, and k as quadruples in the form

$$\mathscr{H} = \{(a, b, c, d) \mid a, b, c, d \in \mathbb{R}\}. \tag{2.38}$$

A unit quaternion has the additional requirement that $a^2 + b^2 + c^2 + d^2 = 1$. The quaternion product of two unit quaternions $H_1 = a_1 + b_1 i + c_1 j + d_1 k$ and $H_2 = a_2 + b_2 i + c_2 j + d_2 k$ gives another unit quaternion (the group property) given by

$$
\begin{aligned}
H_1 \circ H_2 = {} & (a_1 a_2 - b_1 b_2 - c_1 c_2 - d_1 d_2) + (a_1 b_2 + b_1 a_2 + c_1 d_2 - d_1 c_2) i \\
& + (a_1 c_2 - b_1 d_2 + c_1 a_2 + d_1 b_2) j + (a_1 d_2 + b_1 c_2 - c_1 b_2 + d_1 a_2) k.
\end{aligned}
\tag{2.39}
$$

Because the quaternion product is not commutative, we can thus conclude the following:

Property 2.11 The unit quaternion is not a commutative group.

The unit quaternion can be used to represent rotations of a rigid body in the 3-dimensional Euclidean space. Because the unit quaternion representation uses four coordinates instead of three, we avoid the singularity that arises when we use the Euler angle representation.

2.6.2 *Matrix Lie Groups*

Of special interest when it comes to representing transformations of rigid bodies in space are Lie groups that can be written in matrix form. The group operation for matrix Lie groups is matrix multiplication. We note that this operation is continuous because it is simply a combination of multiplication and addition of elements in $\mathbb{R}$.

Matrix multiplication does not commute, so matrix Lie groups are in general not commutative. This is important to bear in mind when working with these groups. Also, the inverse of a matrix is not always defined. As a first requirement, to make sure that a set of matrices satisfies Property 2.6 we need to make sure that an inverse exists. We do this by restricting ourselves to $n \times n$ nonsingular matrices. This is our first matrix group.

2.6.2.1 The General Linear Group

The general linear group of order n consists of all $n \times n$ nonsingular real matrices and is denoted $GL(n, \mathbb{R})$. The manifold of $GL(n, \mathbb{R})$ is thus an open subset of $\mathbb{R}^{n \times n}$ defined by all matrices in $\mathbb{R}^{n \times n}$ except the ones that have determinant equal zero. The identity element is given by the $n \times n$ identity matrix and Property 2.5 is satisfied. As we restrict ourselves to nonsingular matrices the inverse always exists and is given by the matrix inverse. Note that Property 2.6 requires all matrix groups to be subgroups of $GL(n, \mathbb{R})$, i.e., that the inverse exists. As a result a matrix group of $n \times n$ matrices is always a subgroup of $GL(n, \mathbb{R})$.

Property 2.12 The general linear group with matrix multiplication as group operator is not a commutative group.

We can also choose addition as the group operator. In this case the group identity is the zero matrix and the group is commutative. It turns out, however, that choosing matrix multiplication as the group operator allows us to adopt a very nice geometric interpretation of the group elements as points in the configuration space. We will therefore concentrate on matrix groups with multiplication as the group operator.

2.6.2.2 The Orthogonal Group

The orthogonal group is a subgroup of the general linear group defined as

$$O(n) = \{R \in GL(n, \mathbb{R}) \mid R^{\mathsf{T}} R = I\}. \tag{2.40}$$

This group consists of all matrices that, in addition to being nonsingular, preserve the scalar product between n-dimensional vectors. This is an important property that we will use later, for example when we find the corresponding Lie algebra, see Sect. 2.6.4. The fact that these matrices preserve the bilinear form means that the scalar product of two vectors does not change (is invariant) when both vectors are acted on by a matrix in $O(n)$. If we write $x' = Rx$ and $y' = Ry$ we see that

$$x' \cdot y' = x^{\mathsf{T}} R^{\mathsf{T}} R y = x \cdot y \tag{2.41}$$

because $R^{\mathsf{T}} R = I$, which defines this group.

Property 2.13 The orthogonal group with matrix multiplication as group operator is not a commutative group.

2.6.2.3 The Special Orthogonal Group

The special orthogonal group is a subgroup of the orthogonal group defined as

$$SO(n) = \{R \in GL(n, \mathbb{R}) \mid R^{\mathsf{T}} R = I, \det(R) = +1\}. \tag{2.42}$$

Note that for $O(n)$ the determinant is ± 1 as $\det R^{\mathsf{T}} \det R = \det I = 1$. The special orthogonal group $SO(n)$ consists of all elements with determinant $+1$. This group is thus a subset of the orthogonal group with the additional requirement that the determinant is positive, which is what "special" refers to. An element of the special orthogonal group of dimension 3 is a rotation matrix and can be interpreted as pure rotational motion of a rigid body. As $R^{\mathsf{T}} R = I$ we can conclude that the inverse of an element R is the same as the transpose, i.e., $R^{-1} = R^{\mathsf{T}}$. This property is very useful and will simplify the computations in the chapters to come. This property is true for all orthogonal groups, not only the special orthogonal group.

Property 2.14 The special orthogonal group with matrix multiplication as group operator is not a commutative group.

Note that even though $SO(n)$ is not commutative in general, the special case of $SO(2)$ is commutative. This is the group of rotations in the plane which, as we have already seen, is commutative.

2.6.2.4 The Special Euclidean Group

The special Euclidean group $SE(n)$ is the group of *rigid body transformations* on $\mathbb{R}^n$. We are especially interested in the special Euclidean group that act on $\mathbb{R}^3$, denoted $SE(3)$. This is the set of rigid body transformations on $\mathbb{R}^3$ defined as the set of mappings $g : \mathbb{R}^3 \to \mathbb{R}^3$ given by $g(x) = Rx + p$ where $R \in SO(3)$ and $p \in \mathbb{R}^3$. The matrix representation of $SE(3)$ is typically given as

$$g = \begin{bmatrix} R & p \\ 0 & 1 \end{bmatrix}. \tag{2.43}$$

Matrix multiplication of two elements of $SE(3)$ gives

$$g_2 g_1 = \begin{bmatrix} R_2 & p_2 \\ 0 & 1 \end{bmatrix} \begin{bmatrix} R_1 & p_1 \\ 0 & 1 \end{bmatrix} = \begin{bmatrix} R_2 R_1 & R_2 p_1 + p_2 \\ 0 & 1 \end{bmatrix} = g \tag{2.44}$$

where $R_2 R_1 \in SO(3)$ and $R_2 p_1 + p_2 \in \mathbb{R}^3$ and thus $g = g_2 g_1 \in SE(3)$. The inverse is given by the matrix inverse, i.e.,

$$g^{-1} = \begin{bmatrix} R & p \\ 0 & 1 \end{bmatrix}^{-1} = \begin{bmatrix} R^{\mathsf{T}} & -R^{\mathsf{T}} p \\ 0 & 1 \end{bmatrix} \tag{2.45}$$

where we have used that $R^{-1} = R^{\mathsf{T}}$ for orthogonal matrices. We see that because $R^{\mathsf{T}} \in SO(3)$ and $-R^{\mathsf{T}} p \in \mathbb{R}^3$ Property 2.6 is satisfied as $g^{-1} \in SE(3)$.

$SE(3)$ can be also written as a semidirect product of $SO(3)$ and $\mathbb{R}^3$, i.e. as $SO(3) \ltimes \mathbb{R}^3$. This means that $SE(3)$, as a manifold, can be looked upon as the product $SO(3) \times \mathbb{R}^3$, but its group structure includes the action of $SO(3)$ on $\mathbb{R}^3$ which is illustrated by using $\ltimes$ instead of $\times$. Intuitively this is easy to see. Translating a rigid object in space will only change the position of the object, the attitude of the object remains fixed. However, if we rotate an object around an arbitrary line, this will not only change the attitude, but also the position of the rigid body. Hence, the group structure allows for actions of $SO(3)$ on $\mathbb{R}^3$. This can also be seen from the matrix representation of two consecutive transformations in (2.44).

The matrix representation of this group given in (2.43) is an injective homomorphism from $SE(3)$ to $GL(4, \mathbb{R})$, i.e.,

$$(R, p) \to \begin{bmatrix} R & p \\ 0 & 1 \end{bmatrix}. \tag{2.46}$$

We thus obtain a convenient way of writing an element of this group at the expense increasing the number of parameters from 6 to 16 (or 12 assuming the final row as fixed).

Property 2.15 The special Euclidean group $SE(3)$ with matrix multiplication as group operator is not a commutative group.

2.6.2.5 Planar Motion

We can also restrict the motion of a rigid body to the set of rigid transformations in the plane. This is the group $SE(2)$ and can be written in matrix form as

$$g = \begin{bmatrix} R & p \\ 0 & 1 \end{bmatrix} \tag{2.47}$$

where $R \in SO(2)$ and $p \in \mathbb{R}^2$. This representation is thus an injective homomorphism from $SE(2)$ to $GL(3, \mathbb{R})$. Here $R \in SO(2)$ is the rotation matrix representing rotations around the axis orthogonal to the plane, and $p \in \mathbb{R}^2$ represents translational motion in the plane.

Property 2.16 The special Euclidean group $SE(2)$ with matrix multiplication as group operator is not a commutative group.

It is fairly straight forward to see that $SE(2)$ is not commutative also from a geometric point of view. For example, a linear motion in the direction of one of the axes followed by a rotation will give a different transformation than a rotation followed by a linear motion because the axis of the linear motion changes when we chose to perform the rotation before the translation.

2.6.2.6 The Schönflies Group

The Schönflies group describes three degrees of freedom linear motion and one degree of freedom rotational motion. We normally assume that the rotational axis is the z-axis. This is a normal configuration space for pick-and-place tasks in robotic manufacturing. An element of the Schönflies group $\mathscr{X}$ is normally written in matrix form as

$$g = \begin{bmatrix} R & p \\ 0 & 1 \end{bmatrix} \tag{2.48}$$

where $p \in \mathbb{R}^3$ and R is the 3×3 matrix representation of $SO(2)$. We will find these rotation matrices for the three coordinate axes in Sect. 3.2.1.

Table 2.1 The Lie subgroups of $SE(3)$. Only the normal forms are shown. The dimension of each group is given in the parenthesis

Group	Name	Basis	Description
$SE(3)$	Special Euclidean Group	$\{e_1, e_2, e_3, e_4, e_5, e_6\}$	Rigid body motions in $\mathbb{R}^3$ (6)
$\mathscr{X}(z)$	Schönflies Group	$\{e_1, e_2, e_3, e_6\}$	Translations in $\mathbb{R}^3$ and rotations around the z-axis (4)
$\mathbb{R}^3$	Translational Group	$\{e_1, e_2, e_3\}$	Translations in $\mathbb{R}^3$ (3)
$SE(2)$	Planar Group	$\{e_1, e_2, e_6\}$	Planar motion (3)
$SO(3)$	Special Orthogonal Group	$\{e_4, e_5, e_6\}$	Rigid body rotations around a fixed point (3)
$\mathscr{Y}_\rho$		$\{e_1, e_2, e_6 + \rho e_3\}$	Translational motion in the plane and screw motion around the z-axis (3)
$\mathbb{R}^2$	Translational Group	$\{e_1, e_2\}$	Translational motion in the plane (2)
$\mathscr{C}$		$\{e_3, e_6\}$	Translations along and rotations around the z-axis (2)
$\mathbb{R}$	Translational Group	$\{e_1\}$	Linear motion (1)
$SO(2)$	Special orthogonal Group	$\{e_3\}$	Rotational motion in the plane (1)
$\mathscr{H}_\rho$		$\{e_6 + \rho e_3\}$	Screw motion (1)

2.6.2.7 The Lie Subgroups

All the Lie subgroups that we are interested in, i.e., the ones that describe rigid body motion, can be written in terms of the basis of $SE(3)$ and are thus subgroups of $SE(3)$. If we write the basis of the special Euclidean group as $SE(3) = \{e_1, e_2, e_3, e_4, e_5, e_6\}$ we can write the normal form of each subgroup in terms of this basis. Here, e_1, e_2, e_3 represent linear motion in the directions of the x-, y-, and z-axes, respectively, and e_4, e_5, e_6 represent angular motion around the same axes. There are a total of 10 subgroups of $SE(3)$. A short description is given in Table 2.1. We see that there are three 1-dimensional subgroups, these are all screw motions of zero ($SO(2)$), infinite ($\mathbb{R}$), and non-zero ($\mathscr{H}_\rho$) pitch. These are the motions of the most important robotic joints. The higher-dimensional groups are frequently used to describe the configuration space of single rigid bodies or the robot end effector. We see that $SE(3)$ is of dimension 6 and that there are no subgroups of dimension 5. Furthermore there is only one 4-dimensional subgroup and four 3-dimensional subgroups, which includes the three important groups of pure rotational motion, planar motion, and pure translational motion. More details on the normal forms of the Lie subgroups can be found in Meng et al. (2005, 2007).

2.6.3 Local Coordinates of Matrix Lie Groups

We saw in Sect. 2.5.1 that we can find a mapping from a subset of a manifold to the Euclidean space and we studied these maps for $\mathbb{S}^1$ and $\mathbb{S}^2$. We will now look at the mapping from $SO(3)$ to the Euclidean space $\mathbb{R}^3$. As $SO(3)$ is a 3-dimensional space this mapping cannot be illustrated by simple drawings in the same intuitive way as we could with $\mathbb{S}^1$ and $\mathbb{S}^2$, but we need to think of $SO(3)$ as the orientation of a rigid body in space. A point on the manifold $SO(3)$ can be represented by a rotation matrix $R \in \mathbb{R}^{3\times 3}$. It is possible to find an invertible one-to-one mapping

$$\Psi_1 : U_1 \to \mathbb{R}^3 \tag{2.49}$$

from a subset $U_1 \subset SO(3)$ to the Euclidean space $\mathbb{R}^3$. As for $\mathbb{S}^1$ and $\mathbb{S}^2$ the subset U_1 consists of all points on the manifold, except one. $\Theta \in \mathbb{R}^3$ is the Euler angles and is one way to find a minimum representation of $SO(3)$. As there is one point for which this mapping is not well defined we will define sets of the form $\mathbb{R}^3 - \{0, \pm\frac{\pi}{2}, 0\}$ for which we can find a one-to-one mapping to U_1. We see that we have removed one point, similarly to what we did for $\mathbb{S}^1$ and $\mathbb{S}^2$. Depending on how we choose to represent the elements of $SO(3)$, we need to remove the single point on the manifold where the mapping does not exist for the chosen representation. We will see how we arrive at this later in this chapter. The three variables in Θ are normally interpreted as rotations around the coordinate axes. The problem is that no matter what sequence of axes we choose, there always exists one point in one of the variables for which one of the other two variables can be chosen arbitrarily without changing the configuration. If we for example use consecutive rotations around the x, y, and z-axes, the coordinates $(a - \psi, \pm\frac{\pi}{2}, \psi)$ describe the same rotation independently of how ψ is chosen. As a result a continuous one-to-one mapping between $SO(3)$ and Θ cannot be found for a rotation $\pm\frac{\pi}{2}$ around the y-axis. This point represents the well known Euler angles singularity.

It is interesting to note the similarities with the stereographic mappings of $\mathbb{S}^1$ and $\mathbb{S}^2$. Recall that we found two mappings Ψ_1 and Ψ_2 which had different domains $U_1 = \mathbb{S}^n - \{(0, \dots, 0, 1)\}$ and $U_2 = \mathbb{S}^n - \{(0, \dots, 0, -1)\}$. Also for $SO(3)$ we get different domains depending on what mappings we choose. The different mappings are obtained by choosing different sequences of rotation matrices. A different order of rotations leads to singularities at different points and thus also different domains.

2.6.4 Lie Algebra

Lie algebras can be thought of as tangent vectors to the manifold, or more specifically the tangent space of the identity element e. Imagine a path through the identity in a group G defined by $\gamma : \mathbb{R} \to G$ such that $\gamma(0) = e$. The Lie algebra describes the derivatives in some local coordinate system around e. If the parameter is time, i.e., $\gamma(t)$, then the tangent space can be identified with the space of all admissible velocities at the identity.

Taking this approach we can also define the tangent space as an equivalence relation. If we say that two paths are equivalent if their first derivatives are identical at the identity, we see that a tangent vector is the equivalence class that allows us to recognize all such paths. Such a space of equivalence classes is actually a vector space. This is a very important property of the tangent space that we will use frequently in the subsequent sections.

If this vector space $\mathcal{V}$ allows a bilinear operation $\mathcal{V} \times \mathcal{V} \to \mathcal{V}$, called the Lie bracket, which satisfies both skew-symmetry and the Jacobi identities, then this vector space is a Lie algebra. For two elements $v_1, v_2 \in \mathcal{V}$ the Lie bracket is given by

$$[v_1, v_2] = v_1 v_2 - v_2 v_1, \qquad v_1, v_2 \in \mathcal{V}. \tag{2.50}$$

An important property of the Lie bracket is that the resulting product is also an element of $\mathcal{V}$, i.e., $[v_1, v_2] \in \mathcal{V}$. Formally a Lie algebra can be defined as a vector space which allows a Lie bracket:

Definition 2.14 (Lie algebra) A vector space $\mathcal{V}$ is a *Lie algebra* if there exists a bilinear operation $\mathcal{V} \times \mathcal{V} \to \mathcal{V}$, denoted $[\cdot, \cdot]$, satisfying the two properties:

- skew-symmetry, i.e.,

$$[v_1, v_2] = -[v_2, v_1], \qquad \forall v_1, v_2 \in \mathcal{V}, \tag{2.51}$$

- the Jacobi identity, i.e.,

$$\big[v_1, [v_2, v_3]\big] + \big[v_2, [v_3, v_1]\big] + \big[v_3, [v_1, v_2]\big] = 0, \qquad \forall v_1, v_2, v_3 \in \mathcal{V}. \tag{2.52}$$

The Lie bracket $[v_1, v_2]$ plays a very important role in Lie theory as it is a mean to differentiate a vector field v_2 with respect to another vector field v_1. The Lie algebras associated with the groups described in the previous section are described in the following.

2.6.4.1 The General Linear Group

The Lie algebra of $GL(n, \mathbb{R})$ is the set of all $n \times n$ matrices, denoted $gl(n, \mathbb{R})$. It can be shown (Tu 2008) that $[A, B]_{ij} = (AB - BA)_{ij}$, and the Lie bracket is thus given by

$$[A, B] = AB - BA, \qquad A, B \in gl(n, \mathbb{R}). \tag{2.53}$$

2.6.4.2 The Euclidean Space

The Lie algebra associated with the Lie group $\mathbb{R}^n$ is $\mathbb{R}^n$ itself. We will write an element of $\mathbb{R}^n$ as p and an element of the corresponding Lie algebra as v. As we have already seen, this is a very convenient way of writing the velocity variables. Because the position and velocity variables live on the same space we can write the velocity variables as $v = \dot{p}$. We then see that the Lie bracket is given by

$$[v_1, v_2] = 0, \quad \forall v_1, v_2 \in \mathbb{R}^n. \tag{2.54}$$

Thus, for a commutative Lie algebra the Lie bracket is zero. We will make much use of the fact that the velocity variables can be written as the time derivative of the position variables. Recall from Sect. 2.3 that this was in fact the definition of a Euclidean transformation:

Property 2.17 The transformation associated with an element of the Euclidean space $\mathbb{R}^n$ is Euclidean.

2.6.4.3 The Special Orthogonal Group

The Lie algebra of $SO(3)$ is denoted $so(3)$ and can be identified with a skew-symmetric traceless matrix in $\mathbb{R}^{3\times 3}$. We will start by defining the skew-symmetric operator and the opposite operation, the vee operator. For now we will concentrate on $SO(3)$ so we only need the operations on $\mathbb{R}^3$. However, the hat and vee operators are not restricted to $\mathbb{R}^3$ but can be defined for a general vector $\mathbb{R}^n$. We will find these matrices in the next sections.

Definition 2.15 The hat operator $\wedge$ on $\mathbb{R}^3$ maps an element $\omega = [\omega_x\ \omega_y\ \omega_z]^\mathsf{T} \in \mathbb{R}^3$ into a traceless symmetric matrix

$$\widehat{\omega} = \begin{bmatrix} 0 & -\omega_z & \omega_y \\ \omega_z & 0 & -\omega_x \\ -\omega_y & \omega_x & 0 \end{bmatrix}. \tag{2.55}$$

Definition 2.16 The vee operator $\vee$ is the inverse of the hat operator in Definition 2.15. For $so(3)$ it maps a skew-symmetric matrix into a vector in $\mathbb{R}^3$.

We will also use the $\vee$ map to write the vector representation of a matrix in general. As an example, $g^\vee$ is the vector representation of the homogeneous transformation matrix g.

If we write an element of $SO(3)$ as $R(t)$ and recall that $R(t)^\mathsf{T} R(t) = I$, we can differentiate and get

$$\frac{\mathrm{d}}{\mathrm{d}t} R(t)^\mathsf{T} R(t) + R(t)^\mathsf{T} \frac{\mathrm{d}}{\mathrm{d}t} R(t) = 0. \tag{2.56}$$

We see that the constraints that make $R(t)$ an element of $SO(3)$ can be translated into the constraint that $R^\mathsf{T} \dot{R}$ and $\dot{R}^\mathsf{T} R$ are skew-symmetric. We see this if we rewrite (2.56) as $\dot{R}^\mathsf{T} R + (\dot{R}^\mathsf{T} R)^\mathsf{T} = 0$. This shows that $\dot{R}^\mathsf{T} R$ and $R^\mathsf{T} \dot{R}$ must be in the form of (2.55).

Note also that R^{T} takes $\dot{R}$ back to the identity. Thus, if $\dot{R}$ is the tangent space at R then $\widehat{\omega} = R^{\mathsf{T}}\dot{R}$ is the tangent space at the identity. The Lie algebra is defined as the tangent space at the identity which is the set of skew-symmetric 3×3 matrices. An element of $so(3)$ can thus be identified by the vector $\omega = [\omega_x\ \omega_y\ \omega_z]^{\mathsf{T}}$.

Property 2.18 The transformation associated with an element of the special orthogonal group $SO(3)$ is not Euclidean.

This follows directly from the observation that there is no way to write the orientation as a vector in $\mathbb{R}^3$ for which Definition 2.2 is satisfied. We will study this kind of transformations in more detail in Chap. 3.

2.6.4.4 The Special Euclidean Group

The Lie algebra of $SE(3)$ is denoted $se(3)$ and can be identified with the matrix

$$se(3) = \begin{bmatrix} \widehat{\omega} & v \\ 0 & 0 \end{bmatrix}, \tag{2.57}$$

where $\omega, v \in \mathbb{R}^3$ and $\widehat{\omega} \in so(3)$. Elements of $se(3)$ are also referred to in the robotics literature as twists. Physically, twists represent the quasi-velocities of a rigid body and can be written in vector form in terms of twist coordinates as

$$V = \begin{bmatrix} v \\ \omega \end{bmatrix} \tag{2.58}$$

where $v = [v_x\ v_y\ v_z]^{\mathsf{T}}$ represents the linear velocity and $\omega = [\omega_x\ \omega_y\ \omega_z]^{\mathsf{T}}$ represents the angular velocity.

As for $SO(3)$, the constraints on $g \in SE(3)$ are translated into the constraints that $\widehat{V} \in se(3)$ is in the form of (2.57). As we will see in the next chapter when we study rigid body motion in more detail, this is the same as requiring that $\widehat{V}$ equals $\dot{g}$ after it has been transformed back to the identity.

Property 2.19 The transformation associated with an element of the special Euclidean group $SE(3)$ is not Euclidean.

2.6.4.5 The Planar Group

The Lie algebra of $SE(2)$ is denoted $se(2)$ and can be identified with the matrix

$$se(2) = \begin{bmatrix} \widehat{\omega} & v \\ 0 & 0 \end{bmatrix}, \tag{2.59}$$

where $\omega \in \mathbb{R}$, $v \in \mathbb{R}^2$ and $\widehat{\omega} = \left[\begin{smallmatrix} 0 & -\omega \\ \omega & 0 \end{smallmatrix}\right] \in so(2)$. Elements of $se(2)$ can be written in vector form in terms of twist coordinates as

$$V = \begin{bmatrix} v \\ \omega \end{bmatrix} \in \mathbb{R}^3 \tag{2.60}$$

where $v = [v_x \; v_y]^\mathsf{T}$ represents the linear velocity and $\omega = \omega_z$ represents the angular velocity. This is the group of linear motion in the plane and rotation around an axis orthogonal to this plane.

Property 2.20 The transformation associated with an element of the special Euclidean group $SE(2)$ is Euclidean.

Note that this transformation is Euclidean because there exists at least one parameterization for which the variables can be written in the form of Definition 2.2. One such parameterization is when the position variables are written in the normal way as (x, y, ψ) and the velocity variables are written as $(v_x, v_y, \omega_z) = (\dot{x}, \dot{y}, \dot{\psi})$.

2.6.4.6 The Schönflies Group

The Lie algebra of $\mathscr{X}(z)$ is denoted $\varkappa$ and can be identified with the matrix

$$\varkappa = \begin{bmatrix} \widehat{\omega} & v \\ 0 & 0 \end{bmatrix}, \quad \widehat{\omega} = \begin{bmatrix} 0 & -\omega & 0 \\ \omega & 0 & 0 \\ 0 & 0 & 0 \end{bmatrix} \tag{2.61}$$

and thus only angular velocity around the z-axis is allowed. If the rotational motion is in another direction, $\widehat{\omega}$ is modified correspondingly, or we choose the coordinate system so that the z-axis is aligned with the revolute axis. We allow linear velocity in all three directions so $v = [v_x \; v_y \; v_z]^\mathsf{T}$.

Property 2.21 The transformation associated with an element of the Schönflies group $\mathscr{X}(z)$ is Euclidean.

2.6.5 Geometric Interpretation of Lie Group Representations

We have now developed several tools that we can use to represent spaces and transformations on these spaces. In this section we will look into how the different matrix Lie groups are interpreted geometrically. One of the most important tasks in Lie theory is to find suitable matrix representations for the different spaces, i.e., we want to find a set of matrices that satisfy certain properties so that they can be interpreted as geometrically meaningful objects or transformations. We will discuss how these group elements are interpreted geometrically and what they represent.

In general we can say that there are two main interpretations of the elements of a Lie group that are widely used in robotics: either as the configuration of a rigid body or as a transformation of a rigid body. The configuration of a rigid body can be specified with respect to an inertial frame or with respect to another rigid body. The configuration can thus be interpreted as the difference in position and orientation between two different frames.

On the other hand, a transformation can describe the position and orientation of the same rigid body before and after some action. In this case the transformation is to be interpreted as an active transformation of a rigid body, often expressed in the observer frame.

We will also look at how we can use an element of the Lie algebra to describe the velocity of a rigid body to obtain a complete representation of the state space. In the previous section we used twists to represent the velocity of the rigid body. However, as we have already seen, we need to specify the frame in which the velocity is represented. There are two main ways to represent twists that are of importance in robotics, using either body or spatial coordinates.

2.6.5.1 Representations of Matrix Lie Groups

We will now study different ways to represent the location of a rigid body in space. This can be written in terms of three variables (x_{12}, y_{12}, z_{12}) describing the position and three variables $(\phi_{12}, \theta_{12}, \psi_{12})$ describing the orientation, for example by

$$g_{12}^{\vee} = \begin{bmatrix} x_{12} \\ y_{12} \\ z_{12} \\ \phi_{12} \\ \theta_{12} \\ \psi_{12} \end{bmatrix}. \tag{2.62}$$

Here, $g^{\vee}$ denotes the vector representation of the homogeneous transformation matrix g.

If we also know what convention is used for the Euler angles, this vector uniquely describes the location of a frame $\mathscr{F}_2$ in another frame $\mathscr{F}_1$. Suppose now that after we have performed this transformation, we want to perform another transformation from frame $\mathscr{F}_2$ to frame $\mathscr{F}_3$. We thus want to perform two consecutive transformations, the first represented by $g_{12}^{\vee}$ and the second by $g_{23}^{\vee}$. Unfortunately, the final location of the rigid body, represented by $g_{13}^{\vee}$, cannot be found by

$$g_{13}^{\vee} = g_{12}^{\vee} + g_{23}^{\vee} \tag{2.63}$$

because geometrically the transformation represented by $g_{13}^{\vee}$ is not the same as simply adding the transformation corresponding to $g_{12}^{\vee}$ and $g_{23}^{\vee}$ using normal addition. One way to see that this is not correct is to note that by simply adding these, the rotational part of the first transformation will not affect the translational part of the second transformation, as it should (because it is a semi-direct product).

It turns out, however, that if we write each transformation in matrix form as in (2.43), the transformation of g_{12} followed by the transformation g_{23} can be written as

$$g_{13} = g_{12} g_{23} \tag{2.64}$$

where $g \in SE(3)$ is the homogeneous transformation matrix introduced in Sect. 2.6.2. This is the main motivation for writing the transformations in this form. We see that at the cost of increasing the number of variables needed to describe the configuration of a rigid body from 6 to 16 (a 4×4 matrix) we obtain the benefits of this convenient representation.

Another transformation that can be performed in an elegant manner using this matrix representation is the transformation of a point in space. We now write the point in homogeneous coordinates $\bar{q}^b = [(q^b)^\mathsf{T}\ 1]^\mathsf{T}$. Hence, the homogeneous coordinates of a point are obtained by augmenting the coordinates with an identity element. We will now see how this representation allows us to use homogeneous transformation matrices to perform operations on this point. Let $\bar{q}^b$ represent a point that is fixed in frame $\mathscr{F}_b$. Given another frame $\mathscr{F}_a$ that relates to $\mathscr{F}_b$ through $g_{ab}(t)$ this point can be expressed in coordinate frame $\mathscr{F}_a$ by

$$\bar{q}^a(t) = \begin{bmatrix} R_{ab}(t) & p_{ab}(t) \\ 0 & 1 \end{bmatrix} \bar{q}^b. \tag{2.65}$$

This result is obtained by rewriting the well known equations of rotating and translating a point, which is given by $q^a = R_{ab} q^b + p_{ab}$. The same applies to the velocity v of the same point which can be described either in coordinate frame $\mathscr{F}_a$ or in $\mathscr{F}_b$:

$$\bar{v}^a = \begin{bmatrix} R_{ab} & p_{ab} \\ 0 & 1 \end{bmatrix} \bar{v}^b \tag{2.66}$$

where $\bar{v}^b = [(v^b)^\mathsf{T}\ 0]^\mathsf{T}$. Also Eq. (2.66) is only a reformulation of the well known equation $v^a = R_{ab} v^b$. We note that the homogeneous coordinates of a vector are obtained by adding a zero to the end of the vector. Here the velocity v^b expressed in $\mathscr{F}_b$ is transformed so that it is written as observed from frame $\mathscr{F}_a$, denoted v^a. Using this interpretation of the transformation (2.66) the same vector is represented in two different coordinate frames, namely $\mathscr{F}_a$ and $\mathscr{F}_b$, and the transformation thus describes what the same vector looks like for two observers located in two different frames.

Remark 2.2 Note that the homogeneous coordinates of a point q are obtained by adding 1 at the end, i.e., a point q in homogeneous coordinates is written as $\bar{q} = [q^\mathsf{T}\ 1]^\mathsf{T}$. For a vector, however, the homogeneous coordinates are obtained by adding 0 to the vector, i.e., the vector v is written in homogeneous coordinates as $\bar{v} = [v^\mathsf{T}\ 0]^\mathsf{T}$. We note that in this way, adding a point and a vector gives another point, adding two vectors gives a vector, and adding two points is meaningless.

The expression in the form of (2.65) can also be interpreted as a transformation of the point $\bar{q}_a$ to another point $\bar{q}_b$ by first a translation along p_{ab} and then a rotation by R_{ab}. In this case both $\bar{q}_a$ and $\bar{q}_b$ are given with respect to the same reference frame and $\bar{q}_a$ is transformed by g_{ab} to $\bar{q}_b$. The transformation is then written as

$$\bar{q}_b = \begin{bmatrix} R_{ab} & p_{ab} \\ 0 & 1 \end{bmatrix} \bar{q}_a. \tag{2.67}$$

Remark 2.3 Note that a superscript is used to denote in what reference frame a point or vector is represented while a subscript is used to distinguish two different points or vectors: $\bar{q}_a$ and $\bar{q}_b$ are two different points, while $\bar{q}^a$ and $\bar{q}^b$ are two different representations of the same point.

Adapting the interpretation above we can look at an element $g \in SE(3)$ as a transformation from some initial coordinates $\bar{q}(0)$ to the coordinates after a rigid body motion is applied. In this case, both the initial and final coordinates are viewed *from the same coordinate system*. Take a point $\bar{q}(0)$ represented in the inertial frame $\mathscr{F}_0$ and let $g(t, X)$ represent a transformation of t units (or t units of time with constant unitary velocity) along the direction of the twist X. Then, if $\bar{q}(0)$ is transformed into another point by $g(t, X)$ we get

$$\bar{q}(t) = g(t, X)\bar{q}(0) \tag{2.68}$$

where $\bar{q}(0)$ is the point before and $\bar{q}(t)$ is the point after the transformation by $g(t, X)$. Note that both $\bar{q}(0)$ and $\bar{q}(t)$ are represented in $\mathscr{F}_0$. In this case the transformation is not interpreted as a change of observer but rather as an active action on the point $\bar{q}(t)$, in other words a motion. If we want to be specific about what reference frame the point is expressed in, for example the inertial frame $\mathscr{F}_0$, we can denote the point before the transformation as $q^0(0)$ and after the transformation as $q^0(t)$.

Up until now we have only looked at transformations of vectors and points. Similarly, the configuration of a rigid body can be observed from different frames and transformed in space. The configuration of a rigid body is given by the matrix g_{0b}, i.e., the position and orientation of the body frame $\mathscr{F}_b$ with respect to the inertial frame $\mathscr{F}_0$. Using the concept of change of observer's frame we can instead write the position and orientation of the rigid body with respect to an arbitrary frame $\mathscr{F}_i$ as

$$g_{ib} = g_{i0} g_{0b} \tag{2.69}$$

where g_{0i} is the transformation from the inertial frame to frame i and $g_{i0} = g_{0i}^{-1}$. The interpretation of (2.69) is the same as that of (2.65), i.e., a change of observer's frame.

Furthermore, if $g_{0b}(0)$ is the configuration of a rigid body with a reference frame $\mathscr{F}_b$ attached to it with respect to the inertial frame $\mathscr{F}_0$, then the configuration of the same rigid body after a transformation by $g(t, X)$ is given by

$$g_{0b}(t) = g(t, X) g_{0b}(0) \tag{2.70}$$

also with respect to $\mathscr{F}_0$. This is the motion of a rigid body parameterized by t and denoted $g(t, X)$. We will normally parameterize the motion in terms of the time variable t. We then get a time variant homogeneous transformation matrix $g(t)$ that describes the position and orientation of a rigid body as a function of time. We will use both interpretations of the homogeneous transformation matrix frequently in the subsequent chapters.

Summing up, there are two different interpretations of a matrix Lie group. Firstly, a transformation g can be interpreted as a change of the frame in which a set of points is represented. We have already seen that for a point $\bar{q}^b$ represented in coordinate frame $\mathscr{F}_b$ and a homogeneous transformation matrix g_{ab}, this point can be represented in $\mathscr{F}_a$ by the transformation in (2.65). This transformation is referred to as a change of observer's frame. An element g of the Lie group can also be interpreted as an active transformation, i.e., we consider the observer's frame as fixed and let g represent a transformation of a point, a vector, or a rigid body from an initial to a final configuration.

2.6.5.2 Representations of Lie Algebras

Also the velocity of a rigid body can be represented either in vector or matrix form. We will see that as for the position variables, there are several advantages of representing the velocity in matrix form. Consider a point q^b fixed to a rigid body with coordinate frame $\mathscr{F}_b$ that moves corresponding to the 1-parameter subgroup of $SE(3)$. Given a rigid body motion $g_{ab}(t)$ relating $\mathscr{F}_a$ and $\mathscr{F}_b$, this point can be represented with respect to $\mathscr{F}_a$ by

$$\bar{q}^a(t) = g_{ab}(t)\bar{q}^b. \tag{2.71}$$

The velocity of this point is given by

$$\bar{v}_{q^a}(t) = \frac{\mathrm{d}}{\mathrm{d}t}\bar{q}^a(t) = \dot{g}_{ab}(t)\bar{q}^b. \tag{2.72}$$

There is, however a more convenient and compact way to write this relation. First write

$$\bar{v}_{q^a}(t) = \dot{g}_{ab}(t)g_{ab}^{-1}(t)g_{ab}(t)\bar{q}^b. \tag{2.73}$$

Using (2.71) we can rewrite this as

$$\bar{v}_{q^a}(t) = \dot{g}_{ab}(t)g_{ab}^{-1}(t)\bar{q}^a. \tag{2.74}$$

It can be shown that $\dot{g}_{ab}(t)g_{ab}^{-1}(t) \in se(3)$ has a very useful geometric interpretation. If we let $\mathscr{F}_a$ be the inertial frame denoted $\mathscr{F}_0$, and let $\mathscr{F}_b$ be the body frame, we get the spatial velocity variables $\widehat{V}_{0b}^S$ defined by

$$\widehat{V}_{0b}^S = \dot{g}_{0b}g_{0b}^{-1} = \begin{bmatrix} \widehat{\omega}_{0b}^S & v_{0b}^S \\ 0 & 0 \end{bmatrix}. \tag{2.75}$$

The spatial velocity variables $\widehat{V}_{0b}^S$ map the point $\bar{q}^0$ to the velocity of the same point also represented in the inertial frame $\mathscr{F}_0$, i.e. $\bar{v}_{q^0}(t) = \widehat{V}_{0b}^S \bar{q}^0$. Using the terminology of Lie groups, $\widehat{V}_{0b}^S$ is the right translate of $\dot{g}_{0b} \in T_g SE(3)$ which maps $\dot{g}_{0b}$ back to the identity $T_I SE(3)$. $\dot{g}_{0b}$ is thus in the tangent space to $SE(3)$ at configuration g_{0b}, denoted $T_g SE(3)$. If we recall that the Lie algebra was defined as the tangent space at the identity, we can conclude that $\widehat{V}_{0b}^S$ is an element of the Lie algebra $se(3)$. Furthermore, we can write $\widehat{V}_{0b}^S$ in vector form and get

$$V_{0b}^S = \begin{bmatrix} v_{0b}^S \\ \omega_{0b}^S \end{bmatrix} = \begin{bmatrix} -\dot{R}_{0b} R_{0b}^{\mathsf{T}} p_{0b} + \dot{p}_{0b} \\ (\dot{R}_{0b} R_{0b}^{\mathsf{T}})^{\vee} \end{bmatrix}. \tag{2.76}$$

V_{0b}^S is thus the vector representation of the velocity of a reference frame $\mathscr{F}_b$ with respect to $\mathscr{F}_0$ in spatial coordinates, which is what the superscript refers to.

We can also write the velocity in the body frame. Using $\bar{q}^0(t) = g_{0b}(t)\bar{q}^b$ and Eq. (2.74) we get

$$\begin{aligned} \bar{v}_{q^b}(t) &= g_{0b}^{-1}(t)\bar{v}_{q^0}(t) \\ &= g_{0b}^{-1}(t)\dot{g}_{0b}(t)g_{0b}^{-1}(t)\bar{q}^0 \\ &= g_{0b}^{-1}(t)\dot{g}_{0b}(t)\bar{q}^b. \end{aligned} \tag{2.77}$$

Also $g_{0b}^{-1}(t)\dot{g}_{0b}(t) \in se(3)$ has a geometrically meaningful interpretation and represents the same velocity as above, but now as observed from the reference frame attached to the body $\mathscr{F}_b$, i.e., in body coordinates. The body velocity variable $\widehat{V}_{0b}^B$ is defined as

$$\widehat{V}_{0b}^B = g_{0b}^{-1}\dot{g}_{0b} = \begin{bmatrix} \widehat{\omega}_{0b}^B & v_{0b}^B \\ 0 & 0 \end{bmatrix}. \tag{2.78}$$

In this case $\widehat{V}_{0b}^B$ is obtained by left translating $\dot{g}_{0b} \in T_g SE(3)$ back to the identity, and consequently $\widehat{V}_{0b}^B$ is also an element of the Lie algebra. If we write the body velocity variables in vector form we get

$$V_{0b}^B = \begin{bmatrix} v_{0b}^B \\ \omega_{0b}^B \end{bmatrix} = \begin{bmatrix} R_{0b}^{\mathsf{T}} \dot{p}_{0b} \\ (R_{0b}^{\mathsf{T}} \dot{R}_{0b})^{\vee} \end{bmatrix}. \tag{2.79}$$

The map $\widehat{V}_{0b}^B$ thus takes a point represented in the frame attached to the rigid body to the velocity of this point by $\bar{v}_{q^b}(t) = \widehat{V}_{0b}^B \bar{q}^b$, also represented in the body frame. This can also be written as

$$v_{q^b}(t) = \omega_{0b}^B \times q^b + v_{0b}^B \tag{2.80}$$

or using homogeneous coordinates as

$$\bar{v}_{q^b}(t) = \widehat{V}_{0b}^B \bar{q}^b. \tag{2.81}$$

We will study body and spatial velocities in more detail in Chap. 3 where we derive the kinematics of a single rigid body.

2.6.5.3 The Velocity of a Point

Recall from (2.74) that the velocity of a point q^0 (fixed in reference frame $\mathscr{F}_b$), whose motion is given by $\bar{q}^0(t) = g_{0b}(t)\bar{q}^b$ can be written as

$$\bar{v}_{q^0}(t) = \frac{\mathrm{d}}{\mathrm{d}t}\bar{q}^0 = \widehat{V}^S_{0b}\bar{q}^0. \tag{2.82}$$

This gives us the velocity of a point q^0 fixed in the body frame, but expressed in the inertial frame $\mathscr{F}_0$, when the body frame moves with a constant velocity V^S_{0b}. In fact, this is a general result. Given a fixed point q^0 and a matrix representation of the Lie algebra given by $\widehat{X}$, then the velocity of the body-fixed point q^0 represented in the inertial frame $\mathscr{F}_0$ is given by

$$v_{q^0}(t) = \widehat{X}^0_{0b}q^0(t). \tag{2.83}$$

Example 2.7 For $SE(3)$ this becomes

$$\bar{v}_{q^0}(t) = \widehat{V}^S_{0b}\bar{q}^0(t) \tag{2.84}$$

written in spatial velocity variables. If we write the twist coordinates as $V^S_{0b} = [u^S\ v^S\ w^S\ p^S\ q^S\ r^S]^\mathsf{T}$, the twist is defined in matrix form as

$$\widehat{V}^S_{0b} = \begin{bmatrix} 0 & -r^S & q^S & u^S \\ r^S & 0 & -p^S & v^S \\ -q^S & p^S & 0 & w^S \\ 0 & 0 & 0 & 0 \end{bmatrix}. \tag{2.85}$$

In body velocity variables the expression becomes

$$\bar{v}_{q^b}(t) = \widehat{V}^B_{0b}\bar{q}^b(t). \tag{2.86}$$

Example 2.8 Similarly, for pure rotational motion the corresponding expression for the spatial velocity variables is given as

$$\bar{v}_{q^0}(t) = \widehat{\omega}^S_{0b}q^0(t) \tag{2.87}$$

where $\widehat{\omega}$ is the standard skew-symmetric matrix representation of the Lie algebra $so(3)$, so (2.87) can also be written as $v_{q^0}(t) = \omega^S_{0b} \times q^0$. Rotational motion in this form can also be written in terms of body velocity variables as

$$v_{q^b}(t) = \widehat{\omega}^B_{0b}q^b(t) \tag{2.88}$$

which is the same as $v_{q^b}(t) = \omega^B_{0b} \times q^b$.

Example 2.9 Also for planar motion we can obtain the same relation by writing

$$v_{q^0}(t) = \begin{bmatrix} \dot{q}_x^0 \\ \dot{q}_y^0 \\ 0 \end{bmatrix}, \qquad \bar{q}^0(t) = \begin{bmatrix} q_x^0 \\ q_y^0 \\ 1 \end{bmatrix}, \qquad \widehat{V}_{0b}^S = \begin{bmatrix} 0 & -r^S & u^S \\ r^S & 0 & v^S \\ 0 & 0 & 0 \end{bmatrix}. \tag{2.89}$$

$\bar{v}_{q^0}(t) = \widehat{V}_{0b}^S \bar{q}^0(t)$ can be written as

$$\begin{aligned} \dot{q}_x^0 &= -r^S q_y^0 + u^S, \\ \dot{q}_y^0 &= r^S q_x^0 + v^S. \end{aligned} \tag{2.90}$$

We note that the formulation of the twist in (2.89) is just a compact way of writing the more general expression that we found for $se(3)$. That is, we could also write this as an element of $se(3)$ as

$$\widehat{V}_{0b}^S = \begin{bmatrix} 0 & -r^S & 0 & u^S \\ r^S & 0 & 0 & v^S \\ 0 & 0 & 0 & 0 \\ 0 & 0 & 0 & 0 \end{bmatrix}. \tag{2.91}$$

2.6.6 *Actions on Lie Groups*

An element of a Lie group can act on different mathematical spaces, and depending on what space they act on we get different geometrical interpretations. First of all, the elements of a group can act on a vector space: we saw an example of this when a point or a vector was transformed by a homogeneous transformation matrix or a rotation matrix. In this case the group element, for example the rotation matrix, acts on a vector, i.e., it transforms the vector in some way. A group can also act on itself: examples of this are the rigid body motions in Eq. (2.70) and a change of observer's frame. Finally a group element can act on its tangent space. This is useful as it allows us to transform the tangent vectors from the identity to any group element and to represent velocities in different frames. We will discuss such actions in more detail in this section.

2.6.6.1 Conjugations

If we let G be a Lie group and let g and x be elements of G, then we define a *conjugation* as a homomorphism given by

$$h_g(x) = gxg^{-1}. \tag{2.92}$$

This is thus a mapping from the manifold of G to itself, i.e., $h_g : G \to G$ which follows directly from Properties 2.4 and 2.6. If we let X be an element of a Lie algebra, we get a similar transformation from the tangent space at the identity to itself, i.e., $\mathrm{Ad}_g : \mathfrak{g} \to \mathfrak{g}$ given by

$$\mathrm{Ad}_g(X) = g\widehat{X}g^{-1}. \tag{2.93}$$

We see that a conjugation represents the same linear transformation, but as observed from a different reference frame and using a different basis. Let $A \in \mathbb{R}^{n\times n}$ denote a matrix. Then left multiplication $L_A : \mathbb{R}^n \to \mathbb{R}^n$ defines a correspondence between the matrix A and linear functions from $\mathbb{R}^n$ to $\mathbb{R}^n$ given by

$$L_A(X) = AX, \quad X \in \mathbb{R}^n. \tag{2.94}$$

Using this notation we can write the conjugation

$$\mathrm{Ad}_g(X) = g\widehat{X}g^{-1} \tag{2.95}$$

as

$$L_{gXg^{-1}} = L_g \circ L_X \circ L_{g^{-1}}. \tag{2.96}$$

We recall that an element of the Lie algebra X can be interpreted as a linear transformation, for example as in (2.82), of a vector or a point represented in the body frame $\mathscr{F}_b$. If we want to perform the same operation on a point q^a represented in the reference frame $\mathscr{F}_a$ which relates to the body frame through the transformation g_{ab}, we need to perform the following operations, which also gives us an intuitive interpretation of conjugations: First we need to transform the point q^a to frame $\mathscr{F}_b$. This is obtained by a transformation by g_{ab}^{-1} which gives us $q^b = L_{g_{ab}^{-1}} q^a$. Now that the point is represented in the body frame we can perform the desired operation, represented by X, which is applied to q^b by $L_X q^b$. Finally we need to transform the vector back into the frame $\mathscr{F}_a$ by applying $L_{g_{ab}}$. Applying all these transformations gives us $L_{gXg^{-1}} q^a = L_g \circ L_X \circ L_{g^{-1}} q^a$. This is the same transformation as $L_X q^b = \widehat{X} q^b$ but represented in reference frame $\mathscr{F}_a$. A conjugation thus represents a change of basis or change of observer and allows us to represent the linear transformation X but in a different basis or reference frame. In the following we discuss this kind of transformations in more detail.

We can define right multiplication in a similar way to left multiplication. Right multiplications $R_A : \mathbb{R}^n \to \mathbb{R}^n$ define linear functions from $\mathbb{R}^n$ to $\mathbb{R}^n$ given by

$$R_A(X) = XA, \quad X \in \mathbb{R}^n. \tag{2.97}$$

2.6.6.2 The Adjoint Map Ad_g

The Adjoint map Ad_g represents an action of an element of the Lie group on its Lie algebra. The Adjoint map is important as it allows us to represent an element of the Lie algebra in different coordinate frames. The Adjoint map is given by the conjugation by an element g. Take for example the conjugation by g on the identity element, i.e., $geg^{-1} = e$. This operation maps the identity element of the Lie group to itself. The differential of this map is the Jacobian, and maps the tangent space at the identity to itself. More specifically we can write the Adjoint map $\mathrm{Ad} : G \times \mathfrak{g} \to \mathfrak{g}$ as

$$\mathrm{Ad}_g X = g\widehat{X}g^{-1}, \quad \forall g \in G, \widehat{X} \in \mathfrak{g}. \tag{2.98}$$

Ad is thus a function of g for which (2.98) is true. The Adjoint map satisfies the following properties:

Property 2.22 The inverse of the Adjoint map is given as

$$\mathrm{Ad}_g^{-1} = \mathrm{Ad}_{g^{-1}}. \tag{2.99}$$

Property 2.23 The Adjoint map given by the conjugation by gh is found as

$$\mathrm{Ad}_g \mathrm{Ad}_h = \mathrm{Ad}_{gh}. \tag{2.100}$$

This can be seen from Eq. (2.98). We will now look at the adjoint maps of the most important Lie groups used in robotics.

2.6.6.3 The Special Orthogonal Group

Let R be an element of $SO(3)$ and ω an element of $so(3)$. Then the Adjoint map $\mathrm{Ad}_R : so(3) \to so(3)$ is given by

$$\mathrm{Ad}_R \omega = R\widehat{\omega}R^{\mathsf{T}}. \tag{2.101}$$

The Adjoint map on $so(3)$ also admits a 3×3 matrix representation, i.e.,

$$\mathrm{Ad}_R = R \tag{2.102}$$

and we get an alternative action on the vector ω:

$$\mathrm{Ad}_R \omega = R\omega. \tag{2.103}$$

The fact that $\mathrm{Ad}_R = R$ is an "accidental property" of $SO(3)$ and can not be taken as a general result.

Remark 2.4 When the Adjoint map acts on a matrix, it is implicitly understood that it is of the form of (2.101), while when the Adjoint map acts on a vector it is in the form (2.103). Equations (2.101) and (2.103) represent two different formulations of the same transformation.

2.6.6.4 The Special Euclidean Group

Also for $SE(3)$ we can find similar relations. Let $g = (R, p)$ be an element of $SE(3)$ and $V = [v^{\mathsf{T}}\ \omega^{\mathsf{T}}]^{\mathsf{T}}$ an element of $se(3)$, the Adjoint map $\mathrm{Ad}_g : se(3) \to se(3)$ is then defined as

$$\mathrm{Ad}_g V = g\widehat{V}g^{-1}. \tag{2.104}$$

We can also find a 6×6 matrix representation of the Adjoint map that acts on the vector $V \in \mathbb{R}^6$. Let V_{ab}^B be the velocity of a rigid body expressed in the body frame with respect to a fixed frame $\mathscr{F}_a$ and g_{ab} the transformation from $\mathscr{F}_a$ to $\mathscr{F}_b$. Then the linear and angular velocities given in spatial coordinates can be written as

$$\begin{aligned} v_{ab}^S &= p_{ab} \times \left(R_{ab}\omega_{ab}^B\right) + R_{ab}v_{ab}^B, \\ \omega_{ab}^S &= R_{ab}\omega_{ab}^B. \end{aligned} \tag{2.105}$$

This can be written in vector form as

$$V_{ab}^S = \mathrm{Ad}_{g_{ab}} V_{ab}^B = \begin{bmatrix} R_{ab} & \widehat{p}_{ab}R_{ab} \\ 0 & R_{ab} \end{bmatrix} \begin{bmatrix} v_{ab}^B \\ \omega_{ab}^B \end{bmatrix} \tag{2.106}$$

and we thus get

$$\mathrm{Ad}_{g_{ab}} = \begin{bmatrix} R_{ab} & \widehat{p}_{ab}R_{ab} \\ 0 & R_{ab} \end{bmatrix}. \tag{2.107}$$

We see that if we let $\mathscr{F}_a$ be the inertial frame, i.e., $\mathscr{F}_0$, we get the relation between body and spatial velocity variables that we have seen earlier, i.e.,

$$V_{0b}^S = \mathrm{Ad}_{g_{0b}} V_{0b}^B \tag{2.108}$$

or alternatively

$$\begin{bmatrix} v_{0b}^S \\ \omega_{0b}^S \end{bmatrix} = \begin{bmatrix} R_{0b} & \widehat{p}_{0b}R_{0b} \\ 0 & R_{0b} \end{bmatrix} \begin{bmatrix} v_{0b}^B \\ \omega_{0b}^B \end{bmatrix}. \tag{2.109}$$

The Adjoint map can also be used to find the spatial velocities with respect to different frames in the same way.

We can check that the above corresponds to the expressions that we found in (2.76) and (2.79). The right hand side is given by

$$\begin{aligned} \begin{bmatrix} R_{0b} & \widehat{p}_{0b}R_{0b} \\ 0 & R_{0b} \end{bmatrix} \begin{bmatrix} R_{0b}^{\mathsf{T}}\dot{p}_{0b} \\ \left(R_{0b}^{\mathsf{T}}\dot{R}_{0b}\right)^{\vee} \end{bmatrix} &= \begin{bmatrix} R_{0b}R_{0b}^{\mathsf{T}}\dot{p}_{0b} + \widehat{p}_{0b}R_{0b}\left(R_{0b}^{\mathsf{T}}\dot{R}_{0b}\right)^{\vee} \\ R_{0b}\left(R_{0b}^{\mathsf{T}}\dot{R}_{0b}\right)^{\vee} \end{bmatrix} \\ &= \begin{bmatrix} \dot{p}_{0b} + \widehat{p}_{0b}R_{0b}\left(R_{0b}^{\mathsf{T}}\dot{R}_{0b}\right)^{\vee} \\ R_{0b}\left(R_{0b}^{\mathsf{T}}\dot{R}_{0b}\right)^{\vee} \end{bmatrix} \end{aligned}$$

$$
\begin{aligned}
&= \begin{bmatrix} \dot{p}_{0b} + \widehat{p}_{0b} R_{0b} R_{0b}^{\mathsf{T}} \big(\dot{R}_{0b} R_{0b}^{\mathsf{T}}\big)^{\vee} \\ R_{0b} R_{0b}^{\mathsf{T}} \big(\dot{R}_{0b} R_{0b}^{\mathsf{T}}\big)^{\vee} \end{bmatrix} \\
&= \begin{bmatrix} \dot{p}_{0b} + \widehat{p}_{0b} \big(\dot{R}_{0b} R_{0b}^{\mathsf{T}}\big)^{\vee} \\ \big(\dot{R}_{0b} R_{0b}^{\mathsf{T}}\big)^{\vee} \end{bmatrix} \\
&= \begin{bmatrix} \dot{p}_{0b} - \dot{R}_{0b} R_{0b}^{\mathsf{T}} p_{0b} \\ \big(\dot{R}_{0b} R_{0b}^{\mathsf{T}}\big)^{\vee} \end{bmatrix}
\end{aligned} \tag{2.110}
$$

where we have used that

$$
\big(R_{0b}^{\mathsf{T}} \dot{R}_{0b}\big)^{\vee} = \omega_{0b}^{B} = R_{0b}^{\mathsf{T}} \omega_{0b}^{S} = R_{0b}^{\mathsf{T}} \big(\dot{R}_{0b} R_{0b}^{\mathsf{T}}\big)^{\vee} \tag{2.111}
$$

and we get (2.76) as required.

For $SE(3)$ we can show Property 2.22 by straight forward calculation:

$$
\mathrm{Ad}_g^{-1} = \begin{bmatrix} R^{\mathsf{T}} & -\widehat{R^{\mathsf{T}} p} R^{\mathsf{T}} \\ 0 & R^{\mathsf{T}} \end{bmatrix} = \begin{bmatrix} R^{\mathsf{T}} & -R^{\mathsf{T}} \hat{p} \\ 0 & R^{\mathsf{T}} \end{bmatrix} = \mathrm{Ad}_{g^{-1}}. \tag{2.112}
$$

The adjoint map on $SE(3)$ is discussed in detail in Chap. 3.

2.6.6.5 Planar Motion

For planar motion we write

$$
\widehat{\omega}_{ab}^{B} = \begin{bmatrix} 0 & -\omega \\ \omega & 0 \end{bmatrix}, \qquad R_{ab} = \begin{bmatrix} \cos\theta_{ab} & -\sin\theta_{ab} \\ \sin\theta_{ab} & \cos\theta_{ab} \end{bmatrix}. \tag{2.113}
$$

First we note that

$$
\widehat{\omega}_{ab}^{S} = R_{ab} \widehat{\omega}_{ab}^{B} R_{ab}^{\mathsf{T}} = \widehat{\omega}_{ab}^{B} \tag{2.114}
$$

and thus $\widehat{V}_{ab}^{S} = g_{ab} \widehat{V}_{ab}^{B} g_{0b}^{-1}$ can be written as

$$
\begin{aligned}
v_{ab}^{S} &= \widehat{p}_{ab} \omega_{ab}^{B} + R_{ab} v_{ab}^{B}, \\
\omega_{ab}^{S} &= \omega_{ab}^{B}
\end{aligned} \tag{2.115}
$$

where we have defined $\widehat{p}_{ab} = [y_{ab} \ -x_{ab}]^{\mathsf{T}}$. The Adjoint map $\mathrm{Ad}_g : se(2) \to se(2)$ can then be written in vector form as

$$
V_{ab}^{S} = \mathrm{Ad}_{g_{ab}} V_{ab}^{B} = \begin{bmatrix} R_{ab} & \widehat{p}_{ab} \\ 0 & 1 \end{bmatrix} \begin{bmatrix} v_{ab}^{B} \\ \omega_{ab}^{B} \end{bmatrix} \tag{2.116}
$$

and we get

$$\mathrm{Ad}_{g_{ab}} = \begin{bmatrix} R_{ab} & \widehat{p}_{ab} \\ 0 & 1 \end{bmatrix} = \begin{bmatrix} \cos\theta_{ab} & -\sin\theta_{ab} & y_{0b} \\ \sin\theta_{ab} & \cos\theta_{ab} & -x_{0b} \\ 0 & 0 & 1 \end{bmatrix}. \tag{2.117}$$

2.6.6.6 The Adjoint Map ad_X

As we have seen, there exists a commutator, or a Lie bracket, on every Lie algebra which maps two elements on the Lie algebra to itself. In this section we will look into how this transformation, denoted $\mathrm{ad}_X Y = [\widehat{X}, \widehat{Y}] = \widehat{X}\widehat{Y} - \widehat{Y}\widehat{X}$ can be interpreted geometrically. This will lead us to the definition of another action on a Lie group, the *adjoint map* ad.

Close to the identity, a group element $g \in G$ can be approximated by $g = I + t\widehat{X} + o(t^2)$ and its inverse by $g^{-1} = I - t\widehat{X} + o(t^2)$ where $o(t^2)$ guarantees that g is still an element in G. Recall that conjugation of an element of the Lie algebra is given by $g\widehat{Y}g^{-1}$. Substituting the expression for g gives

$$g\widehat{Y}g^{-1} = \big(I + t\widehat{X} + o(t^2)\big)\widehat{Y}\big(I - t\widehat{X} + o(t^2)\big) = \widehat{Y} + t(\widehat{X}\widehat{Y} - \widehat{Y}\widehat{X}) + o\big(t^2\big). \tag{2.118}$$

Differentiating and setting $t = 0$ allows us to write this locally in terms of the Lie algebra element $\widehat{X}$ as (recall that $\widehat{Y}$ is fixed)

$$\frac{\mathrm{d}}{\mathrm{d}t} g\widehat{Y}g^{-1} = \widehat{X}\widehat{Y} - \widehat{Y}\widehat{X} = [\widehat{X}, \widehat{Y}]. \tag{2.119}$$

Note that this is the derivative of the Adjoint map Ad_g, i.e., we define

$$\mathrm{ad}_X Y = \frac{\mathrm{d}}{\mathrm{d}t}\mathrm{Ad}_g Y. \tag{2.120}$$

The adjoint map ad can also be written in terms of the important Lie bracket:

Definition 2.17 The adjoint map ad_X is defined as

$$\mathrm{ad}_X Y = [\widehat{X}, \widehat{Y}] = \widehat{X}\widehat{Y} - \widehat{Y}\widehat{X} \tag{2.121}$$

where $[\cdot, \cdot]$ is the Lie bracket.

Again we note the slight abuse of notation because Y is a matrix on the left hand side, but a vector on the right. We will assume that it is implicitly understood that the adjoint map ad_X is given by an $n \times n$-matrix when acting on a vector and by the expression in Definition 2.17 when acting on a matrix. The two formulations will of course give us the same result.

The Lie bracket is thus an action of the Lie algebra on itself. We note that the Lie bracket is linear, i.e.,

$$[a\widehat{X}_1 + b\widehat{X}_2, \widehat{Y}] = a[\widehat{X}_1, \widehat{Y}] + b[\widehat{X}_2, \widehat{Y}]. \tag{2.122}$$

This means that all the elements of the Lie algebra can be found from the pairs of the basis elements of the same Lie algebra.

Equation (2.119) provides us with an intuition of the action imposed by the adjoint map ad_g: Given an element $\widehat{X} \in \mathfrak{g}$. This is an element of the Lie algebra and gives us a 1-parameter subgroup of G around the identity parameterized by t. Locally this is written as $g(t) = I + t\widehat{X} + o(t^2)$. The action of g on another element of the Lie group Y is given by $\mathrm{Ad}_g Y$ and transforms a (fixed) element of the Lie algebra Y by g. The time derivative of this transformation, denoted $\mathrm{ad}_X Y$, thus gives us how Y changes with time as a result of the action imposed by X.

To get a deeper understanding of the geometrical interpretation of the Lie bracket, we first write the following corollary:

Corollary 2.1 *For $\widehat{X}, \widehat{Y} \in \mathfrak{g}$ and small t the following relation is true*:

$$e^{-t\widehat{Y}} e^{-t\widehat{X}} e^{t\widehat{Y}} e^{t\widehat{X}} = e^{t^2[\widehat{X},\widehat{Y}] + o(t^3)}. \tag{2.123}$$

Equation (2.123) provides us with a geometric interpretation of the Lie bracket. Recall that $\mathrm{Ad}_g Y$ tells us how Y is transformed by an element of the Lie group g. The differential of this, namely $\mathrm{ad}_X Y$, thus tells us how Y changes with X near the identity. Assume a curve given by $e^{-t\widehat{Y}} e^{-t\widehat{X}} e^{t\widehat{Y}} e^{t\widehat{X}}$, then the Lie bracket $[\widehat{X}, \widehat{Y}]$ represents the tangent vector to this curve at the identity element as illustrated in Fig. 2.7. If the initial point on the manifold is x, the Lie bracket is given by

$$[X, Y](x) = \lim_{t \to 0^+} \frac{e^{-t\widehat{Y}} e^{-t\widehat{X}} e^{t\widehat{Y}} e^{t\widehat{X}} x - x}{t} \tag{2.124}$$

which is illustrated in Fig. 2.7. Exponential maps of this kind are discussed in more detail in Sect. 2.8.

We can also look at the Lie bracket as a measure of how X fails to commute with Y. More specifically, the Lie bracket tells us how elements of G near the identity in the direction of X fail to commute with elements in the direction of Y. For instance, if X and Y commute, then $[X, Y] = 0$. We also see that if $[X, Y] \neq 0$ then this tells us that if we first follow the trajectories of X followed by Y we do not get the same result as if we first follow the trajectories of Y followed by X. To gain some more insight into the structure of the Lie bracket we include an example taken from Bullo and Lewis (2000).

Example 2.10 Consider the vector fields $X = \frac{\partial}{\partial y}$ and $Y = \frac{\partial}{\partial x} + y\frac{\partial}{\partial z}$ on a manifold $\mathscr{M} = \mathbb{R}^3$. The Lie bracket is then given by $[X, Y] = \frac{\partial}{\partial z}$ which means that if we

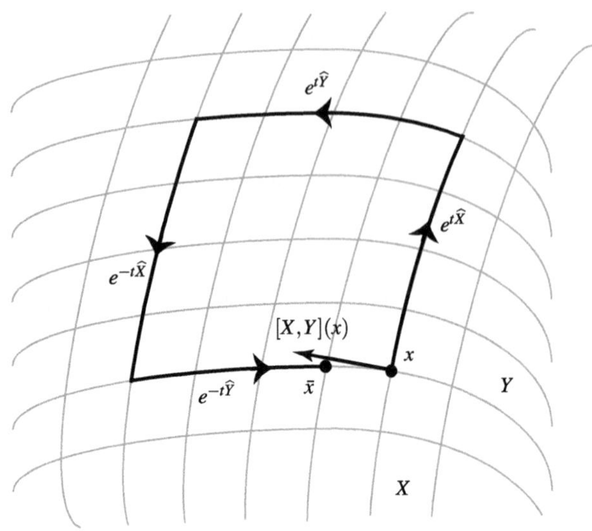

Fig. 2.7 A geometric interpretation of the Lie bracket $[X, Y]$. The path that takes us from an initial point x to a final point $\bar{x} = e^{-t\widehat{Y}} e^{-t\widehat{X}} e^{t\widehat{Y}} e^{t\widehat{X}} x$. The Lie bracket tells us how Y changes in the direction of X

follow a flow first along X, then along Y, then along $-X$ and finally along $-Y$, all for the same amount of time, then this motion will move us upwards in the direction of the z-axis.

We will now look at how we find the adjoint map ad_X for the most important groups.

2.6.6.7 Euclidean Space

The Lie algebra associated with the Euclidean space $\mathbb{R}^n$ is $\mathbb{R}^n$ itself. As this is commutative for all n the Lie bracket is trivially zero. The adjoint map ad_v corresponding to the Euclidean space of dimension n is thus an $n \times n$ zero matrix.

2.6.6.8 The Special Orthogonal Group

The Lie algebra of the special orthogonal group can be written in terms of the basis elements

$$E_x = \begin{bmatrix} 0 & 0 & 0 \\ 0 & 0 & -1 \\ 0 & 1 & 0 \end{bmatrix}, \quad E_y = \begin{bmatrix} 0 & 0 & 1 \\ 0 & 0 & 0 \\ -1 & 0 & 0 \end{bmatrix}, \quad E_z = \begin{bmatrix} 0 & -1 & 0 \\ 1 & 0 & 0 \\ 0 & 0 & 0 \end{bmatrix}. \tag{2.125}$$

We can find an adjoint representation of E_x, E_y, and E_z in a very simple form. Take for example the adjoint action of E_x on a vector $Y = [a\ b\ c]^\mathsf{T} \in \mathbb{R}^3$, then

$$
\begin{aligned}
E_x\widehat{Y} - \widehat{Y}E_x &= \begin{bmatrix} 0 & 0 & 0 \\ 0 & 0 & -1 \\ 0 & 1 & 0 \end{bmatrix}\begin{bmatrix} 0 & -c & b \\ c & 0 & -a \\ -b & a & 0 \end{bmatrix} \\
&\quad - \begin{bmatrix} 0 & -c & b \\ c & 0 & -a \\ -b & a & 0 \end{bmatrix}\begin{bmatrix} 0 & 0 & 0 \\ 0 & 0 & -1 \\ 0 & 1 & 0 \end{bmatrix} \\
&= \begin{bmatrix} 0 & -b & -c \\ b & 0 & 0 \\ c & 0 & 0 \end{bmatrix}.
\end{aligned} \tag{2.126}
$$

If we write the skew-symmetric matrix $\widehat{Y}$ as a vector Y we can find the adjoint map ad_{E_x} through the relation

$$
\mathrm{ad}_{E_x}Y = \begin{bmatrix} 0 & 0 & 0 \\ 0 & 0 & -1 \\ 0 & 1 & 0 \end{bmatrix}\begin{bmatrix} a \\ b \\ c \end{bmatrix} = \begin{bmatrix} 0 \\ -c \\ b \end{bmatrix} \tag{2.127}
$$

whose skew-symmetric form is the same as (2.126). The same can be shown for E_y and E_z, i.e.,

$$
\mathrm{ad}_{E_x} = E_x, \qquad \mathrm{ad}_{E_y} = E_y, \qquad \mathrm{ad}_{E_z} = E_z. \tag{2.128}
$$

To find the adjoint map of an element ω in $so(3)$ we write $\omega = [p\ q\ r]^\mathsf{T}$, and write the adjoint map as

$$
\mathrm{ad}_\omega = p\mathrm{ad}_{E_x} + q\mathrm{ad}_{E_y} + r\mathrm{ad}_{E_z} = pE_x + qE_y + rE_z = \widehat{\omega} \tag{2.129}
$$

where

$$
\widehat{\omega} = \begin{bmatrix} 0 & -r & q \\ r & 0 & -p \\ -q & p & 0 \end{bmatrix}. \tag{2.130}
$$

For $SO(3)$ we thus have the rather special case that the adjoint representation ad_ω equals the matrix representation of the Lie algebra $\widehat{\omega}$ of the angular velocities. This property is only true for $SO(3)$, however, and cannot be taken as a general rule. This will be clear when we find the adjoint representation of an element of $se(3)$.

2.6.6.9 The Special Euclidean Group

The Lie algebra of the Special Euclidean group can be written in terms of the basis elements

$$
\begin{gathered}
\widehat{v}_x = \begin{bmatrix} 0 & 0 & 0 & 1 \\ 0 & 0 & 0 & 0 \\ 0 & 0 & 0 & 0 \\ 0 & 0 & 0 & 0 \end{bmatrix}, \quad \widehat{v}_y = \begin{bmatrix} 0 & 0 & 0 & 0 \\ 0 & 0 & 0 & 1 \\ 0 & 0 & 0 & 0 \\ 0 & 0 & 0 & 0 \end{bmatrix}, \\
\widehat{v}_z = \begin{bmatrix} 0 & 0 & 0 & 0 \\ 0 & 0 & 0 & 0 \\ 0 & 0 & 0 & 1 \\ 0 & 0 & 0 & 0 \end{bmatrix}, \quad \widehat{\omega}_x = \begin{bmatrix} 0 & 0 & 0 & 0 \\ 0 & 0 & -1 & 0 \\ 0 & 1 & 0 & 0 \\ 0 & 0 & 0 & 0 \end{bmatrix}, \\
\widehat{\omega}_y = \begin{bmatrix} 0 & 0 & 1 & 0 \\ 0 & 0 & 0 & 0 \\ -1 & 0 & 0 & 0 \\ 0 & 0 & 0 & 0 \end{bmatrix}, \quad \widehat{\omega}_z = \begin{bmatrix} 0 & -1 & 0 & 0 \\ 1 & 0 & 0 & 0 \\ 0 & 0 & 0 & 0 \\ 0 & 0 & 0 & 0 \end{bmatrix}.
\end{gathered} \tag{2.131}
$$

For two twists $X = [v_1^{\mathsf{T}}\ \omega_1^{\mathsf{T}}]^{\mathsf{T}}$ and $Y = [v_2^{\mathsf{T}}\ \omega_2^{\mathsf{T}}]^{\mathsf{T}}$ we find

$$
\widehat{X}\widehat{Y} - \widehat{Y}\widehat{X} = \begin{bmatrix} \widehat{\omega_1 \times \omega_2} & \omega_1 \times v_2 - \omega_2 \times v_1 \\ 0 & 0 \end{bmatrix}. \tag{2.132}
$$

If we compare this to the vector representation of the adjoint map we will find the basis elements of the corresponding adjoint maps

$$
\begin{gathered}
\mathrm{ad}_{v_x} = \begin{bmatrix} 0 & E_x \\ 0 & 0 \end{bmatrix}, \quad \mathrm{ad}_{v_y} = \begin{bmatrix} 0 & E_y \\ 0 & 0 \end{bmatrix}, \quad \mathrm{ad}_{v_z} = \begin{bmatrix} 0 & E_z \\ 0 & 0 \end{bmatrix}, \\
\mathrm{ad}_{\omega_x} = \begin{bmatrix} E_x & 0 \\ 0 & E_x \end{bmatrix}, \quad \mathrm{ad}_{\omega_y} = \begin{bmatrix} E_y & 0 \\ 0 & E_y \end{bmatrix}, \quad \mathrm{ad}_{\omega_z} = \begin{bmatrix} E_z & 0 \\ 0 & E_z \end{bmatrix}.
\end{gathered}
$$

Let $V = \begin{bmatrix} u\ v\ w\ p\ q\ r \end{bmatrix}^{\mathsf{T}}$ and write

$$
\mathrm{ad}_V = u\mathrm{ad}_{v_x} + v\mathrm{ad}_{v_y} + w\mathrm{ad}_{v_z} + p\mathrm{ad}_{\omega_x} + q\mathrm{ad}_{\omega_y} + r\mathrm{ad}_{\omega_z} \tag{2.133}
$$

which gives us the adjoint map as

$$\mathrm{ad}_V = \begin{bmatrix} 0 & -r & q & 0 & -w & v \\ r & 0 & -p & w & 0 & -u \\ -q & p & 0 & -v & u & 0 \\ 0 & 0 & 0 & 0 & -r & q \\ 0 & 0 & 0 & r & 0 & -p \\ 0 & 0 & 0 & -q & p & 0 \end{bmatrix}. \tag{2.134}$$

For a twist V_2 the adjoint map $\mathrm{ad}_{V_1} V_2$ tells us how the twist V_2 changes with V_1 in the vicinity of the identity element.

2.6.6.10 Planar Motion

The basis elements of the planar group are given by

$$\widehat{v}_x = \begin{bmatrix} 0 & 0 & 1 \\ 0 & 0 & 0 \\ 0 & 0 & 0 \end{bmatrix}, \qquad \widehat{v}_y = \begin{bmatrix} 0 & 0 & 0 \\ 0 & 0 & 1 \\ 0 & 0 & 0 \end{bmatrix}, \qquad \widehat{\omega}_z = \begin{bmatrix} 0 & -1 & 0 \\ 1 & 0 & 0 \\ 0 & 0 & 0 \end{bmatrix}. \tag{2.135}$$

We can find an adjoint representation for the planar group similar to the ones that we found for $SO(3)$ and $SE(3)$. The adjoint maps corresponding to each degree of freedom is given by

$$\mathrm{ad}_{v_x} = \begin{bmatrix} 0 & 0 & 0 \\ 0 & 0 & -1 \\ 0 & 0 & 0 \end{bmatrix}, \qquad \mathrm{ad}_{v_y} = \begin{bmatrix} 0 & 0 & 1 \\ 0 & 0 & 0 \\ 0 & 0 & 0 \end{bmatrix}, \qquad \mathrm{ad}_{\omega_z} = \begin{bmatrix} 0 & -1 & 0 \\ 1 & 0 & 0 \\ 0 & 0 & 0 \end{bmatrix}, \tag{2.136}$$

and we see that

$$\mathrm{ad}_{v_x} = -\widehat{v}_y, \qquad \mathrm{ad}_{v_y} = \widehat{v}_x, \qquad \mathrm{ad}_{\omega_z} = \widehat{\omega}_z. \tag{2.137}$$

The adjoint representation is thus given by

$$\mathrm{ad}_V = \begin{bmatrix} 0 & -r & v \\ r & 0 & -u \\ 0 & 0 & 0 \end{bmatrix} \tag{2.138}$$

where $V = [u\ v\ r]^\mathsf{T}$.

2.6.6.11 The Schönflies Group

The Schönflies group allows translation in three degrees of freedom and rotation around one of the coordinate axes, normally the z-axis. The Lie algebra is therefore given by

$$\widehat{X} = \begin{bmatrix} 0 & -\omega & 0 & v_x \\ \omega & 0 & 0 & v_y \\ 0 & 0 & 0 & v_z \\ 0 & 0 & 0 & 0 \end{bmatrix}. \tag{2.139}$$

We can now use that $\mathrm{ad}_X Y = \hat{X}\hat{Y} - \hat{Y}\hat{X}$ to find the adjoint map. This is not normally shown in standard text books on robotics, so we show the derivation in some more detail. If we write $X = [X_x \; X_y \; X_z \; X_\omega]^\mathsf{T}$ and $Y = [Y_x \; Y_y \; Y_z \; Y_\omega]^\mathsf{T}$ we find the adjoint map as

$$\begin{aligned} \mathrm{ad}_X Y &= \begin{bmatrix} 0 & -X_\omega & 0 & X_x \\ X_\omega & 0 & 0 & X_y \\ 0 & 0 & 0 & X_z \\ 0 & 0 & 0 & 0 \end{bmatrix} \begin{bmatrix} 0 & -Y_\omega & 0 & Y_x \\ Y_\omega & 0 & 0 & Y_y \\ 0 & 0 & 0 & Y_z \\ 0 & 0 & 0 & 0 \end{bmatrix} \\ &\quad - \begin{bmatrix} 0 & -Y_\omega & 0 & Y_x \\ Y_\omega & 0 & 0 & Y_y \\ 0 & 0 & 0 & Y_z \\ 0 & 0 & 0 & 0 \end{bmatrix} \begin{bmatrix} 0 & -X_\omega & 0 & X_x \\ X_\omega & 0 & 0 & X_y \\ 0 & 0 & 0 & X_z \\ 0 & 0 & 0 & 0 \end{bmatrix} \\ &= \begin{bmatrix} -X_\omega Y_\omega & 0 & 0 & -X_\omega Y_y \\ 0 & -X_\omega Y_\omega & 0 & X_\omega Y_x \\ 0 & 0 & 0 & 0 \\ 0 & 0 & 0 & 0 \end{bmatrix} \\ &\quad - \begin{bmatrix} -Y_\omega X_\omega & 0 & 0 & -Y_\omega X_y \\ 0 & -Y_\omega X_\omega & 0 & Y_\omega X_x \\ 0 & 0 & 0 & 0 \\ 0 & 0 & 0 & 0 \end{bmatrix} \\ &= \begin{bmatrix} 0 & 0 & 0 & -X_\omega Y_y + Y_\omega X_y \\ 0 & 0 & 0 & X_\omega Y_x - Y_\omega X_x \\ 0 & 0 & 0 & 0 \\ 0 & 0 & 0 & 0 \end{bmatrix}. \end{aligned} \tag{2.140}$$

We can now extract the matrix ad_X as

$$\mathrm{ad}_X = \begin{bmatrix} 0 & -X_\omega & 0 & X_y \\ X_\omega & 0 & 0 & -X_x \\ 0 & 0 & 0 & 0 \\ 0 & 0 & 0 & 0 \end{bmatrix} \tag{2.141}$$

because

$$\begin{bmatrix} -X_\omega Y_y + Y_\omega X_y \\ X_\omega Y_x - Y_\omega X_x \\ 0 \\ 0 \end{bmatrix} = \begin{bmatrix} 0 & -X_\omega & 0 & X_y \\ X_\omega & 0 & 0 & -X_x \\ 0 & 0 & 0 & 0 \\ 0 & 0 & 0 & 0 \end{bmatrix} \begin{bmatrix} Y_x \\ Y_y \\ Y_z \\ Y_\omega \end{bmatrix}. \tag{2.142}$$

The basis of the adjoint map ad for the Schönflies group is thus given by

$$\begin{aligned} \mathrm{ad}_{v_x} &= \begin{bmatrix} 0 & 0 & 0 & 0 \\ 0 & 0 & 0 & -1 \\ 0 & 0 & 0 & 0 \\ 0 & 0 & 0 & 0 \end{bmatrix}, \qquad \mathrm{ad}_{v_y} = \begin{bmatrix} 0 & 0 & 0 & 1 \\ 0 & 0 & 0 & 0 \\ 0 & 0 & 0 & 0 \\ 0 & 0 & 0 & 0 \end{bmatrix}, \\ \mathrm{ad}_{v_z} &= \begin{bmatrix} 0 & 0 & 0 & 0 \\ 0 & 0 & 0 & 0 \\ 0 & 0 & 0 & 0 \\ 0 & 0 & 0 & 0 \end{bmatrix}, \qquad \mathrm{ad}_{\omega} = \begin{bmatrix} 0 & -1 & 0 & 0 \\ 1 & 0 & 0 & 0 \\ 0 & 0 & 0 & 0 \\ 0 & 0 & 0 & 0 \end{bmatrix}. \end{aligned} \tag{2.143}$$

The adjoint map ad_X of the Schönflies group is normally written as

$$\mathrm{ad}_X = \begin{bmatrix} 0 & -r & 0 & v \\ r & 0 & 0 & -u \\ 0 & 0 & 0 & 0 \\ 0 & 0 & 0 & 0 \end{bmatrix}. \tag{2.144}$$

2.7 Tangent Spaces, Vector Fields, and Integral Curves

When we are to derive the dynamic equations of a mechanical system we want to know what the state space of the system looks like. In this section we will look at the tangent bundle which associates a set of admissible velocities with each position that the mechanical system can take. We will see that it is not meaningful to talk about velocity without regard to the configuration: we need to define the velocity at a certain configuration of the mechanical system. Thus, at each point on the manifold we can assign a tangent vector. If we do this in a smooth way we obtain a vector field which gives us information about the flow, or motion, of the system. Finally, a specific motion, i.e., a specified initial condition and flow, allows us to find a curve for which the tangent vector corresponds to the vector field for every point on the curve. These curves are called integral curves.

2.7.1 Tangent Spaces

At any point x on a manifold $\mathscr{M}$ we can identify a tangent vector as an equivalence relation between two paths that pass through x. This equivalence relation tells us

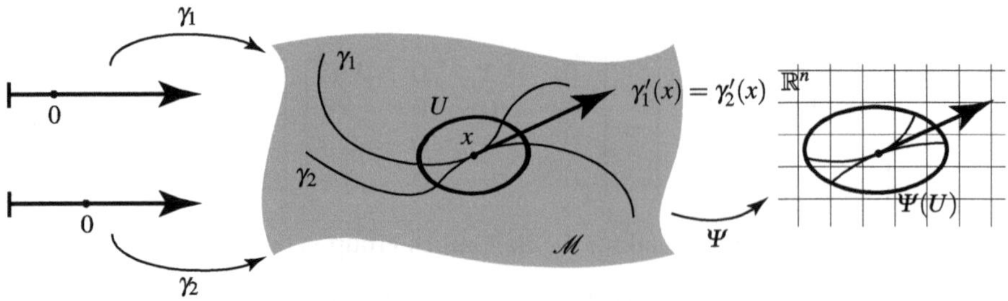

Fig. 2.8 The definition of a tangent vector at a point x on a manifold $\mathcal{M}$

that the first order derivative of the two paths are equal at x. Intuitively we can imagine a particle that moves along a path on a surface. Then, at every point in the path the velocity vector will be a tangent vector for that path (and possibly other paths that pass through the same point).

We will use the definition presented in Bullo and Lewis (2000) and illustrated in Fig. 2.8. Define a curve $\gamma : \mathbb{R} \to \mathcal{M}$ as a mapping from an interval on $\mathbb{R}$ which contains 0 to the manifold $\mathcal{M}$ where $\gamma(0) = x$. Then two curves γ_1 and γ_2 are equivalent at x if, for some local coordinate system, γ_1 and γ_2 have the same derivative at 0. Intuitively this means that the two curves move in the same direction when they pass through the point x. The collection of all possible directions at x is called the *tangent space* at x and is denoted $T\mathcal{M}_x$.

We are mainly concerned with manifolds that are submanifolds of the Euclidean space: a sphere $\mathbb{S}^2$ can for example be thought of as a submanifold of $\mathbb{R}^3$. In this case we can visualize the tangent space at each point as a plane tangent to the sphere at that point.

We can also define a tangent vector in a somewhat more intuitive way:

Definition 2.18 Define $\gamma : \mathbb{R} \to \mathcal{M}$ so that $\gamma(t)$ is a curve in a manifold $\mathcal{M}$. The tangent vector $\dot{\gamma}$ to the curve $\gamma(t)$ at a point $\gamma(t_0)$ is defined by

$$v = \dot{\gamma}(t_0) = \lim_{t \to t_0} \frac{\gamma(t) - \gamma(t_0)}{t - t_0}. \tag{2.145}$$

We see the definition of the tangent vector corresponds well with the classical definition of the velocity of a particle when γ is a function of time t.

At every point in $\mathcal{M}$ we can define such a tangent space. The tangent bundle is the collection of all the tangent spaces at every point in the manifold (Kwatny and Blankenship 2000):

Definition 2.19 (Tangent bundle) The union of all the tangent spaces to $\mathscr{M}$ is called the *tangent bundle* $T\mathscr{M}$ and is defined as

$$T\mathscr{M} = \bigcup_{x \in \mathscr{M}} T\mathscr{M}_x. \tag{2.146}$$

The tangent bundle plays a very important role in mechanics. A point in the tangent bundle $T\mathscr{M}$ is a pair (x, v) where $x \in \mathscr{M}$ and $v \in T\mathscr{M}_x$. We see that the velocity vector therefore is position dependent, i.e., it is taken from the tangent space at position x, i.e., $T\mathscr{M}_x$. This is a manifold of dimension $\dim T\mathscr{M} = 2\dim \mathscr{M}$. *The tangent bundle is normally referred to as the state space in mechanics.* We can write the local coordinates on $\mathscr{M}$ as $(x_1, x_2, \dots, x_n)$ and the components of the tangent vectors on $T\mathscr{M}_x$ as $(v_1, v_2, \dots, v_n)$. The coordinates of the tangent bundle completely describe the state space of the system in coordinates $(x_1, x_2, \dots, x_n, v_1, v_2, \dots, v_n)$.

Given the local coordinates as a chart (U, Ψ) we get the following important result (Kwatny and Blankenship 2000):

Definition 2.20 The components of the tangent vector v to the curve $\gamma(t)$ on $\mathscr{M}$ in the local coordinate chart (U, Ψ) are the numbers $v_1, v_2, \dots, v_m$ where

$$v_i = \frac{\mathrm{d}\Psi_i}{\mathrm{d}t}. \tag{2.147}$$

We will study in great detail what these velocity variables look like for different mechanical systems.

2.7.2 *Vector Fields, Flows, and Integral Curves*

A vector field tells us in what direction our system moves at each position. Using the concept of tangent spaces, a vector field smoothly assigns a tangent vector to each point on the manifold. The vector field thus presents us with a solution to an ordinary differential equation for a given initial condition. Figure 2.9 shows a vector field F and the solution to the corresponding differential equation with initial condition $\gamma(0)$.

The solution $\gamma(t)$ to the differential equation is called an integral curve:

Definition 2.21 A differentiable curve $\gamma(t)$ is called an integral curve at $x \in \mathscr{M}$ for a vector field F if

$$\dot{\gamma}(t) = F\big(\gamma(t)\big), \qquad \gamma(0) = x. \tag{2.148}$$

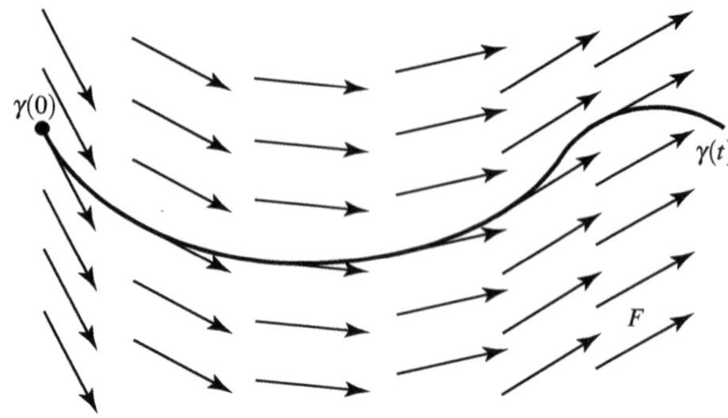

Fig. 2.9 A vector field F and the solution to a differential equation with initial condition $\gamma(0)$

We see that the vector field tells us what the tangent vector looks like for the curve $\gamma(t)$ at each point on the curve (Fig. 2.10).

A vector field is a mapping $F : \mathcal{M} \to T\mathcal{M}$ that for every point $x \in \mathcal{M}$ assigns a tangent vector $F(x) \in T\mathcal{M}$. A Lie algebra element is a tangent vector at the identity and we can therefore find a one-to-one correspondence between the Lie algebra and a vector field. In order to do this we need to define a tangent vector at every point on the manifold. This can be obtained by right translating the tangent vector at the identity to every element of the group. We write the group elements as g and a corresponding tangent vector at g as $\widehat{X}g$. This follows directly from Eq. (2.78). In this case we have $F(g) = R_g\widehat{X} = \widehat{X}g$ and the integral curve is thus found by solving the differential equation

$$\frac{\mathrm{d}g(t)}{\mathrm{d}t} = \widehat{X}g(t). \tag{2.149}$$

The solution to this differential equation is very important and will be treated in the next section when we look at the exponential map.

Recall that the tangent space at a point $g(t)$ in a Lie group G, denoted $\dot{g}(t)$ is transformed back to the identity, i.e., to the Lie algebra $\widehat{X}$, by

$$\widehat{X} = \dot{g}(t)g(t)^{-1}. \tag{2.150}$$

The Lie algebras are in this case interpreted as spatial velocity variables. If we instead use left translation the Lie algebra is in the form $\widehat{X} = g^{-1}\dot{g}$ and interpreted as body frame velocities.

Example 2.11 An element of the Lie algebra $so(2)$ can be represented as

$$\widehat{\omega} = \begin{bmatrix} 0 & -1 \\ 1 & 0 \end{bmatrix} \in so(2). \tag{2.151}$$

A differentiable path $R(t)$ can be written in the normal way as

$$R(t) = \begin{bmatrix} \cos t & -\sin t \\ \sin t & \cos t \end{bmatrix} \in SO(2). \tag{2.152}$$

Fig. 2.10 The derivative of the integral curve is given by the vector field as $\dot{\gamma}(t) = F(\gamma(t))$

For a point y_0 we can study the velocity of this point at time $t = 0$ due to the vector field by looking at the Lie algebra. The instantaneous velocity of this point is given by $L_X y_0 = \widehat{X} y_0$. A point on the positive y-axis represented by $y_0 = [0\ 1]^\mathsf{T}$ will for example move to the left at time $t = 0$ because

$$L_X y_0 = \begin{bmatrix} 0 & -1 \\ 1 & 0 \end{bmatrix} \begin{bmatrix} 0 \\ 1 \end{bmatrix} = \begin{bmatrix} -1 \\ 0 \end{bmatrix}. \tag{2.153}$$

However, because $y(t) = L_{R(t)} y_0$ is an integral curve of the vector field L_X, $\widehat{X}$ tells us how y moves for all time, not just at $t = 0$. This vector field is given by $L_X R(t)$:

$$F\big(R(t)\big) = L_X R(t) y_0 = \begin{bmatrix} 0 & -1 \\ 1 & 0 \end{bmatrix} \begin{bmatrix} \cos t & -\sin t \\ \sin t & \cos t \end{bmatrix} \begin{bmatrix} 0 \\ 1 \end{bmatrix} = -\begin{bmatrix} \cos t \\ \sin t \end{bmatrix}. \tag{2.154}$$

We see that if we associate y_0 with the identity $R = I$ at time $t = 0$ we can find the tangent vector to the curve $R(t)$ at any time t by Eq. (2.154). In Fig. 2.11 this is illustrated for $t = \frac{\pi}{2}$ which corresponds to a clockwise rotation of $\frac{\pi}{2}$ radians around the origin.

2.8 The Exponential Map

One of the most important tools that we will use when deriving the kinematics is the fact that there exists an onto map from the Lie algebra to the Lie group. In the context of robotics the exponential map was introduced by Brockett (1984) who derived the forward kinematics of a robotic manipulator with 1-DoF joints using the exponential map. In Chap. 4 we will show how to use the exponential map to write the forward kinematics of a robotic manipulator in terms of the joint twists.

We have seen that an integral curve of a vector field $F(g)$, where g is a point on the manifold $\mathscr{M}$, is a curve $\gamma(t)$ for which $\gamma(0) = I$ and $\dot{\gamma}(0) = X$. In other words, the derivative at $t = 0$ is the Lie algebra X. In this section we will look at the solution to differential equations in the form

$$\dot{g}(t) = F\big(g(t)\big) \tag{2.155}$$

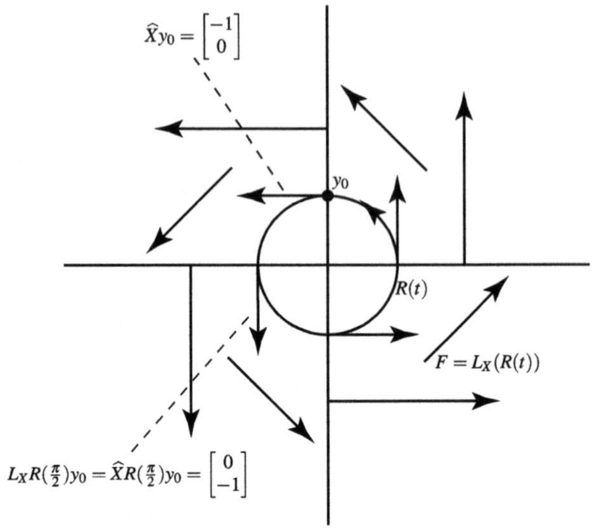

Fig. 2.11 The vector field $F = L_X(R(t))$ and an integral curve $R(t)$. The Lie algebra $\widehat{X} \in so(2)$ tells us what the tangent vector to the integral curve looks like for all t

where $g(t)$ is a path on a manifold $\mathcal{M}$ and $F(g(t)) : \mathcal{M} \to T\mathcal{M}$ is a vector field in the form $F(g(t)) = L_X g(t)$. We will study the exponential map which presents us with a solution to the initial value problem on left invariant vector fields of the Lie algebras.

We will use the exponential map extensively when deriving the dynamics of multibody systems. There are two main interpretations of the exponential map. First of all the exponential map allows us to map an element of the Lie algebra to the corresponding Lie group. Thus, for a given twist, there is a neighborhood around 0 in the Lie algebra for which the exponential map maps this twist homeomorphically to a neighborhood of the identity in the Lie group. The Lie algebra therefore gives us much information about how the Lie group behaves locally.

We can also think of the exponential mapping as a solution to a left invariant vector field. The solution to differential equations of the form $\dot{g}(t) = L_X g(t)$ form 1-parameter subgroups that can be described in terms of the Lie algebra. These paths are often called optimal paths and have the nice property that $g(t)$ is the integral curve of the vector field $F(g(t)) = L_X g(t)$.

2.8.1 The Exponential of a Matrix

For the group of $n \times n$ nonsingular real matrices $GL(n)$ the exponential map is given by the power series of the Lie algebra elements (Murray et al. 1994). Let $A \in gl(n)$, where $gl(n)$ is the Lie algebra associated with $GL(n)$. Then the exponential map exp(A) is given by

$$e^A = I + A + \frac{A^2}{2!} + \frac{A^3}{3!} + \cdots = \sum_{n=0}^{\infty} \frac{A^n}{n!} \tag{2.156}$$

where I is the identity matrix. This expression is valid for all subgroups of $SE(3)$ and $SE(3)$ itself by replacing A with the matrix representation of the Lie algebra associated with the Lie group. We denote the matrix representation of the corresponding Lie algebra by $\widehat{X}$ and thus get

$$e^{\widehat{X}} = I + \widehat{X} + \frac{\widehat{X}^2}{2!} + \frac{\widehat{X}^3}{3!} + \cdots = \sum_{n=0}^{\infty} \frac{\widehat{X}^n}{n!}. \tag{2.157}$$

We found $\widehat{X}$ for the different Lie groups in Sect. 2.6.4. Later in this section we will look at what the exponential map looks like for the most important Lie groups in robotics. We will first look at how we can use the exponential map to solve differential equations and how these present us with a solution to the natural paths on the Lie groups.

2.8.2 *Differential Equations*

Assume a standard homogeneous differential equation in the form

$$\frac{\mathrm{d}y}{\mathrm{d}t} = ky \tag{2.158}$$

where k is a constant and y is the variable. We know that the solution to this differential equation is given by

$$y(t) = y(0)e^{kt} \tag{2.159}$$

where $y(0)$ is the initial condition. We can see this by substituting this expression back into (2.158):

$$\frac{\mathrm{d}y}{\mathrm{d}t} = \frac{\mathrm{d}}{\mathrm{d}t}\big(y(0)e^{kt}\big) = y(0)\frac{\mathrm{d}e^{kt}}{\mathrm{d}t} = y(0)ke^{kt} = k\big(y(0)e^{kt}\big) = ky. \tag{2.160}$$

Similarly, if y is a vector we have

$$\frac{\mathrm{d}y}{\mathrm{d}t} = Ky \tag{2.161}$$

where K is a matrix and y the vector of variables, we get the solution

$$y(t) = e^{Kt}y(0) \tag{2.162}$$

where $e^{Kt} = Ae^{\Lambda t}A^{-1}$ for a matrix Λ with the Eigenvalues and a matrix A with the Eigenvectors of K (Jordan and Smith 2004).

We will see that we can find a similar solution when the underlying space is a general manifold.

2.8.3 The Natural Path in a Matrix Group

The following proposition is valid for any Lie group $\mathfrak{g}$ and will play an important role in the remaining of the book (Tapp 2005):

Proposition 2.1 *Let $X \in \mathfrak{g}$ and $\gamma(t) = e^{t\widehat{X}}$.*

1. *For all $y_0 \in \mathbb{R}^n$, $y(t) = L_{\gamma(t)}y_0$ is an integral curve of L_X.*
2. *$\gamma(t)$ is itself an integral curve of the vector field on G. The value of the vector field at g is $g\widehat{X}$.*

Part 1 is illustrated in Fig. 2.11. We need to choose $\gamma(t)$ so that $y(t) = L_{\gamma(t)}y_0$ is an integral curve of the vector field L_X. In other words, we need $y(0) = y_0$ to satisfy the initial condition and $\dot{y}(t) = L_X y(t)$ for $y(t)$ to be an integral curve of L_X. Following the approach in Tapp (2005) we can show this by writing the path $y(t)$ by the power series

$$y(t) = c_0 + c_1 t + c_2 t^2 + c_3 t^3 + \cdots \tag{2.163}$$

and its derivative as

$$\dot{y}(t) = c_1 + 2c_2 t + 3c_3 t^2 + \cdots . \tag{2.164}$$

Because $\dot{y}(t) = L_X y(t)$ we get

$$\begin{aligned}\left(c_1 + 2c_2 t + 3c_3 t^2 + \cdots\right) &= \widehat{X}\left(c_0 + c_1 t + c_2 t^2 + c_3 t^3 + \cdots\right)\\ &= \left(\widehat{X}c_0 + \widehat{X}c_1 t + \widehat{X}c_2 t^2 + \widehat{X}c_3 t^3 + \cdots\right)\end{aligned} \tag{2.165}$$

We first note that $y(0) = L_{\gamma(0)}y_0 = y_0$ and (2.163) then gives us $c_0 = y(0) = y_0$. From (2.165) we then find $c_1 = \widehat{X}y_0$ so that the terms with no time variables are correct. Next, the terms with t give us $c_2 = \frac{1}{2}\widehat{X}^2 y_0$. Continuing in this way gives us the recursive formula $c_i = \frac{1}{i!}\widehat{X}c_{i-1}$. If we now write the integral curve in terms of these c_i's we get

$$y(t) = y_0 + \widehat{X}t y_0 + \frac{1}{2!}(\widehat{X}t)^2 y_0 + \frac{1}{3!}(\widehat{X}t)^3 y_0 + \cdots \tag{2.166}$$

and we have found an expression for the integral curve in terms of the Lie algebra X. The final step is to recognize that this power series can be written as an exponential map of the form

$$\begin{aligned} y(t) &= y_0 + \widehat{X} t y_0 + \frac{1}{2!}(\widehat{X}t)^2 y_0 + \frac{1}{3!}(\widehat{X}t)^3 y_0 + \cdots \\ &= \left(I + \widehat{X}t + \frac{1}{2!}(\widehat{X}t)^2 + \frac{1}{3!}(\widehat{X}t)^3 + \cdots\right) y_0 \\ &= e^{\widehat{X}t} y_0 \\ &= L_{e^{\widehat{X}t}} y_0. \end{aligned} \tag{2.167}$$

We have thus shown that for $\gamma(t) = e^{t\widehat{X}}$, then $y(t) = L_{\gamma(t)} y_0$ is an integral curve of L_X.

The second part of Proposition 2.1 says that also $\gamma(t)$ is an integral curve of a vector field with values $g\widehat{X}$ at g. For this to be true we need $\gamma(0) = I$ and $\dot{\gamma}(0) = \widehat{X}$, and also that $\dot{\gamma}(t) = L_g \widehat{X}$. First write

$$\gamma(t) = e^{\widehat{X}t} = I + \widehat{X}t + \frac{1}{2!}(\widehat{X}t)^2 + \frac{1}{3!}(\widehat{X}t)^3 + \cdots \tag{2.168}$$

and

$$\dot{\gamma}(t) = \frac{\mathrm{d}}{\mathrm{d}t} e^{\widehat{X}t} = \widehat{X} + \widehat{X}^2 t + \frac{1}{2!}\widehat{X}^3 t^2 + \cdots \tag{2.169}$$

which gives $\gamma(0) = I$ and $\dot{\gamma}(0) = \widehat{X}$ as required. We also see that if we factor $\widehat{X}$ out on the right we get

$$\begin{aligned} \dot{\gamma}(t) &= \left(I + \widehat{X}t + \frac{1}{2!}(\widehat{X}t)^2 + \cdots\right)\widehat{X} \\ &= \gamma(t)\widehat{X} \end{aligned} \tag{2.170}$$

which shows that $\gamma(t)$ is an integral curve to the vector field generated by the Lie algebra X.

Remark 2.5 We note that we could also have factored $\widehat{X}$ out on the left hand side in Eq. (2.170). This is a solution to the differential equation $\dot{\gamma} = R_g \widehat{X}$. Even though the vector fields L_X and R_X are in general different, they agree along the trajectory of $\gamma(t)$.

What we have shown here is that a curve $\gamma(t) = e^{\widehat{X}t}$ is actually a 1-parameter subgroup of G, i.e., the exponential mapping maps $\widehat{X}t \in \mathfrak{g}$ to a 1-parameter subgroup $\gamma(t) = e^{\widehat{X}t} \in G$. More specifically it maps it to the 1-parameter subgroups that are tangent to X at I (because $\dot{\gamma}(0) = X$). These 1-parameter subgroups play a very important role in the derivation of the dynamic equations of mechanical systems such as robotic manipulators.

We are now ready to present the following important theorem:

Theorem 2.1 *If* $X \in \mathfrak{g}$, *then* $e^{\widehat{X}t} \in G$.

The exponential map is thus a mapping from the Lie algebra $\widehat{X}$ to the corresponding Lie group. We will not prove this for a general Lie group, but refer to Tapp (2005) for the proof. In the following sections we will, however, show that this is true for the most important Lie groups and find the explicit expressions whenever possible.

2.8.4 Rigid Body Motion in Terms of Exponential Coordinates

Recall that a point $\bar{q}^a(0)$ can be transformed into another point by a homogeneous transformation $g(t, X)$ by

$$\bar{q}^a(t) = g(t, X)\bar{q}^a(0) \tag{2.171}$$

where $\bar{q}^a(0)$ is the point before and $\bar{q}^a(t)$ is the point after the transformation, both expressed in the same coordinate frame $\mathscr{F}_a$. For a constant twist, this transformation can be written in terms of exponential coordinates as

$$\bar{q}^a(t) = e^{\widehat{X}t}\bar{q}^a(0). \tag{2.172}$$

We have seen that if $\widehat{X} \in se(3)$, then $e^{\widehat{X}t} \in SE(3)$, so this expression is in the form of (2.171). Equation (2.172) represents a 1-parameter transformation of $\bar{q}^a(0)$ and can for example give us information about how a one degree of freedom revolute or translational joint acts on a point or a vector.

Similarly, we can look at the transformation of a rigid body with initial position $g_{0b}(0)$ transformed by $g(t, X)$ as in (2.70):

$$g_{0b}(t) = g(t, X)g_{0b}(0). \tag{2.173}$$

Also this transformation can be written in terms of exponential coordinates as

$$g_{0b}(t) = e^{\widehat{X}t}g_{0b}(0). \tag{2.174}$$

If for example $g_{0b}(0)$ denotes the position of a rigid body, then $g_{0b}(t)$ denotes the position after a 1-parameter transformation $e^{\widehat{X}t}$. We note that both $g_{0b}(0)$ and $g_{0b}(t)$ are represented in the same inertial reference frame $\mathscr{F}_0$ so Eq. (2.174) can be interpreted as a rigid body motion. We will see that we can derive the kinematics of robots using exponential coordinates in this way.

2.8.5 *The Exponential Map of the Most Important Lie Groups*

The exponential map is used to represent motion in a fixed direction, i.e., for a given Lie algebra. In this section we will look at what this motion looks like for the most important Lie groups. We will see that these motions are 1-parameter subgroups of $SE(3)$.

2.8.5.1 The Euclidean Group $\mathbb{R}^n$

We have seen that the Lie algebra of $\mathbb{R}^n$ is $\mathbb{R}^n$ itself. The integral curve of the constant vector field $F(x) = v$ is thus given by $\gamma(t) = vt$. For the Euclidean group $\mathbb{R}^n$ we can find the solution to the exponential map by simple reasoning. From Eq. (2.174) we see that the exponential map gives us the transformation by an element $\widehat{X}$ of the Lie algebra by t units of time given an initial condition $g_{0b}(0)$. Correspondingly, for the Euclidean group we write the elements of the Lie group as $x(t)$ with initial condition $x(0)$ and the elements of the Lie algebra as v. We then know that this transformation is given by

$$x(t) = x(0) + vt \tag{2.175}$$

for the Euclidean group with addition as the group operator and a constant vector field $F(x) = v$.

2.8.5.2 The Special Orthogonal Group $SO(2)$

For rotational motion in the plane the exponential map can be written as in (2.175). This is because, even though $\mathbb{S}^1$ and $\mathbb{R}$ strictly speaking are not topologically the same, we can treat 1-DoF rotations as Euclidean. We saw this in Sect. 2.5. The exponential map of $SO(2)$ is thus the same as for $\mathbb{R}^n$.

2.8.5.3 The Special Orthogonal Group $SO(3)$

For rotational motion in the three dimensional Euclidean space we have the following important result:

Theorem 2.2 *Given an element $\widehat{\omega} \in so(3)$. Then the exponential map is given by*

$$e^{\widehat{\omega}t} = I + \widehat{\omega}\sin t + \widehat{\omega}^2(1 - \cos t) \tag{2.176}$$

for $\|\omega\| = 1$ and

$$e^{\widehat{\omega}t} = I + \frac{\widehat{\omega}}{\|\omega\|}\sin\|\omega\|t + \frac{\widehat{\omega}^2}{\|\omega\|^2}\big(1 - \cos\|\omega\|t\big) \tag{2.177}$$

for $\|\omega\| \neq 1$. Equation (2.176) is normally referred to as Rodrigues' formula.

Following the proof found in Murray et al. (1994) we will show how we arrive at Rodrigues' formula when $\|\omega\| = 1$. First write

$$e^{\widehat{\omega}t} = I + t\widehat{\omega} + \frac{t^2}{2!}\widehat{\omega}^2 + \frac{t^3}{3!}\widehat{\omega}^3 + \cdots. \tag{2.178}$$

It can be shown (see Murray et al. 1994) that for $\widehat{\omega} \in so(3)$, this can be written as

$$e^{\widehat{\omega}t} = I + \left(t - \frac{t^3}{3!} + \frac{t^5}{5!} \cdots\right)\widehat{\omega} + \left(\frac{t^2}{2!} - \frac{t^4}{4!} + \frac{t^6}{6!} \cdots\right)\widehat{\omega}^2. \tag{2.179}$$

If we use the relations $\sin t = t - \frac{t^3}{3!} + \frac{t^5}{5!} \cdots$ and $\cos t = 1 - \frac{t^2}{2!} + \frac{t^4}{4!} - \frac{t^6}{6!} \cdots$ we get an expression for the exponential map on $so(3)$ given by

$$e^{\widehat{\omega}t} = I + \widehat{\omega}\sin t + \widehat{\omega}^2(1 - \cos t). \tag{2.180}$$

The exponential map of an element of $so(3)$ produces an element of $SO(3)$:

Theorem 2.3 *For $\widehat{\omega} \in so(3)$ and $t \in \mathbb{R}$ we have*

$$e^{\widehat{\omega}t} \in SO(3). \tag{2.181}$$

In other words, the exponential map of $\widehat{\omega}$ gives us a rotation matrix that corresponds to t units of rotation around the axis ω. To prove Theorem 2.3 we follow the approach in Murray et al. (1994) and show that $e^{\widehat{\omega}t}$ satisfies the group properties $R^{\mathsf{T}}R = I$ and $\det R = 1$. First note that

$$\left(e^{\widehat{\omega}t}\right)^{-1} = e^{-\widehat{\omega}t} = e^{\widehat{\omega}^{\mathsf{T}}t} = \left(e^{\widehat{\omega}t}\right)^{\mathsf{T}}. \tag{2.182}$$

This shows that $R^{-1} = R^{\mathsf{T}}$ and we have shown that $R^{\mathsf{T}}R = I$. Further, we know that the determinant of a matrix exponential is given simply by the exponential of the trace of the matrix, i.e., $\det e^{\widehat{\omega}} = e^{\mathrm{Tr}(\widehat{\omega})}$. As the trace of $\widehat{\omega} \in so(3)$ is zero, the determinant of $e^{\widehat{\omega}t}$ equals 1. $e^{\widehat{\omega}t}$ is therefore an element of $SO(3)$. Finally we note that the exponential map on $so(3)$ is surjective.

Using the expression in (2.172) we can use the exponential map to transform a point corresponding to a 1-parameter motion represented by a rotation of t radians around the axis ω which is then in the form

$$q^a(t) = e^{\widehat{\omega}t}q^a(0) = R(\omega, t)q^a(0) \tag{2.183}$$

where $R(\omega, t)$ is the rotation matrix that represents a rotation of t radians around ω.

2.8.5.4 The Special Euclidean Group $SE(3)$

We can find a theorem for $SE(3)$ similar to Theorem 2.2 for $SO(3)$:

Theorem 2.4 *Given an element* $V = [v^{\mathsf{T}}\ \omega^{\mathsf{T}}]^{\mathsf{T}} \in se(3)$. *Then the exponential map is given by*

$$e^{\widehat{V}t} = \begin{bmatrix} I & vt \\ 0 & 1 \end{bmatrix}, \quad \omega = 0 \tag{2.184}$$

$$e^{\widehat{V}t} = \begin{bmatrix} e^{\widehat{\omega}t} & (I - e^{\widehat{\omega}t})(\omega \times v) + \omega\omega^{\mathsf{T}} vt \\ 0 & 1 \end{bmatrix}, \quad \omega \neq 0 \tag{2.185}$$

where $e^{\widehat{\omega}t}$ *is given as in* (2.176) *and we have assumed* $\|\omega\| = 1$.

We refer to Tapp (2005) or Murray et al. (1994) for the proofs. Also for the Special Euclidean group it can be shown that the exponential map of an element of $se(3)$ produces an element of $SE(3)$:

Theorem 2.5 *For* $\widehat{V} \in se(3)$ *and* $t \in \mathbb{R}$ *we have*

$$e^{\widehat{V}t} \in SE(3). \tag{2.186}$$

The proof follows directly by observing that (2.184) and (2.185) are elements of $SE(3)$. This is a very important result as it allows us to derive the kinematics of multibody systems in terms of exponential coordinates by Eq. (2.174).

Again, using the expression in (2.172) we can use the exponential map to transform a point corresponding to a 1-parameter motion represented by the constant twist V which is then given by

$$\bar{q}^a(t) = e^{\widehat{V}t}\bar{q}^a(0) = g(V, t)\bar{q}^a(0) \tag{2.187}$$

where $g(V, t)$ is the homogeneous transformation matrix that corresponds to a 1-parameter motion of t units in the direction of V.

2.8.5.5 The Planar Group $SE(2)$

For consistency we include the following theorem from $SE(2)$:

Theorem 2.6 *For* $\widehat{X} \in se(2)$ *and* $t \in \mathbb{R}$ *we have*

$$e^{\widehat{X}t} \in SE(2) \tag{2.188}$$

which is the homogeneous transformation matrix in the plane.

This theorem follows from Theorem 2.4 by selecting the basis $\hat{v}_x$, $\hat{v}_y$, and $\hat{\omega}_z$ corresponding to $SE(2)$ in (2.131).

2.8.5.6 The General Linear Group $GL(n)$

For the General Linear Group with a corresponding Lie algebra $X \in \mathfrak{g}$ the exponential map is given by the series expansion

$$e^{\widehat{X}t} = \sum_{i=0}^{\infty} \frac{t^i \widehat{X}^i}{i!}. \tag{2.189}$$

We note that the expressions found for $SO(3)$, $SE(3)$ and $SE(2)$ are specific formulas of this general expression. It is straight forward to show that $e^{t\widehat{X}}$ is an integral curve of the vector field L_X:

$$\frac{\mathrm{d}}{\mathrm{d}t} e^{\widehat{X}t} = \sum_{i=1}^{\infty} \frac{t^{i-1}\widehat{X}^i}{(i-1)!} = \sum_{i=1}^{\infty} \frac{t^{i-1}\widehat{X}^{i-1}}{(i-1)!}\widehat{X} = \sum_{i=0}^{\infty} \frac{t^i \widehat{X}^i}{i!}\widehat{X} = e^{\widehat{X}t}\widehat{X}. \tag{2.190}$$

We also have the following general result:

Theorem 2.7 *For $\widehat{X} \in \mathfrak{gl}(n)$ and $t \in \mathbb{R}$ we have*

$$e^{\widehat{X}t} \in GL(n) \tag{2.191}$$

where $GL(n)$ is the general linear group and $\mathfrak{gl}(n)$ the corresponding Lie algebra.

2.8.6 Charts and Exponential Maps

As we have seen, the exponential map can be interpreted as a mapping from the Lie algebra to the corresponding Lie group, e.g., the mapping $\exp : se(3) \to SE(3)$. There is, however, a slightly different interpretation of the exponential map which leads to a different way of utilizing the results from the previous section. Recall that a chart Ψ maps an element on the manifold, for example $SE(3)$, onto the Euclidean space with the same dimension. In this section we will look at the inverse of this mapping, i.e., the mapping Φ which maps en element of $\mathbb{R}^n$—which in our case are the exponential coordinates—to an element of the Lie group. As the Lie algebra also lives on the Euclidean space it is thus natural to ask whether we can use the exponential map as a mapping from $\mathbb{R}^n$ (where $\mathbb{R}^n$ denotes the Euclidean space in the neighborhood of some point) to the manifold, for example $SE(3)$. From the previous section it is quite intuitive that the transformation represented by a quantity t in the direction of a Lie algebra X can be *interpreted as a displacement*, and not only as an element of the Lie algebra.

We will denote local coordinates in the Euclidean space as $\varphi \in \mathbb{R}^n$. These local coordinates are the exponential coordinates that we found in the previous section, just interpreted a little differently. We interpret these coordinates as tiny displacements in the neighborhood of some point on $\mathbb{R}^n$. From (2.168) we see that we can interpret vt as position coordinates for small t. We can therefore replace vt with φ in the exponential map. The exponential map thus provides us with a transformation between local and global variables. The local variables are represented in $\mathbb{R}^n$ and the global variables are defined on the manifold itself.

Recall that the time derivative of an element of a Lie group does not have a physical meaningful interpretation. That is, we cannot simply write $\dot{g}(t)$ to get the velocity that corresponds to a path $g(t)$. However, as the local position variables are written as $\varphi \in \mathbb{R}^n$ the time derivative can be written simply as $\dot{\varphi} = \frac{\mathrm{d}\varphi}{\mathrm{d}t}$. Because $\varphi \in \mathbb{R}^n$ this is a physically meaningful quantity and gives us the local velocity variables. We can then use this to find a relation between the local and global velocity variables.

A transformation can be uniquely described by a matrix $Q \in GL(n)$. We have seen several examples of such matrices: a rotation matrix ($Q = R$) can uniquely describe a ball joint, a homogeneous transformation matrix ($Q = g$) can represent unconstrained rigid body motion in $\mathbb{R}^3$ and a scalar ($Q = \mathbb{R}$) can represent a 1-dimensional linear or revolute transformation. Let's assume that Q is constant and denote this by $\bar{Q}$. The exponential map allows us to express the dynamics in exponential coordinates φ so that locally every state $\bar{Q}$ is described by a set of Euclidean coordinates $\varphi \in \mathbb{R}^n$. Thus, in the neighborhood of $\bar{Q}$ there exist a function $\Phi(\bar{Q}, \varphi)$ that defines a local diffeomorphism between a neighborhood of $0 \in \mathbb{R}^n$ and a neighborhood of $\bar{Q}$. $\bar{Q}$ is locally described by $Q = \Phi(\bar{Q}, \varphi)$ with $\Phi(\bar{Q}, 0) = \bar{Q}$. Using the results from the previous section, the mapping $\Phi(\bar{Q}, \varphi)$ can be expressed in terms of the exponential map.

In this way the position of a rigid body in space is given globally by a matrix Lie group Q, but locally around an element $\bar{Q}$ it is also described by the local coordinates $\varphi \in \mathbb{R}^n$. Also the velocity variables can be defined both locally and globally. Locally the velocity variables are given simply by $\dot{\varphi} = \frac{\mathrm{d}\varphi}{\mathrm{d}t}$ and globally by a vector $v \in \mathbb{R}^n$. The vector v uniquely describes the joint twist as $V_{0b}^B = H(Q)v$ where $H(Q)$ is called a selection matrix. The selection matrix maps the n-dimensional velocity vector to a 6-dimensional twist. The term *selection matrix* arises because in most cases $H(Q) = H$ consists of only ones and zeros, and selects what components of the twist the elements of the vector v correspond to.

We will now look at some examples of how we can use the coordinate charts to describe the neighborhood of a point $\bar{Q}$ on a manifold in terms of local coordinates φ.

2.8.6.1 The Euclidean Space $\mathbb{R}$

For the Euclidean space the allowed velocity is given as an element of the tangent space of the Lie group and is uniquely described by a vector $v \in \mathbb{R}^n$. For Euclidean 1-DoF transformations, such as translational and rotational joints, we have $Q \in \mathbb{R}$ and $v = \dot{Q} \in \mathbb{R}$. The coordinate mapping is given by $\Phi(\bar{Q}, \varphi) = \bar{Q} + \varphi$ with $\varphi \in \mathbb{R}$. We see that if we replace tv with φ—which we can do because both are representations of displacements—this corresponds to the exponential map of $\mathbb{R}^n$ that we found in the previous section.

Also other transformations can be written in this way, for example pure translational motion in $\mathbb{R}^2$ or $\mathbb{R}^3$. In this case we get $Q \in \mathbb{R}^n$ and $v = \dot{Q} \in \mathbb{R}^n$. The coordinate mapping is given by

$$\Phi(\bar{Q}, \varphi) = \bar{Q} + \varphi \tag{2.192}$$

with $\varphi \in \mathbb{R}^n$.

2.8.6.2 Matrix Lie Groups *GL*(*n*)

An important group of transformations consists of all transformations with a Lie group topology. Φ is then given by the exponential map, i.e.

$$\Phi(\bar{Q}, \varphi) = \bar{Q} e^{\sum_{i=1}^{n} b_i \varphi_i} \tag{2.193}$$

where b_i represents the basis elements of the Lie algebra found in (2.131). n is the dimension of the manifold and the basis elements determine in what directions we allow displacements. For example, for pure rotational motion b_i represents the three axes that we can rotate around and φ_i determines the size of the rotation in the direction of b_i in the normal way. In this case we have $b_1 = \hat{\omega}_x$, $b_2 = \hat{\omega}_y$, and $b_3 = \hat{\omega}_z$.

Equation (2.193) can be interpreted as first a transformation $g_{0\bar{Q}}$ from the inertial frame to the configuration represented by $\bar{Q}$ followed by a transformation $g_{\bar{Q}Q}$ from $\bar{Q}$ to Q. This transformation is only valid locally, i.e., in the neighborhood of $\bar{Q}$ and can therefore be written in terms of the local coordinates φ. For $SE(3)$ Eq. (2.193) can thus be written as

$$g_{0Q} = g_{0\bar{Q}} g_{\bar{Q}Q}. \tag{2.194}$$

g_{0Q} is a homogeneous transformation matrix $g_{0Q} \in SE(3)$ and we can find the spatial velocities from

$$\widehat{V}^S_{0Q} = \dot{g}_{0Q} g^{-1}_{0Q} = \begin{bmatrix} \widehat{\omega}^S_{0Q} & v^S_{0Q} \\ 0 & 0 \end{bmatrix}. \tag{2.195}$$

More generally, for a general manifold with a matrix representation Q we can write

$$\widehat{v}^B_{0b} = \Phi^{-1}(\bar{Q}, \varphi)\dot{\Phi}(\bar{Q}, \varphi) \tag{2.196}$$

in body coordinates, or in spatial coordinates as

$$\widehat{v}^S_{0b} = \dot{\Phi}(\bar{Q}, \varphi)\Phi^{-1}(\bar{Q}, \varphi). \tag{2.197}$$

2.9 Local Coordinates and Velocity Transformation Matrices

We have seen that $SE(3)$ is topologically different from the Euclidean space and that it is not possible to continuously and globally cover it using six coordinates. As a direct result of this, there is no way to write the configuration of a rigid body as a vector in $\mathbb{R}^6$ in a way that is globally valid. There are, however, several representation methods that describe $SE(3)$ either only locally continuously using six numbers, or globally continuously using more than six numbers.

In this section we will describe the location of a rigid body locally using vectors in $\mathbb{R}^6$ and we will see how we can use the relation in (2.197) to find a velocity transformation matrix in terms of these local coordinates. This will give us the global velocity variables in terms of the local velocity variables in $\mathbb{R}^6$ and the velocity transformation matrix which depends only on the local position variables. This velocity transformation matrix is important when we are to derive the dynamics of the system in the subsequent chapters.

2.9.1 Velocity Transformation Matrices in Dynamics

In general, the topology of a Lie group is not Euclidean. When deriving the dynamic equations for vehicles such as ships (Fossen 2002), AUVs (Antonelli 2006), and spacecraft (Hughes 2002), this is normally dealt with by introducing a transformation matrix that relates the velocity variables represented in the different frames. However, forcing the dynamics into a vector representation in this way, without taking the topology of the configuration space into account, leads to singularities in the representation. To preserve the topology of the configuration space we will use quasi-coordinates, i.e., velocity coordinates that are not given by the time-derivative of position coordinates, but by a linear relation. Thus, there exist differentiable matrices S_i such that we can write $v_i = S_i(Q_i, \varphi_i)\dot{\varphi}_i$ for every Q_i. For Euclidean joints this relation is given simply by the identity map while for joints with a Lie group topology we can use the exponential map to derive this relation.

We will represent the configuration of the multibody system as a set of configuration states $Q = \{Q_i\}$. The configuration state Q_i of joint i is then the matrix representation of the Lie group corresponding to the topology of the joint. The corresponding block $S_i(Q_i, \varphi_i)$ relating the velocity variables is well known from the Lie theory and can be found in terms of the Lie bracket or the exponential map (Rossmann 2002). For standard revolute and prismatic joints Q_i becomes a scalar $Q_i = q_i$ and $S_i = 1$ while for joints or transformations with a Lie group topology, Q_i is the matrix Lie group. We will now look at what form $S_i(Q_i, \varphi_i)$ will take in this case.

2.9.2 The Velocity Transformation Matrix in Terms of Exponential Coordinates

We have seen that the velocity transformation matrix, i.e., the mapping from the velocity variables to the time derivative of the position variables is not always well defined. In this section we will show that we can rewrite the expression in (2.197) to find a matrix in this form. We will start by stating the following important theorem:

Proposition 2.2 *It is always possible to find a matrix* $S(\bar{Q}, \varphi)$ *such that*

$$v = S(\bar{Q}, \varphi)\dot{\varphi}, \tag{2.198}$$

i.e., there is a linear relation between v *and* $\dot{\varphi}$.

Proof (Duindam 2006) Note that when $g(\bar{Q}, \varphi)$ is in the form of (2.193), i.e., $\bar{Q}$ is constant, then the time derivative of the transformation $g(\bar{Q}, \varphi)$ is a function $\dot{g}(\bar{Q}, \varphi, \dot{\varphi})$ which is linear in $\dot{\varphi}$. There is also a linear relation from $\dot{g}(\bar{Q}, \varphi, \dot{\varphi})$ to the twist, given by the conjugation $\widehat{V}^c_{ab} = g_{ca}\dot{g}_{ab}g_{bc}$. If we now write the vector v in terms of the twist as $V^c_{ab} = H(\bar{Q})v$ we conclude that there exists a linear relation between v and $\dot{\varphi}$. □

We will see several examples of typical expressions for the relation in (2.198). For the cases we are interested in we can find simplified expressions for this velocity transformation map. In general we will use a series representation of the exponential map. This is because, when we are to derive the dynamics later on, we will not use the exponential coordinates themselves, but rather their differential properties. After differentiating we will also evaluate the functions at $\varphi = 0$. We see that if we differentiate the map given in (2.193) and evaluate at $\varphi = 0$ all higher order terms of φ vanish. From this point of view a series representation is easier to work with than analytical expressions. We will now look at what the expression in (2.193) looks like for the most important Lie groups.

2.9.2.1 The Euclidean Space $\mathbb{R}^n$

We note that for a Euclidean transformation the Jacobian is given simply by setting $S(Q, \varphi) = I$. We can see this by simple inspection or from (2.192). Equation (2.198) thus simplifies to

$$v = \dot{\varphi}. \tag{2.199}$$

2.9.2.2 The Special Orthogonal Group $SO(3)$

For $SO(3)$ the exponential map of an element $\widehat{\varphi} \in so(3)$ is given in the normal way as

$$e^{\widehat{\varphi}} = I + \widehat{\varphi} + \frac{1}{2!}\widehat{\varphi}^2 + \frac{1}{3!}\widehat{\varphi}^3 + o(\varphi^4) \tag{2.200}$$

where $o(\varphi^n)$ collects the terms with order n or higher. The time derivative is given by

$$\begin{aligned}\frac{\mathrm{d}}{\mathrm{d}t}e^{\widehat{\varphi}} &= \frac{\mathrm{d}}{\mathrm{d}t}I + \frac{\mathrm{d}}{\mathrm{d}t}\widehat{\varphi} + \frac{1}{2!}\frac{\mathrm{d}}{\mathrm{d}t}\widehat{\varphi}^2 + \frac{1}{3!}\frac{\mathrm{d}}{\mathrm{d}t}\widehat{\varphi}^3 + o(\varphi^3) \\ &= \widehat{\dot{\varphi}} + \frac{1}{2}(\widehat{\varphi}\widehat{\dot{\varphi}} + \widehat{\dot{\varphi}}\widehat{\varphi}) + \frac{1}{6}(\widehat{\dot{\varphi}}\widehat{\varphi}^2 + \widehat{\varphi}\widehat{\dot{\varphi}}\widehat{\varphi} + \widehat{\varphi}^2\widehat{\dot{\varphi}}) + o(\varphi^3)\end{aligned} \tag{2.201}$$

and the inverse is given by

$$\begin{aligned}\left(e^{\widehat{\varphi}}\right)^{-1} = \left(e^{-\widehat{\varphi}}\right) &= I + (-\widehat{\varphi}) + \frac{1}{2!}(-\widehat{\varphi})^2 + \frac{1}{3!}(-\widehat{\varphi})^3 + o(\varphi^4) \\ &= I - \widehat{\varphi} + \frac{1}{2!}\widehat{\varphi}^2 - \frac{1}{3!}\widehat{\varphi}^3 + o(\varphi^4).\end{aligned} \tag{2.202}$$

We can now write the *body frame* angular velocities in (2.196) as

$$\begin{aligned}\widehat{\omega}_{0b}^B &= \Phi^{-1}(\bar{Q},\varphi)\dot{\Phi}(\bar{Q},\varphi) = \left(\bar{Q}e^{\widehat{\varphi}}\right)^{-1}\frac{\mathrm{d}}{\mathrm{d}t}\left(\bar{Q}e^{\widehat{\varphi}}\right) = \left(e^{\widehat{\varphi}}\right)^{-1}\bar{Q}^{\mathsf{T}}Q\frac{\mathrm{d}}{\mathrm{d}t}\left(e^{\widehat{\varphi}}\right) \\ &= \left(I - \widehat{\varphi} + \frac{1}{2}\widehat{\varphi}^2 + o(\varphi^3)\right) \\ &\quad \times \left(\widehat{\dot{\varphi}} + \frac{1}{2}(\widehat{\varphi}\widehat{\dot{\varphi}} + \widehat{\dot{\varphi}}\widehat{\varphi}) + \frac{1}{6}(\widehat{\dot{\varphi}}\widehat{\varphi}^2 + \widehat{\varphi}\widehat{\dot{\varphi}}\widehat{\varphi} + \widehat{\varphi}^2\widehat{\dot{\varphi}}) + o(\varphi^3)\right) \\ &= \widehat{\dot{\varphi}} + \frac{1}{2}(\widehat{\varphi}\widehat{\dot{\varphi}} + \widehat{\dot{\varphi}}\widehat{\varphi}) - \widehat{\varphi}\widehat{\dot{\varphi}} - \frac{1}{2}(\widehat{\varphi}^2\widehat{\dot{\varphi}} - \widehat{\varphi}\widehat{\dot{\varphi}}\widehat{\varphi}) + \frac{1}{2}(\widehat{\varphi}\widehat{\dot{\varphi}} + \widehat{\dot{\varphi}}\widehat{\varphi}) \\ &\quad + \frac{1}{6}(\widehat{\dot{\varphi}}\widehat{\varphi}^2 + \widehat{\varphi}\widehat{\dot{\varphi}}\widehat{\varphi} + \widehat{\varphi}^2\widehat{\dot{\varphi}}) + \frac{1}{2}\widehat{\varphi}\widehat{\dot{\varphi}}\widehat{\varphi} + o(\varphi^3) \\ &= \widehat{\dot{\varphi}} - \frac{1}{2}(\widehat{\varphi}\widehat{\dot{\varphi}} - \widehat{\dot{\varphi}}\widehat{\varphi}) + \frac{1}{6}(\widehat{\dot{\varphi}}\widehat{\varphi}^2 - 2\widehat{\varphi}\widehat{\dot{\varphi}}\widehat{\varphi} + \widehat{\varphi}^2\widehat{\dot{\varphi}}) + o(\varphi^3).\end{aligned} \tag{2.203}$$

We will use that for two vectors $a, b \in \mathbb{R}^3$ we have the relation $\widehat{(\widehat{a}b)} = \widehat{a}\widehat{b} - \widehat{b}\widehat{a}$ so that the second term is simplified by

$$(\widehat{\varphi}\widehat{\dot{\varphi}} - \widehat{\dot{\varphi}}\widehat{\varphi}) = \widehat{\widehat{\varphi}\dot{\varphi}} \tag{2.204}$$

and the body angular velocity can now be written in matrix form as

$$\widehat{\omega}_{0b}^B = \widehat{\dot{\varphi}} - \frac{1}{2}\widehat{\widehat{\varphi}\dot{\varphi}} + o(\varphi^2) \tag{2.205}$$

which can also be written in vector form as

$$\omega_{0b}^B = \left(I - \frac{1}{2}\widehat{\varphi} + o(\varphi^2)\right)\dot{\varphi}. \tag{2.206}$$

We do not calculate the higher order terms because these vanish when we differentiate and evaluate the expressions at $\varphi = 0$. The velocity transformation matrix relating the time derivative of the exponential coordinates and the body frame angular velocities can now be written in terms of the exponential coordinates as

$$S(\varphi) = I - \frac{1}{2}\widehat{\varphi} + o(\varphi^2). \tag{2.207}$$

We note that this expression does not depend on the current configuration $\bar{Q}$, which is reasonable because the transformation represents a mapping to the *body* velocity twist.

2.9.2.3 The Special Euclidean Group $SE(3)$

For the special Euclidean group the exponential coordinates $\varphi \in \mathbb{R}^6$ are chosen using the standard basis so that the first three coordinates represent translation and the last three rotational motion. We can find a formulation for the special Euclidean group similar to the one that we found for the special orthogonal group by using the adjoint representation ad_φ. In fact the twist written in the body frame for $SE(3)$ given by (2.196) becomes

$$V_{0b}^B = \sum_{i=0}^{\infty} \frac{(-1)^i}{(i+1)!}\mathrm{ad}_\varphi^i \dot{\varphi}, \tag{2.208}$$

which, if we write the first three terms explicitly, becomes

$$V_{0b}^B = \left(I - \frac{1}{2}\mathrm{ad}_\varphi + \frac{1}{6}\mathrm{ad}_\varphi^2 + o(\varphi^3)\right)\dot{\varphi}. \tag{2.209}$$

We will not show it here, but these expressions can be found in the same way as we did for the special orthogonal group. We refer to Rossmann (2002) for the proofs.

2.10 Geometric Integrators

Numerical integration methods are normally designed to give us the evolution of a differential equation whose configuration space is a vector space $\mathbb{R}^n$. These approaches cannot be directly applied to cases where the configuration space is a curved manifold because there is no guarantee that we stay on the manifold. When the domain is a Lie group, for example, it is important that we can guarantee that the computed solution remains in the same group. Geometric integration techniques are a group of integrators that guarantee that the computed solution stays on a certain manifold. There will always be a numerical error, but this is an error on the manifold itself.

There are two important families of geometric integrators: embedded and intrinsic integrators (Munthe-Kaas 1998). Firstly, embedded integrators require the manifold to be embedded in $\mathbb{R}^n$ and integrates using a standard integrator such as Euler's method or Runge-Kutta. The problem with embedding the manifold in $\mathbb{R}^n$ is that these classical integration method do not in general stay on the right manifold, and the computed solution therefore has to be projected onto the manifold after each iteration.

Intrinsic methods, on the other hand, compute the flow and therefore guarantee that we stay on the manifold. This is normally performed by computing the Lie algebra at each iteration step and then find the computed solution by the exponential map (Iserles 1984). We have already seen that the exponential map maps an element of the Lie algebra to the Lie group, and we can therefore conclude that we do not leave the manifold.

An early and comprehensive discussion on geometric integrators is found in Crouch and Grossman (1993) and a more recent treatment in McLachlan and Quispel (2006). An overview of different methods is given in Engø and Marthinsen (1997) and Hairer (2001). Geometric integrators in the context of multibody systems is discussed in Park and Chung (2005) and the special case of attitude dynamics of a rigid body is discussed in Lee et al. (2005, 2011).

References

Absil, P.-A., Mahony, R., & Sepulchre, R. (2008). *Optimization algorithms on matrix manifolds*. Princeton: Princeton University Press.

Antonelli, G. (2006). *Underwater robots. Motion and force control of vehicle-manipulator systems*. Berlin: Springer.

Brockett, R. W. (1984). Robotic manipulators and the product of exponentials formula. In *Proceedings of mathematical theory of networks and systems*, Beer Sheva, Israel (pp. 120–129).

Bullo, F., & Lewis, A. D. (2000). *Geometric control of mechanical systems: modeling, analysis, and design for simple mechanical control systems*. New York: Springer.

Cartan, E., & Adam, P. (2000). *Sur la structure des groupes de transformations finis et continus*. Nony.

Cayley, A. (1854). On the theory of groups, as depending on the symbolic equation $\theta^n = 1$. *Philosophical Magazine*, *7*(47), 40–47.

Crouch, P. E., & Grossman, R. (1993). Numerical integration of ordinary differential equations on manifolds. *Nonlinear Science*, *3*(1), 1–33.

Duindam, V. (2006). *Port-based modeling and control for efficient bipedal walking robots*. Ph.D. thesis, University of Twente.

Engø, K., & Marthinsen, A. (1997). Application of geometric integration to some mechanical problems. *Multibody System Dynamics*, *2*(1), 71–88.

Fossen, T. I. (2002). *Marine control systems*. Trondheim: Marine Cybernetics AS. 3rd printing.

Gingsberg, J. (2007). *Engineering dynamics*. New York: Cambridge University Press.

Hairer, E. (2001). Geometric integration of ordinary differential equations on manifolds. *BIT Numerical Mathematics*, *41*(2), 996–1007.

Hughes, P. C. (2002). *Spacecraft attitude dynamics*. Mineola: Dover Publications.

Iserles, A. (1984). Solving linear ordinary differential equations by exponentials of iterated commutators. *Numerische Mathematik*, *45*(2), 183–199.

Jordan, D. W., & Smith, P. (2004). *Nonlinear ordinary differential equations—an introduction to dynamical systems*. Oxford: Oxford University Press.

Killing, W. (1888). *Mathematische Annalen: Vol. 31. Die Zusammensetzung der stetigen/endlichen Transformationsgruppen*

Kwatny, H. G., & Blankenship, G. (2000). *Nonlinear control and analytical mechanics a computational approach*. Boston: Birkhäuser.

Lee, T., Leok, M., & McClamroch, N. (2011). Geometric numerical integration for complex dynamics of tethered spacecraft. In *Proceedings of American control conference*, San Francisco, California, USA (pp. 1885–1891).

Lee, T., McClamroch, N., & Leok, M. (2005). A lie group variational integrator for the attitude dynamics of a rigid body with applications to the 3d pendulum. In *Proceedings of IEEE conference on control applications*, Toronto, Canada (pp. 962–967).

Lie, S. (1888). *Encyclopædia Britannica. Theorie der Transformationsgruppen I.*

Lie, S. (1890). *Encyclopædia Britannica. Theorie der Transformationsgruppen II.*

Lie, S. (1893). *Encyclopædia Britannica. Theorie der Transformationsgruppen III.*

McLachlan, R. I., & Quispel, G. R. W. (2006). Geometric integrators for ODEs. *Journal of Physics. A, Mathematical and General*, *39*(19), 5251–5285.

Meng, J., Liu, G., & Li, Z. (2005). A geometric theory for synthesis and analysis of sub-6 DoF parallel manipulators. In *Proceedings of the IEEE international conference on robotics and automation*, Barcelona, Spain (pp. 3342–3347).

Meng, J., Liu, G., & Li, Z. (2007). A geometric theory for analysis and synthesis of sub-6 DoF parallel manipulators. *IEEE Transactions on Robotics*, *23*(4), 625–649.

Munthe-Kaas, H. (1998). Runge-Kutta methods on lie groups. *BIT Numerical Mathematics*, *38*(1), 92–111.

Murray, R. M., Li, Z., & Sastry, S. S. (1994). *A mathematical introduction to robotic manipulation*. Boca Raton: CRC Press.

Park, J., & Chung, W.-K. (2005). Geometric integration on Euclidean group with application to articulated multibody systems. *IEEE Transactions on Robotics*, *21*(5), 850–863.

Rao, A. (2006). *Dynamics of particles and rigid bodies—a systematic approach*. Cambridge: Cambridge University Press.

Rossmann, W. (2002). *Lie groups—an introduction through linear algebra*. Oxford: Oxford Science Publications.

Study, E. (1903). *Geometrie der Dynamen. Die Zusammensetzung von kräften und verwandte gegenstände der Geometrie*. Ithaca: Cornell University Library.

Tapp, K. (2005). *Matrix groups for undergraduates*. New York: Am. Math. Soc.

Tu, L. W. (2008). *An introduction to manifolds*. New York: Springer.

Chapter 3
Rigid Body Kinematics

Any analysis of the dynamics of a mechanical system starts with a kinematic study which defines the admissible configurations and velocities of the system. The kinematics of a rigid body is the study of its possible configurations and admissible velocities without taking into account what causes the motion. The relation between the motion of the system and the forces and torques that cause the motion is referred to as dynamics and is treated in Chaps. 6–8.

The location of a rigid body is completely described by its position and orientation. The position is given by the location of the origin of the coordinate frame attached to the rigid body with respect to an inertial coordinate frame, and the orientation is given by representing the coordinate axes of the frame attached to the rigid body in the inertial coordinate frame. The main goal of this chapter is thus to describe the position and orientation of a reference frame attached to a rigid body with respect to an inertial reference frame. In the 3-dimensional Euclidean space, the position and orientation have three degrees of freedom each.

We start the study of rigid body kinematics by looking at how to represent the position of a point in space. We then study the velocity and acceleration of this point, often referred to as differential kinematics. We then look at the orientation of a rigid body and how to represent the angular velocity and angular acceleration.

3.1 Translational Motion in $\mathbb{R}^3$

Translational motion is the study of the position, linear velocity, and linear acceleration of the origin of one coordinate frame with respect to another. We refer to Rao (2006) for details on this topic. Given an inertial coordinate frame $\mathscr{F}_0$ with basis $\{e_x, e_y, e_z\}$ and a coordinate frame $\mathscr{F}_b$ attached to a rigid body, the position of this rigid body can be completely described in Cartesian coordinates by the vector

$$p_{0b} = \begin{bmatrix} x_{0b} \\ y_{0b} \\ z_{0b} \end{bmatrix} \in \mathbb{R}^3 \tag{3.1}$$

P.J. From et al., *Vehicle-Manipulator Systems*, Advances in Industrial Control, 91
DOI 10.1007/978-1-4471-5463-1_3,

such that the vector from the origin of $\mathscr{F}_0$, denoted $\mathscr{O}_0$, to the origin of $\mathscr{F}_b$, denoted $\mathscr{O}_b$, is given in Cartesian coordinates as

$$p_{0b} = x_{0b}e_x + y_{0b}e_y + z_{0b}e_z. \tag{3.2}$$

The position of a rigid body is thus uniquely defined by a 3-dimensional vector and consequently has three degrees of freedom. As we have seen in Sect. 2.2.1 the velocity of the rigid body observed from the inertial frame can be given by

$$v_{0b}^0 = \frac{\mathrm{d}p_{0b}}{\mathrm{d}t} = \dot{x}_{0b}e_x + \dot{y}_{0b}e_y + \dot{z}_{0b}e_z. \tag{3.3}$$

We will use the following notation: v_{ij}^k denotes the velocity of $\mathscr{O}_j$ (the origin of frame $\mathscr{F}_j$) with respect to reference frame $\mathscr{F}_i$ expressed in frame $\mathscr{F}_k$. The velocity of a rigid body can thus be written in compact vector form as

$$v_{0b}^0 = \dot{p}_{0b} = \begin{bmatrix} \dot{x}_{0b} \\ \dot{y}_{0b} \\ \dot{z}_{0b} \end{bmatrix} \in \mathbb{R}^3 \tag{3.4}$$

when it is observed from the inertial frame $\mathscr{F}_0$. We can also write the velocity as observed from the body frame $\mathscr{F}_b$. This is denoted the *body velocity* and written as v_{0b}^b.

The acceleration of the rigid body can be written as

$$\dot{v}_{0b}^0 = \frac{\mathrm{d}^2 p_{0b}}{\mathrm{d}t^2} = \ddot{x}_{0b}e_x + \ddot{y}_{0b}e_y + \ddot{z}_{0b}e_z \tag{3.5}$$

and in vector form as

$$\dot{v}_{0b}^0 = \ddot{p}_{0b} = \begin{bmatrix} \ddot{x}_{0b} \\ \ddot{y}_{0b} \\ \ddot{z}_{0b} \end{bmatrix} \in \mathbb{R}^3. \tag{3.6}$$

Also the acceleration can be written with respect to the inertial frame $\mathscr{F}_0$, as above, or as body acceleration $\dot{v}_{0b}^b$ when observed from the body frame $\mathscr{F}_b$.

An important property of translational motion in the Euclidean space is that the velocity variables v_{0b}^0 can be written as the time derivative of the position variables p_{0b} and we conclude the following:

Property 3.1 The transformation between two rigid bodies represented by pure translational motion in $\mathbb{R}^3$ is Euclidean.

Note 3.1 We make a brief remark regarding the difference in notation between the position vector on one hand and the velocity and acceleration vectors on the other. As we learned in Sect. 2.2.1 the velocity and acceleration vectors given by (3.3) and (3.5) depend on the reference frames in which they are expressed and hence the notation v_{0b}^0 and $\dot{v}_{0b}^0$ with the superscript denoting the reference frame in which the vectors are observed, in this case the inertial frame $\mathscr{F}_0$. In general they can, of course, be observed from a different reference frame than the inertial frame, and we get the notation v_{ij}^k for velocity. Velocities represented in different frames are in general different and this is the reason that we need to distinguish between velocities v_{0b}^0 represented in $\mathscr{F}_0$ on one hand and velocities v_{0b}^b represented in $\mathscr{F}_b$ on the other. For position however, it is customary to write the velocity simply as p_{0b} where it is implicitly understood that we mean the position of $\mathscr{O}_b$ with respect to $\mathscr{F}_0$, as observed from $\mathscr{F}_0$. If we want to represent the position in another reference frame $\mathscr{F}_a$ we write p_{0b}^a explicitly.

3.1.1 *Spatial Linear Velocities*

We have seen that we can represent the linear velocities in different reference frames, for example $\mathscr{F}_0$ and $\mathscr{F}_b$, for which the relation between the two representation is given by change of basis, i.e. by $v_{0b}^0 = R_{0b} v_{0b}^b$. There is, however another important way to represent the velocity of a rigid body, i.e., in terms of spatial velocity variables (Murray et al. 1994; Selig 2000). Spatial velocity variables have a somewhat unintuitive geometric interpretation:

Definition 3.1 (Spatial Linear Velocity) The spatial linear velocity v_{0b}^S of a rigid body with reference frame $\mathscr{F}_b$ with respect to an inertial frame $\mathscr{F}_0$ represents the velocity of a (possibly) imaginary point located at (possibly an imaginary extension of) the rigid body when this point travels through the origin of $\mathscr{F}_0$.

For pure translational motion the spatial velocity is given simply by $v_{0b}^S = v_{0b}^0$. However, when we allow for rotational motion in the next sections we will see that it is important to distinguish between the spatial velocity v_{0b}^S and the velocity as observed from the inertial frame $\mathscr{F}_0$, denoted v_{0b}^0. The geometrical interpretation of the spatial velocity variable is illustrated in Fig. 3.1.

3.2 Rotational Motion in $\mathbb{R}^3$

We now show how to represent the orientation of a rigid body with respect to an inertial frame where the origin of the two frames coincide. In this case the orientation of a rigid body can be given by representing the three coordinate axes of the body frame $\mathscr{F}_b$ in the inertial reference frame $\mathscr{F}_0$. This transformation has three degrees of freedom. Wen and Kreutz-Delgado (1991) gives a good overview of representation methods for rotational motion.

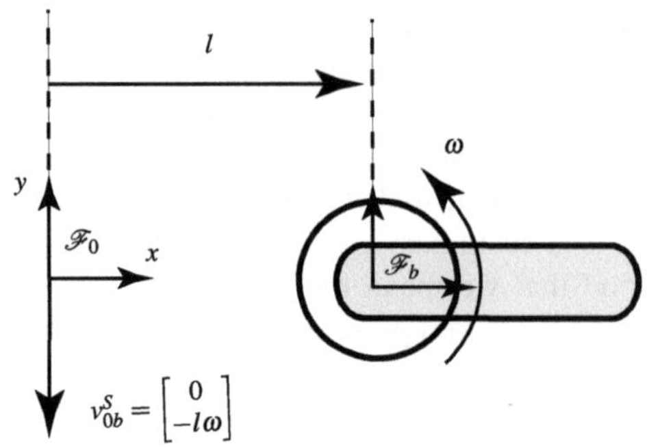

Fig. 3.1 The spatial velocity variable: A rigid body with reference frame $\mathscr{F}_b$ rotates around the origin of $\mathscr{F}_b$ with constant velocity ω. This results in a spatial velocity $v^S_{0b} = [0\ -l\omega]^\mathsf{T}$ which is the velocity of an imaginary point attached to the rigid body that travels through the origin of the inertial frame $\mathscr{F}_0$. The spatial velocity therefore depends on the distance l between the inertial frame $\mathscr{F}_0$ and the body frame $\mathscr{F}_b$

3.2.1 The Euler Angles

One representation that is commonly used to describe the orientation of a rigid body is the *Euler angles representation*. The orientation of the rigid body is then given by a sequence of three rotations about the coordinate axes. There are several different sequences of rotations, in total 12, but the most common ones are the ZYX, ZXZ, and ZYZ Euler angles, which are often referred to as Euler angles of type I, II, and III, respectively. The ZYX Euler angles are also referred to as the roll, pitch and yaw angles and are commonly used to describe the motion of aircraft and marine vessels.

If the motion is restricted to rotations around one of the coordinate axes, we can represent the rotations as direction cosine matrices of a rather simple form. For a rotation ϕ around the x-axis, for example, the relation between the original basis $\{e_x, e_y, e_z\}$ and the new basis $\{e'_x, e'_y, e'_z\}$ is given by Siciliano et al. (2011)

$$e'_x = e_x, \tag{3.7}$$

$$e'_y = \cos\phi e_y - \sin\phi e_z, \tag{3.8}$$

$$e'_z = \sin\phi e_y + \cos\phi e_z. \tag{3.9}$$

The direction cosine matrix, i.e., the matrix that rotates a rigid body from an orientation aligned with $\{e_x, e_y, e_z\}$ to an orientation aligned with $\{e'_x, e'_y, e'_z\}$ is then given by

$$R_x(\phi) = \begin{bmatrix} 1 & 0 & 0 \\ 0 & \cos\phi & -\sin\phi \\ 0 & \sin\phi & \cos\phi \end{bmatrix}. \tag{3.10}$$

Similarly, the direction cosine matrices for rotations about the y- and z-axes are given by

$$R_y(\theta) = \begin{bmatrix} \cos\theta & 0 & \sin\theta \\ 0 & 1 & 0 \\ -\sin\theta & 0 & \cos\theta \end{bmatrix}, \tag{3.11}$$

$$R_z(\psi) = \begin{bmatrix} \cos\psi & -\sin\psi & 0 \\ \sin\psi & \cos\psi & 0 \\ 0 & 0 & 1 \end{bmatrix}. \tag{3.12}$$

We see that the orientation of a rigid body with coordinate frame $\mathscr{F}_b$ with respect to the inertial frame $\mathscr{F}_0$ can be described in terms of the vector

$$\eta_2 = \begin{bmatrix} \phi_{0b} \\ \theta_{0b} \\ \psi_{0b} \end{bmatrix}. \tag{3.13}$$

In this chapter when we consider only single rigid bodies, and whenever there is no ambiguity as to what transformation we refer to, we will suppress the subscript and write $\eta_2 = [\phi\ \theta\ \psi]^{\mathsf{T}}$.

3.2.2 The Rotation Matrix

If we write the basis of the rigid body as $\{e'_x, e'_y, e'_z\}$, this can be written in terms of the inertial coordinate frame $\{e_x, e_y, e_z\}$ as

$$e'_x = r_{xx}e_x + r_{xy}e_y + r_{xz}e_z, \tag{3.14}$$

$$e'_y = r_{yx}e_x + r_{yy}e_y + r_{yz}e_z, \tag{3.15}$$

$$e'_z = r_{zx}e_x + r_{zy}e_y + r_{zz}e_z. \tag{3.16}$$

Alternatively we can write

$$\begin{bmatrix} e'_x \\ e'_y \\ e'_z \end{bmatrix} = [r_1 \quad r_2 \quad r_3] \begin{bmatrix} e_x \\ e_y \\ e_z \end{bmatrix} \tag{3.17}$$

where the vectors r_1, r_2 and r_3 are the coordinates of the basis $\{e'_x, e'_y, e'_z\}$ relative to $\{e_x, e_y, e_z\}$, as illustrated in Fig. 3.2. The matrix

$$R = [r_1 \quad r_2 \quad r_3] \tag{3.18}$$

is denoted a *rotation matrix*. There are many ways to find the rotation matrix. We can for example use the ZYX Euler angles described in the previous section. We then start with (1) a rotation ϕ about the x-axis, represented by $R_x(\phi)$; (2) a rotation

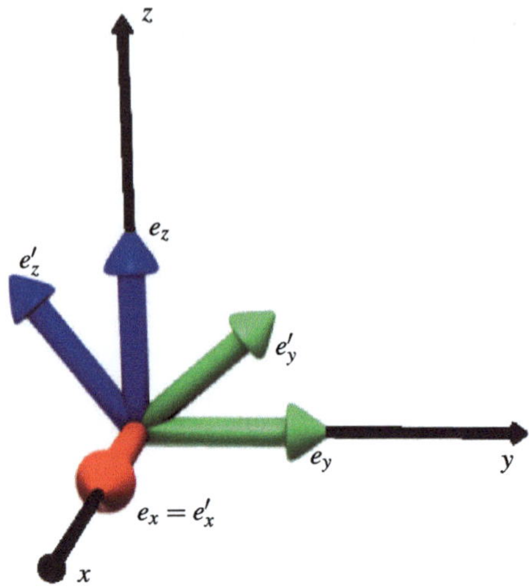

(a) Rotation around the x-axis

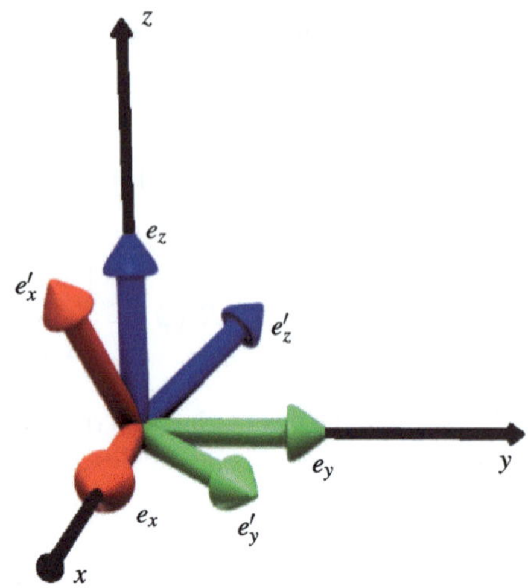

(b) A rotation that can only be obtained by three consecutive rotations around the coordinate axes

Fig. 3.2 The orientation of a rigid body represented by reference frame $\mathscr{F}_b$ with basis $\{e'_x, e'_y, e'_z\}$ with respect to an inertial reference frame $\mathscr{F}_0$ with basis $\{e_x, e_y, e_z\}$ for rotations with one and three degrees of freedom

θ about the new (rotated) y-axis, represented by $R_y(\theta)$; and (3) a rotation ψ about the rotated z-axis, represented by $R_z(\psi)$. We note that if we apply these rotations to a vector by left multiplication, we multiply these "backwards" as

$$R = R_z(\psi)R_y(\theta)R_x(\phi) \in \mathbb{R}^{3\times 3}. \tag{3.19}$$

The rotation matrix gives a complete description of the orientation of a rigid body with reference frame $\mathscr{F}_b$ with respect to the inertial frame $\mathscr{F}_0$. We will denote such a rotation matrix R_{0b}. The rotation matrix can also be used to rotate a vector p^i observed in coordinate frame i into the same vector p^j observed in coordinate frame j by

$$p^j = R_{ji}p^i. \tag{3.20}$$

The superscripts mean that p^i and p^j are the same vector, but viewed from two different coordinate frames.

The column vectors of the rotation matrix represent the unit vectors of an orthonormal frame, i.e.,

$$r_i^{\mathsf{T}} r_j = \begin{cases} 0 & \text{for } i \neq j, \\ 1 & \text{for } i = j. \end{cases}$$

The orthogonality and unity conditions each add three constraints to the rotation matrix. As the rotation matrix has nine elements and we have a total of six constraints, we see that the orientation of a rigid body has three degrees of freedom.

The conditions can also be written as

$$RR^{\mathsf{T}} = R^{\mathsf{T}}R = I, \tag{3.21}$$

from which, if we postmultiply by R^{-1}, we get

$$R^{\mathsf{T}} = R^{-1}. \tag{3.22}$$

This is an important property that we will use frequently in the following. Furthermore, as we assume a right-handed coordinate system, we require that $\det R = 1$. We get the following definition of a rotation matrix:

Definition 3.2 A rotation matrix $R \in \mathbb{R}^{n\times n}$ is defined by

$$SO(n) = \{R \in \mathbb{R}^{n\times n} \mid RR^{\mathsf{T}} = I, \det R = +1\}. \tag{3.23}$$

We are particularly interested in rotation matrices where $n = 3$—which describe rotations in the 3-dimensional Euclidean space—and rotation matrices where $n = 2$—which describe rotations in the plane. For an appropriate choice of coordinate frames a rotation in the plane can always be represented by one of the matrices given in (3.10)–(3.12).

For rotations of a rigid body in $\mathbb{R}^3$ we adopt the roll-pitch-yaw Euler angles. The rotation matrix is then given by (3.19) which, if we write it out, gives

$$R = \begin{bmatrix} \cos\psi\cos\theta & \cos\psi\sin\theta\sin\phi - \sin\psi\cos\phi & \cos\psi\cos\phi\sin\theta + \sin\psi\sin\phi \\ \sin\psi\cos\theta & \sin\phi\sin\theta\sin\psi + \cos\psi\cos\phi & \sin\theta\sin\psi\cos\phi - \cos\psi\sin\phi \\ -\sin\theta & \cos\theta\sin\phi & \cos\theta\cos\phi \end{bmatrix} \tag{3.24}$$

3.2.3 *The Quaternion Representation*

The unit quaternion can also be used to represent rotations in $\mathbb{R}^3$ (Egeland and Gravdahl 2003; Hanson 2006; Kuipers 2002). Recall that a unit quaternion is written in terms of the basis 1, i, j, and k as quadruples in the form

$$\mathscr{H} = \{(a, b, c, d) \mid a, b, c, d \in \mathbb{R}, a^2 + b^2 + c^2 + d^2 = 1\} \tag{3.25}$$

where a is normally referred to as the scalar part and (b, c, d) as the vector part. A rotation of t units around an axis ω can be written as a rotation matrix in terms of the exponential coordinates as $R = e^{\widehat{\omega} t}$. The corresponding unit quaternion, i.e., the quaternion that corresponds to the same rotation, is given by

$$H = \begin{bmatrix} \cos\frac{t}{2} & \omega\sin\frac{t}{2} \end{bmatrix}. \tag{3.26}$$

We see that we can determine the size of the rotation from the scalar part and the axis that we rotate around from the vector part of the quaternion. The inverse is obtained by substituting t with $-t$ which gives:

$$H^{-1} = \begin{bmatrix} \cos\frac{t}{2} & -\omega \sin\frac{t}{2} \end{bmatrix}. \tag{3.27}$$

This is normally denoted the conjugate of the quaternion and corresponds to the rotation matrix $R = e^{-\widehat{\omega}t}$. Two consecutive rotations can be represented by the quaternion product given in (2.39) in the same way that the rotation obtained by two consecutive rotations represented by rotation matrices is given by matrix multiplication.

3.2.4 Angular Velocity and Angular Acceleration

A natural starting point when considering the angular velocity of a rigid body is to look at the time derivative of the rotation matrix. We will see, however, that $\dot{R}$ does not provide us with a very meaningful geometric interpretation. We will thus define a set of variables ω that are easier to interpret geometrically.

Let R_{ab} represent the orientation of a reference frame $\mathscr{F}_b$ with respect to a reference frame $\mathscr{F}_a$. In general the rotation matrix is time varying and we can differentiate Eq. (3.21) with respect to time which gives

$$\dot{R}_{ab}(t)R_{ab}^{\mathsf{T}}(t) + R_{ab}(t)\dot{R}_{ab}^{\mathsf{T}}(t) = 0. \tag{3.28}$$

Obviously, as R needs to satisfy certain constraints, so must $\dot{R}$. In fact, the constraints that $\dot{R}$ needs to satisfy are given by (3.28). It is, however, possible to write these constraints in a more convenient form. First, define the matrix

$$\widehat{\omega}_{ab}^{S}(t) = \dot{R}_{ab}(t)R_{ab}^{\mathsf{T}}(t) \tag{3.29}$$

which allows us to write

$$\begin{aligned} \widehat{\omega}_{ab}^{S}(t) + \widehat{\omega}_{ab}^{S}(t)^{\mathsf{T}} &= \dot{R}_{ab}(t)R_{ab}^{\mathsf{T}}(t) + \big(\dot{R}_{ab}(t)R_{ab}^{\mathsf{T}}(t)\big)^{\mathsf{T}} \\ &= \dot{R}_{ab}(t)R_{ab}^{\mathsf{T}}(t) + R_{ab}(t)\dot{R}_{ab}^{\mathsf{T}}(t) = 0. \end{aligned} \tag{3.30}$$

We can conclude that $\widehat{\omega}_{ab}^{S}(t)$ is skew-symmetric. From (3.29) we can write the time derivative of $R_{ab}(t)$ as a function of $R_{ab}(t)$ itself as

$$\dot{R}_{ab}(t) = \widehat{\omega}_{ab}^{S}(t)R_{ab}(t) \tag{3.31}$$

where we have used that $R_{ab}^{\mathsf{T}}R_{ab} = I$. This relation may seem somewhat constructed, but $\widehat{\omega}_{ab}^{S}$ actually has a geometrically meaningful interpretation: Let p^b be fixed in $\mathscr{F}_b$ and the relative motion between $\mathscr{F}_a$ and $\mathscr{F}_b$ be given by R_{ab}. Then the same vector, but as seen from $\mathscr{F}_a$ is given by $p^a(t) = R_{ab}(t)p^b$. R_{ab} thus maps the position coordinates of a point in $\mathscr{F}_b$ into the position coordinates of the same point in the frame $\mathscr{F}_a$. The time derivative of $p^a(t)$ is given by

$$\begin{aligned}\dot{p}^a(t) &= \dot{R}_{ab}(t)p^b \\ &= \widehat{\omega}^S_{ab}(t)R_{ab}(t)p^b \\ &= \widehat{\omega}^S_{ab}(t)p^a. \end{aligned} \tag{3.32}$$

We see that $\omega^S_{ab}(t)$ has a very clear geometric interpretation as it maps the position of a point p^a, observed in frame $\mathscr{F}_a$, into the velocities of the same point, also represented in $\mathscr{F}_a$. The angular velocity is thus given by the three variables in $\omega^S_{ab}(t)$. Recall that the skew-symmetric operator maps the vector $\omega = [\omega_x\ \omega_y\ \omega_z]^\mathsf{T}$ to the matrix representation in the following way:

$$\widehat{\omega} = \omega\times = \begin{bmatrix} 0 & -\omega_z & \omega_y \\ \omega_z & 0 & -\omega_x \\ -\omega_y & \omega_x & 0 \end{bmatrix}. \tag{3.33}$$

We choose $\mathscr{F}_a$ to be the inertial frame $\mathscr{F}_0$ and define the *spatial angular velocities* $\widehat{\omega}^S_{0b}$ as

$$\widehat{\omega}^S_{0b} = \dot{R}_{0b}R^{-1}_{0b} = \dot{R}_{0b}R^\mathsf{T}_{0b}. \tag{3.34}$$

We will denote the spatial angular velocity in vector form as

$$\omega^S_{0b} = \begin{bmatrix} p^S \\ q^S \\ r^S \end{bmatrix}. \tag{3.35}$$

The body angular velocity represents the velocity of the same rigid body, but as observed from the body frame $\mathscr{F}_b$. The *body angular velocity* $\widehat{\omega}^B_{0b}$ is defined as

$$\widehat{\omega}^B_{0b} = R^{-1}_{0b}\dot{R}_{0b} = R^\mathsf{T}_{0b}\dot{R}_{0b}. \tag{3.36}$$

We will denote the body angular velocity in vector form as

$$\omega^B_{0b} = \begin{bmatrix} p \\ q \\ r \end{bmatrix}. \tag{3.37}$$

From Sect. 2.6.6.2 we know that we can find the relation between the body and spatial velocity variables in terms of the Adjoint map by $\widehat{\omega}^S_{0b} = \mathrm{Ad}_{R_{0b}}\widehat{\omega}^B_{0b}$. We find the Adjoint map by combining (3.34) and (3.36) which gives

$$\widehat{\omega}^S_{0b} = R_{0b}\widehat{\omega}^B_{0b}R^\mathsf{T}_{0b}, \tag{3.38}$$

or alternatively

$$\omega^S_{0b} = R_{0b}\omega^B_{0b}. \tag{3.39}$$

Recall that for translations in $\mathbb{R}^3$ we could find a set of position variables whose rate of change corresponds to the velocity. For angular motion on the other hand, we cannot find such a set of coordinates. In other words, we cannot find the orientation of the rigid body by integrating the angular velocity in the standard way. From Sect. 2.4 we know that we can choose a set of three coordinates $(\gamma_x, \gamma_y, \gamma_z)$, which relate to the angular velocities by

$$\dot{\gamma}_x = \omega_x, \tag{3.40}$$

$$\dot{\gamma}_y = \omega_y, \tag{3.41}$$

$$\dot{\gamma}_z = \omega_z. \tag{3.42}$$

However, the set $(\gamma_x, \gamma_y, \gamma_z)$ does not have a physical interpretation in the same way as the integral of the velocity variables for linear motion. If we use Euler angles, for example, then $(\gamma_x, \gamma_y, \gamma_z) \neq (\phi, \theta, \psi)$. The parameter γ is often referred to as quasi-coordinates and will be discussed in detail in Sect. 3.3.3.

Once the angular velocity of the rigid body is resolved, it is straight forward to derive the angular accelerations. Given the body frame angular velocity ω_{0b}^B the body frame angular acceleration is given simply by

$$\ddot{\omega}_{0b}^B = \frac{\mathrm{d}\dot{\omega}_{0b}^B}{\mathrm{d}t} = \begin{bmatrix} \dot{p} \\ \dot{q} \\ \dot{r} \end{bmatrix} \tag{3.43}$$

and the spatial angular accelerations are given by

$$\ddot{\omega}_{0b}^S = \frac{\mathrm{d}\dot{\omega}_{0b}^S}{\mathrm{d}t} = \begin{bmatrix} \dot{p}^S \\ \dot{q}^S \\ \dot{r}^S \end{bmatrix}. \tag{3.44}$$

3.2.5 The Relation Between Euler Angle Derivatives and Twists

As we have seen we can represent the velocity of a rigid body in terms of twists, defined either in spatial or body coordinates. We can also write the angular velocity as the time derivative of the Euler angles. In this section we will look at the relation between the derivative of the Euler angles and the body and spatial angular velocities defined in the previous section. This relation represents the kinematics of a rigid body. We start with a simple example (Murray et al. 1994):

Example 3.1 Assume a rigid body that rotates around a fixed axis so that its motion is characterized by

$$R(t) = \begin{bmatrix} \cos\psi(t) & -\sin\psi(t) & 0 \\ \sin\psi(t) & \cos\psi(t) & 0 \\ 0 & 0 & 1 \end{bmatrix} \tag{3.45}$$

where $\psi(t)$ is the rotation around the z-axis. The instantaneous spatial and body velocities are given by (3.34) and (3.36), respectively:

$$\begin{aligned}
\widehat{\omega}^S_{0b}(t) &= \dot{R}(t)R^\mathsf{T}(t) \\
&= \begin{bmatrix} -\dot{\psi}(t)\sin\psi(t) & -\dot{\psi}(t)\cos\psi(t) & 0 \\ -\dot{\psi}(t)\cos\psi(t) & -\dot{\psi}(t)\sin\psi(t) & 0 \\ 0 & 0 & 0 \end{bmatrix} \begin{bmatrix} \cos\psi(t) & \sin\psi(t) & 0 \\ -\sin\psi(t) & \cos\psi(t) & 0 \\ 0 & 0 & 1 \end{bmatrix} \\
&= \begin{bmatrix} 0 & -\dot{\psi}(t) & 0 \\ \dot{\psi}(t) & 0 & 0 \\ 0 & 0 & 0 \end{bmatrix}
\end{aligned} \tag{3.46}$$

$$\begin{aligned}
\widehat{\omega}^B_{0b}(t) &= R^\mathsf{T}(t)\dot{R}(t) \\
&= \begin{bmatrix} \cos\psi(t) & \sin\psi(t) & 0 \\ -\sin\psi(t) & \cos\psi(t) & 0 \\ 0 & 0 & 1 \end{bmatrix} \begin{bmatrix} -\dot{\psi}(t)\sin\psi(t) & -\dot{\psi}(t)\cos\psi(t) & 0 \\ -\dot{\psi}(t)\cos\psi(t) & -\dot{\psi}(t)\sin\psi(t) & 0 \\ 0 & 0 & 0 \end{bmatrix} \\
&= \begin{bmatrix} 0 & -\dot{\psi}(t) & 0 \\ \dot{\psi}(t) & 0 & 0 \\ 0 & 0 & 0 \end{bmatrix}
\end{aligned} \tag{3.47}$$

and we have that $\omega^S_{0b} = \omega^B_{0b} = [0\ 0\ \dot{\psi}^\mathsf{T}]$.

The example above shows that for a one degree of freedom rotational motion, both the spatial and body angular velocities are exactly the time derivative of the angle of rotation. This is true for one degree of freedom rotational motion, but cannot be taken as a general result. This is illustrated in the next example:

Example 3.2 Consider a rigid body motion that can be parameterized first by a rotation θ around the y-axis, followed by a rotation ψ around the z-axis. Let $R_z(\psi)$ and $R_y(\theta)$ be the elementary rotation matrices. The motion of the rigid body is then given by

$$R_{0b} = R_z(\psi)R_y(\theta) = \begin{bmatrix} \cos\psi\cos\theta & -\sin\psi & \cos\psi\sin\theta \\ \sin\psi\cos\theta & \cos\psi & \sin\psi\sin\theta \\ -\sin\theta & 0 & \cos\theta \end{bmatrix} \tag{3.48}$$

and the time derivative is found to be

$$\dot{R}_{0b} = \begin{bmatrix} -\sin\psi\cos\theta\dot{\psi} - \cos\psi\sin\theta\dot{\theta} & -\cos\psi\dot{\psi} & -\sin\psi\sin\theta\dot{\psi} + \cos\psi\cos\theta\dot{\theta} \\ \cos\psi\cos\theta\dot{\psi} - \sin\psi\sin\theta\dot{\theta} & -\sin\psi\dot{\psi} & \cos\psi\sin\theta\dot{\psi} + \sin\psi\cos\theta\dot{\theta} \\ -\cos\theta\dot{\theta} & 0 & -\sin\theta\dot{\theta} \end{bmatrix}. \tag{3.49}$$

In this case the spatial and body velocities are given by

$$\omega_{0b}^{S} = \left(\dot{R}R^{\mathsf{T}}\right)^{\vee} = \begin{bmatrix} -\sin\psi\dot{\theta} \\ \cos\psi\dot{\theta} \\ \dot{\psi} \end{bmatrix}, \tag{3.50}$$

$$\omega_{0b}^{B} = \left(R^{\mathsf{T}}\dot{R}\right)^{\vee} = \begin{bmatrix} -\sin\theta\dot{\psi} \\ \dot{\theta} \\ \cos\theta\dot{\psi} \end{bmatrix}. \tag{3.51}$$

Here, the vee map $\vee$ maps the matrix representation of the Lie group into the corresponding vector representation. We see that $\omega_{0b}^{S} \neq \omega_{0b}^{B}$ so in the general case the body and spatial velocities are different. We can easily verify that $\widehat{\omega}_{0b}^{S} = R_{0b}\widehat{\omega}_{0b}^{B}$, i.e.,

$$\begin{bmatrix} -\sin\psi\dot{\theta} \\ \cos\psi\dot{\theta} \\ \dot{\psi} \end{bmatrix} = \begin{bmatrix} \cos\psi\cos\theta & -\sin\psi & \cos\psi\sin\theta \\ \sin\psi\cos\theta & \cos\psi & \sin\psi\sin\theta \\ -\sin\theta & 0 & \cos\theta \end{bmatrix} \begin{bmatrix} -\sin\theta\dot{\psi} \\ \dot{\theta} \\ \cos\theta\dot{\psi} \end{bmatrix}. \tag{3.52}$$

We can also find the relation between the body and spatial velocities (3 DoF) and the time derivative of the Euler angles using the same approach. The rotation matrix is then given by Fossen (1994, 2002)

$$\begin{aligned} R_{0b} &= R_z(\psi)R_y(\theta)R_x(\phi) \\ &= \begin{bmatrix} \cos\psi\cos\theta & \cos\psi\sin\theta\sin\phi - \sin\psi\cos\phi & \cos\psi\cos\phi\sin\theta + \sin\psi\sin\phi \\ \sin\psi\cos\theta & \sin\phi\sin\theta\sin\psi + \cos\psi\cos\phi & \sin\theta\sin\psi\cos\phi - \cos\psi\sin\phi \\ -\sin\theta & \cos\theta\sin\phi & \cos\theta\cos\phi \end{bmatrix} \end{aligned} \tag{3.53}$$

and we get the spatial and body velocities given by

$$\omega_{0b}^{S} = \left(\dot{R}R^{\mathsf{T}}\right)^{\vee} = \begin{bmatrix} -\sin\psi\dot{\theta} + \cos\theta\cos\psi\dot{\phi} \\ \cos\psi\dot{\theta} + \cos\theta\sin\psi\dot{\phi} \\ \dot{\psi} - \sin\theta\dot{\phi} \end{bmatrix}, \tag{3.54}$$

$$\omega_{0b}^{B} = \left(R^{\mathsf{T}}\dot{R}\right)^{\vee} = \begin{bmatrix} \dot{\phi} - \sin\theta\dot{\psi} \\ \cos\theta\dot{\psi}\sin\phi + \dot{\theta}\cos\phi \\ \cos\theta\dot{\psi}\cos\phi - \dot{\theta}\sin\phi \end{bmatrix}. \tag{3.55}$$

We can now find the relations between the spatial and angular velocities and the Euler angles:

$$\omega_{0b}^{S} = \begin{bmatrix} \cos\theta\cos\psi & -\sin\psi & 0 \\ \cos\theta\sin\psi & \cos\psi & 0 \\ -\sin\theta & 0 & 1 \end{bmatrix} \begin{bmatrix} \dot{\phi} \\ \dot{\theta} \\ \dot{\psi} \end{bmatrix}, \tag{3.56}$$

$$\omega_{0b}^{B}=\begin{bmatrix}1 & 0 & -\sin\theta\\ 0 & \cos\phi & \cos\theta\sin\phi\\ 0 & -\sin\phi & \cos\theta\cos\phi\end{bmatrix}\begin{bmatrix}\dot\phi\\ \dot\theta\\ \dot\psi\end{bmatrix}. \tag{3.57}$$

The inverse of this mapping is given by

$$\begin{bmatrix}\dot\phi\\ \dot\theta\\ \dot\psi\end{bmatrix}=\begin{bmatrix}\frac{\cos\psi}{\cos\theta} & \frac{\sin\psi}{\cos\theta} & 0\\ -\sin\psi & \cos\psi & 0\\ \frac{\cos\psi\sin\theta}{\cos\theta} & \frac{\sin\psi\sin\theta}{\cos\theta} & 1\end{bmatrix}\begin{bmatrix}p^S\\ q^S\\ r^S\end{bmatrix}, \tag{3.58}$$

$$\begin{bmatrix}\dot\phi\\ \dot\theta\\ \dot\psi\end{bmatrix}=\begin{bmatrix}1 & \frac{\sin\theta\sin\phi}{\cos\theta} & \frac{\sin\theta\cos\phi}{\cos\theta}\\ 0 & \cos\phi & -\sin\phi\\ 0 & \frac{\sin\phi}{\cos\theta} & \frac{\cos\phi}{\cos\theta}\end{bmatrix}\begin{bmatrix}p\\ q\\ r\end{bmatrix}. \tag{3.59}$$

Based on the relations in (3.58) and (3.59) we can conclude the following property of angular transformations on $SO(3)$:

Property 3.2 The transformation of a rigid body represented by pure rotational motion in $\mathbb{R}^3$ is non-Euclidean.

A matrix that gives the relation between velocities in this way is denoted a Jacobian matrix, or a velocity transformation matrix. The Jacobian matrix relating the body velocities and the time derivatives of the Euler angles such that $\dot\eta_2 = J_{b,o}(\eta_2)\omega_{0b}^{B}$ is given by

$$J_{b,o}(\eta_2)=T_{0b}(\eta_2)=\begin{bmatrix}1 & \frac{\sin\theta\sin\phi}{\cos\theta} & \frac{\sin\theta\cos\phi}{\cos\theta}\\ 0 & \cos\phi & -\sin\phi\\ 0 & \frac{\sin\phi}{\cos\theta} & \frac{\cos\phi}{\cos\theta}\end{bmatrix} \tag{3.60}$$

where $\eta_2=[\phi\ \theta\ \psi^{\mathsf T}]\in\mathbb{R}^3$.

The inverse of the Jacobian matrix mapping the time derivatives of the Euler angles to the body velocities such that $\omega_{0b}^{B}=J_{b,o}^{-1}(\eta_2)\dot\eta_2$ is given by

$$J_{b,o}^{-1}(\eta_2)=T_{0b}^{-1}(\eta_2)=\begin{bmatrix}1 & 0 & -\sin\theta\\ 0 & \cos\phi & \cos\theta\sin\phi\\ 0 & -\sin\phi & \cos\theta\cos\phi\end{bmatrix}. \tag{3.61}$$

The Jacobian matrices relating the spatial angular velocities and the Euler angles are given in the same way as $J_{b,o}^{S}(\eta_2)=T_{0b}^{S}(\eta_2)$ as found in (3.58) and $(J_{b,o}^{S})^{-1}(\eta_2)=(T_{0b}^{S})^{-1}(\eta_2)$ as found in (3.56).

Velocity transformation matrices are very important when deriving the kinematics and dynamics of both single rigid bodies and multibody systems. In fact, the transformation matrix is what determines the kinematics of a rigid body because it gives us the necessary relation between the velocity variables on one hand, and (the time derivative of) the position variables on the other.

3.3 Rigid Body Transformations

The location of a rigid body in space can be described by the position and orientation of a reference frame attached to the rigid body with respect to the inertial frame, as illustrated in Fig. 3.3. We find the configuration of a reference frame $\mathscr{F}_b$ with respect to $\mathscr{F}_0$ in two steps: (1) we represent the *position* of the origin $\mathscr{O}_b$ of $\mathscr{F}_b$ with respect to the inertial frame $\mathscr{F}_0$ as seen from $\mathscr{F}_0$; and (2) write the *orientation* of $\mathscr{F}_b$ with respect to $\mathscr{F}_0$ as if the two origins were coinciding, i.e., as if $\mathscr{O}_0 = \mathscr{O}_b$. We note that if we perform step 2 before step 1 so that the position is represented in $\mathscr{F}_b$ instead of $\mathscr{F}_0$, we get a different representation of the position of $\mathscr{O}_b$, but the same orientation.

As we have seen, both the position and orientation of a rigid body have three degrees of freedom each. Since the configuration of a rigid body has six degrees of freedom it is reasonable to represent the configuration as a vector in $\mathbb{R}^6$. One way to do this is to first write the position of the rigid body as a vector $\eta_1 = p_{0b} = [x_{0b}\ y_{0b}\ z_{0b}]^{\mathsf{T}}$ as described in Sect. 3.1, and the orientation, also as a vector in $\mathbb{R}^3$ by the Euler angles $\eta_2 = [\phi\ \theta\ \psi]^{\mathsf{T}}$, as described in Sect. 3.2.1. The location of a rigid body in space is then given by a vector

$$g_{0b}^{\vee} = \begin{bmatrix} \eta_1 \\ \eta_2 \end{bmatrix} = \begin{bmatrix} x_{0b} \\ y_{0b} \\ z_{0b} \\ \phi \\ \theta \\ \psi \end{bmatrix}. \tag{3.62}$$

The problem with this representation is that $SE(3)$ has a different topology than the Euclidean space $\mathbb{R}^6$. It is thus not possible to cover $SE(3)$ continuously and globally using six coordinates in this way. To obtain a continuous and global covering of $SE(3)$ we need to use more than six real numbers. One representation that we can use is the unit quaternion, which requires four parameters to describe the orientation. We can also combine the rotation matrix and the position vector into one matrix, called the homogeneous transformation matrix, using 16 real numbers. In this way we choose a representation that allows us to describe the topology of $SE(3)$ globally and continuously at the expense of increasing the number of variables used to describe the configuration of the system to 7 or 16.

Before we look at homogeneous transformations in more detail, we will derive the kinematic relations using vectors in $\mathbb{R}^6$. This is a commonly found formulation in for example modeling of ships and spacecraft.

3.3.1 Vector Representation

Assume that we choose to represent the configuration of the rigid body as a vector in $\mathbb{R}^6$ as in (3.62). In the same way as for rotational motion we need to find the

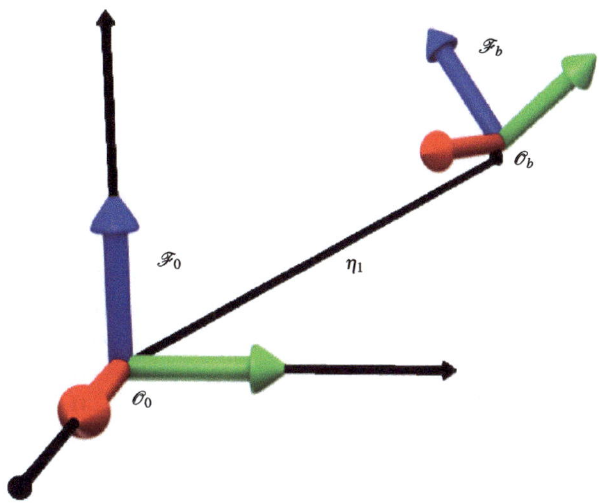

Fig. 3.3 The position and orientation of a coordinate frame $\mathscr{F}_b$ with origin $\mathscr{O}_b$ with respect to an inertial reference frame $\mathscr{F}_0$

relation between the velocity variables and the time derivative of the position variables. The position variables are given as $\eta = [\eta_1^\mathsf{T}\ \eta_2^\mathsf{T}]^\mathsf{T} = [x_{0b}\ y_{0b}\ z_{0b}\ \phi\ \theta\ \psi]^\mathsf{T}$ and the velocity variables are given by stacking the spatial or body linear and angular velocity variables that we found in Sects. 3.1 and 3.2, respectively. For $SE(3)$ the body and spatial velocity variables are given by

$$V_{0b}^B = \begin{bmatrix} u \\ v \\ w \\ p \\ q \\ r \end{bmatrix}, \qquad V_{0b}^S = \begin{bmatrix} u^S \\ v^S \\ w^S \\ p^S \\ q^S \\ r^S \end{bmatrix}. \tag{3.63}$$

We see that we need to specify in what reference frame both the linear and angular velocities are represented.

We can now find the mapping from the body velocities of the rigid body to the time derivative of the position variables as

$$\dot{\eta} = J_b(\eta_2) V_{0b}^B. \tag{3.64}$$

$J_b(\eta_2)$ is defined as

$$J_b(\eta_2) = \begin{bmatrix} R_{0b}(\eta_2) & 0 \\ 0 & T_{0b}(\eta_2) \end{bmatrix} \tag{3.65}$$

where $R_{0b}(\eta_2)$ is the rotation matrix and depends only on the orientation of the body represented by the Euler angles $\eta_2 = [\phi\ \theta\ \psi]^\mathsf{T}$. This is the same as the rotation matrix that we found in Sect. 3.2.2. $T_{0b}(\eta_2)$ is given by (ZYX-sequence)

$$T_{0b}(\eta_2) = \begin{bmatrix} 1 & \sin\phi\tan\theta & \cos\phi\tan\theta \\ 0 & \cos\phi & -\sin\phi \\ 0 & \frac{\sin\phi}{\cos\theta} & \frac{\cos\phi}{\cos\theta} \end{bmatrix} \tag{3.66}$$

and was found in Sect. 3.2.5. The velocity transformation matrix is then given by

$$\begin{bmatrix} \dot{x}_{0b} \\ \dot{y}_{0b} \\ \dot{z}_{0b} \\ \dot{\phi} \\ \dot{\theta} \\ \dot{\psi} \end{bmatrix} = \begin{bmatrix} c\psi c\theta & c\psi s\theta s\phi - s\psi c\phi & c\psi c\phi s\theta + s\psi s\phi & 0 & 0 & 0 \\ s\psi c\theta & s\phi s\theta s\psi + c\psi c\phi & s\theta s\psi c\phi - c\psi s\phi & 0 & 0 & 0 \\ -s\theta & c\theta s\phi & c\theta c\phi & 0 & 0 & 0 \\ 0 & 0 & 0 & 1 & \frac{s\phi s\theta}{c\theta} & \frac{c\phi s\theta}{c\theta} \\ 0 & 0 & 0 & 0 & c\phi & -s\phi \\ 0 & 0 & 0 & 0 & \frac{s\phi}{c\theta} & \frac{c\phi}{c\theta} \end{bmatrix} \begin{bmatrix} u \\ v \\ w \\ p \\ q \\ r \end{bmatrix} \tag{3.67}$$

where $s\theta$ means $\sin\theta$ etc. We note that $T_{0b}(\eta_2)$, and thus also $J_b(\eta_2)$, is not defined for $\theta = \pm\frac{\pi}{2}$. This is the well known Euler angle singularity which for the ZYX-sequence appears as θ approaches $\frac{\pi}{2}$. Equation (3.67) describes the kinematics of a rigid body motion in $\mathbb{R}^6$ when the velocity variables are represented in the body frame.

If we instead choose to represent the velocity variables in the inertial frame $\mathscr{F}_0$, the relation $\dot{\eta} = J_b^0(\eta_2)V_{0b}^0$ is given by a velocity transformation matrix in the form

$$J_b^0(\eta_2) = \begin{bmatrix} I & 0 \\ 0 & T_{0b}(\eta_2)R_{0b}^{-1}(\eta_2) \end{bmatrix}. \tag{3.68}$$

Similarly, in spatial coordinates the mapping $\dot{\eta} = J_b^S(\eta_2)V_{0b}^S$ is given by

$$J_b^S(\eta) = J_b(\eta_2)\mathrm{Ad}_{g_{0b}}^{-1}(\eta). \tag{3.69}$$

Finally we find the accelerations in the same way that we did in the previous sections. The body accelerations are given by $\dot{V}_{0b}^B = [\dot{u}\ \dot{v}\ \dot{w}\ \dot{p}\ \dot{q}\ \dot{r}]^\mathsf{T}$ and the spatial accelerations as $\dot{V}_{0b}^S = [\dot{u}^S\ \dot{v}^S\ \dot{w}^S\ \dot{p}^S\ \dot{q}^S\ \dot{r}^S]^\mathsf{T}$.

3.3.2 Singularities in the Representation

It is a well known fact that the kinematics of a rigid body contains singularities if the Euler angles are used to represent the orientation of the body and the topology is not taken into account. This is easily seen from the velocity transformation matrix in (3.67) which has a singularity at $\theta = \pm\frac{\pi}{2}$. Another way to see this is to consider the Euler angles described by three consecutive rotations around the x-, y-, and z-axes. In this case the coordinates $(a - \psi, \pm\frac{\pi}{2}, \psi)$ describe the same orientation for all values of ψ. Thus, for a smooth change in the orientation of the rigid body, we can have a non-smooth change in the coordinates. This is undesirable when it comes to modeling and implementing the equations in a simulation software. Depending on what sequence of Euler angles we use, these singularities and discontinuities will appear at different points in the configuration space. This is always true, and it is not possible to find an Euler angle representation for which this problem does not arise at one given configuration (Murray et al. 1994; Duindam and Stramigioli 2007, 2008; From et al. 2010).

The reason that these singularities appear is that we try to write the kinematics in vector form while the configuration space in fact is $SE(3)$. From the following lemma, taken from Duindam (2006), we see that the $SE(3)$ is in fact different from $\mathbb{R}^6$:

Lemma 3.1 *The space of all possible configurations of a rigid body in the 3-dimensional space, relative to the inertial frame, is the 6-dimensional space* $SE(3)$, *which is topologically equivalent to the set* $\mathbb{R}^1 \times \mathbb{R}^1 \times \mathbb{R}^1 \times \mathbb{S}^2 \times \mathbb{S}^1$.

Proof A rigid body (for example a reference frame) is described using three non-collinear reference points. The space of rigid body motions in the 3-dimensional Euclidean space is thus the space needed to position these three points in space. The first point can be positioned freely in space (a point has no orientation), i.e., there are three degrees of freedom used to position this point. These are translational degrees of freedom and denoted $\mathbb{R}^1 \times \mathbb{R}^1 \times \mathbb{R}^1$. Because we are dealing with rigid bodies, the second point needs to be positioned so that its distance to the first point is constant. This is obtained by placing the second point at a sphere $\mathbb{S}^2$ with the first point at the center. Finally the last point needs to be positioned so that it remains at a constant distance from the first two points. This means that the final point needs to be placed at a circle $\mathbb{S}^1$. A rigid body motion in the 3-dimensional Euclidean space thus has six degrees of freedom and is described by $\mathbb{R}^1 \times \mathbb{R}^1 \times \mathbb{R}^1 \times \mathbb{S}^2 \times \mathbb{S}^1$. More details can be found in Duindam (2006). □

One solution to solve the singularity problem is to use a non-minimal representation such as the unit quaternion to represent the orientation. However, because this representation is not a minimal set of generalized coordinates, the unit quaternion cannot be used in Lagrange's equations, as will be clear in Chap. 6. This means that when deriving the kinematics, the quaternion representation is a good choice, but when we move on to deriving the dynamic equations in Chaps. 6–8 from a Lagrangian starting point, we tend to use other representations.

As noted, the conventional Lagrangian framework requires the state variables to be written in the form of a set of generalized coordinates and a corresponding set of generalized velocities given by the time derivative of the generalized coordinates. One major motivation for using the formulation that we do in this book is to derive the dynamics of systems for which the conventional Lagrangian approach cannot be applied. As we have seen, the configuration space of a single rigid body cannot be written in terms of generalized coordinates and velocities of this kind, and as a result vehicle-manipulator dynamics cannot be derived from the Lagrangian in the conventional way. In the following we will thus derive a set of state variables for which it is possible to derive the dynamics of both single rigid bodies and multibody systems by rewriting the Lagrangian. More specifically, we will find a set of state variables which allows us to find a linear mapping from the velocity to the time derivative of the position variables. This will allow us to formulate the dynamics in terms of generalized coordinates and quasi-velocities instead of generalized coordinates and generalized velocities.

Because we will adopt a Lagrangian approach—although a somewhat modified version of the classical formulation in generalized coordinates and generalized velocities—we will stick with the Euler angle formulation using rotation matrices and not address the quaternion representation to any extent in the subsequent chapters.

3.3.3 Configuration States as Matrix Lie Groups

In this section we derive the kinematics of a rigid body using the framework introduced in Sect. 2.6, i.e., we write the configuration of a rigid body in the 3-dimensional Euclidean space as a homogeneous transformation matrix and the velocity as a twist. Because all configuration spaces that we are interested in are subgroups of $SE(3)$ or $SE(3)$ itself, is it always possible to describe the location of a rigid body in space by the homogeneous transformation matrix $g \in SE(3)$. For configuration spaces that have dimensions lower than 6, we then need to impose certain constraints on the homogeneous transformation matrices. In this case we will write the configuration in terms of the *configuration states* Q, which may be a scalar, a vector in $\mathbb{R}^m$, or a matrix Lie group (Tu 2008; Rossmann 2002).

Because all configuration spaces of interest are subgroups of $SE(3)$, the configuration can always be written in the form of a homogeneous transformation matrix with the appropriate restrictions, in which case we will write $g = g(Q)$. Writing the configurations in the form of a homogeneous transformation matrix allows us to treat all transformations in the same way, and will also simplify the notation when we look at multibody kinematics in the next chapter. The configuration state Q guarantees that the configuration state has the right dimension and constraints.

Similarly, the velocity of a rigid body is written in its most general form as a twist V. For a subspace of $SE(3)$ the configuration space only allows for velocities in certain directions and we write the velocity as $v = HV$ where H is the *selection matrix* which selects the elements of V which are not identically equal to zero.

Because all rigid body transformations can be described using $SE(3)$—possibly with some geometric restriction to one of its subspaces—we will start by studying this group in detail. We then go on to study the global parameterization of the configuration states and describe the subspaces of $SE(3)$ in Sect. 3.3.5.

Because all rigid body motions can be represented by a curve on a manifold, we will make frequent use of the most important property of manifolds: namely that manifolds are locally Euclidean. Locally, both the position variables and the velocity variables can therefore be represented as vector spaces and we can find a well defined mapping between the two. This local structure allows us to perform differentiation on a vector space while the group structure tells us what the manifold looks like globally. If we put all of this together we get a well defined and global representation of the kinematics of a rigid body. By representing the position and orientation of the rigid body as a matrix Lie group and the velocity as a twist and define local charts that allow us to find a linear and well-defined relation between the time derivative of the local position variables and the twists, we obtain a geometrically meaningful representation of the state space without the presence of singularities. This is studied in detail in Sect. 3.3.6. Geometric methods in robotics and control have been studied in great detail by several authors such as Bullo and Lewis (2000), Selig (2000), Marsden and Ratiu (1999); and Gallier (2001).

3.3.4 The Special Euclidean Group

Any rigid body, no matter what space it lives on, can be described by $SE(3)$. The special Euclidean group therefore plays a very special role in rigid body kinematics. This is because, as we have seen, any rigid body lives on $SE(3)$ or one of its subspaces. A subgroup of the special Euclidean group is a special case of $SE(3)$ for which some dimensions (or directions) are removed. In this section we will therefore study the configuration space $SE(3)$ in some more detail, and we will look at the subgroups in the next sections.

The configuration of a rigid body in the 3-dimensional Euclidean space can be determined by three variables representing the position and three variables representing the orientation. As we saw in Sect. 2.6.2, we can also write the configuration of a rigid body in matrix form, for example as

$$g_{0b} = \begin{bmatrix} R_{0b} & p_{0b} \\ 0 & 1 \end{bmatrix} \tag{3.70}$$

where $R_{0b} \in SO(3)$ and $p_{0b} \in \mathbb{R}^3$. The main advantage of writing the configuration of a rigid body in this way is that any element $g \in SE(3)$ written in this form represents a configuration that corresponds to an actual physical location of the rigid body in space. Furthermore, a smooth variation in the homogeneous transformation matrix corresponds to a smooth change in the configuration of the rigid body. This representation of the configuration of a rigid body describes the same configuration as in Sect. 3.3, but we will now adapt a more geometric interpretation of this matrix as an element of a manifold. This manifold is of course $SE(3)$ and we avoid the problems that we encountered in the previous section when the state space was represented in $\mathbb{R}^6$.

To describe the location of a rigid body in space there are certain properties that our representation needs. First of all we need a reference. This is often chosen as the inertial reference frame, which in our case is the identity matrix $g_I = I$. We can easily verify that the identity matrix satisfies the properties of the identity element because $gg_I = g_I g = g$ for all g.

Furthermore we need to be able to represent an opposite transformation. For a single rigid body this is given by the matrix inverse:

Lemma 3.2 *The inverse of an element of $SE(3)$ is given by the matrix inverse, i.e.,*

$$g^{-1} = \begin{bmatrix} R & p \\ 0 & 1 \end{bmatrix}^{-1} = \begin{bmatrix} R^\top & -R^\top p \\ 0 & 1 \end{bmatrix} \tag{3.71}$$

where we have used that $R^\top = R^{-1}$ for orthogonal matrices. We see that as $R^\top \in SO(3)$ and $-R^\top p \in \mathbb{R}^3$ we have $g^{-1} \in SE(3)$.

The fact that g^{-1} is an element of $SE(3)$ is a necessary property, because this guarantees that also opposite transformations can be interpreted as rigid body motions.

The final property that needs to be satisfied for $SE(3)$ to represent the space of rigid body transformations is that two consecutive transformations can be interpreted as a rigid body transformation:

Lemma 3.3 *Matrix multiplication of two elements of $SE(3)$ gives*

$$g_2 g_1 = \begin{bmatrix} R_2 & p_2 \\ 0 & 1 \end{bmatrix} \begin{bmatrix} R_1 & p_1 \\ 0 & 1 \end{bmatrix} = \begin{bmatrix} R_2 R_1 & R_2 p_1 + p_2 \\ 0 & 1 \end{bmatrix} = g \tag{3.72}$$

where $R_2 R_1 \in SO(3)$ and $R_2 p_1 + p_2 \in \mathbb{R}^3$ and thus $g = g_2 g_1 \in SE(3)$. g can thus be interpreted as a rigid body motion.

We can now conclude the following:

Theorem 3.1 *An element $g(t) \in SE(3)$ can be interpreted as a rigid body motion in the 3-dimensional Euclidean space. The identity element is given by the identity matrix, the opposite transformation is given by the matrix inverse, and the transformation that corresponds to two or more consecutive transformations is obtained by matrix multiplication.*

Proof This theorem follow directly from Lemmas 3.2 and 3.3. □

One of the main advantages of using homogeneous transformation matrices is that we are guaranteed that each element corresponds in a smooth and one-to-one manner to a physical configuration of a rigid body in space. We obtain this nice property because we choose a redundant representation. What guarantees that this representation describes a 6-DoF motion, is the set of constraints imposed on the matrix. As 16 real numbers are used to describe a 6 degrees of freedom motion, constraints are imposed to remove the remaining 10 degrees of freedom, i.e., R must be orthogonal and with determinant 1 and the bottom row must be equal to [0 0 0 1]. For the matrix operations that we will use, these constrains are always true. This is easily verified for matrix multiplication and matrix inverses.

There are two important interpretations of the homogeneous transformation matrix. Firstly, if the relative transformation between two coordinate frames is known, any quantity computed in one reference frame can be transformed to the other frame by simple matrix multiplication. This allows us to observe the same computed quantity from two different coordinate frames. Secondly, we can describe the way a reference frame moves in space as a time-varying homogeneous transformation matrix. This time-dependent homogeneous transformation matrix therefore gives us the time-evolution of a rigid body in space, as observed from one single reference frame.

Now that we have defined a set of matrices that correspond to certain configurations in space, we want to describe the velocities of the rigid bodies in the same way. However, as $g(t)$ is constrained to be a homogeneous matrix, we need to impose certain constraints also on $\dot{g}(t)$. The constraints on $\dot{g}(t)$ will depend on the current value of $g(t)$, as we saw for the rotation matrix. There are two ways to describe the velocity variables in this way, by spatial or body velocities.

We can find the spatial velocity variables $\widehat{V}_{0b}^S$ by

$$\widehat{V}_{0b}^S = \dot{g}_{0b} g_{0b}^{-1} = \begin{bmatrix} \widehat{\omega}_{0b}^S & v_{0b}^S \\ 0 & 0 \end{bmatrix} \tag{3.73}$$

which, as we have seen, maps a point represented in the inertial frame to the velocity of the same point, also in the inertial frame. Using the terminology of Lie groups, $\widehat{V}_{0b}^S$ is the right translate of the tangent element $\dot{g}_{0b} \in T_g SE(3)$ at g_{0b} to the identity, in other words the right translate of $\dot{g}$ is an element of $se(3)$ (recall that the Lie algebra is the tangent space at the identity). If we write $\widehat{V}_{0b}^S$ in vector form we get

$$V_{0b}^S = \begin{bmatrix} \dot{p}_{0b} - \dot{R}_{0b} R_{0b}^{\mathsf{T}} p_{0b} \\ (\dot{R}_{0b} R_{0b}^{\mathsf{T}})^{\vee} \end{bmatrix}. \tag{3.74}$$

We can also write the velocity in the body frame as

$$\widehat{V}_{0b}^B = g_{0b}^{-1} \dot{g}_{0b} = \begin{bmatrix} \widehat{\omega}_{0b}^B & v_{0b}^b \\ 0 & 0 \end{bmatrix}. \tag{3.75}$$

Also $g_{ab}^{-1}(t)\dot{g}_{ab}(t) \in se(3)$ has a geometrically meaningful interpretation and represents the velocity of the rigid body as observed from the body-fixed frame. In this case $\widehat{V}_{0b}^B$ is the left translate of the tangent element $\dot{g}_{0b} \in T_g SE(3)$ at g_{0b} which maps $\widehat{V}^B_{0b}$ back to the identity $T_I SE(3)$. Also $\widehat{V}^B_{0b}$ is thus an element of the Lie algebra $se(3)$. If we write the body velocity variables in vector form we get

$$V^B_{0b} = \begin{bmatrix} R^{\mathsf{T}}_{0b}\dot{p}_{0b} \\ (R^{\mathsf{T}}_{0b}\dot{R}_{0b})^{\vee} \end{bmatrix}. \tag{3.76}$$

The relation in Eq. (3.73) gives the kinematics of the system in terms of the spatial velocity variables. This describes the same relation as we found in (3.69), but we note that when we introduce the matrix representation to describe the location of the rigid body instead of the vector representation, we avoid the singularities that appeared in (3.69). In the same way, the relation in Eq. (3.75) presents us with a singularity free formulation of the kinematics that we found in (3.64), in terms of the body velocities.

By representing the velocities in this way using twists, the problem of imposing constraints on $\dot{g}(t)$ is avoided. Casting the constraints on g into constraints on $\dot{g}$ is not straight forward, but we avoid this by writing the velocity state as $\hat{V}^S_{0b}$ or $\hat{V}^B_{0b}$ which, as we have seen, is a very efficient way to cast the constraints on the position variables into constraints on the velocity state. The time derivative of the homogeneous transformation matrix can be recovered from the twist. This is given by

$$\dot{g}_{0b} = g_{0b}\widehat{V}^B_{0b}g_{bb} = g_{0b}\widehat{V}^B_{0b}, \tag{3.77}$$

for the body velocity twist, and

$$\dot{g}_{0b} = g_{00}\widehat{V}^S_{0b}g_{0b} = \widehat{V}^S_{0b}g_{0b}. \tag{3.78}$$

for the spatial velocity twist.

We see that twists give us a very effective way to represent a constrained space. In this case the space that we want to describe is the manifold of rigid body motion embedded in the 3-dimensional Euclidean space. Using the formulation of twists we constrain the motion of the rigid body to a configuration space that corresponds to the possible positions and admissible velocities that the rigid body can take in the real world. The formulation presented here is not only a mathematically convenient way to represent a constrained higher-dimensional space, twists also have very useful geometric interpretations as the velocities of rigid bodies:

The constraints on the time derivative of the homogeneous transformation matrix can be avoided by representing the velocity variables as twist. The velocities are then written in either body or spatial coordinates:

- The body velocity twist V^B_{0b} represents the linear and angular velocities of a rigid body with a frame $\mathscr{F}_b$ attached to it with respect to the inertial frame $\mathscr{F}_0$ as seen for an observer in the body frame $\mathscr{F}_b$. Thus, at every time instant the velocity is described in the body frame $\mathscr{F}_b$.

- The spatial velocity twist V_{0b}^S represents the linear and angular velocities of a rigid body with a frame $\mathscr{F}_b$ attached to it, with respect to the inertial frame $\mathscr{F}_0$ in spatial coordinates. I.e., ω_{0b}^S describes the angular velocities of $\mathscr{F}_b$ with respect to $\mathscr{F}_0$ as observed from $\mathscr{F}_0$ whereas v_{0b}^S describes the linear velocities of a (possibly) imaginary point located at (possibly an imaginary extension of) the rigid body when this point travels through the origin of $\mathscr{F}_0$. Thus, at every time instant we measure the velocity of the point which, at that time instant, travels through the origin of $\mathscr{F}_0$. The spatial velocities are illustrated in Fig. 3.1.

We will now look at a different type of constrained space, i.e., rigid body motions that are constrained to a subspace of $SE(3)$. Also these spaces are easily constrained by choosing the correct formulation of the twist variables.

3.3.5 Configuration Spaces as Subgroups of $SE(3)$

The most important configuration spaces in mechanics, and also the ones that we will encounter throughout this book, are subgroups of the special Euclidean group, or the special Euclidean group itself (Hervé 1978; Meng et al. 2007). As we have seen, the configuration of a rigid body in space can be written as a homogeneous transformation matrix with 6 degrees of freedom. For many transformations, however, the configuration space is of a lower dimension. This is the case for standard 1-degree of freedom robotic joints, pure rotational motion, and planar motion, as a few examples. We will now look at how we can represent the configuration space of these transformations in a more compact form, and how this compact form relates to the homogeneous transformation matrix described above.

The kinematics of the system can be naturally described in terms of the state variables g_{0b} for position/orientation and V_{0b}^B or V_{0b}^S for velocity. To allow for more general systems, and also multibody systems such as robotic manipulators, we will write the configuration of a rigid body as Q, where it is implicitly understood that $Q = q \in \mathbb{R}^m$ if the configuration space is Euclidean and where Q is a matrix Lie group if the configuration space is non-Euclidean. The velocity variable is written as a vector $v = \dot{q} \in \mathbb{R}^m$ for a Euclidean configuration space and as $v = \tilde{V}_{0b}^B \in \mathbb{R}^m$ if it is non-Euclidean. Using this formalism we obtain a global parameterization of a rigid idealized joint (Duindam 2006):

Definition 3.3 A globally parametrized rigid body transformation is a kinematic restriction on the allowed relative twist of two rigid bodies i and j to a linear subspace of dimension m, where the relative motion of the bodies is described by two sets of states, namely

- a matrix Q, parameterizing the relative configuration as $g_{ij} = g_{ij}(Q)$,
- a vector $v \in \mathbb{R}^m$, parameterizing the relative twists as $V_{ij}^k = H(Q)v$.

We can thus parameterize a transformation, for example of a rigid body with reference frame $\mathscr{F}_b$, as an m degree of freedom motion with a matrix Q (with m degrees of freedom) representing the location of the rigid body with respect to an inertial frame $\mathscr{F}_0$, and a vector v with m degrees of freedom, representing the velocity. For convenience, both for representation and for implementation, we will normally write the configuration as a homogeneous transformation matrix and the velocity as a twist. Because all configuration spaces that are of interest to us are subgroups of $SE(3)$, we can always write the state space as an element of the tangent bundle in this way.

We see that it is important that we adapt a formalism that satisfies the restrictions of the configuration space also when the configuration space has less than six degrees of freedom. For rigid bodies with a configuration space other than $SE(3)$ and with dimension $m < 6$ we only need m parameters to define the velocity state. Hence, in the case of $m < 6$ we can define a selection matrix in the following way:

Definition 3.4 A *selection matrix* $H \in \mathbb{R}^{6\times m}$ represents a mapping from a velocity state in $\mathbb{R}^m$ to the twist in $\mathbb{R}^6$ such that the velocity twist is given by

$$V_{0b}^B = Hv. \tag{3.79}$$

$v \in \mathbb{R}^m$ fully determines the velocity state of the rigid body.

For Lie groups we can also write $v = H^\top V_{0b}^B$ where H selects the elements of V_{0b}^B that are not trivially zero. In the special case when we want to represent the configuration space of a single rigid body, or a vehicle, we will write

$$V_{0b}^B = H\tilde{V}_{0b}^B \tag{3.80}$$

where the tilde in $\tilde{V}_{0b}^B \in \mathbb{R}^m$ indicates that we have selected the m entries of V_{0b}^B needed to parameterize the m degree of freedom motion. v is thus a general velocity variable while $\tilde{V}$ represents the velocity of a single rigid body. In this chapter, v and $\tilde{V}$ can thus be used interchangeably, whereas in the next chapters we will use v to describe the velocity of a VM system and $\tilde{V}$ to specify that we mean only the vehicle, or a single rigid body. The spatial velocity can now be expressed in terms of the body velocity as:

$$V_{0b}^S = \mathrm{Ad}_{g_{0b}} V_{0b}^B = \mathrm{Ad}_{g_{0b}} H\tilde{V}_{0b}^B. \tag{3.81}$$

3.3.5.1 Translational motion

A transformation by a linear motion is Euclidean and the configuration space can be written as

$$Q = p_{0b} = \begin{bmatrix} x_{0b} \\ y_{0b} \\ z_{0b} \end{bmatrix}, \qquad v = \tilde{V}_{0b}^{B} = \dot{p}_{0b} = \begin{bmatrix} \dot{x}_{0b} \\ \dot{y}_{0b} \\ \dot{z}_{0b} \end{bmatrix} = \begin{bmatrix} u \\ v \\ w \end{bmatrix}. \tag{3.82}$$

From Definition 3.3 we can write the transformation as a homogeneous transformation matrix and the velocity as a twist as

$$g_{0b}(Q) = \begin{bmatrix} 1 & 0 & 0 & x_{0b} \\ 0 & 1 & 0 & y_{0b} \\ 0 & 0 & 1 & x_{0b} \\ 0 & 0 & 0 & 1 \end{bmatrix}, \qquad V_{0b}^{S} = V_{0b}^{B} = \begin{bmatrix} \dot{x}_{0b} \\ \dot{y}_{0b} \\ \dot{z}_{0b} \\ 0 \\ 0 \\ 0 \end{bmatrix} = \begin{bmatrix} u \\ v \\ w \\ 0 \\ 0 \\ 0 \end{bmatrix}. \tag{3.83}$$

Writing the position and velocity variables in this form effectively restricts the motion to pure translations in $\mathbb{R}^3$. We see that the velocity is written as

$$V_{0b}^{B} = H\tilde{V}_{0b}^{B} \tag{3.84}$$

with the selection matrix

$$H = \begin{bmatrix} I_{3\times 3} \\ 0_{3\times 3} \end{bmatrix}. \tag{3.85}$$

An important transformation that we will make frequent use of when we move on to modeling robotic manipulators is 1-DoF transformations. If, for example, the kinematic constraints of a robotic joint restricts its motion to translation along the y-axis, this joint can be globally parameterized by

$$Q = y_{0b}, \qquad v = \dot{y}_{0b}, \tag{3.86}$$

which gives the following homogeneous transformation matrix and velocity variable:

$$g_{0b}(Q) = \begin{bmatrix} 1 & 0 & 0 & 0 \\ 0 & 1 & 0 & y_{0b} \\ 0 & 0 & 1 & 0 \\ 0 & 0 & 0 & 1 \end{bmatrix}, \qquad V_{0b}^{S} = V_{0b}^{B} = \begin{bmatrix} 0 \\ \dot{y}_{0b} \\ 0 \\ 0 \\ 0 \\ 0 \end{bmatrix} = \begin{bmatrix} 0 \\ v \\ 0 \\ 0 \\ 0 \\ 0 \end{bmatrix}. \tag{3.87}$$

Both these have one degree of freedom as required.

3.3.5.2 Rotational motion

Rotational motion is parameterized using the rotation matrix and the angular velocities as

$$Q = R_{0b}, \qquad v = \omega_{0b}^{B}. \tag{3.88}$$

For rotational motion the state space can then be written as

$$g_{0b}(Q) = \begin{bmatrix} R_{0b} & 0 \\ 0 & 1 \end{bmatrix}, \qquad V_{0b}^{B} = \begin{bmatrix} 0 \\ 0 \\ 0 \\ p \\ q \\ r \end{bmatrix}. \tag{3.89}$$

The velocity twist is written in terms of ω_{0b}^{B} as

$$V_{0b}^{B} = H\omega_{0b}^{B} \tag{3.90}$$

with the selection matrix

$$H = \begin{bmatrix} 0_{3\times 3} \\ I_{3\times 3} \end{bmatrix}. \tag{3.91}$$

The most commonly found transformation in robotics is the one of 1-DoF rotational motion. We will thus make much use of the homogeneous transformation matrix that constrains the motion to simple rotations of this kind. For a rotation around the z-axis, for example, we have $Q = \psi \in \mathbb{R}$ and $v = \dot{\psi} \in \mathbb{R}$ and the homogeneous transformation matrix and velocity twists are parameterized by

$$g_{0b}(Q) = \begin{bmatrix} \cos\psi & -\sin\psi & 0 & 0 \\ \sin\psi & \cos\psi & 0 & 0 \\ 0 & 0 & 1 & 0 \\ 0 & 0 & 0 & 1 \end{bmatrix}, \qquad V_{0b}^{B} = \begin{bmatrix} 0 \\ 0 \\ 0 \\ 0 \\ 0 \\ \dot{\psi} \end{bmatrix}.$$

For a revolute robotic joint $q = \psi$ is often referred to as the joint variable.

3.3.5.3 Planar Motion

For planar motion, i.e., free translation in the plane and rotation around the axis orthogonal to the plane, we write

$$Q = \begin{bmatrix} \cos\psi & -\sin\psi & x_{0b} \\ \sin\psi & \cos\psi & y_{0b} \\ 0 & 0 & 1 \end{bmatrix}, \qquad v = \begin{bmatrix} \dot{x}_{0b} \\ \dot{y}_{0b} \\ \dot{\psi} \end{bmatrix}. \tag{3.92}$$

The position variables for planar motion in the xy-plane are thus given by $\left[\begin{smallmatrix} q_1 \\ q_2 \end{smallmatrix}\right] = \left[\begin{smallmatrix} x_{0b} \\ y_{0b} \end{smallmatrix}\right]$ and the orientation is given by $q_3 = \psi$. It is important to note, however, that even though the state space is thus fully parameterized by $(q_1, q_2, q_3) = (x_{0b}, y_{0b}, \psi)$ the topology of the configuration space is not taken into account using this parameterization because the cyclic behavior and multi-covering of ψ is not considered. In (3.92), however, the cyclic motion is taken into account by the trigonometric functions.

Also planar motion can be written as a homogeneous matrix with the imposed constraints:

$$g_{0b}(Q) = \begin{bmatrix} \cos\psi & -\sin\psi & 0 & x_{0b} \\ \sin\psi & \cos\psi & 0 & y_{0b} \\ 0 & 0 & 1 & 0 \\ 0 & 0 & 0 & 1 \end{bmatrix},$$

$$V_{0b}^S = \begin{bmatrix} u^S \\ v^S \\ 0 \\ 0 \\ 0 \\ r^S \end{bmatrix} = \begin{bmatrix} \dot{x}_{0b} + y_{0b}\dot{\psi} \\ \dot{y}_{0b} - x_{0b}\dot{\psi} \\ 0 \\ 0 \\ 0 \\ \dot{\psi} \end{bmatrix}, \qquad V_{0b}^B = \begin{bmatrix} u \\ v \\ 0 \\ 0 \\ 0 \\ r \end{bmatrix}. \tag{3.93}$$

The velocities are now found by transforming the variables to the body frame $\mathscr{F}_b$. For $SE(2)$, for example, the body velocities are given by

$$v = \tilde{V}_{0b}^B = \begin{bmatrix} u \\ v \\ r \end{bmatrix} = \begin{bmatrix} R_{0b}^{-1} \left[\begin{smallmatrix} u^0 \\ v^0 \end{smallmatrix}\right] \\ r^S \end{bmatrix} = \begin{bmatrix} R_{0b}^{-1} \left[\begin{smallmatrix} \dot{x}_{0b} \\ \dot{y}_{0b} \end{smallmatrix}\right] \\ \dot{\psi} \end{bmatrix} \tag{3.94}$$

where R_{0b} is the 2×2 rotation matrix in (3.92), i.e.,

$$v = \tilde{V}_{0b}^B = \begin{bmatrix} u \\ v \\ r \end{bmatrix} = \begin{bmatrix} \dot{x}_{0b}\cos\psi + \dot{y}_{0b}\sin\psi \\ -\dot{x}_{0b}\sin\psi + \dot{y}_{0b}\cos\psi \\ \dot{\psi} \end{bmatrix}. \tag{3.95}$$

For rigid body motion in the plane, the selection matrix H selects two linear velocity variables, normally in the direction of the x- and y-axes, and one variable representing the rotation around the z-axis. We write the body velocities as $\tilde{V}_{0b}^B = [u\ v\ r]^\mathsf{T}$ so that

$$V_{0b}^B = H\tilde{V}_{0b}^B, \qquad H = \begin{bmatrix} 1 & 0 & 0 \\ 0 & 1 & 0 \\ 0 & 0 & 0 \\ 0 & 0 & 0 \\ 0 & 0 & 0 \\ 0 & 0 & 1 \end{bmatrix}. \tag{3.96}$$

3.3.5.4 Rigid Bodies with Configuration Space $\mathscr{X}(z)$

The Schönflies group represents linear motion in three degrees of freedom and rotation around one of the coordinate axes, normally the z-axis. The configuration state is written as a homogeneous transformation matrix with rotational motion only around the z-axis and full translational freedom, i.e., as

$$g_{0b}(Q) = \begin{bmatrix} \cos\psi & -\sin\psi & 0 & x_{0b} \\ \sin\psi & \cos\psi & 0 & y_{0b} \\ 0 & 0 & 1 & z_{0b} \\ 0 & 0 & 0 & 1 \end{bmatrix}. \tag{3.97}$$

The body and spatial velocities are found in the normal way as

$$\begin{aligned} \hat{V}_{0b}^S = \dot{g}_{0b} g_{0b}^{-1} &= \begin{bmatrix} -\sin\psi\dot{\psi} & -\cos\psi\dot{\psi} & 0 & \dot{x}_{0b} \\ \cos\psi\dot{\psi} & -\sin\psi\dot{\psi} & 0 & \dot{y}_{0b} \\ 0 & 0 & 0 & \dot{z}_{0b} \\ 0 & 0 & 0 & 0 \end{bmatrix} \\ &\times \begin{bmatrix} \cos\psi & \sin\psi & 0 & -x_{0b}\cos\psi - y_{0b}\sin\psi \\ -\sin\psi & \cos\psi & 0 & x_{0b}\sin\psi - y_{0b}\cos\psi \\ 0 & 0 & 1 & z_{0b} \\ 0 & 0 & 0 & 1 \end{bmatrix} \\ &= \begin{bmatrix} 0 & -\dot{\psi} & 0 & \dot{x}_{0b} + y_{0b}\dot{\psi} \\ \dot{\psi} & 0 & 0 & \dot{y}_{0b} - x_{0b}\dot{\psi} \\ 0 & 0 & 0 & \dot{z}_{0b} \\ 0 & 0 & 0 & 0 \end{bmatrix} \end{aligned} \tag{3.98}$$

$$\begin{aligned} \hat{V}_{0b}^B = g_{0b}^{-1} \dot{g}_{0b} &= \begin{bmatrix} \cos\psi & \sin\psi & 0 & -x_{0b}\cos\psi - y_{0b}\sin\psi \\ -\sin\psi & \cos\psi & 0 & x_{0b}\sin\psi - y_{0b}\cos\psi \\ 0 & 0 & 1 & z_{0b} \\ 0 & 0 & 0 & 1 \end{bmatrix} \\ &\times \begin{bmatrix} -\sin\psi\dot{\psi} & -\cos\psi\dot{\psi} & 0 & \dot{x}_{0b} \\ \cos\psi\dot{\psi} & -\sin\psi\dot{\psi} & 0 & \dot{y}_{0b} \\ 0 & 0 & 0 & \dot{z}_{0b} \\ 0 & 0 & 0 & 0 \end{bmatrix} \\ &= \begin{bmatrix} 0 & -\dot{\psi} & 0 & \dot{x}_{0b}\cos\psi + \dot{y}_{0b}\sin\psi \\ \dot{\psi} & 0 & 0 & -\dot{x}_{0b}\sin\psi + \dot{y}_{0b}\cos\psi \\ 0 & 0 & 0 & \dot{z}_{0b} \\ 0 & 0 & 0 & 0 \end{bmatrix} \end{aligned} \tag{3.99}$$

The velocity state is thus fully described by $\tilde{V}_{0b}^B = [u\ v\ w\ r]^\mathsf{T}$ so the selection matrix is given by

Table 3.1 The position and velocity variables used to describe the configuration states for different configuration manifolds

Configuration space	Position variable	Velocity variable
$SE(3)$	$Q = g_{0b} = \begin{bmatrix} R_{0b} & p_{0b} \\ 0 & 1 \end{bmatrix} \in SE(3)$	$\hat{V}_{0b}^B = \begin{bmatrix} \hat{\omega}_{0b}^B & v_{0b}^B \\ 0 & 0 \end{bmatrix} \in se(3)$
	$R_{0b} \in SO(3), \quad p_{0b} \in \mathbb{R}^3$	$v = V_{0b}^B = \begin{bmatrix} v_{0b}^B \\ \omega_{0b}^B \end{bmatrix}$
		$v_{0b}^B \in \mathbb{R}^3, \quad \omega_{0b}^B \in \mathbb{R}^3$
$\mathscr{X}(z)$	$Q = g_{0b} = \begin{bmatrix} R_{0b} & p_{0b} \\ 0 & 1 \end{bmatrix} \in \mathscr{X}(z)$	$\hat{V}_{0b}^B = \begin{bmatrix} \hat{\omega}_{0b}^B & v_{0b}^B \\ 0 & 0 \end{bmatrix} \in \varkappa(z)$
	$R_{0b} \in SO(2), \quad p_{0b} \in \mathbb{R}^3$	$v = \tilde{V}_{0b}^B = \begin{bmatrix} v_{0b}^B \\ \omega_{0b}^B \end{bmatrix}$
		$v_{0b}^B \in \mathbb{R}^3, \quad \omega_{0b}^B \in \mathbb{R}$
$\mathbb{R}^3$	$Q = p_{0b} \in \mathbb{R}^3$	$v = \tilde{V}_{0b}^B = v_{0b}^B \in \mathbb{R}^3$
$SE(2)$	$Q = g_{0b} = \begin{bmatrix} R_{0b} & p_{0b} \\ 0 & 1 \end{bmatrix} \in SE(2)$	$\hat{V}_{0b}^B = \begin{bmatrix} \hat{\omega}_{0b}^B & v_{0b}^B \\ 0 & 0 \end{bmatrix} \in se(2)$
	$R_{0b} \in SO(2), \quad p_{0b} \in \mathbb{R}^2$	$v = V_{0b}^B = \begin{bmatrix} v_{0b}^B \\ \omega_{0b}^B \end{bmatrix}$
		$v_{0b}^B \in \mathbb{R}^2, \quad \omega_{0b}^B \in \mathbb{R}$
$SO(3)$	$Q = R_{0b} \in SO(3)$	$\hat{V}_{0b}^B = \hat{\omega}_{0b}^B \in so(3)$ $v = \tilde{V}_{0b}^B = \omega_{0b}^B \in \mathbb{R}^3$
$\mathbb{R}^2$	$Q = p_{0b} \in \mathbb{R}^2$	$v = \tilde{V}_{0b}^B = v_{0b}^B \in \mathbb{R}^2$
$C(1)$	$Q = g_{0b} \in C(1)$	$v = \tilde{V}_{0b}^B = \begin{bmatrix} v_{0b}^B \\ \omega_{0b}^B \end{bmatrix}$
	$R_{0b} \in SO(2), \quad p_{0b} \in \mathbb{R}$	$v_{0b}^B \in \mathbb{R}, \quad \omega_{0b}^B \in \mathbb{R}$
$\mathbb{R}$	$Q = p_{0b} \in \mathbb{R}$	$v = \tilde{V}_{0b}^B = v_{0b}^B \in \mathbb{R}$
$SO(2)$	$Q = R_{0b} \in SO(2)$	$v = \tilde{V}_{0b}^B = \omega_{0b}^B \in \mathbb{R}$
$H(1)$	$Q = q_{0b} \in H(1)$	$v = \tilde{V}_{0b}^B \in h(1)$

$$V_{0b}^B = H\tilde{V}_{0b}^B, \qquad H = \begin{bmatrix} 1 & 0 & 0 & 0 \\ 0 & 1 & 0 & 0 \\ 0 & 0 & 1 & 0 \\ 0 & 0 & 0 & 0 \\ 0 & 0 & 0 & 0 \\ 0 & 0 & 0 & 1 \end{bmatrix} \tag{3.100}$$

and the spatial and body velocities are written as

$$V_{0b}^S = \begin{bmatrix} \dot{x}_{0b} + y_{0b}\dot{\psi} \\ \dot{y}_{0b} - x_{0b}\dot{\psi} \\ \dot{z}_{0b} \\ \dot{\psi} \end{bmatrix}, \qquad V_{0b}^B = \begin{bmatrix} \dot{x}_{0b}\cos\psi + \dot{y}_{0b}\sin\psi \\ -\dot{x}_{0b}\sin\psi + \dot{y}_{0b}\cos\psi \\ \dot{z}_{0b} \\ \dot{\psi} \end{bmatrix}. \tag{3.101}$$

Table 3.2 The selection matrices H for different Lie subgroups of $SE(3)$. Note that the selection matrices must be chosen so that the resulting twist represents velocities in the right direction, not only of the right dimension. It is, however, always possible to choose the coordinate frames so that the selection matrix is in the form above. This is referred to as the nominal form of the group

Lie Group	H	Lie Group	H
$SE(3)$	$\begin{bmatrix} 1&0&0&0&0&0 \\ 0&1&0&0&0&0 \\ 0&0&1&0&0&0 \\ 0&0&0&1&0&0 \\ 0&0&0&0&1&0 \\ 0&0&0&0&0&1 \end{bmatrix}$	$\mathscr{X}(z)$	$\begin{bmatrix} 1&0&0&0 \\ 0&1&0&0 \\ 0&0&1&0 \\ 0&0&0&0 \\ 0&0&0&0 \\ 0&0&0&1 \end{bmatrix}$
$\mathbb{R}^3$	$\begin{bmatrix} 1&0&0 \\ 0&1&0 \\ 0&0&1 \\ 0&0&0 \\ 0&0&0 \\ 0&0&0 \end{bmatrix}$	$SE(2)$	$\begin{bmatrix} 1&0&0 \\ 0&1&0 \\ 0&0&0 \\ 0&0&0 \\ 0&0&0 \\ 0&0&1 \end{bmatrix}$
$SO(3)$	$\begin{bmatrix} 0&0&0 \\ 0&0&0 \\ 0&0&0 \\ 1&0&0 \\ 0&1&0 \\ 0&0&1 \end{bmatrix}$	$\mathbb{R}^2, C(1)$	$\begin{bmatrix} 1&0 \\ 0&1 \\ 0&0 \\ 0&0 \\ 0&0 \\ 0&0 \end{bmatrix}, \begin{bmatrix} 0&0 \\ 0&0 \\ 1&0 \\ 0&0 \\ 0&0 \\ 0&1 \end{bmatrix}$
$\mathbb{R}, SO(2)$	$\begin{bmatrix} 1 \\ 0 \\ 0 \\ 0 \\ 0 \\ 0 \end{bmatrix}, \begin{bmatrix} 0 \\ 0 \\ 0 \\ 0 \\ 0 \\ 1 \end{bmatrix}$	H	$\begin{bmatrix} 0 \\ 0 \\ 1 \\ 0 \\ 0 \\ \rho \end{bmatrix}$

We have now found the state variables of the most important configuration spaces and the corresponding selection matrices. The state variables in Table 3.1 together with the selection matrices and velocity transformation matrices in Table 3.2 define the kinematics of a single rigid body for the most important configuration spaces in robotics.

3.3.6 Local Coordinates

In this section we will study in detail how we can use the local structure of manifolds to find an expression for the spatial and body velocities in terms of the local position variables. Even though a manifold itself is in general not Euclidean, we know that the neighborhood of each point on the manifold is locally Euclidean. We will introduce the local position variables, i.e., a set of position variables that describe the vicinity of a point, by the variable $\varphi \in \mathbb{R}^m$. It is then possible to choose φ so that $\dot{\varphi}$ is related to the body velocity twist through the mapping

$$V_{0b}^B = S(Q, \varphi)\dot{\varphi} \tag{3.102}$$

and similarly for spatial velocity. This kinematic relation will become very useful when we are to derive the dynamics of rigid body systems in the subsequent chapters: instead of letting Q represent the configuration variable of the rigid body, we will regard φ as this variable. This will simplify the expressions because both φ and $\dot{\varphi}$ can be written in $\mathbb{R}^m$. We do, however, want our *final equations* to depend on

Table 3.3 The velocity transformation matrices S and the adjoint maps ad_X used to compute the velocity transformation matrices for different Lie subgroups of $SE(3)$. Note that the adjoint maps vanish for the Abelian groups. The velocity transformation matrices are shown for body velocity variables

Lie Group	S	ad_X
$SE(3)$	$I - \frac{1}{2}\mathrm{ad}_{\varphi_V} + \frac{1}{6}\mathrm{ad}^2_{\varphi_V} - \cdots$	$\begin{bmatrix} 0 & -r & q & 0 & -w & v \\ r & 0 & -p & w & 0 & -u \\ -q & p & 0 & -v & u & 0 \\ 0 & 0 & 0 & 0 & -r & q \\ 0 & 0 & 0 & -r & 0 & p \\ 0 & 0 & 0 & -q & p & 0 \end{bmatrix}$
$\mathscr{X}(z)$	$I - \frac{1}{2}\mathrm{ad}_{\varphi_V} + \frac{1}{6}\mathrm{ad}^2_{\varphi_V} - \cdots$	$\begin{bmatrix} 0 & -r & 0 & v \\ r & 0 & 0 & -u \\ 0 & 0 & 0 & 0 \\ 0 & 0 & 0 & 0 \end{bmatrix}$
$\mathbb{R}^3$	$I_{3\times 3}$	$\begin{bmatrix} 0 & 0 & 0 \\ 0 & 0 & 0 \\ 0 & 0 & 0 \end{bmatrix}$
$SE(2)$	$I - \frac{1}{2}\mathrm{ad}_{\varphi_V} + \frac{1}{6}\mathrm{ad}^2_{\varphi_V} - \cdots$	$\begin{bmatrix} 0 & -r & v \\ r & 0 & -u \\ 0 & 0 & 0 \end{bmatrix}$
$SO(3)$	$I - \frac{1}{2}\widehat{\varphi}_\omega + \frac{1}{6}\widehat{\varphi}^2_\omega - \cdots$	$\begin{bmatrix} 0 & -r & q \\ r & 0 & -p \\ -q & p & 0 \end{bmatrix}$
$\mathbb{R}^2, C(1)$	$I_{2\times 2}$	$\begin{bmatrix} 0 & 0 \\ 0 & 0 \end{bmatrix}$
$\mathbb{R}, H, SO(2)$	$I_{1\times 1}$	0

Q and not φ to get a globally valid formulation. We do this by differentiating with respect to φ and setting $\varphi = 0$. Because $\Phi(Q, \varphi) = Q$ for $\varphi = 0$ this gives us the dynamic equations in terms of the global position variable Q and not φ. We will study this formulation in more detail when we derive the dynamic equations in the subsequent chapters.

For now we will write the kinematics of a rigid body in terms of the local position variables φ and the spatial and body velocities (twists). This kinematic relation is given by $S(Q, \varphi)$ and gives us a well-defined mapping from the time derivative of the position variable φ to the velocity twist V^B_{0b}. Based on the results in Sect. 2.9 we can find this mapping for different configuration spaces of interest, represented by the different matrix Lie groups. We also refer to Sect. 2.9 for the derivations and the proofs of the expressions in the next sections.

In the subsequent sections we will find the velocity transformation matrices S for several different configuration spaces. This is summarized in Table 3.3.

3.3.6.1 Rigid Bodies with Configuration Space $SE(3)$

The configuration space of a free-floating rigid body, such as an AUV or an airplane, can be described by the matrix Lie group $SE(3)$. In this case we have the mapping (Rossmann 2002; Duindam 2006)

$$V^B_{0b} = \left(I - \frac{1}{2}\mathrm{ad}_{\varphi_V} + \frac{1}{6}\mathrm{ad}^2_{\varphi_V} - \cdots\right)\dot{\varphi}_V \tag{3.103}$$

with $\mathrm{ad}_p = \begin{bmatrix} \widehat{p}_{4\dots6} & \widehat{p}_{1\dots3} \\ 0 & \widehat{p}_{4\dots6} \end{bmatrix} \in \mathbb{R}^{6\times6}$ for $p \in \mathbb{R}^6$ relating the local and global velocity variables, and $\tilde{V}_{0b}^B = V_{0b}^B$ as $H = I$. The velocity transformation matrix from the time derivative of the local position variables to the globally valid velocity variables V_{0b}^B is given by

$$S(Q,\varphi) = \left(I - \frac{1}{2}\mathrm{ad}_{\varphi_V} + \frac{1}{6}\mathrm{ad}_{\varphi_V}^2 - \cdots \right) \in \mathbb{R}^{6\times6}. \tag{3.104}$$

3.3.6.2 Rigid Bodies with Configuration Space $SO(3)$

The velocity state of a rigid body in $SO(3)$ is fully determined by only three variables and we choose H so that $V_{0b}^B = H\omega_{0b}^B$. We then get

$$\tilde{V}_{0b}^B = \omega_{0b}^B = \left(I - \frac{1}{2}\widehat{\varphi}_\omega + \frac{1}{6}\widehat{\varphi}_\omega^2 - \cdots \right)\dot{\varphi}_\omega \tag{3.105}$$

where $\widehat{\varphi}$ is the skew-symmetric matrix representation of the vector $\varphi \in \mathbb{R}^3$ given by

$$\widehat{\varphi} = \begin{bmatrix} 0 & -\varphi_3 & \varphi_2 \\ \varphi_3 & 0 & -\varphi_1 \\ -\varphi_2 & \varphi_1 & 0 \end{bmatrix}. \tag{3.106}$$

The velocity transformation matrix for $SO(3)$ is given by

$$S(Q,\varphi) = \left(I - \frac{1}{2}\widehat{\varphi}_\omega + \frac{1}{6}\widehat{\varphi}_\omega^2 - \cdots \right) \in \mathbb{R}^{3\times3}. \tag{3.107}$$

3.3.6.3 Rigid Bodies with Configuration Space $SE(2)$

For planar motion the adjoint map ad_V is given by

$$\mathrm{ad}_V = \begin{bmatrix} 0 & -r & v \\ r & 0 & -u \\ 0 & 0 & 0 \end{bmatrix}. \tag{3.108}$$

The velocity transformation matrix $S(Q,\varphi) \in \mathbb{R}^{3\times3}$ is found in the same way as for $SE(3)$ and $SO(2)$ above with

$$\mathrm{ad}_\varphi = \begin{bmatrix} 0 & -\varphi_3 & \varphi_2 \\ \varphi_3 & 0 & -\varphi_1 \\ 0 & 0 & 0 \end{bmatrix}. \tag{3.109}$$

3.3.6.4 Rigid Bodies with Configuration Space $\mathscr{X}(z)$

For the Schönflies group the body velocities are written as $\tilde{V}_{0b}^B = [u\ v\ w\ r]^\mathsf{T}$ and the adjoint map ad_X is found in the normal way with From (2012a, 2012b)

$$\mathrm{ad}_X = \begin{bmatrix} 0 & -r & 0 & v \\ r & 0 & 0 & -u \\ 0 & 0 & 0 & 0 \\ 0 & 0 & 0 & 0 \end{bmatrix}. \tag{3.110}$$

The body velocity transformation matrix is found by substituting

$$\mathrm{ad}_\varphi = \begin{bmatrix} 0 & -\varphi_4 & 0 & \varphi_2 \\ \varphi_4 & 0 & 0 & -\varphi_1 \\ 0 & 0 & 0 & 0 \\ 0 & 0 & 0 & 0 \end{bmatrix} \tag{3.111}$$

into Eq. (3.104).

3.3.6.5 Rigid Bodies with Configuration Space $T(n)$, $C(1)$, $SO(2)$ and $H(1)$

The velocity state of a rigid body with a Euclidean configuration space is fully determined by m variables $Q = q \in \mathbb{R}^m$ where m is the dimension of the manifold. Also the velocity state can be written in this way as $v = \dot{q} \in \mathbb{R}^m$. Again we have

$$V_{0b}^B = H\tilde{V}_{0b}^B \tag{3.112}$$

where $H \in \mathbb{R}^{6\times m}$. For a rigid body with a Euclidean configuration space we have the very simple relation

$$\tilde{V}_{0b}^B = \dot{\varphi}_v \tag{3.113}$$

because we can write $Q = q = \varphi \in \mathbb{R}^m$ and $v = \dot{q} = \dot{\varphi} \in \mathbb{R}^m$. The corresponding velocity transformation matrix is given by the identity matrix, i.e.,

$$S(Q,\varphi) = I \in \mathbb{R}^{m\times m}. \tag{3.114}$$

References

Bullo, F., & Lewis, A. D. (2000). *Geometric control of mechanical systems: modeling, analysis, and design for simple mechanical control systems*. New York: Springer.

Duindam, V. (2006). *Port-based modeling and control for efficient bipedal walking robots*. Ph.D. Thesis, University of Twente.

Duindam, V., & Stramigioli, S. (2007). Lagrangian dynamics of open multibody systems with generalized holonomic and nonholonomic joints. In *Proceedings of the IEEE/RSJ international conference on intelligent robots and systems*, San Diego, CA, USA (pp. 3342–3347).

Duindam, V., & Stramigioli, S. (2008). Singularity-free dynamic equations of open-chain mechanisms with general holonomic and nonholonomic joints. *IEEE Transactions on Robotics*, *24*(3), 517–526.
Egeland, O., & Gravdahl, J. T. (2003). *Modeling and simulation for automatic control*. Trondheim: Marine Cybernetics AS
Fossen, T. I. (1994). *Guidance and control of Ocean vehicles*. Chichester: Wiley.
Fossen, T. I. (2002). *Marine control systems*. Trondheim: Marine Cybernetics AS. 3rd printing.
From, P. J. (2012a). An explicit formulation of singularity-free dynamic equations of mechanical systems in Lagrangian form—part one: single rigid bodies. *Modeling, Identification and Control*, *33*(2), 45–60.
From, P. J. (2012b). An explicit formulation of singularity-free dynamic equations of mechanical systems in Lagrangian form—part two: multibody systems. *Modeling, Identification and Control*, *33*(2), 61–68.
From, P. J., Duindam, V., Pettersen, K. Y., Gravdahl, J. T., & Sastry, S. (2010). Singularity-free dynamic equations of vehicle-manipulator systems. *Simulation Modelling Practice and Theory*, *18*(6), 712–731.
Gallier, J. (2001). *Texts in applied mathematics series. Geometric methods and applications: for computer science and engineering*. Berlin: Springer.
Hanson, A. J. (2006). *Visualizing quaternions*. San Francisco: Morgan Kaufmann.
Hervé, J. M. (1978). Analyse structurelle des mécanismes par groupe des déplacements. *Mechanism and Machine Theory*, *13*(4), 435–437.
Kuipers, J. B. (2002). *Quaternions and rotation sequences*. Princeton: Princeton University Press.
Marsden, J. E., & Ratiu, T. S. (1999). *Texts in applied mathematics. Introduction to mechanics and symmetry* (2nd ed.). New York: Springer.
Meng, J., Liu, G., & Li, Z. (2007). A geometric theory for analysis and synthesis of sub-6 DoF parallel manipulators. *IEEE Transactions on Robotics*, *23*(4), 625–649.
Murray, R. M., Li, Z., & Sastry, S. S. (1994). *A mathematical introduction to robotic manipulation*. Boca Raton: CRC Press.
Rao, A. (2006). *Dynamics of particles and rigid bodies—a systematic approach*. Cambridge: Cambridge University Press.
Rossmann, W. (2002). *Lie groups—an introduction through linear algebra*. Oxford: Oxford Science Publications.
Selig, J. M. (2000). *Geometric fundamentals of robotics*. New York: Springer.
Siciliano, B., Sciavicco, L., Villani, L., & Oriolo, G. (2011). *Advanced textbooks in control and signal processing. Robotics: modelling, planning and control*. Berlin: Springer.
Tu, L. W. (2008). *An introduction to manifolds*. New York: Springer.
Wen, J. T.-Y., & Kreutz-Delgado, K. (1991). The attitude control problem. *IEEE Transactions on Automatic Control*, *30*(10), 1148–1162.

Chapter 4
Kinematics of Manipulators on a Fixed Base

A robotic manipulator is a collection of rigid bodies whose relative motion is constrained by the admissible velocities of the joins connecting two consecutive rigid bodies in the system. The rigid bodies are called the links of the robot. For most robots we can assume that the links are rigid, but there also exist robots with flexible links, particularly robots used in space, where weight is more important than stiffness. The stiffness of the links not only increases the lifetime and bandwidth of the robot, it also simplifies the modeling and control because each link can be modeled as a single rigid body using the framework of Chap. 3. An ideal joint constrains the motion between two links so that only certain relative velocities are allowed, independently of the torques and forces applied to the links. For our purpose we can assume that all joints are ideal (Selig 2000; Hervé and Sparacino 1991; Angeles 1982; Karger and Novák 1985).

The kinematics of a robotic manipulator can be described as the mapping from joint to operational space. In general, we can divide the kinematics problem into two types: static and differential kinematics. Static kinematics is a mapping from the joint positions to the end-effector location. This is thus a mapping from one manifold to another and we will make much use of the properties of the different manifolds introduced in Sect. 2.5. We will find this mapping either by a sequence of homogeneous transformation matrices or by the product of exponentials formula. The differential kinematics is a mapping from the joint velocities to the end-effector velocities. This mapping is given by the geometric Jacobian if the end-effector velocity is given as a twist, and the analytical Jacobian if the end-effector velocity is given as the time derivative of the configuration vector η.

In this chapter we address fixed-base manipulators with 1-DoF joints, i.e., revolute or prismatic joints with one degree of freedom. The vast majority of robotic manipulators are built with this kind of joints because they are easy to drive and give very favorable characteristics to the robot's workspace. One degree of freedom joints are also easy to model: Firstly, the homogeneous transformation matrices can be written in a very simple form and we can also write the forward kinematics in terms of the exponential map by using the product of exponentials formula. Sec-

P.J. From et al., *Vehicle-Manipulator Systems*, Advances in Industrial Control,
DOI 10.1007/978-1-4471-5463-1_4,

ondly, the differential kinematics is simplified, and we can for example write the columns of the geometric Jacobian as (position-dependent) joint twists.

The problem of finding the joint positions from a given end-effector configuration is called the inverse kinematics problem. The inverse kinematics problem is treated in detail in most textbooks on robotics so we will not address this problem in any detail in this book (Siciliano et al. 2011). We will, however, discuss the inverse of the differential kinematics throughout the book. This is a considerably easier problem than the inverse kinematics problem, as the joint velocities can be found from the end-effector velocities through the inverse of the Jacobian. Robot kinematics has been studied in great detail in several textbooks such as Siciliano et al. (2011), Spong and Vidyasagar (2008), Murray et al. (1994)

4.1 Static Kinematics

The forward kinematics map of a fixed-base manipulator gives us the transformation from the base frame $\mathscr{F}_0$ inertially fixed to the base of the robot to the end-effector frame $\mathscr{F}_e$. This transformation is given by the homogeneous transformation $g_{0e}(q)$ and depends on the joint positions $q \in \mathbb{R}^n$. The first step in finding the forward kinematics map is to define the reference frames in which the robot parameters and joint variables are defined.

4.1.1 Reference Frames

In addition to identify an inertial reference frame $\mathscr{F}_0$ with the robot base and a non-inertial frame $\mathscr{F}_e$ with the end effector, we will attach a reference frame to each link of the robot, denoted $\mathscr{F}_i$ for $i = 1, \dots, n$, in such a way that link 1 is closest to the base and link i comes directly after link $i-1$ in the chain. The links are connected by joints in such a way that the joint connecting link i and link $i+1$ is denoted the $(i+1)$'th joint. We note that for a serial chain of links denoted in this way and when only the kinematics is taken into account, a change in the joint position of joint i will affect the motion of link i but not the motion of the links closer to the base, i.e., the links j for $j < i$ are not affected by the motion of joint i. We will only consider 1 degree of freedom rotational or prismatic joints for which the joint position is represented by $Q_i = q_i \in \mathbb{R}$. The transformation from one link to the next is thus Euclidean and we can describe the position of the n joints by the vector $q \in \mathbb{R}^n$ and the velocity as $v = \dot{q} \in \mathbb{R}^n$. The configuration of frame $\mathscr{F}_i$ attached to each link i with respect to the inertial frame $\mathscr{F}_0$ is given by $g_{0i}(q)$.

It is common to define a reference position called the *home position* of the robot and identify this position with $q = 0$. This position can be chosen arbitrarily, but is often chosen when the robot is stretched out or when it is positioned as in Fig. 4.1. This configuration is also often referred to as the zero pose position in the robotics literature.

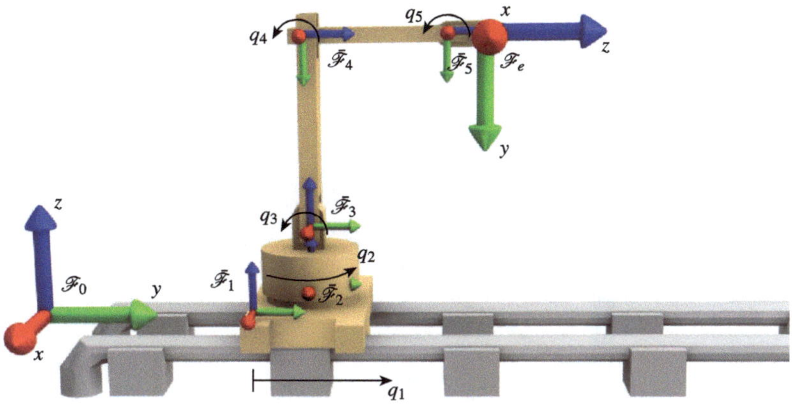

Fig. 4.1 One example of the home position for a robotic manipulator with five joints. The inertial frame $\mathscr{F}_0$ and the end-effector frame $\mathscr{F}_e$ are shown with large coordinate frames and the coordinate frames of joints 1–5 are shown with small coordinate frames

When we are to derive the kinematic and dynamic equations of robotic manipulators and other mechanical systems we can simplify the expressions and computations substantially by making a well-considered choice when choosing the location of the coordinate frames in which the different parameters and variables are represented. Unfortunately, there is a conflict of interest when it comes to choosing the reference frames:

- When deriving the *kinematics* of a robotic system we can greatly simplify the equations by choosing the reference frames of each link so that the origin is located at the joint axis (the axis of rotation for revolute joints or the axis of translation for prismatic joints) and such that one of the coordinate axes is aligned with the joint axis. We will denote a frame defined in this way as $\bar{\mathscr{F}}$ (with bar).
- When deriving the *dynamics* of a robotic system we can greatly simplify the equations by choosing the reference frames of each link so that the origin is located at the center of gravity of the link and the coordinate axes are aligned with the link's principal axes of inertia. We will denote a frame defined in this way as $\mathscr{F}$ (without bar).

In most cases it is not possible to choose the frames so that both these requirements are fulfilled. In this chapter we will therefore, as a general rule, study the motion of reference frames $\bar{\mathscr{F}}$ with the origin at the link axes in order to simplify the kinematic expressions. We see that if we choose the reference frames in this

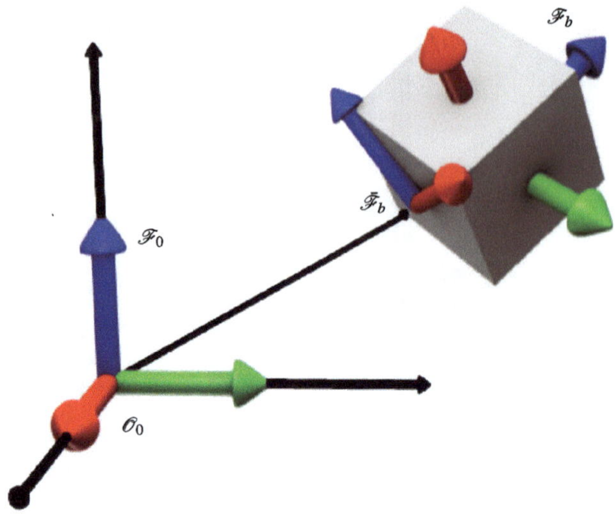

Fig. 4.2 The definition of the different coordinate frames for a robotic link. The coordinate frame $\mathscr{F}_i$ is chosen so $\mathscr{O}_i$ is at the center of gravity and the x-, y-, and z-axes are aligned with the principal axes of inertia of the link. The coordinate frame $\bar{\mathscr{F}}_b$ is defined so that $\bar{\mathscr{O}}_i$ is located at the joint axis and one of the coordinate axes is aligned with the joint axis

way, the origin of $\bar{\mathscr{F}}$ does not move for revolute joints. Furthermore, if we choose one of the coordinate axes to point in the direction of the joint axis, this coordinate axis will point in the same direction independently of the joint position for both prismatic and revolute joints.

When we need to describe the position or velocity of the center of gravity of a link we will attach a frame $\mathscr{F}_i$ to this point. This will simplify the dynamic equations because the kinetic energy can be written in terms of the generalized inertia matrix which is constant and diagonal. We will discuss the reference frames $\mathscr{F}$ located in the center of gravity in more detail in the chapters on dynamics.

We will write the frames located at the joint axes as $\bar{\mathscr{F}}$ (with bar) and the frames attached to the center of gravity as $\mathscr{F}$ (without the bar) to make it clear what reference frame we refer to. It is important to note that even though both reference frames are attached to the same rigid body, their motion will in general not be identical. The difference between the reference frames is illustrated in Fig. 4.2.

It is also important that we distinguish between $\bar{\mathscr{F}}$ and $\mathscr{F}$ when referring to velocities and transformations. For velocities we will use the following notation:

- $V^S_{0\bar{i}}$ is the spatial velocity of frame $\bar{\mathscr{F}}_i$ with respect to the inertial frame $\mathscr{F}_0$,
- V^S_{0i} is the spatial velocity of frame $\mathscr{F}_i$ with respect to the inertial frame $\mathscr{F}_0$,

and similarly for body velocities. We will denote transformations in the same way:

- $g_{0\bar{i}}$ denotes the location of frame $\bar{\mathscr{F}}_i$ in $\mathscr{F}_0$,
- g_{0i} denotes the location of frame $\mathscr{F}_i$ in $\mathscr{F}_0$.

4.1.2 Homogeneous Transformations

Using standard notation we can describe the pose of each frame $\mathscr{F}_i$ relative to $\mathscr{F}_0$ as a homogeneous transformation matrix $g_{0i} \in SE(3)$ of the form

$$g_{0i} = \begin{bmatrix} R_{0i} & p_{0i} \\ 0 & 1 \end{bmatrix} \in \mathbb{R}^{4\times 4} \tag{4.1}$$

with rotation matrix $R_{0i} \in SO(3)$ and translation vector $p_{0i} \in \mathbb{R}^3$. This pose can also be described using the vector of joint coordinates q as

$$g_{0i} = g_{0(i-1)} g_{(i-1)i} = g_{0(i-1)}(q) g_{(i-1)i}(q_i). \tag{4.2}$$

The transformation from frame $\mathscr{F}_{i-1}$ to frame $\mathscr{F}_i$ depends only on the joint position of joint q_i. We can thus write the transformation from the inertial frame $\mathscr{F}_0$ to the end-effector frame $\mathscr{F}_e$ as a series of homogeneous transformations

$$g_{0e}(q) = g_{01}(q_1) g_{12}(q_2) \cdots g_{(n-1)n}(q_n) g_{ne} \tag{4.3}$$

where g_{ne} is the constant transformation from the last link to the end effector.

We can also derive the forward kinematics map in terms of the frames $\bar{\mathscr{F}}_i$ located at joint axis i. The forward kinematics map is then given in the same way by

$$g_{0e}(q) = g_{0\bar{1}}(q_1) g_{\bar{1}\bar{2}}(q_2) \cdots g_{\overline{(n-1)}\bar{n}}(q_n) g_{\bar{n}e}. \tag{4.4}$$

Choosing the coordinate frames in this way allows us to divide the transformation from one joint to the next into one transformation that depends on the kinematic structure of the manipulator and a configuration-dependent part:

The transformation from one frame to the next frame in a robotic chain can be written in terms of two transformations that take a rather simple form:

- first a constant transformation g_i^0 from frame $\bar{\mathscr{F}}_{i-1}$ to the next frame $\bar{\mathscr{F}}_i^0$ when the manipulator is in the home position $q_i = 0$ (also called zero pose position and thus the zero in the superscript in $\bar{\mathscr{F}}^0$); and
- a transformation that depends on the position of the joint, i.e., the transformation $g_i(q_i)$ from $\bar{\mathscr{F}}_i^0$ to $\bar{\mathscr{F}}_i$.

The transformation from frame $\bar{\mathscr{F}}_{i-1}$ to the subsequent frame $\bar{\mathscr{F}}_i$ is then given by

$$g_{\overline{(i-1)}\bar{i}}(q_i) = g_i^0 g_i(q_i). \tag{4.5}$$

As a general rule we can say that the first transformation g_i^0 is obtained from the manipulator structure, i.e., the length of the links, the link offsets, the link positions

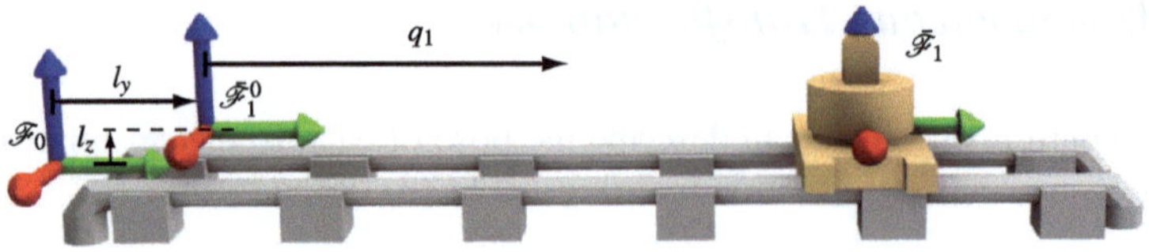

Fig. 4.3 The transformations of a prismatic joint first from the inertial frame $\mathscr{F}_0$ to the joint frame $\bar{\mathscr{F}}_1^0$ at the home position, and then to the frame at joint position at q_1 represented by $\bar{\mathscr{F}}_1$

at home position, etc. The second transformation $g_i(q_i)$ is given by the current joint position q_i and gives us the motion of link i with respect to $\bar{\mathscr{F}}_i^0$.

Example 4.1 We will illustrate this convention for a prismatic joint illustrated in Fig. 4.3. We first define the transformation from frame $\mathscr{F}_0$ (in this case the inertial frame) to link 1 when the manipulator is at the home position. For the manipulator in Fig. 4.3 this is given by

$$g_1^0 = \begin{bmatrix} 1 & 0 & 0 & 0 \\ 0 & 1 & 0 & l_y \\ 0 & 0 & 1 & l_z \\ 0 & 0 & 0 & 1 \end{bmatrix} \tag{4.6}$$

where l_y and l_z give the position of $\bar{\mathscr{F}}_1^0$ in $\mathscr{F}_0$. The position of the joint with respect to the home position when the joint motion is taken into account is given by

$$g_1(q_1) = \begin{bmatrix} 1 & 0 & 0 & 0 \\ 0 & 1 & 0 & q_1 \\ 0 & 0 & 1 & 0 \\ 0 & 0 & 0 & 1 \end{bmatrix}. \tag{4.7}$$

The transformation from $\mathscr{F}_0$ to $\bar{\mathscr{F}}_1$ is denoted $g_{0\bar{1}}(q_1)$ and is given by

$$g_{0\bar{1}}(q_1) = g_1^0 g_1(q_1) = \begin{bmatrix} 1 & 0 & 0 & 0 \\ 0 & 1 & 0 & l_y + q_1 \\ 0 & 0 & 1 & l_z \\ 0 & 0 & 0 & 1 \end{bmatrix}. \tag{4.8}$$

Example 4.2 For a joint with one degree of rotational freedom around the x-axis as in Fig. 4.4, the homogeneous transformation matrix that describes the position of the joint in home position is given, by

$$g_1^0 = \begin{bmatrix} 1 & 0 & 0 & 0 \\ 0 & 1 & 0 & l_y \\ 0 & 0 & 1 & l_z \\ 0 & 0 & 0 & 1 \end{bmatrix} \tag{4.9}$$

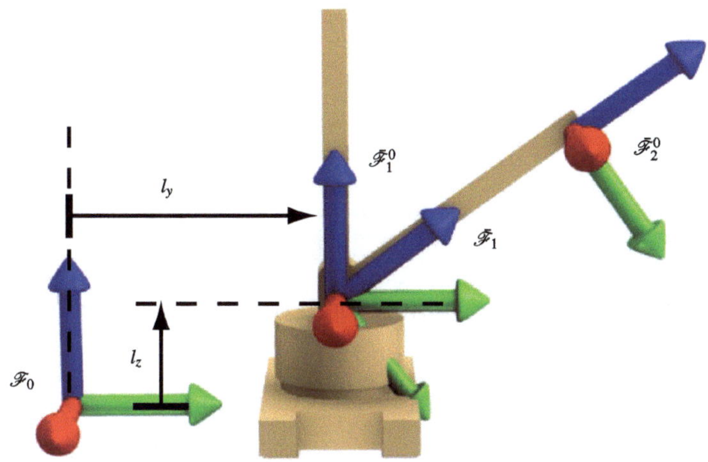

Fig. 4.4 The transformations of a revolute joint first from the inertial frame $\mathscr{F}_0$ to the joint frame $\bar{\mathscr{F}}_1^0$ at the home position, and then to the frame at joint position at q_1 represented by $\bar{\mathscr{F}}_1$. The frame $\bar{\mathscr{F}}_2^0$ of joint 2 is also shown

which is the same as for the prismatic joint, and the transformation that depends on the joint position is given by

$$g_1(q_1) = \begin{bmatrix} 1 & 0 & 0 & 0 \\ 0 & \cos q_1 & -\sin q_1 & 0 \\ 0 & \sin q_1 & \cos q_1 & 0 \\ 0 & 0 & 0 & 1 \end{bmatrix}. \tag{4.10}$$

The transformation from $\mathscr{F}_0$ to $\bar{\mathscr{F}}_1$ is denoted $g_{0\bar{1}}(q_1)$ and is given by

$$g_{0\bar{1}}(q_i) = g_1^0 g_1(q_1) = \begin{bmatrix} 1 & 0 & 0 & 0 \\ 0 & \cos q_1 & -\sin q_1 & l_y \\ 0 & \sin q_1 & \cos q_1 & l_z \\ 0 & 0 & 0 & 1 \end{bmatrix}. \tag{4.11}$$

Now that we can find the transformation corresponding to one single joint we can apply Eq. (4.4) to find the forward kinematics.

Example 4.3 Consider the manipulator illustrated in Fig. 4.5. The forward kinematics map can be solved by finding the homogeneous transformation associated with each joint and substitute this into (4.4). To find the end-effector configuration we need to find the homogeneous transformation matrices that give us the transformations between any two subsequent reference frames. The reference frames of the manipulator in Fig. 4.5 are shown in Fig. 4.6 for the home position. To simplify the expressions we will assume that the end effector frame $\mathscr{F}_e$ coincides with the last

Fig. 4.5 The five degree of freedom manipulator in Example 4.3. The first joint is prismatic and the last four joints are revolute

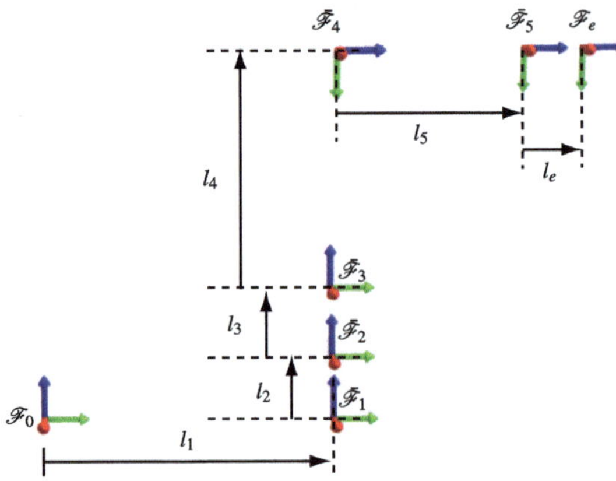

Fig. 4.6 The coordinate frames of the manipulator in Fig. 4.5

link $\bar{\mathcal{F}}_5$, i.e., $l_e = 0$. We first find the transformation from the inertial frame to the prismatic joint. The manipulator moves on a rail and $g_{0\bar{1}}$ therefore depends on the position q_1 of the first joint and we get a homogeneous transformation matrix

$$g_{0\bar{1}}(q_1) = \begin{bmatrix} 1 & 0 & 0 & 0 \\ 0 & 1 & 0 & l_1 + q_1 \\ 0 & 0 & 1 & 0 \\ 0 & 0 & 0 & 1 \end{bmatrix} \tag{4.12}$$

that represents a pure translation along the y-axis and a constant offset also in the direction of the y-axis. The transformation from the first to the second joint is given by a pure rotational motion around the z-axis and a constant offset in the direction of the z-axis:

$$g_{\bar{1}\bar{2}}(q_2) = \begin{bmatrix} \cos q_2 & -\sin q_2 & 0 & 0 \\ \sin q_2 & \cos q_2 & 0 & 0 \\ 0 & 0 & 1 & l_2 \\ 0 & 0 & 0 & 1 \end{bmatrix}.$$

In a similar way we find the transformation matrices for the other joints in the manipulator as:

$$g_{\overline{23}}(q_3) = \begin{bmatrix} 1 & 0 & 0 & 0 \\ 0 & \cos q_3 & -\sin q_3 & 0 \\ 0 & \sin q_3 & \cos q_3 & l_3 \\ 0 & 0 & 0 & 1 \end{bmatrix},$$

$$g_{\overline{34}}(q_4) = \begin{bmatrix} 1 & 0 & 0 & 0 \\ 0 & \cos(q_4 - \frac{\pi}{2}) & -\sin(q_4 - \frac{\pi}{2}) & 0 \\ 0 & \sin(q_4 - \frac{\pi}{2}) & \cos(q_4 - \frac{\pi}{2}) & l_4 \\ 0 & 0 & 0 & 1 \end{bmatrix}.$$

$$g_{\overline{45}}(q_5) = \begin{bmatrix} 1 & 0 & 0 & 0 \\ 0 & \cos q_5 & -\sin q_5 & 0 \\ 0 & \sin q_5 & \cos q_5 & l_5 \\ 0 & 0 & 0 & 1 \end{bmatrix}, \qquad g_{\overline{5}e} = \begin{bmatrix} 1 & 0 & 0 & 0 \\ 0 & 1 & 0 & 0 \\ 0 & 0 & 1 & l_e \\ 0 & 0 & 0 & 1 \end{bmatrix}.$$

The forward kinematics map is then found by

$$g_{0e} = g_{0\overline{1}}(q_1) g_{\overline{12}}(q_2) g_{\overline{23}}(q_3) g_{\overline{34}}(q_4) g_{\overline{45}}(q_5) g_{\overline{5}e} \tag{4.13}$$

which can be written as

$$g_{0e} = \begin{bmatrix} R_{0e} & p_{0e} \\ 0 & 1 \end{bmatrix} \in \mathbb{R}^{4\times 4} \tag{4.14}$$

with (assume $l_e = 0$)

$$R_{0e}(q) = \begin{bmatrix} \cos q_2 & -\sin q_2 \cos q_{35\pi} & \sin q_2 \sin q_{35\pi} \\ \sin q_2 & \cos q_2 \cos q_{35\pi} & -\cos q_2 \sin q_{35\pi} \\ 0 & \sin q_{35\pi} & \cos q_{35\pi} \end{bmatrix} \tag{4.15}$$

$$p_{0e}(q) = \begin{bmatrix} l_5 \sin q_2 \sin q_{34\pi} + l_4 \sin q_2 \sin q_3 \\ -l_5 \cos q_2 \sin q_{34\pi} - l_4 \cos q_2 \sin q_3 + l_1 + q_1 \\ l_5 \cos q_{34\pi} + l_4 \cos q_3 + l_2 + l_3 \end{bmatrix} \tag{4.16}$$

where $q_{34\pi} = q_3 + q_4 - \frac{\pi}{2}$ and $q_{35\pi} = q_3 + q_4 + q_5 - \frac{\pi}{2}$.

4.1.3 Product of Exponentials Formula

The forward kinematics of a serial manipulator with one degree of freedom joints can also be written in terms of the joint twists. First we define a set of joint angles that represent the home position of the robot and identify this configuration with $q = 0$. We then transform the configuration of the end effector from the home position, denoted by $g_{0e}(0)$, to the configuration for the new joint position q, denoted

$g_{0e}(q)$. The product of exponentials formula was introduced in Brockett (1984) and also discussed in Park (1994).

For a one joint mechanism we see from Eq. (2.174) that the new end-effector configuration can be written as

$$g_{0e}(q_1) = e^{\widehat{X}_1 q_1} g_{0e}(0). \tag{4.17}$$

Recall that this expression is interpreted as an *active* transformation, i.e., a transformation of a reference frame $\mathscr{F}_e$ from one configuration to another. The transformed reference frame is observed from the same inertial reference frame, in our case $\mathscr{F}_0$, both before and after the transformation. This is also how we will interpret this transformation when used to resolve the forward kinematics problem, namely a transformation of the end-effector frame $\mathscr{F}_e$ from the home position to the position after a change q in the joint position or orientation.

Lemma 4.1 *Given a one joint robotic manipulator with a joint twist $\widehat{X}_1$ when the manipulator is in home position $q_1 = 0$. Denote further the end-effector configuration at home position by $g_{0e}(0)$. Then the end-effector configuration for the joint position q_1 is given by*

$$g_{0e}(q_1) = e^{\widehat{X}_1 q_1} g_{0e}(0). \tag{4.18}$$

This follows directly from the geometric interpretation described in Sect. 2.8 and above.

Similarly, for a manipulator with two joints, we can perform this operation first by moving the second joint and assuming the first joint as fixed and then move the first joint and assume the second joint as fixed. The rotation of the second joint will not affect the twist of the first joint and we can therefore move the first joint corresponding to the twist in home position. In other words, both twists are constant. The total displacement then becomes

$$g_{0e}(q) = e^{\widehat{X}_1 q_1} e^{\widehat{X}_2 q_2} g_{0e}(0). \tag{4.19}$$

It is important to note that $\widehat{X}_1$ and $\widehat{X}_2$ are constant twists. This is easily seen if we interpret (4.19) as a sequence of transformations starting at the end of the chain and moving its way towards the base. We do, however, get the same result if we start with the joint closest to the base (Murray et al. 1994): Assume a two joint robotic manipulator with joint twists $\widehat{X}_1$ and $\widehat{X}_2$ at home position $q = 0$. Then if we move to a joint position q_1 for the first joint the twist of the second joint becomes

$$X_2' = \mathrm{Ad}_{e^{\hat{X}_1 q_1}} X_2 \tag{4.20}$$

where X_2' is the spatial joint twist after a transformation by q_1. Conjugation gives

$$e^{\widehat{X}_2' q_2} = e^{\widehat{X}_1 q_1}\left(e^{\widehat{X}_2 q_2}\right)e^{-\widehat{X}_1 q_1} \tag{4.21}$$

which is the same as the conjugation in (2.92) written in terms of the exponential map. The new end-effector configuration after transforming first corresponding to joint 1, and then afterwards corresponding to joint 2, is then given by

$$\begin{aligned} g_{0e}(q) &= e^{\widehat{X}_2' q_2} e^{\widehat{X}_1 q_1} g_{0e}(0) \\ &= e^{\widehat{X}_1 q_1}\left(e^{\widehat{X}_2 q_2}\right)e^{-\widehat{X}_1 q_1} e^{\widehat{X}_1 q_1} g_{0e}(0) \\ &= e^{\widehat{X}_1 q_1} e^{\widehat{X}_2 q_2} g_{0e}(0) \end{aligned} \tag{4.22}$$

which is the same as (4.19). The forward kinematics problem can thus be solved using the following theorem:

Theorem 4.1 *The forward kinematics map of an n degree of freedom manipulator with joint twists $\widehat{X}_i$ for $i = 1, \dots, n$ and an end-effector configuration $g_{0e}(0)$ at home position $q = 0$ is given by*

$$g_{0e}(q) = e^{\widehat{X}_1 q_1} e^{\widehat{X}_2 q_2} \cdots e^{\widehat{X}_n q_n} g_{0e}(0). \tag{4.23}$$

The forward kinematics problem is thus solved if we find the joint twist of each joint in the manipulator. We see that because we choose $\mathscr{F}_0$ as the reference frame we need to represent the twists X_i with respect to this frame. There are, however, many different ways to represent the twists. This is discussed in the next section and we show how to find the joint twists of the robot links for prismatic and revolute joints.

4.2 The Manipulator Twists

We see that the forward kinematics map for a robotic manipulator with 1-DoF joints is completely determined by the joint twists. To find the configuration of the end-effector frame $\mathscr{F}_e$ with respect to the inertial frame $\mathscr{F}_0$ it is thus sufficient to find the twists of the manipulator for a given position, denoted the home position. Because we want to describe the end-effector frame with respect to the inertial frame we must use the spatial twists. We will first find the body joint twists X_i^i and then find the spatial joint twist X_i by the Adjoint transformation in the normal way. We will also see that for pure rotational and translational motion the spatial twists take a very simple form and are particularly easy to compute.

The joint body twists describe the allowed motion of the joints as seen from the frame attached to the joint, i.e., the motion of $\bar{\mathscr{F}}_i$ with respect to $\bar{\mathscr{F}}_i^0$. This is in other words always constant and independent of the manipulator configuration. A standard robotic manipulator is composed of prismatic and revolute joints: For a prismatic joint the joint twist describes the admissible linear velocities, i.e., the direction p_i^i for which motion is allowed. The body joint twist of a prismatic joint is thus given by

$$X_i^i = \begin{bmatrix} v_i^i \\ \omega_i^i \end{bmatrix} = \begin{bmatrix} p_i^i \\ 0 \end{bmatrix} \tag{4.24}$$

where p_i^i is represented in the body frame $\bar{\mathscr{F}}_i$. Note the difference between p_i^i (single subscript) which describes the direction of the motion of joint i, and p_{0i} (double subscript) which is the position of frame $\mathscr{F}_i$ in $\mathscr{F}_0$.

Similarly, the body joint twist of a rotational joint that rotates around the axis p_i^i can be written as

$$X_i^i = \begin{bmatrix} v_i^i \\ \omega_i^i \end{bmatrix} = \begin{bmatrix} 0 \\ p_i^i \end{bmatrix}. \tag{4.25}$$

Because p_i^i is written in the body frame $\bar{\mathscr{F}}_i$, the body twists for both revolute and prismatic joints are constant.

In the previous section we found the forward kinematics map in terms of the spatial joint twists at the home position. The spatial twists are found from the body twist by the Adjoint map. Because the Adjoint map is configuration dependent, the spatial twist is in general not constant. However, because the twists used in the forward kinematics map are found at the home position these are constant both for revolute and prismatic motion. For a prismatic joint the spatial joint twists are thus given by $X_i = \mathrm{Ad}_{g_{0\bar{i}}} X_i^i$:

$$X_i = \begin{bmatrix} v_i \\ \omega_i \end{bmatrix} = \begin{bmatrix} p_i \\ 0 \end{bmatrix} = \begin{bmatrix} R_{0\bar{i}} & \hat{p}_{0\bar{i}} R_{0\bar{i}} \\ 0 & R_{0\bar{i}} \end{bmatrix} \begin{bmatrix} p_i^i \\ 0 \end{bmatrix} = \begin{bmatrix} R_{0\bar{i}} p_i^i \\ 0 \end{bmatrix} \tag{4.26}$$

where p_i is represented in the inertial frame $\mathscr{F}_0$ when the manipulator is in the home position. $\mathrm{Ad}_{g_{0\bar{i}}}$ thus takes us from the body twist in frame $\bar{\mathscr{F}}_i$ to the spatial twist in the inertial frame $\mathscr{F}_0$.

We see that a prismatic joint only affects the linear part of the spatial twist, as expected. For a revolute joint, however, the joint motion affects both the linear and rotational part of the spatial twist. This can be seen from the Adjoint map $X_i = \mathrm{Ad}_{g_{0\bar{i}}} X_i^i$ where X_i^i is given as in (4.25). It is, however, possible to find the spatial twist without applying the Adjoint map: Let p_i denote the axis that joint i rotates around when it is in the home position and let $\bar{l}_i \in \mathbb{R}^3$ be a point on this axis, both represented in $\mathscr{F}_0$. Then the rotational part of the twist is given by p_i and the translational part is given by $-p_i \times \bar{l}_i$. The twist of a rotational joint can be written as

$$X_i = \begin{bmatrix} v_i \\ \omega_i \end{bmatrix} = \begin{bmatrix} -p_i \times \bar{l}_i \\ p_i \end{bmatrix}. \tag{4.27}$$

where both p_i and $\bar{l}_i$ are written in $\mathscr{F}_0$. We see that we have some freedom in choosing the point $\bar{l}_i$: all that is required is that it lies on the axis p_i when at the home position $q = 0$. $\bar{l}_i$ tells us how frame $\bar{\mathscr{F}}_i^0$ is located in $\mathscr{F}_0$, i.e., the collection of all $\bar{l}_i$'s and p_i's for $i = 1, \ldots, n$ gives us the manipulator kinematics (the structure of the manipulator).

To show that the expression in (4.27) is in fact identical to $X_i = \mathrm{Ad}_{g_{0\bar{i}}} X_i^i$ is fairly straight forward. We let $R_{0\bar{i}}$ be the orientation of $\bar{\mathscr{F}}_i^0$ with respect to $\mathscr{F}_0$ and $p_{0\bar{i}} = \bar{l}_i$ be the location of the origin of $\bar{\mathscr{F}}_i^0$ in $\mathscr{F}_0$ in the normal way. Then we can write

$$\begin{bmatrix} v_i \\ \omega_i \end{bmatrix} = \begin{bmatrix} R_{0\bar{i}} & \hat{p}_{0\bar{i}} R_{0\bar{i}} \\ 0 & R_{0\bar{i}} \end{bmatrix} \begin{bmatrix} 0 \\ p_i^i \end{bmatrix} = \begin{bmatrix} \hat{p}_{0\bar{i}} R_{0\bar{i}} p_i^i \\ R_{0\bar{i}} p_i^i \end{bmatrix} = \begin{bmatrix} \hat{p}_{0\bar{i}} p_i \\ p_i \end{bmatrix} = \begin{bmatrix} \bar{l}_i \times p_i \\ p_i \end{bmatrix} = \begin{bmatrix} -p_i \times \bar{l}_i \\ p_i \end{bmatrix}. \tag{4.28}$$

The reason that we can choose $\bar{l}_i$ as *any* point on the axis p_i is that the component on this axis will always vanish from the final expressions. If the joint rotates around the z-axis, for example, we see that we can choose the z-coordinate of l_i freely because

$$p_i \times \bar{l}_i = \begin{bmatrix} 0 & -1 & 0 \\ 1 & 0 & 0 \\ 0 & 0 & 0 \end{bmatrix} \begin{bmatrix} l_{i,x} \\ l_{i,y} \\ l_{i,z} \end{bmatrix} \tag{4.29}$$

is not affected by the choice of $l_{i,z}$.

We can now define the body and spatial joint twists for prismatic and revolute joints in the following way:

Given a joint i with axis p_i^i represented in frame $\bar{\mathscr{F}}_i$. Then the body joint twist is given by

$$X_i^i = \begin{bmatrix} p_i^i \\ 0 \end{bmatrix} \tag{4.30}$$

for a prismatic joint and by

$$X_i^i = \begin{bmatrix} 0 \\ p_i^i \end{bmatrix} \tag{4.31}$$

for a revolute joint.

The spatial twist describes the same motion but in spatial coordinates and needs to be transformed by the Adjoint map in the normal way:

Given a joint i with axis p_i and some point $\bar{l}_i$ on this axis, both represented in frame $\mathscr{F}_0$. Then the spatial joint twist is given by

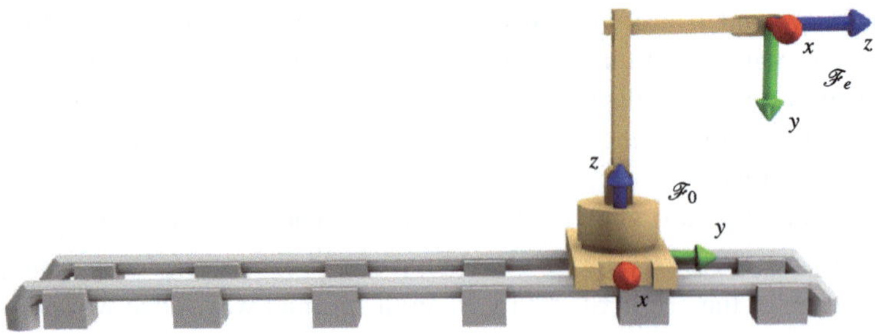

Fig. 4.7 The five degree of freedom manipulator in Example 4.4 with one translational and four rotational joints. The inertial frame $\mathcal{F}_0$ is chosen so that it coincides with the base of the manipulator at the home position

$$X_i = \mathrm{Ad}_{g_{0\bar{i}}} X_i^i = \begin{bmatrix} p_i \\ 0 \end{bmatrix} \tag{4.32}$$

for a prismatic joint and by

$$X_i = \mathrm{Ad}_{g_{0\bar{i}}} X_i^i = \begin{bmatrix} -p_i \times \bar{l}_i \\ p_i \end{bmatrix} \tag{4.33}$$

for a revolute joint, where $p_i = R_{0\bar{i}} p_i^i$.

Because we can choose the inertial reference frame freely, we can define the spatial joint twists with respect to any frame. If we for example need to define the spatial joint twist with respect to the frame $\mathcal{F}_e$ attached to the end effector, we will denote the spatial twist X_i^e. It is important to note that this is the spatial joint twist with respect to the end-effector frame, i.e., the admissible velocities in spatial coordinates, which is different from the admissible velocities as seen from the end-effector frame. If not otherwise stated, we will use the expression spatial joint twist to describe the twist in spatial coordinates with respect to an inertial reference frame.

Example 4.4 Consider the robot in Fig. 4.7 and identify the joint variables with $q = 0$ at the position shown, called home position. We choose the frames attached to each link as in Fig. 4.6, but with $l_1 = 0$. We further define $l_{23} = l_2 + l_3$ and $l_{24} = l_2 + l_3 + l_4$. The motion of the first joint is generated by translating along the rail and the twist is thus given by

$$X_1 = X_1^1 = \begin{bmatrix} 0 \\ 1 \\ 0 \\ 0 \\ 0 \\ 0 \end{bmatrix}. \tag{4.34}$$

The second joint revolutes about z-axis that goes through the origin of the base frame. We need to choose a point on this axis, and we choose $\bar{l}_i = [0\,0\,0]^\mathsf{T}$ for simplicity. The spatial joint twist can then be written in a simple form as

$$X_2 = \begin{bmatrix} 0 \\ 0 \\ 0 \\ 0 \\ 0 \\ 1 \end{bmatrix}. \tag{4.35}$$

We can also find X_2 from the Adjoint map $X_2 = \mathrm{Ad}_{g_{0\bar{2}}} X_{\bar{2}}^2$ which, if we write it explicitly, gives (4.35). The third joint does not revolute around an axis that goes through the origin, but through a point $\bar{l}_3 = [0\,0\,l_{23}]^\mathsf{T}$. Similarly joints four and five revolute around axes that go through the points $\bar{l}_4 = [0\,0\,l_{24}]^\mathsf{T}$ and $\bar{l}_5 = [0\,l_5\,l_{24}]^\mathsf{T}$. The twists are then given by

$$X_3 = \begin{bmatrix} -\begin{bmatrix} 1 \\ 0 \\ 0 \end{bmatrix} \times \begin{bmatrix} 0 \\ 0 \\ l_{23} \end{bmatrix} \\ \begin{bmatrix} 1 \\ 0 \\ 0 \end{bmatrix} \end{bmatrix} = \begin{bmatrix} 0 \\ l_{23} \\ 0 \\ 1 \\ 0 \\ 0 \end{bmatrix}, \quad X_4 = \begin{bmatrix} -\begin{bmatrix} 1 \\ 0 \\ 0 \end{bmatrix} \times \begin{bmatrix} 0 \\ 0 \\ l_{24} \end{bmatrix} \\ \begin{bmatrix} 1 \\ 0 \\ 0 \end{bmatrix} \end{bmatrix} = \begin{bmatrix} 0 \\ l_{24} \\ 0 \\ 1 \\ 0 \\ 0 \end{bmatrix}.$$

$$X_5 = \begin{bmatrix} -\begin{bmatrix} 1 \\ 0 \\ 0 \end{bmatrix} \times \begin{bmatrix} 0 \\ l_5 \\ l_{24} \end{bmatrix} \\ \begin{bmatrix} 1 \\ 0 \\ 0 \end{bmatrix} \end{bmatrix} = \begin{bmatrix} 0 \\ l_{24} \\ -l_5 \\ 1 \\ 0 \\ 0 \end{bmatrix}. \tag{4.36}$$

Now that we have found the twist of each joint, we can find the forward kinematics map which is given by

$$g_{0e}(q) = e^{\widehat{X}_1 q_1} e^{\widehat{X}_2 q_2} e^{\widehat{X}_3 q_3} e^{\widehat{X}_4 q_4} e^{\widehat{X}_5 q_5} g_{0e}(0). \tag{4.37}$$

$g_{0e}(0)$ is given by a rotation of $-\frac{\pi}{2}$ around the x-axis and the translation vector $p_{0e}(0) = [0\,l_5\,l_{24}]^\mathsf{T}$, i.e.,

$$g_{0e}(0) = \begin{bmatrix} 1 & 0 & 0 & 0 \\ 0 & \cos\frac{\pi}{2} & \sin\frac{\pi}{2} & l_5 \\ 0 & -\sin\frac{\pi}{2} & \cos\frac{\pi}{2} & l_{24} \\ 0 & 0 & 0 & 1 \end{bmatrix} = \begin{bmatrix} 1 & 0 & 0 & 0 \\ 0 & 0 & 1 & l_5 \\ 0 & -1 & 0 & l_{24} \\ 0 & 0 & 0 & 1 \end{bmatrix}. \tag{4.38}$$

The exponential maps of each joint is given by (2.185)

$$e^{\widehat{X}_i q_i} = \begin{bmatrix} e^{\widehat{\omega}_i q_i} & (I - e^{\widehat{\omega}_i q_i})(\omega_i \times v_i) + \omega_i \omega_i^\mathsf{T} v_i q_i \\ 0 & 1 \end{bmatrix}. \tag{4.39}$$

We see that $e^{\widehat{\omega}_i q_i}$ is simply the rotation matrix of each joint looked at separately which has a rather simple form when the rotation is about the coordinate axes. The linear terms are found by the upper right part of (4.39), which for joint 3 becomes

$$\begin{aligned}
&\left(I - e^{\widehat{\omega}_3 q_3}\right)(\omega_3 \times v_3) + \omega_3 \omega_3^{\mathsf{T}} v_3 q_3 \\
&= \begin{bmatrix} 0 & 0 & 0 \\ 0 & 1-\cos q_3 & \sin q_3 \\ 0 & -\sin q_3 & 1-\cos q_3 \end{bmatrix} \left(\begin{bmatrix} 0 & 0 & 0 \\ 0 & 0 & -1 \\ 0 & 1 & 0 \end{bmatrix} \begin{bmatrix} 0 \\ l_{23} \\ 0 \end{bmatrix} \right) \\
&\quad + \left(\begin{bmatrix} 1 \\ 0 \\ 0 \end{bmatrix} \begin{bmatrix} 1 & 0 & 0 \end{bmatrix} \right) \begin{bmatrix} 0 \\ l_{23} \\ 0 \end{bmatrix} q_3 \\
&= \begin{bmatrix} 0 & 0 & 0 \\ 0 & 1-\cos q_3 & \sin q_3 \\ 0 & -\sin q_3 & 1-\cos q_3 \end{bmatrix} \begin{bmatrix} 0 \\ 0 \\ l_{23} \end{bmatrix} + \begin{bmatrix} 1 & 0 & 0 \\ 0 & 0 & 0 \\ 0 & 0 & 0 \end{bmatrix} \begin{bmatrix} 0 \\ l_{23} \\ 0 \end{bmatrix} q_3 \\
&= \begin{bmatrix} 0 \\ l_{23} \sin q_3 \\ l_{23}(1-\cos q_3) \end{bmatrix} + \begin{bmatrix} 0 \\ 0 \\ 0 \end{bmatrix} q_3 = \begin{bmatrix} 0 \\ l_{23} \sin q_3 \\ l_{23}(1-\cos q_3) \end{bmatrix}
\end{aligned} \tag{4.40}$$

and similarly for the other joints. The exponential map of the individual joints are then found to be

$$\begin{aligned}
e^{\widehat{X}_1 q_1} &= \begin{bmatrix} 1 & 0 & 0 & 0 \\ 0 & 1 & 0 & q_1 \\ 0 & 0 & 1 & 0 \\ 0 & 0 & 0 & 1 \end{bmatrix}, \\
e^{\widehat{X}_2 q_2} &= \begin{bmatrix} \cos q_2 & -\sin q_2 & 0 & 0 \\ \sin q_2 & \cos q_2 & 0 & 0 \\ 0 & 0 & 1 & 0 \\ 0 & 0 & 0 & 1 \end{bmatrix}, \\
e^{\widehat{X}_3 q_3} &= \begin{bmatrix} 1 & 0 & 0 & 0 \\ 0 & \cos q_3 & -\sin q_3 & l_{23} \sin q_3 \\ 0 & \sin q_3 & \cos q_3 & l_{23}(1-\cos q_3) \\ 0 & 0 & 0 & 1 \end{bmatrix}, \\
e^{\widehat{X}_4 q_4} &= \begin{bmatrix} 1 & 0 & 0 & 0 \\ 0 & \cos q_4 & -\sin q_4 & l_{24} \sin q_4 \\ 0 & \sin q_4 & \cos q_4 & l_{24}(1-\cos q_4) \\ 0 & 0 & 0 & 1 \end{bmatrix}, \\
e^{\widehat{X}_5 q_5} &= \begin{bmatrix} 1 & 0 & 0 & 0 \\ 0 & \cos q_5 & -\sin q_5 & l_5(1-\cos q_5) + l_{24} \sin q_5 \\ 0 & \sin q_5 & \cos q_5 & l_{24}(1-\cos q_5) - l_5 \sin q_5 \\ 0 & 0 & 0 & 1 \end{bmatrix}.
\end{aligned} \tag{4.41}$$

We can now find the forward kinematics map by substituting this into Eq. (4.37) which gives

$$R_{0e}(q) = \begin{bmatrix} \cos q_2 & -\sin q_2 \sin q_{35} & -\sin q_2 \cos q_{35} \\ \sin q_2 & \cos q_2 \sin q_{35} & \cos q_2 \cos q_{35} \\ 0 & -\cos q_{35} & \sin q_{35} \end{bmatrix} \tag{4.42}$$

$$p_{0e}(q) = \begin{bmatrix} -l_5 \sin q_2 \cos q_{34} + l_4 \sin q_2 \sin q_3 \\ l_1 + q_1 + l_5 \cos q_2 \cos q_{34} - l_4 \cos q_2 \sin q_3 \\ l_2 + l_3 + l_5 \sin q_{34} + l_4 \cos q_3 \end{bmatrix} \tag{4.43}$$

To arrive at these expressions does, however, require quite a few computations, so we will show how we can simplify the computations by choosing the inertial frame more wisely.

Example 4.5 We can simplify the computations slightly by choosing the inertial reference frame $\mathscr{F}_{\tilde{0}}$ (where we have denoted the inertial reference frame by $\mathscr{F}_{\tilde{0}}$ to distinguish it from $\mathscr{F}_0$) so that it coincides with the end-effector frame for the home position, as shown in Fig. 4.8. In this case $g_{\tilde{0}e}(0)$ is trivially given by the identity matrix because $\mathscr{F}_{\tilde{0}}$ coincides with $\mathscr{F}_e$. We must, however, choose the twists differently as these are expressed in the inertial frame. If we formulate the problem in this way, the twists are given by

$$X_1^e = \begin{bmatrix} 0 \\ 0 \\ 1 \\ 0 \\ 0 \\ 0 \end{bmatrix}, \quad X_2^e = \begin{bmatrix} -\begin{bmatrix} 0 \\ -1 \\ 0 \end{bmatrix} \times \begin{bmatrix} 0 \\ l_3+l_4 \\ -l_5 \end{bmatrix} \\ \begin{bmatrix} 0 \\ -1 \\ 0 \end{bmatrix} \end{bmatrix} = \begin{bmatrix} -l_5 \\ 0 \\ 0 \\ 0 \\ -1 \\ 0 \end{bmatrix},$$

$$X_3^e = \begin{bmatrix} -\begin{bmatrix} 1 \\ 0 \\ 0 \end{bmatrix} \times \begin{bmatrix} 0 \\ l_4 \\ -l_5 \end{bmatrix} \\ \begin{bmatrix} 1 \\ 0 \\ 0 \end{bmatrix} \end{bmatrix} = \begin{bmatrix} 0 \\ -l_5 \\ -l_4 \\ 1 \\ 0 \\ 0 \end{bmatrix}, \tag{4.44}$$

$$X_4^e = \begin{bmatrix} -\begin{bmatrix} 1 \\ 0 \\ 0 \end{bmatrix} \times \begin{bmatrix} 0 \\ 0 \\ -l_5 \end{bmatrix} \\ \begin{bmatrix} 1 \\ 0 \\ 0 \end{bmatrix} \end{bmatrix} = \begin{bmatrix} 0 \\ -l_5 \\ 0 \\ 1 \\ 0 \\ 0 \end{bmatrix},$$

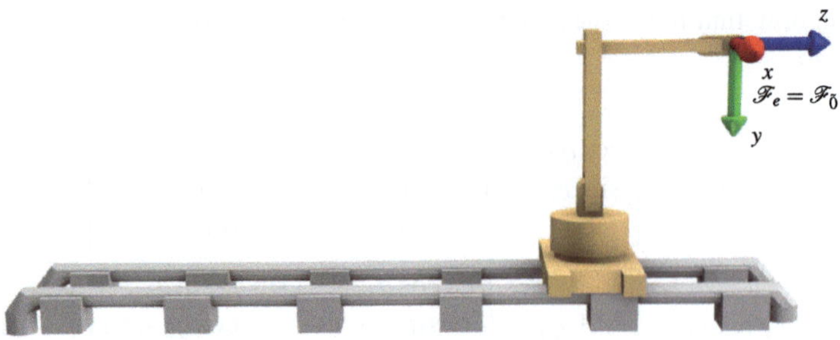

Fig. 4.8 A five degree of freedom manipulator with one translational and four rotational joints. The inertial frame is chosen so that it coincides with the end-effector frame at the home position

$$X_5^e = \begin{bmatrix} 0 \\ 0 \\ 0 \\ 1 \\ 0 \\ 0 \end{bmatrix}.$$

We note that if we choose the same direction as positive rotation as we did in Example 4.4 for the second joint, we need to use an axis of rotation $p_2 = [0\ -1\ 0]^{\mathsf{T}}$ instead of $p_2 = [0\ 1\ 0]^{\mathsf{T}}$ (recall that p_i is defined in $\mathscr{F}_{\tilde{0}}$). The exponential map of the individual joints are then slightly different from the ones we found in Example 4.4:

$$e^{\widehat{X}_1^e q_1} = \begin{bmatrix} 1 & 0 & 0 & 0 \\ 0 & 1 & 0 & q_1 \\ 0 & 0 & 1 & 0 \\ 0 & 0 & 0 & 1 \end{bmatrix},$$

$$e^{\widehat{X}_2^e q_2} = \begin{bmatrix} \cos q_2 & 0 & -\sin q_2 & -l_5 \sin q_2 \\ 0 & 1 & 0 & 0 \\ \sin q_2 & 0 & \cos q_2 & -l_5(1-\cos q_2) \\ 0 & 0 & 0 & 1 \end{bmatrix},$$

$$e^{\widehat{X}_3^e q_3} = \begin{bmatrix} 1 & 0 & 0 & 0 \\ 0 & \cos q_3 & -\sin q_3 & l_4(1-\cos q_3) - l_5 \sin q_3 \\ 0 & \sin q_3 & \cos q_3 & -l_5(1-\cos q_3) - l_4 \sin q_3 \\ 0 & 0 & 0 & 1 \end{bmatrix}, \tag{4.45}$$

$$e^{\widehat{X}_4^e q_4} = \begin{bmatrix} 1 & 0 & 0 & 0 \\ 0 & \cos q_4 & -\sin q_4 & -l_5 \sin q_4 \\ 0 & \sin q_4 & \cos q_4 & -l_5(1-\cos q_4) \\ 0 & 0 & 0 & 1 \end{bmatrix},$$

$$e^{\widehat{X}_5^e q_5} = \begin{bmatrix} 1 & 0 & 0 & 0 \\ 0 & \cos q_5 & -\sin q_5 & 0 \\ 0 & \sin q_5 & \cos q_5 & 0 \\ 0 & 0 & 0 & 1 \end{bmatrix}.$$

We can now find the forward kinematics by starting with the last joint: Because $g_{\tilde{0}e} = I$ we have $g_{\tilde{0}e}(q_5) = e^{\widehat{X}_5^e q_5}$. Continuing with the fourth joint we get

$$\begin{aligned} g_{\tilde{0}e}(q_4, q_5) &= e^{\widehat{X}_4^e q_4} g_{\tilde{0}e}(q_5) \\ &= \begin{bmatrix} 1 & 0 & 0 & 0 \\ 0 & \cos q_4 & -\sin q_4 & -l_5 \sin q_4 \\ 0 & \sin q_4 & \cos q_4 & -l_5(1-\cos q_4) \\ 0 & 0 & 0 & 1 \end{bmatrix} \begin{bmatrix} 1 & 0 & 0 & 0 \\ 0 & \cos q_5 & -\sin q_5 & 0 \\ 0 & \sin q_5 & \cos q_5 & 0 \\ 0 & 0 & 0 & 1 \end{bmatrix} \\ &= \begin{bmatrix} 1 & 0 & 0 & 0 \\ 0 & \cos q_{45} & -\sin q_{45} & -l_5 \sin q_4 \\ 0 & \sin q_{45} & \cos q_{45} & -l_5(1-\cos q_4) \\ 0 & 0 & 0 & 1 \end{bmatrix} \end{aligned} \tag{4.46}$$

where $q_{45} = q_4 + q_5$ as usual. This gives us the end-effector position and orientation with respect to the end-effector position and orientation at home position when only the last two joints move. We can include the action of the third joint in the same way:

$$\begin{aligned} & g_{\tilde{0}e}(q_3, q_4, q_5) \\ &= e^{\widehat{X}_3^e q_3} g_{\tilde{0}e}(q_4, q_5) \\ &= \begin{bmatrix} 1 & 0 & 0 & 0 \\ 0 & \cos q_3 & -\sin q_3 & l_4(1-\cos q_3) - l_5 \sin q_3 \\ 0 & \sin q_3 & \cos q_3 & l_5(\cos q_3 - 1) - l_4 \sin q_3 \\ 0 & 0 & 0 & 1 \end{bmatrix} \\ &\quad \times \begin{bmatrix} 1 & 0 & 0 & 0 \\ 0 & \cos q_{45} & -\sin q_{45} & -l_5 \sin q_4 \\ 0 & \sin q_{45} & \cos q_{45} & -l_5(1-\cos q_4) \\ 0 & 0 & 0 & 1 \end{bmatrix} \\ &= \begin{bmatrix} 1 & 0 & 0 & 0 \\ 0 & \cos q_{35} & -\sin q_{35} & \left[\begin{matrix} -l_5 \cos q_3 \sin q_4 + l_5 \sin q_3 (1-\cos q_4) \\ + l_4(1-\cos q_3) - l_5 \sin q_3 \end{matrix} \right] \\ 0 & \sin q_{35} & \cos q_{35} & \left[\begin{matrix} -l_5 \sin q_3 \sin q_4 - l_5 \cos q_3 (1-\cos q_4) \\ - l_5(1-\cos q_3) - l_4 \sin q_3 \end{matrix} \right] \\ 0 & 0 & 0 & 1 \end{bmatrix} \end{aligned}$$

$$
= \begin{bmatrix} 1 & 0 & 0 & 0 \\ 0 & \cos q_{35} & -\sin q_{35} & -l_5 \sin q_{34} + l_4(1-\cos q_3) \\ 0 & \sin q_{35} & \cos q_{35} & l_5 \cos q_{34} - l_5 - l_4 \sin q_3 \\ 0 & 0 & 0 & 1 \end{bmatrix}. \tag{4.47}
$$

The action of the second joint on the end-effector configuration is given by a rotation around the y-axis of the end-effector frame in the home position. Note that if the direction of rotation is chosen as in the previous examples, this is a negative rotation, so

$$
\begin{aligned}
& g_{\tilde{0}e}(q_2, q_3, q_4, q_5) \\
&\quad = e^{\widehat{X}^e_2 q_2} g_{\tilde{0}e}(q_3, q_4, q_5) \\
&\quad = \begin{bmatrix} \cos q_2 & 0 & -\sin q_2 & -l_5 \sin q_2 \\ 0 & 1 & 0 & 0 \\ \sin q_2 & 0 & \cos q_2 & l_5(\cos q_2 - 1) \\ 0 & 0 & 0 & 1 \end{bmatrix} \\
&\quad \times \begin{bmatrix} 1 & 0 & 0 & 0 \\ 0 & \cos q_{35} & -\sin q_{35} & -l_5 \sin q_{34} + l_4(1-\cos q_3) \\ 0 & \sin q_{35} & \cos q_{35} & -l_4 \sin q_3 - l_5(1 - \cos q_{34}) \\ 0 & 0 & 0 & 1 \end{bmatrix} \\
&\quad = \begin{bmatrix} \cos q_2 & -\sin q_2 \sin q_{35} & -\sin q_2 \cos q_{35} & -l_5 \sin q_2 \cos q_{34} + l_4 \sin q_2 \sin q_3 \\ 0 & \cos q_{35} & -\sin q_{35} & -l_5 \sin q_{34} + l_4(1-\cos q_3) \\ \sin q_2 & \cos q_2 \sin q_{35} & \cos q_2 \cos q_{35} & -l_4 \cos q_2 \sin q_3 + l_5 \cos q_2 \cos q_{34} - l_5 \\ 0 & 0 & 0 & 1 \end{bmatrix}.
\end{aligned} \tag{4.48}
$$

Finally, if we include the first joint, which is a pure linear motion, the end-effector orientation can be written as above, and we add the linear motion in the direction of the z-axis which gives

$$
\begin{aligned}
& g_{\tilde{0}e}(q) \\
&\quad = \begin{bmatrix} \cos q_2 & -\sin q_2 \sin q_{35} & -\sin q_2 \cos q_{35} & -l_5 \sin q_2 \cos q_{34} + l_4 \sin q_2 \sin q_3 \\ 0 & \cos q_{35} & -\sin q_{35} & -l_5 \sin q_{34} + l_4(1-\cos q_3) \\ \sin q_2 & \cos q_2 \sin q_{35} & \cos q_2 \cos q_{35} & q_1 + l_4 \cos q_2 \sin q_3 + l_5 \cos q_2 \cos q_{34} - l_5 \\ 0 & 0 & 0 & 1 \end{bmatrix}.
\end{aligned} \tag{4.49}
$$

$g_{\tilde{0}e}(q)$ gives us the end-effector configuration when the robot joints are at position q with respect to the end-effector frame at $q = 0$. In Examples 4.3 and 4.4 we found the end-effector configuration with respect to the inertial frame $\mathscr{F}_0$. We can show that $g_{0e}(q)$ in Examples 4.3 and 4.4 represents the same configuration as $g_{\tilde{0}e}(q)$ here. We show this by transforming $g_{\tilde{0}e}(q)$ back to the inertial frame $\mathscr{F}_0$:

$$
g_{0e}(q) = g_{0\tilde{0}} g_{\tilde{0}e}(q). \tag{4.50}
$$

where

$$g_{0\tilde{0}} = \begin{bmatrix} 1 & 0 & 0 & 0 \\ 0 & 0 & 1 & l_1 + l_5 \\ 0 & -1 & 0 & l_2 + l_3 + l_4 \\ 0 & 0 & 0 & 1 \end{bmatrix} \tag{4.51}$$

which gives

$$R_{0e}(q) = \begin{bmatrix} \cos q_2 & -\sin q_2 \sin q_{35} & -\sin q_2 \cos q_{35} \\ \sin q_2 & \cos q_2 \sin q_{35} & \cos q_2 \cos q_{35} \\ 0 & -\cos q_{35} & \sin q_{35} \end{bmatrix} \tag{4.52}$$

$$p_{0e}(q) = \begin{bmatrix} -l_5 \sin q_2 \cos q_{34} + l_4 \sin q_2 \sin q_3 \\ l_1 + q_1 - l_4 \cos q_2 \sin q_3 + l_5 \cos q_2 \cos q_{34} \\ l_2 + l_3 + l_5 \sin q_{34} + l_4 \cos q_3 \end{bmatrix}. \tag{4.53}$$

Now, if we take the expression in Example 4.3, and use that $\cos(x - \frac{\pi}{2}) = \sin(x)$ and $\sin(x - \frac{\pi}{2}) = -\cos(x)$ to substitute $-\cos(q_{45}) \Leftarrow \sin(q_{45\pi})$ and $\sin(q_{45}) \Leftarrow \cos(q_{45\pi})$ the transformation (4.15)–(4.16) can be written as

$$\begin{aligned} R_{0e}(q) &= \begin{bmatrix} \cos q_2 & -\sin q_2 \cos q_{35\pi} & \sin q_2 \sin q_{35\pi} \\ \sin q_2 & \cos q_2 \cos q_{35\pi} & -\cos q_2 \sin q_{35\pi} \\ 0 & \sin q_{35\pi} & \cos q_{35\pi} \end{bmatrix} \\ &= \begin{bmatrix} \cos q_2 & -\sin q_2 \sin q_{35} & -\sin q_2 \cos q_{35} \\ \sin q_2 & \cos q_2 \sin q_{35} & \cos q_2 \cos q_{35} \\ 0 & -\cos q_{35} & \sin q_{35} \end{bmatrix} \end{aligned} \tag{4.54}$$

$$\begin{aligned} p_{0e}(q) &= \begin{bmatrix} l_5 \sin q_2 \sin q_{34\pi} + l_4 \sin q_2 \sin q_3 \\ l_1 + q_1 - l_5 \cos q_2 \sin q_{34\pi} - l_4 \cos q_2 \sin q_3 \\ l_2 + l_3 + l_5 \cos q_{34\pi} + l_4 \cos q_3 \end{bmatrix} \\ &= \begin{bmatrix} -l_5 \sin q_2 \cos q_{34} + l_4 \sin q_2 \sin q_3 \\ l_1 + q_1 + l_5 \cos q_2 \cos q_{34} - l_4 \cos q_2 \sin q_3 \\ l_2 + l_3 + l_5 \sin q_{34} + l_4 \cos q_3 \end{bmatrix} \end{aligned} \tag{4.55}$$

and we have shown that we get the same result with the product of exponentials formula as we did with multiplications of the homogeneous transformation matrices in Example 4.3.

4.3 Manipulator Velocities

We have seen that the manipulator twists define the admissible velocities of each joint defined in spatial or body coordinates, or in the frame attached to the joint itself. By collecting the effects of all the joints we can use the manipulator twists to describe the velocities of all the links in the mechanism and in particular the end

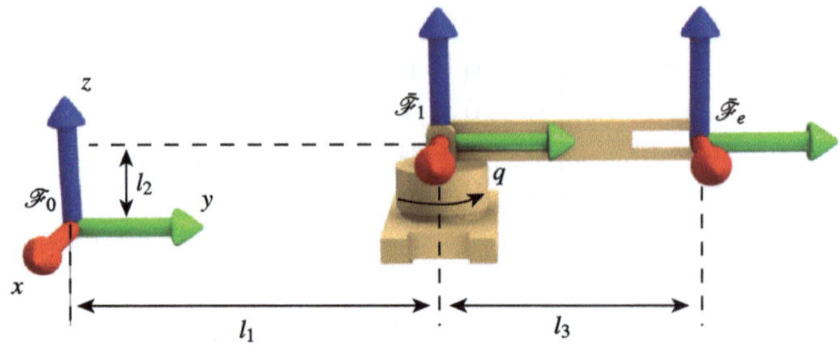

Fig. 4.9 The 1-DoF manipulator with one rotational joint described in Example 4.6

effector. One of the most important problems in robotics is to find the end-effector velocity from the joint velocities. This is given by the Jacobian map and is discussed in detail in the next section. We will start by looking at the different representations of this velocity.

In the same way that the manipulator twists can be described in different ways, the end-effector velocities can be represented in spatial or body coordinates. Firstly, the body velocity V_{0e}^B describes the end-effector velocity as seen from the end-effector frame $\mathcal{F}_e$. The linear part v_{0e}^B thus describes the velocity of the frame $\mathcal{F}_e$ attached to the end-effector with respect to the inertial frame, as seen from the frame $\mathcal{F}_e$. We note that we need to choose the instantaneous body frame $\mathcal{F}_e$, i.e, the frame $\mathcal{F}_e$ at the current configuration. Similarly, the rotational part ω_{0e}^B gives us the angular velocities of the body frame $\mathcal{F}_e$ in the inertial frame $\mathcal{F}_0$, also as seen from the instantaneous end-effector frame $\mathcal{F}_e$.

We can also describe the velocity in terms of spatial coordinates. The rotational part of the spatial velocity ω_{0e}^S has the straight forward interpretation as the angular velocity of the end-effector frame $\mathcal{F}_e$ with respect to the inertial frame $\mathcal{F}_0$, represented in the inertial frame $\mathcal{F}_0$. Unfortunately, the interpretation of the linear part of the spatial velocity twist v_{0e}^S is not as intuitive as this is not simply the linear velocity of $\mathcal{F}_e$ in $\mathcal{F}_0$. The geometric interpretation of the spatial linear velocity is rather different: the linear part of the spatial twist describes the instantaneous velocity of a point moving through the origin of $\mathcal{F}_0$. This can be a point attached to the end effector or an imaginary point moving with the end effector. We will show the difference through a simple example (Murray et al. 1994):

Example 4.6 We will now find the spatial and body velocity of a one-link manipulator with a rotational joint as seen in Fig. 4.9. The transformation from the inertial frame to the end-effector frame is found in the normal way as

$$g_{0e}(t) = \begin{bmatrix} \cos q & -\sin q & 0 & -l_3 \sin q \\ \sin q & \cos q & 0 & l_1 + l_3 \cos q \\ 0 & 0 & 1 & l_2 \\ 0 & 0 & 0 & 1 \end{bmatrix}. \tag{4.56}$$

The velocity of the rigid body can be described in terms of body or spatial coordinates. In body coordinates the velocity is given as (3.76)

$$V_{0e}^{B} = \begin{bmatrix} v_{0e}^{B} \\ \omega_{0e}^{B} \end{bmatrix} = \begin{bmatrix} R_{0e}^{\mathsf{T}} \dot{p}_{0e} \\ (R_{0e}^{\mathsf{T}} \dot{R}_{0e})^{\vee} \end{bmatrix} = \begin{bmatrix} -l_3\dot{q} \\ 0 \\ 0 \\ 0 \\ 0 \\ \dot{q} \end{bmatrix}. \tag{4.57}$$

If we imagine that we are attached to the end-effector frame $\mathscr{F}_e$, a positive rotation q will result in a linear velocity in the direction of the negative x-axis of the body frame. The angular velocity ω_{0e}^{B} similarly describes a rotational motion around the z-axis of the body frame. We see that the body velocity twist V_{0e}^{B} gives us the velocity of the end-effector with respect to the inertial frame as seen from the end-effector frame, as expected.

The spatial velocity is given by (3.74)

$$V_{0e}^{S} = \begin{bmatrix} v_{0e}^{S} \\ \omega_{0e}^{S} \end{bmatrix} = \begin{bmatrix} -\dot{R}_{0e} R_{0e}^{\mathsf{T}} p_{0e} + \dot{p}_{0e} \\ (\dot{R}_{0e} R_{0e}^{\mathsf{T}})^{\vee} \end{bmatrix} = \begin{bmatrix} l_1\dot{q} \\ 0 \\ 0 \\ 0 \\ 0 \\ \dot{q} \end{bmatrix}. \tag{4.58}$$

The spatial angular velocity gives us the instantaneous angular velocity as seen from the inertial frame which for pure angular motion in the plane is identical to the body velocity. The geometric interpretation of the spatial linear velocity, however, is not the linear velocity of the end-effector frame as seen from the inertial frame. It is rather the velocity of an imaginary point situated at the origin of $\mathscr{F}_0$, but moving with the end-effector, or possibly a rigid body extension of the end effector. The velocity of this point is always in the direction of the positive x-axis of $\mathscr{F}_0$ for a positive q and the velocity is given by $l_1\dot{q}$.

We can, however, also find the velocity of the end-effector frame with respect to the inertial frame as seen from the inertial frame $\mathscr{F}_0$. Note that this is not the same as the spatial velocity. We will denote this velocity the *inertial velocity twist* V_{0e}^{0} and write

$$\begin{aligned} V_{0e}^{0} &= \begin{bmatrix} R_{0e} & 0 \\ 0 & R_{0e} \end{bmatrix} \begin{bmatrix} v_{0e}^{B} \\ \omega_{0e}^{B} \end{bmatrix} = \begin{bmatrix} R_{0e} R_{0e}^{\mathsf{T}} \dot{p}_{0e} \\ R_{0e}(R_{0e}^{\mathsf{T}} \dot{R}_{0e})^{\vee} \end{bmatrix} \\ &= \begin{bmatrix} \dot{p}_{0e} \\ R_{0e}(R_{0e}^{\mathsf{T}} \dot{R}_{0e})^{\vee} \end{bmatrix} = \begin{bmatrix} -l_3\dot{q}\cos q \\ -l_3\dot{q}\sin q \\ 0 \\ 0 \\ 0 \\ \dot{q} \end{bmatrix}. \end{aligned} \tag{4.59}$$

This is the linear and angular velocity of the end-effector frame $\mathscr{F}_e$ with respect to the inertial frame $\mathscr{F}_0$ for an observer located in the fixed inertial frame $\mathscr{F}_0$. The geometric interpretation of the velocity V_{0e}^0 is thus somewhat more intuitive than the spatial velocity V_{0e}^S. Note, however, that this intuitive interpretation comes at the cost of valuable information: if we use V_{0e}^0 instead of V_{0e}^S we see that we lose the information describing the location (translational part) of the frame $\bar{\mathscr{F}}_1$ with respect to the reference frame $\mathscr{F}_0$ and also the nice interpretation as an element of the Lie algebra as described above.

Example 4.7 We have seen that we can also use the body and spatial velocity twists to find the velocity of a point. Consider once again the 1-DoF manipulator in Fig. 4.9. The velocity of a point $q^e = [0\,0\,0]^\mathrm{T}$ in the body frame $\mathscr{F}_e$ (in this case $\mathscr{O}_e$) is given by

$$\bar{v}_{q^e} = \hat{V}_{0e}^B \bar{q}^e = \begin{bmatrix} 0 & -\dot{q} & 0 & -l_3\dot{q} \\ \dot{q} & 0 & 0 & 0 \\ 0 & 0 & 0 & 0 \\ 0 & 0 & 0 & 0 \end{bmatrix} \begin{bmatrix} 0 \\ 0 \\ 0 \\ 1 \end{bmatrix} = \begin{bmatrix} -l_3\dot{q} \\ 0 \\ 0 \\ 0 \end{bmatrix}. \tag{4.60}$$

We can also find the spatial velocity of the same point

$$q^0 = \begin{bmatrix} -l_3 \sin q \\ l_1 + l_3 \cos q \\ l_2 \end{bmatrix} \tag{4.61}$$

represented in $\mathscr{F}_0$ by

$$\begin{aligned} \bar{v}_{q^0} = \hat{V}_{0e}^S \bar{q}^0 &= \begin{bmatrix} 0 & -\dot{q} & 0 & l_1\dot{q} \\ \dot{q} & 0 & 0 & 0 \\ 0 & 0 & 0 & 0 \\ 0 & 0 & 0 & 0 \end{bmatrix} \begin{bmatrix} -l_3 \sin q \\ l_1 + l_3 \cos q \\ l_2 \\ 1 \end{bmatrix} \\ &= \begin{bmatrix} -l_3\dot{q}\cos q - l_1\dot{q} + l_1\dot{q} \\ -l_3\dot{q}\sin q \\ 0 \\ 0 \end{bmatrix} = \begin{bmatrix} -l_3\dot{q}\cos q \\ -l_3\dot{q}\sin q \\ 0 \\ 0 \end{bmatrix} \end{aligned} \tag{4.62}$$

and we see that the velocity of a point in spatial coordinates has a more intuitive geometric interpretation than the spatial velocity twist in the previous example.

We see that the spatial velocity depends on the choice of inertial frame $\mathscr{F}_0$. The linear part of the spatial velocity describes the velocity of an imaginary point at the origin of the inertial frame $\mathscr{F}_0$, which is obtained by including the upper right part of the Adjoint map:

$$\mathrm{Ad}_g = \begin{bmatrix} R & \hat{p}R \\ 0 & R \end{bmatrix}. \tag{4.63}$$

Therefore, if we want to find the spatial velocity with respect to a different frame we also need to use the Adjoint map (and not the transformation in (4.59)) so that we describe the velocity of a point traveling through the origin of this new frame. Let for example V_{jk}^S be the spatial velocity with respect to the frame $\mathscr{F}_j$. Then the spatial velocity with respect to a different frame $\mathscr{F}_i$ which relates to $\mathscr{F}_j$ by g_{ij} is given by $V_{ik}^S = \mathrm{Ad}_{g_{ij}} V_{jk}^S$. If g_{ij} is time-varying, we also need to include the velocity of frame $\mathscr{F}_j$ with respect to $\mathscr{F}_i$ and we get $V_{ik}^S = V_{ij}^S + \mathrm{Ad}_{g_{ij}} V_{jk}^S$.

The Adjoint map is thus used to find the relation between the body and the spatial velocities of a rigid body and to find the relation between the spatial velocities of a rigid body described with respect to different coordinate frames.

4.4 Manipulator Jacobian

In Sect. 4.1 we found a mapping from the joint positions to the end-effector position. Another useful relation is the mapping from the joint velocities to the end-effector velocity. It is often easier to work in the velocity space, particularly when resolving the inverse problem, i.e., it is in most cases easier to find the joint velocities from the end-effector velocity than it is to find the joint positions from the end-effector position and orientation. There are two approaches to find this relation in the velocity space, either by the geometric or the analytical Jacobian.

Firstly, the geometric Jacobian maps a set of joint velocities to the end-effector velocities in the form $V_{0e}^B = [(v_{0e}^B)^\mathsf{T}\ (\omega_{0e}^B)^\mathsf{T}]^\mathsf{T} \in \mathbb{R}^6$. This mapping is given by

$$V_{0e}^B = J_{m,g}(q)\dot{q} \tag{4.64}$$

and maps the joint velocities to the end-effector twist expressed in the end-effector frame, i.e., the body velocity. We will denote this Jacobian the body geometric Jacobian. Similarly we can find a mapping from the joint velocities to the spatial end-effector velocity twist by the spatial geometric Jacobian as

$$V_{0e}^S = J_{m,g}^S(q)\dot{q}. \tag{4.65}$$

The geometric Jacobian is often referred to as the manipulator Jacobian in the robotics literature. We will not use this term as we will sometimes use the term manipulator Jacobian to make it clear that we talk about the Jacobian of the manipulator and not the whole system (including the vehicle).

Secondly, the analytical Jacobian is the Jacobian of the forward kinematics map $f : \mathbb{R}^n \to \mathbb{R}^6$ given by $J_{m,a} = \frac{\partial f}{\partial q}$. $J_{m,a}$ gives the relation

$$\dot{\eta}_{0e} = J_{m,a}(q)\dot{q} \tag{4.66}$$

where η_{0e} represents the position and orientation of the end effector in the inertial frame.

The Jacobian mappings are important tools in robotics and will be frequently used in the remainder of the book. It is therefore important to distinguish the different Jacobians from each other and to know the properties associated with each of them. We will now look at how we can derive the different Jacobians and the relationships between them, and we will study their structural properties in detail.

4.4.1 Body Geometric Jacobian

The velocity of the end effector expressed in the end-effector frame, often referred to as the body velocity in the robotics literature, is given by

$$\widehat{V}_{0e}^{B} = g_{0e}^{-1}(q)\dot{g}_{0e}(q) \tag{4.67}$$

and depends on the joint positions $q = [q_1\ q_2\ \cdots\ q_n]^{\mathsf{T}}$. The time derivative of $g_{0e}(q)$ is given by the chain rule as

$$\dot{g}_{0e}(q) = \sum_{i=1}^{n} \left(\frac{\partial g_{0e}}{\partial q_i} \dot{q}_i \right). \tag{4.68}$$

If we recall that, from Definition 2.15, the vee-map allows us to write a matrix $\widehat{V} \in se(3)$ (twist) as a vector V (twist coordinates), Eq. (4.67) can be written as (Duindam 2006; Stramigioli 2001)

$$\begin{aligned} V_{0e}^{B} &= \left(g_{0e}^{-1} \frac{\partial g_{0e}}{\partial q_1} \dot{q}_1 \right)^{\vee} + \left(g_{0e}^{-1} \frac{\partial g_{0e}}{\partial q_2} \dot{q}_2 \right)^{\vee} + \cdots + \left(g_{0e}^{-1} \frac{\partial g_{0e}}{\partial q_n} \dot{q}_n \right)^{\vee} \\ &= \underbrace{\left[\left(g_{0e}^{-1} \frac{\partial g_{0e}}{\partial q_1} \right)^{\vee} \ \left(g_{0e}^{-1} \frac{\partial g_{0e}}{\partial q_2} \right)^{\vee} \ \cdots \ \left(g_{0e}^{-1} \frac{\partial g_{0e}}{\partial q_n} \right)^{\vee} \right]}_{J_{m,g}} \begin{bmatrix} \dot{q}_1 \\ \dot{q}_2 \\ \vdots \\ \dot{q}_n \end{bmatrix} \\ &= J_{m,g}\dot{q}, \end{aligned} \tag{4.69}$$

where the matrix on the left is called the *body geometric Jacobian*. It turns out that the column vectors of the body geometric Jacobian have a very nice geometric interpretation. To see this we write each column of the Jacobian in matrix form, which from Murray et al. (1994) and Theorem 4.1 gives

$$
\begin{aligned}
g_{0e}^{-1}\frac{\partial g_{0e}}{\partial q_i} &= g_{0e}^{-1} e^{\widehat{X}_1 q_1} \cdots e^{\widehat{X}_{i-1} q_{i-1}} \frac{\partial}{\partial q_i}\big(e^{\widehat{X}_i q_i}\big) e^{\widehat{X}_{i+1} q_{i+1}} \cdots e^{\widehat{X}_n q_n} g_{0e}(0) \\
&= g_{0e}^{-1} e^{\widehat{X}_1 q_1} \cdots e^{\widehat{X}_{i-1} q_{i-1}} \big(\widehat{X}_i e^{\widehat{X}_i q_i}\big) e^{\widehat{X}_{i+1} q_{i+1}} \cdots e^{\widehat{X}_n q_n} g_{0e}(0) \\
&= \underbrace{g_{0e}^{-1}(0) e^{-\widehat{X}_n q_n} \cdots e^{-\widehat{X}_1 q_1}}_{g_{0e}^{-1}} e^{\widehat{X}_1 q_1} \cdots e^{\widehat{X}_{i-1} q_{i-1}} \big(\widehat{X}_i e^{\widehat{X}_i q_i}\big) \\
&\quad \times e^{\widehat{X}_{i+1} q_{i+1}} \cdots e^{\widehat{X}_n q_n} g_{0e}(0) \\
&= g_{0e}^{-1}(0) e^{-\widehat{X}_n q_n} \cdots e^{-\widehat{X}_i q_i} \widehat{X}_i e^{\widehat{X}_i q_i} e^{\widehat{X}_{i+1} q_{i+1}} \cdots e^{\widehat{X}_n q_n} g_{0e}(0),
\end{aligned}
\tag{4.70}
$$

where we have used that

$$
g_{0e} = g_{0e}(q) = e^{\widehat{X}_1 q_1} \cdots e^{\widehat{X}_n q_n} g_{0e}(0), \tag{4.71}
$$

which gives

$$
g_{0e}^{-1}(q) = g_{0e}^{-1}(0) e^{-\widehat{X}_n q_n} \cdots e^{-\widehat{X}_1 q_1}. \tag{4.72}
$$

The twists in (4.70) are thus given by conjugations in the form of (2.98) and we have

$$
Ad_{g^{-1}} \widehat{X}_i = g^{-1} \widehat{X}_i (g^{-1})^{-1} = g^{-1} \widehat{X}_i g \tag{4.73}
$$

where

$$
g = e^{\widehat{X}_i q_i} e^{\widehat{X}_{i+1} q_{i+1}} \cdots e^{\widehat{X}_n q_n} g_{0e}(0). \tag{4.74}
$$

The geometric interpretation of the twists in the body geometric Jacobian is obtained by dividing the actions of g in (4.74) in two steps. Firstly, $g_{0e}(0)$ transforms the spatial joint twist X_i to the end-effector joint twist X_i^e through the adjoint map $\mathrm{Ad}_{g_{0e}(0)}^{-1}$ at home position. Secondly, the terms $e^{\widehat{X}_i q_i} e^{\widehat{X}_{i+1} q_{i+1}} \cdots e^{\widehat{X}_n q_n}$ takes us from the end-effector joint twist in home position (X_i^e) to the end-effector twist at the current configuration ($X_i^\dagger(q)$). We note that because the twist is written in the end-effector frame this is only affected by joints $(i, i+1, \ldots, n)$.

The columns of the body geometric Jacobian thus correspond to the joint twists written with respect to the tool frame at configuration q. We denote by $X_i^\dagger$ the joint twist as seen from $\mathscr{F}_e$ at configuration q given by

$$
X_i^\dagger = \mathrm{Ad}^{-1}_{e^{\widehat{X}_i q_i} \cdots e^{\widehat{X}_n q_n} g_{0e}(0)} X_i \tag{4.75}
$$

or alternatively

$$
X_i^\dagger = \mathrm{Ad}^{-1}_{g_{\bar{i}e}(q)} X_i^i = \mathrm{Ad}_{g_{\bar{i}e}^{-1}(q)} X_i^i. \tag{4.76}
$$

The definition and geometrical interpretation of the body geometric Jacobian can now be given as follows:

Theorem 4.2 *The body geometric Jacobian $J_{m,g}(q)$ maps the joint velocities to the body velocity twist expressed in the end-effector frame $\mathscr{F}_e$ by*

$$V_{0e}^B = J_{m,g}(q)\dot{q} \tag{4.77}$$

and is defined as

$$J_{m,g}(q) = \begin{bmatrix} X_1^\dagger & X_2^\dagger & \cdots & X_n^\dagger \end{bmatrix} \tag{4.78}$$

$$= \left[\left(g_{0e}^{-1} \frac{\partial g_{0e}}{\partial q_1} \right)^\vee \quad \left(g_{0e}^{-1} \frac{\partial g_{0e}}{\partial q_2} \right)^\vee \quad \cdots \quad \left(g_{0e}^{-1} \frac{\partial g_{0e}}{\partial q_n} \right)^\vee \right] \tag{4.79}$$

where the column vectors $(g_{0e}^{-1} \frac{\partial g_{0e}}{\partial q_i})^\vee$ are the joint twists $X_i^\dagger(q)$ as observed from the end-effector frame $\mathscr{F}_e$ at configuration q.

Example 4.8 Consider the manipulator structure in Fig. 4.7. We will now find the body geometric Jacobian by Eq. (4.76). The body geometric Jacobian is then given by

$$\begin{aligned} J_{m,g}(q) &= \begin{bmatrix} X_1^\dagger & X_2^\dagger & X_3^\dagger & X_4^\dagger & X_5^\dagger \end{bmatrix} \\ &= \begin{bmatrix} \mathrm{Ad}_{g_{\bar{1}e}^{-1}} X_1^1 & \mathrm{Ad}_{g_{\bar{2}e}^{-1}} X_2^2 & \mathrm{Ad}_{g_{\bar{3}e}^{-1}} X_3^3 & \mathrm{Ad}_{g_{\bar{4}e}^{-1}} X_4^4 & \mathrm{Ad}_{g_{\bar{5}e}^{-1}} X_5^5 \end{bmatrix}. \end{aligned} \tag{4.80}$$

Here $g_{\bar{i}e}$ denotes the transformation from frame $\bar{\mathscr{F}}_i^0$ to the end-effector frame $\mathscr{F}_e$. We use that the Adjoint transformations can be written as

$$\mathrm{Ad}_{g_{e\bar{i}}} = \mathrm{Ad}_{g_{\bar{i}e}^{-1}} = \begin{bmatrix} R_{\bar{i}e}^\mathsf{T} & -R_{\bar{i}e}^\mathsf{T} \hat{p}_{\bar{i}e} \\ 0 & R_{\bar{i}e}^\mathsf{T} \end{bmatrix} \tag{4.81}$$

and find each column of the body geometric Jacobian as

$$\mathrm{Ad}_{g_{e\bar{1}}} X_1^1 = \begin{bmatrix} cq_2 & sq_2 & 0 & * & * & * \\ -sq_2 sq_{35} & cq_2 sq_{35} & -cq_{35} & * & * & * \\ -sq_2 cq_{35} & cq_2 cq_{35} & sq_{35} & * & * & * \\ 0 & 0 & 0 & * & * & * \\ 0 & 0 & 0 & * & * & * \\ 0 & 0 & 0 & * & * & * \end{bmatrix} \begin{bmatrix} 0 \\ 1 \\ 0 \\ 0 \\ 0 \\ 0 \end{bmatrix} = \begin{bmatrix} sq_2 \\ cq_2 cq_{35} \\ -cq_2 sq_{35} \\ 0 \\ 0 \\ 0 \end{bmatrix},$$

$$
\begin{aligned}
\mathrm{Ad}_{g_{e\bar{2}}} X_2^2 &= \begin{bmatrix} cq_2 & sq_2 & 0 & * & * & -l_5cq_{34}+l_4sq_3 \\ -sq_2sq_{35} & cq_2sq_{35} & -cq_{35} & * & * & 0 \\ -sq_2cq_{35} & cq_2cq_{35} & sq_{35} & * & * & 0 \\ 0 & 0 & 0 & * & * & 0 \\ 0 & 0 & 0 & * & * & -cq_{35} \\ 0 & 0 & 0 & * & * & sq_{35} \end{bmatrix} \begin{bmatrix} 0 \\ 0 \\ 0 \\ 0 \\ 0 \\ 1 \end{bmatrix} \\
&= \begin{bmatrix} l_5cq_{34}+l_4sq_3 \\ 0 \\ 0 \\ 0 \\ -cq_{35} \\ sq_{35} \end{bmatrix}, \\
\mathrm{Ad}_{g_{e\bar{3}}} X_3^3 &= \begin{bmatrix} 1 & 0 & 0 & 0 & * & * \\ 0 & sq_{35} & -cq_{35} & -l_5cq_5-l_4sq_{45} & * & * \\ 0 & cq_{35} & sq_{35} & l_5sq_5-l_4cq_{45} & * & * \\ 0 & 0 & 0 & 1 & * & * \\ 0 & 0 & 0 & 0 & * & * \\ 0 & 0 & 0 & 0 & * & * \end{bmatrix} \begin{bmatrix} 0 \\ 0 \\ 0 \\ 1 \\ 0 \\ 0 \end{bmatrix} \\
&= \begin{bmatrix} 0 \\ -l_5cq_5-l_4sq_{45} \\ l_5sq_5-l_4cq_{45} \\ 1 \\ 0 \\ 0 \end{bmatrix}, \\
\mathrm{Ad}_{g_{e\bar{4}}} X_4^4 &= \begin{bmatrix} 1 & 0 & 0 & 0 & l_5\cos q_4 & l_5\sin q_4 \\ 0 & \cos q_{45} & \sin q_{45} & -l_5\cos q_5 & 0 & 0 \\ 0 & -\sin q_{45} & \cos q_{45} & l_5\sin q_5 & 0 & 0 \\ 0 & 0 & 0 & 1 & 0 & 0 \\ 0 & 0 & 0 & 0 & \cos q_{45} & \sin q_{45} \\ 0 & 0 & 0 & 0 & -\sin q_{45} & \cos q_{45} \end{bmatrix} \begin{bmatrix} 0 \\ 0 \\ 0 \\ 1 \\ 0 \\ 0 \end{bmatrix} \\
&= \begin{bmatrix} 0 \\ -l_5\cos q_5 \\ l_5\sin q_5 \\ 1 \\ 0 \\ 0 \end{bmatrix}, \\
\mathrm{Ad}_{g_{e\bar{5}}} X_5^5 &= \begin{bmatrix} 1 & 0 & 0 & 0 & 0 & 0 \\ 0 & \cos q_5 & \sin q_5 & 0 & 0 & 0 \\ 0 & -\sin q_5 & \cos q_5 & 0 & 0 & 0 \\ 0 & 0 & 0 & 1 & 0 & 0 \\ 0 & 0 & 0 & 0 & \cos q_5 & \sin q_5 \\ 0 & 0 & 0 & 0 & -\sin q_5 & \cos q_5 \end{bmatrix} \begin{bmatrix} 0 \\ 0 \\ 0 \\ 1 \\ 0 \\ 0 \end{bmatrix} = \begin{bmatrix} 0 \\ 0 \\ 0 \\ 1 \\ 0 \\ 0 \end{bmatrix}
\end{aligned}
\tag{4.82}
$$

and the body Jacobian is given by

$$J_{m,g}(q) = \begin{bmatrix} \mathrm{Ad}_{g_{\bar{1}e}^{-1}} X_1^1 & \mathrm{Ad}_{g_{\bar{2}e}^{-1}} X_2^2 & \mathrm{Ad}_{g_{\bar{3}e}^{-1}} X_3^3 & \mathrm{Ad}_{g_{\bar{4}e}^{-1}} X_4^4 & \mathrm{Ad}_{g_{\bar{5}e}^{-1}} X_5^5 \end{bmatrix}$$

$$= \begin{bmatrix} \sin q_2 & -l_5 c q_{34} + l_4 \sin q_3 & 0 & 0 & 0 \\ \cos q_2 \sin q_{35} & 0 & -l_5 \cos q_5 - l_4 \sin q_{45} & -l_5 \cos q_5 & 0 \\ \cos q_2 \cos q_{35} & 0 & l_5 \sin q_5 - l_4 \cos q_{45} & l_5 \sin q_5 & 0 \\ 0 & 0 & 1 & 1 & 1 \\ 0 & -\cos q_{35} & 0 & 0 & 0 \\ 0 & \sin q_{35} & 0 & 0 & 0 \end{bmatrix}. \tag{4.83}$$

Example 4.9 Consider once again the manipulator structure in Fig. 4.8. In Example 4.5 we found the manipulator twists X_i^e expressed in the end-effector frame $\mathscr{F}_e$. We can now find the body geometric Jacobian in terms of these twists by first writing the twists in $\mathscr{F}_e$ and associate this pose with $q = 0$, and then transforming the body joint twists to the current robot configuration. We then write

$$\begin{aligned} J_{m,g}(q) &= \begin{bmatrix} X_1^\dagger & X_2^\dagger & X_3^\dagger & X_4^\dagger & X_5^\dagger \end{bmatrix} \\ &= \Big[\mathrm{Ad}^{-1}_{e^{\hat{X}_1^e q_1} \cdots e^{\hat{X}_5^e q_5}} X_1^e \quad \mathrm{Ad}^{-1}_{e^{\hat{X}_2^e q_2} \cdots e^{\hat{X}_5^e q_5}} X_2^e \quad \mathrm{Ad}^{-1}_{e^{\hat{X}_3^e q_3} \cdots e^{\hat{X}_5^e q_5}} X_3^e \\ &\qquad \mathrm{Ad}^{-1}_{e^{\hat{X}_4^e q_4} e^{\hat{X}_5^e q_5}} X_4^e \quad \mathrm{Ad}^{-1}_{e^{\hat{X}_5^e q_5}} X_5^e \Big] \end{aligned} \tag{4.84}$$

where we have chosen $g_{0e}(0) = I$, which means that we assume that the inertial frame $\mathscr{F}_0$ coincides with the end-effector frame $\mathscr{F}_e$ for $q = 0$. The transformation due to each joint is found from (4.45) in Example 4.5. We start with the last joint, which gives $X_5^\dagger = X_5^e$. To find the contribution to the end-effector velocity for the fourth joint, we first calculate

$$e^{\hat{X}_4^e q_4} e^{\hat{X}_5^e q_5} = \begin{bmatrix} 1 & 0 & 0 & 0 \\ 0 & \cos q_{45} & -\sin q_{45} & -l_5 \sin q_4 \\ 0 & \sin q_{45} & \cos q_{45} & -l_5(1 - \cos q_4) \\ 0 & 0 & 0 & 1 \end{bmatrix} \tag{4.85}$$

which gives the Adjoint transformation

$$\mathrm{Ad}^{-1}_{e^{\hat{X}_4^e q_4} e^{\hat{X}_5^e q_5}}$$

$$= \begin{bmatrix} 1 & 0 & 0 & 0 & -l_5(1-\cos q_4) & l_5 \sin q_4 \\ 0 & \cos q_{45} & \sin q_{45} & l_5 \cos q_{45} - l_5 \cos q_5 & 0 & 0 \\ 0 & -\sin q_{45} & \cos q_{45} & -l_5 \sin q_{45} + l_5 \sin q_5 & 0 & 0 \\ 0 & 0 & 0 & 1 & 0 & 0 \\ 0 & 0 & 0 & 0 & \cos q_{45} & \sin q_{45} \\ 0 & 0 & 0 & 0 & -\sin q_{45} & \cos q_{45} \end{bmatrix}. \tag{4.86}$$

The joint twist is given by $X_4^e = \begin{bmatrix} 0 & -l_5 & 0 & 1 & 0 & 0 \end{bmatrix}^{\mathrm{T}}$ which gives the fourth column in the body geometric Jacobian as

$$\mathrm{Ad}^{-1}_{e^{\hat{X}^e_4 q_4} e^{\hat{X}^e_5 q_5}} X^e_4 = \begin{bmatrix} 0 & -l_5 \cos q_5 & l_5 \sin q_5 & 1 & 0 & 0 \end{bmatrix}^\mathsf{T}. \tag{4.87}$$

Similarly we can find the remaining columns of the body Jacobian as

$$X_3^\dagger = \begin{bmatrix} 1 & 0 & 0 & 0 & * & * \\ 0 & cq_{35} & sq_{35} & -l_5(cq_{35} - cq_5) - l_4(sq_{45} - sq_{35}) & * & * \\ 0 & -sq_{35} & cq_{35} & -l_5(sq_{35} - sq_5) - l_4(cq_{45} - cq_{35}) & * & * \\ 0 & 0 & 0 & 1 & * & * \\ 0 & 0 & 0 & -cq_{35} & * & * \\ 0 & 0 & 0 & -sq_{35} & * & * \end{bmatrix} \begin{bmatrix} 0 \\ -l_5 \\ -l_4 \\ 1 \\ 0 \\ 0 \end{bmatrix}$$

$$= \begin{bmatrix} 0 \\ -l_5 cq_5 - l_4 sq_{45} \\ l_5 sq_5 - l_4 cq_{45} \\ 1 \\ 0 \\ 0 \end{bmatrix}$$

$$X_2^\dagger = \begin{bmatrix} cq_2 & 0 & sq_2 & * & -l_4 sq_3 + l_5 cq_{34} - l_5 cq_2 & * \\ -sq_2 sq_{35} & cq_{35} & cq_2 sq_{35} & * & l_5 sq_2 sq_{35} & * \\ -sq_2 cq_{35} & -sq_{35} & cq_2 cq_{35} & * & l_5 sq_2 cq_{35} & * \\ 0 & 0 & 0 & * & 0 & * \\ 0 & 0 & 0 & * & cq_{35} & * \\ 0 & 0 & 0 & * & -sq_{35} & * \end{bmatrix} \begin{bmatrix} -l_5 \\ 0 \\ 0 \\ 0 \\ -1 \\ 0 \end{bmatrix} \tag{4.88}$$

$$= \begin{bmatrix} l_4 sq_3 - l_5 cq_{34} \\ 0 \\ 0 \\ 0 \\ -cq_{35} \\ sq_{35} \end{bmatrix}$$

$$X_1^\dagger = \begin{bmatrix} cq_2 & 0 & sq_2 & * & -l_4 sq_3 + l_5 cq_{34} - l_5 cq_2 & * \\ -sq_2 sq_{35} & cq_{35} & cq_2 sq_{35} & * & l_5 sq_2 sq_{35} & * \\ -sq_2 cq_{35} & -sq_{35} & cq_2 cq_{35} & * & l_5 sq_2 cq_{35} & * \\ 0 & 0 & 0 & * & 0 & * \\ 0 & 0 & 0 & * & cq_{35} & * \\ 0 & 0 & 0 & * & -sq_{35} & * \end{bmatrix} \begin{bmatrix} 0 \\ 0 \\ 1 \\ 0 \\ 0 \\ 0 \end{bmatrix}$$

$$= \begin{bmatrix} sq_2 \\ cq_2 sq_{35} \\ cq_2 cq_{35} \\ 0 \\ 0 \\ 0 \end{bmatrix}$$

which gives the body geometric Jacobian (4.83) in Example 4.8.

4.4.2 Spatial Geometric Jacobian

The spatial velocity of the end effector is given by

$$\widehat{V}^S_{0e} = \dot{g}_{0e}(q)g^{-1}_{0e}(q). \tag{4.89}$$

Similarly to the body velocity, we can find the spatial velocity in (4.89) as

$$\begin{aligned}
\widehat{V}^S_{0e} &= \left(\frac{\partial g_{0e}}{\partial q_1}\dot{q}_1 g^{-1}_{0e}\right)^\vee + \left(\frac{\partial g_{0e}}{\partial q_2}\dot{q}_2 g^{-1}_{0e}\right)^\vee + \cdots + \left(\frac{\partial g_{0e}}{\partial q_n}\dot{q}_n g^{-1}_{0e}\right)^\vee \\
&= \underbrace{\left[\left(\frac{\partial g_{0e}}{\partial q_1} g^{-1}_{0e}\right)^\vee \quad \left(\frac{\partial g_{0e}}{\partial q_2} g^{-1}_{0e}\right)^\vee \quad \cdots \quad \left(\frac{\partial g_{0e}}{\partial q_n} g^{-1}_{0e}\right)^\vee\right]}_{J^S_{m,g}} \begin{bmatrix} \dot{q}_1 \\ \dot{q}_2 \\ \vdots \\ \dot{q}_n \end{bmatrix},
\end{aligned} \tag{4.90}$$

where the matrix on the left is called the *spatial geometric Jacobian*. It turns out that also the column vectors of the spatial geometric Jacobian have a very nice geometric interpretation. Following more or less the same computations as for the body velocities, we find

$$\frac{\partial g_{0e}}{\partial q_i} g^{-1}_{0e} = e^{\widehat{X}_1 q_1} \cdots e^{\widehat{X}_{i-1} q_{i-1}} \frac{\partial}{\partial q_i}\left(e^{\widehat{X}_i q_i}\right) e^{\widehat{X}_{i+1} q_{i+1}} \cdots e^{\widehat{X}_n q_n} g_{0e}(0) g^{-1}_{0e} \tag{4.91}$$

$$= e^{\widehat{X}_1 q_1} \cdots e^{\widehat{X}_{i-1} q_{i-1}} \left(\widehat{X}_i e^{\widehat{X}_i q_i}\right) e^{\widehat{X}_{i+1} q_{i+1}} \cdots e^{\widehat{X}_n q_n} g_{0e}(0) g^{-1}_{0e} \tag{4.92}$$

$$= e^{\widehat{X}_1 q_1} \cdots e^{\widehat{X}_{i-1} q_{i-1}} \widehat{X}_i e^{-\widehat{X}_{i-1} q_{i-1}} e^{-\widehat{X}_{i-2} q_{i-2}} \cdots e^{-\widehat{X}_1 q_1} \tag{4.93}$$

where we have used that if we first write

$$g_{0e}(q) = e^{\widehat{X}_1 q_1} \cdots e^{\widehat{X}_{i-1} q_{i-1}} e^{\widehat{X}_i q_i} e^{\widehat{X}_{i+1} q_{i+1}} \cdots e^{\widehat{X}_n q_n} g_{0e}(0) \tag{4.94}$$

we get the relation

$$\begin{aligned}
e^{\widehat{X}_1 q_1} \cdots e^{\widehat{X}_{i-1} q_{i-1}} e^{\widehat{X}_i q_i} e^{\widehat{X}_{i+1} q_{i+1}} \cdots e^{\widehat{X}_n q_n} g_{0e}(0) g^{-1}_{0e} &= I \\
e^{\widehat{X}_i q_i} e^{\widehat{X}_{i+1} q_{i+1}} \cdots e^{\widehat{X}_n q_n} g_{0e}(0) g^{-1}_{0e} &= e^{-\widehat{X}_{i-1} q_{i-1}} \cdots e^{-\widehat{X}_1 q_1}
\end{aligned} \tag{4.95}$$

which we substituted into (4.92) to get (4.93).

The columns of the spatial geometric Jacobian thus correspond to the joint twists written with respect to the base frame at configuration q and can be written in a similar form to the body geometric Jacobian. X'_i is the configuration dependent joint twist as seen from $\mathscr{F}_0$ and is given by

$$X'_i = \mathrm{Ad}_{e^{\widehat{X}_1 q_1} \dots e^{\widehat{X}_{i-1} q_{i-1}}} X_i \tag{4.96}$$

or alternatively

$$X'_i = \mathrm{Ad}_{g_{0\bar{i}}(q)} X^i_i. \tag{4.97}$$

Theorem 4.3 *The spatial geometric Jacobian $J^S_{m,g}(q)$ maps the joint velocities to the spatial velocity twist expressed in the inertial frame $\mathscr{F}_0$ by*

$$V^S_{0e} = J^S_{m,g}(q)\dot{q} \tag{4.98}$$

and is defined as

$$J^S_{m,g}(q) = \begin{bmatrix} X'_1 & X'_2 & \cdots & X'_n \end{bmatrix} \tag{4.99}$$

$$= \left[\left(\frac{\partial g_{0e}}{\partial q_1} g_{0e}^{-1} \right)^\vee \quad \left(\frac{\partial g_{0e}}{\partial q_2} g_{0e}^{-1} \right)^\vee \quad \cdots \quad \left(\frac{\partial g_{0e}}{\partial q_n} g_{0e}^{-1} \right)^\vee \right] \tag{4.100}$$

where the column vectors $(\frac{\partial g_{0e}}{\partial q_2} g_{0e}^{-1})^\vee$ are the joint twists X'_i as observed from the inertial frame $\mathscr{F}_0$.

Note that X'_i is the twist of joint i as seen from $\mathscr{F}_0$ and therefore depends only on the position of the previous links $q_1 \dots q_{(i-1)}$. We can therefore write

$$X'_i = \mathrm{Ad}_{g_{0\bar{i}}}(q_1 \dots q_{(i-1)}) X^i_i. \tag{4.101}$$

Example 4.10 Once again consider the manipulator structure in Fig. 4.7. The spatial geometric Jacobian is then given by

$$\begin{aligned} J^S_{m,g}(q) &= \begin{bmatrix} X'_1 & X'_2 & X'_3 & X'_4 & X'_5 \end{bmatrix} \\ &= \begin{bmatrix} \mathrm{Ad}_{g_{0\bar{1}}} X^1_1 & \mathrm{Ad}_{g_{0\bar{2}}} X^2_2 & \mathrm{Ad}_{g_{0\bar{3}}} X^3_3 & \mathrm{Ad}_{g_{0\bar{4}}} X^4_4 & \mathrm{Ad}_{g_{0\bar{5}}} X^5_5 \end{bmatrix}. \end{aligned} \tag{4.102}$$

The configuration dependent spatial twists are given by

$$\mathrm{Ad}_{g_{0\bar{1}}} X^1_1 = \begin{bmatrix} 1 & 0 & 0 & 0 & 0 & q_1 \\ 0 & 1 & 0 & 0 & 0 & 0 \\ 0 & 0 & 1 & -q_1 & 0 & 0 \\ 0 & 0 & 0 & 1 & 0 & 0 \\ 0 & 0 & 0 & 0 & 1 & 0 \\ 0 & 0 & 0 & 0 & 0 & 1 \end{bmatrix} \begin{bmatrix} 0 \\ 1 \\ 0 \\ 0 \\ 0 \\ 0 \end{bmatrix} = \begin{bmatrix} 0 \\ 1 \\ 0 \\ 0 \\ 0 \\ 0 \end{bmatrix},$$

$$\mathrm{Ad}_{g_{0\bar{2}}} X_2^{\bar{2}} = \begin{bmatrix} 1 & 0 & 0 & 0 & -l_2 & q_1 \\ 0 & 1 & 0 & l_2 & 0 & 0 \\ 0 & 0 & 1 & -q_1 & 0 & 0 \\ 0 & 0 & 0 & 1 & 0 & 0 \\ 0 & 0 & 0 & 0 & 1 & 0 \\ 0 & 0 & 0 & 0 & 0 & 1 \end{bmatrix} \begin{bmatrix} 0 \\ 0 \\ 0 \\ 0 \\ 0 \\ 1 \end{bmatrix} = \begin{bmatrix} q_1 \\ 0 \\ 0 \\ 0 \\ 0 \\ 1 \end{bmatrix},$$

$$\mathrm{Ad}_{g_{0\bar{3}}} X_3^{\bar{3}} = \begin{bmatrix} \cos q_2 & -\sin q_2 & 0 & -l_{23}\sin q_2 & * & * \\ \sin q_2 & \cos q_2 & 0 & l_{23}\cos q_2 & * & * \\ 0 & 0 & 1 & -q_1\cos q_2 & * & * \\ 0 & 0 & 0 & \cos q_2 & * & * \\ 0 & 0 & 0 & \sin q_2 & * & * \\ 0 & 0 & 0 & 0 & * & * \end{bmatrix} \begin{bmatrix} 0 \\ 0 \\ 0 \\ 1 \\ 0 \\ 0 \end{bmatrix} = \begin{bmatrix} -l_{23}\sin q_2 \\ l_{23}\cos q_2 \\ -q_1\cos q_2 \\ \cos q_2 \\ \sin q_2 \\ 0 \end{bmatrix},$$

$$\mathrm{Ad}_{g_{0\bar{4}}} X_4^{\bar{4}} = \begin{bmatrix} cq_2 & -sq_2sq_3 & -sq_2cq_3 & -(l_{23}+l_4cq_3)sq_2 & * & * \\ sq_2 & cq_2sq_3 & cq_2cq_3 & (l_{23}+l_4cq_3)cq_2 & * & * \\ 0 & -cq_3 & sq_3 & l_4sq_3 - q_1cq_2 & * & * \\ 0 & 0 & 0 & cq_2 & * & * \\ 0 & 0 & 0 & sq_2 & * & * \\ 0 & 0 & 0 & 0 & * & * \end{bmatrix} \begin{bmatrix} 0 \\ 0 \\ 0 \\ 1 \\ 0 \\ 0 \end{bmatrix} \tag{4.103}$$

$$= \begin{bmatrix} -(l_{23}+l_4cq_3)sq_2 \\ (l_{23}+l_4cq_3)cq_2 \\ l_4sq_3 - q_1cq_2 \\ cq_2 \\ sq_2 \\ 0 \end{bmatrix},$$

$$\mathrm{Ad}_{g_{0\bar{5}}} X_5^{\bar{5}} = \begin{bmatrix} cq_2 & -sq_2sq_{34} & -sq_2cq_{34} & -(l_{23}+l_4cq_3+l_5sq_{34})sq_2 & * & * \\ sq_2 & cq_2sq_{34} & cq_2cq_{34} & (l_{23}+l_4cq_3+l_5sq_{34})cq_2 & * & * \\ 0 & -cq_{34} & sq_{34} & l_4sq_3 - l_5cq_{34} - q_1cq_2 & * & * \\ 0 & 0 & 0 & cq_2 & * & * \\ 0 & 0 & 0 & sq_2 & * & * \\ 0 & 0 & 0 & 0 & * & * \end{bmatrix}$$

$$\times \begin{bmatrix} 0 \\ 0 \\ 0 \\ 1 \\ 0 \\ 0 \end{bmatrix}$$

$$= \begin{bmatrix} -(l_{23} + l_4 cq_3 + l_5 sq_{34}) sq_2 \\ (l_{23} + l_4 cq_3 + l_5 sq_{34}) cq_2 \\ l_4 sq_3 - l_5 cq_{34} - q_1 cq_2 \\ cq_2 \\ sq_2 \\ 0 \end{bmatrix}.$$

The spatial geometric Jacobian of the manipulator in Fig. 4.7 is thus given by

$$J^S_{m,g}(q) = \begin{bmatrix} 0 & q_1 & 0 & -(l_{23} + l_4 cq_3) sq_2 & -(l_{23} + l_4 cq_3 + l_5 sq_{34}) sq_2 \\ 1 & 0 & l_{23} & (l_{23} + l_4 cq_3) cq_2 & (l_{23} + l_4 cq_3 + l_5 sq_{34}) cq_2 \\ 0 & 0 & -q_1 & l_4 sq_3 - q_1 cq_2 & l_4 sq_3 - l_5 cq_{34} - q_1 cq_2 \\ 0 & 0 & cq_2 & cq_2 & cq_2 \\ 0 & 0 & sq_2 & sq_2 & sq_2 \\ 0 & 1 & 0 & 0 & 0 \end{bmatrix}. \tag{4.104}$$

Example 4.11 We can also find the spatial geometric Jacobian by Eq. (4.96) as

$$\begin{aligned} J^S_{m,g}(q) &= \begin{bmatrix} X'_1 & X'_2 & X'_3 & X'_4 & X'_5 \end{bmatrix} \\ &= \Big[X_1 \quad \mathrm{Ad}_{e^{\hat{X}_1 q_1}} X_2 \quad \mathrm{Ad}_{e^{\hat{X}_1 q_1} e^{\hat{X}_2 q_2}} X_3 \\ &\qquad \mathrm{Ad}_{e^{\hat{X}_1 q_1} \ldots e^{\hat{X}_3 q_3}} X_4 \quad \mathrm{Ad}_{e^{\hat{X}_1 q_1} \ldots e^{\hat{X}_4 q_4}} X_5 \Big]. \end{aligned} \tag{4.105}$$

In this case the spatial twists are found by

$$\mathrm{Ad}_{e^{\hat{X}_1 q_1}} X_2 = \begin{bmatrix} 1 & 0 & 0 & 0 & 0 & q_1 \\ 0 & 1 & 0 & 0 & 0 & 0 \\ 0 & 0 & 1 & -q_1 & 0 & 0 \\ 0 & 0 & 0 & 1 & 0 & 0 \\ 0 & 0 & 0 & 0 & 1 & 0 \\ 0 & 0 & 0 & 0 & 0 & 1 \end{bmatrix} \begin{bmatrix} 0 \\ 0 \\ 0 \\ 0 \\ 0 \\ 1 \end{bmatrix} = \begin{bmatrix} q_1 \\ 0 \\ 0 \\ 0 \\ 0 \\ 1 \end{bmatrix},$$

$$\mathrm{Ad}_{e^{\hat{X}_1 q_1} e^{\hat{X}_2 q_2}} X_3 = \begin{bmatrix} \cos q_2 & -\sin q_2 & 0 & 0 & * & * \\ \sin q_2 & \cos q_2 & 0 & 0 & * & * \\ 0 & 0 & 1 & -q_1 \cos q_2 & * & * \\ 0 & 0 & 0 & \cos q_2 & * & * \\ 0 & 0 & 0 & \sin q_2 & * & * \\ 0 & 0 & 0 & 0 & * & * \end{bmatrix} \begin{bmatrix} 0 \\ l_{23} \\ 0 \\ 1 \\ 0 \\ 0 \end{bmatrix}$$

$$= \begin{bmatrix} -l_{23} \sin q_2 \\ l_{23} \cos q_2 \\ -q_1 \cos q_2 \\ \cos q_2 \\ \sin q_2 \\ 0 \end{bmatrix},$$

$$\operatorname{Ad}_{e^{\hat{X}_1 q_1}\ldots e^{\hat{X}_3 q_3}} X_4$$

$$= \begin{bmatrix} * & -\sin q_2 \cos q_3 & * & -l_{23}\sin q_2 + l_{23}\sin q_2 \cos q_3 & * & * \\ * & \cos q_2 \cos q_3 & * & l_{23}\cos q_2 - l_{23}\cos q_2 \cos q_3 & * & * \\ * & \sin q_3 & * & -l_{23}\sin q_3 - q_1 \cos q_2 & * & * \\ * & 0 & * & \cos q_2 & * & * \\ * & 0 & * & \sin q_2 & * & * \\ * & 0 & * & 0 & * & * \end{bmatrix} \begin{bmatrix} 0 \\ l_{24} \\ 0 \\ 1 \\ 0 \\ 0 \end{bmatrix}$$

$$= \begin{bmatrix} -(l_{23} + l_4 c q_3) s q_2 \\ (l_{23} + l_4 c q_3) c q_2 \\ l_4 s q_3 - q_1 c q_2 \\ c q_2 \\ s q_2 \\ 0 \end{bmatrix}, \tag{4.106}$$

$$\operatorname{Ad}_{e^{\hat{X}_1 q_1}\ldots e^{\hat{X}_4 q_4}} X_5$$

$$= \begin{bmatrix} c q_2 & -s q_2 c q_{34} & s q_2 s q_{34} & l_{24} s q_2 c q_{34} - l_3 s q_2 - l_4 s q_2 c q_3 & * & * \\ s q_2 & c q_2 c q_{34} & -c q_2 s q_{34} & l_{24} c q_2 c q_{34} + l_3 c q_2 + l_4 c q_2 c q_3 & * & * \\ 0 & s q_{34} & -c q_{34} & -l_{24} s q_{34} + l_4 s q_3 + q_1 c q_2 & * & * \\ 0 & 0 & 0 & c q_2 & * & * \\ 0 & 0 & 0 & s q_2 & * & * \\ 0 & 0 & 0 & 0 & * & * \end{bmatrix}$$

$$\times \begin{bmatrix} 0 \\ l_{34} \\ -l_5 \\ 1 \\ 0 \\ 0 \end{bmatrix}$$

$$= \begin{bmatrix} -(l_{23} + l_4 c q_3 + l_5 s q_{34}) s q_2 \\ (l_{23} + l_4 c q_3 + l_5 s q_{34}) c q_2 \\ l_4 s q_3 - l_5 c q_{34} - q_1 c q_2 \\ c q_2 \\ s q_2 \\ 0 \end{bmatrix},$$

where $l_{23} = l_2 + l_3$. The spatial geometric Jacobian of the manipulator in Fig. 4.5 is thus given by

$$J_{m,g}^S(q) = \begin{bmatrix} 0 & q_1 & 0 & -(l_{23}+l_4cq_3)sq_2 & -(l_{23}+l_4cq_3+l_5sq_{34})sq_2 \\ 1 & 0 & l_{23} & (l_{23}+l_4cq_3)cq_2 & (l_{23}+l_4cq_3+l_5sq_{34})cq_2 \\ 0 & 0 & -q_1 & l_4sq_3-q_1cq_2 & l_4sq_3-l_5cq_{34}-q_1cq_2 \\ 0 & 0 & cq_2 & cq_2 & cq_2 \\ 0 & 0 & sq_2 & sq_2 & sq_2 \\ 0 & 1 & 0 & 0 & 0 \end{bmatrix} \tag{4.107}$$

The mapping from the body velocities to the spatial velocities is given by the Adjoint map

$$V_{0e}^S = \mathrm{Ad}_{g_{0e}} V_{0e}^B. \tag{4.108}$$

We can use this relation to find a mapping from the body geometric Jacobian to the spatial geometric Jacobian. Substituting (4.64) and (4.65) into (4.108) we get

$$J_{m,g}^S \dot{q} = \mathrm{Ad}_{g_{0e}} J_{m,g} \dot{q}, \tag{4.109}$$

and the mapping from the body geometric Jacobian to the spatial geometric Jacobian is given by

$$J_{m,g}^S = \mathrm{Ad}_{g_{0e}} J_{m,g}. \tag{4.110}$$

4.4.3 The Geometric Jacobian of the Manipulator Links

We have found the velocity of the end effector with respect to the base link $\mathscr{F}_0$ as $V_{0e}^S = J_{m,g}^S(q)\dot{q}$. In the same way we can find the mapping from the joint velocity to the velocity of each link. This relation is given by the link Jacobian. We find the link Jacobian by first observing that the velocity of link i only depends on joints $1 \ldots i$: joints further out in the chain ($> i$) have no effect on the velocity of link i for open chain manipulators.

The Jacobian $\bar{J}_i$ of link i is given by

$$\bar{J}_i(q) = \begin{bmatrix} \mathrm{Ad}_{g_{0\bar{1}}} X_1^1 & \mathrm{Ad}_{g_{0\bar{2}}} X_2^2 & \mathrm{Ad}_{g_{0\bar{3}}} X_3^3 & \cdots & \mathrm{Ad}_{g_{0\bar{i}}} X_i^i & 0_{(n-i)\times 6} \end{bmatrix}. \tag{4.111}$$

The mapping from the joint velocities to the velocity of a frame $\bar{\mathscr{F}}_i$ attached to the joint axis of link i is now given by

$$V_{0\bar{i}}^S = \bar{J}_i(q)\dot{q}. \tag{4.112}$$

Recall that the spatial velocity twist $V^S_{0\bar{i}}$ is to be interpreted as the velocity of an imaginary point "attached" to the rigid body with frame $\bar{\mathscr{F}}_i$ as it travels through $\mathscr{O}_0$. It does therefore not make any difference where on the rigid body i we choose to attach the frame $\bar{\mathscr{F}}_i$. From this we can conclude the following

Proposition 4.1 *The velocities V^S_{0i} and $V^S_{0\bar{i}}$ of two frames $\mathscr{F}_i$ and $\bar{\mathscr{F}}_i$ attached to the same rigid body, represented with respect to an inertial frame $\mathscr{F}_0$ are equal when written in spatial coordinates*:

$$V^S_{0\bar{i}} = V^S_{0i}. \tag{4.113}$$

As a result the spatial Jacobian of a frame $\mathscr{F}_i$ attached to the center of gravity of a link i in a robotic chain is found as

$$J_i(q) = \bar{J}_i(q) \tag{4.114}$$

where $\bar{J}_i(q)$ is found in Eq. (4.111).

This is a very important property that we will use frequently in the subsequent chapters. It is important to note, however, that the same is not true for velocities written in body coordinates:

The body Jacobian $\bar{J}^B_i$ for a frame $\bar{\mathscr{F}}_i$ attached to link i is given by

$$\bar{J}^B_i(q) = \begin{bmatrix} \mathrm{Ad}_{g^{-1}_{\bar{1}\bar{i}}} X^1_1 & \mathrm{Ad}_{g^{-1}_{\bar{2}\bar{i}}} X^2_2 & \mathrm{Ad}_{g^{-1}_{\bar{3}\bar{i}}} X^3_3 & \cdots & \mathrm{Ad}_{g^{-1}_{\bar{i}\bar{i}}} X^i_i & 0_{(n-i)\times 6} \end{bmatrix}. \tag{4.115}$$

The mapping from the joint velocities to the velocity of frame $\bar{\mathscr{F}}_i$ attached to the joint axis of link i is now given by

$$V^B_{0\bar{i}} = \bar{J}^B_i(q)\dot{q}. \tag{4.116}$$

If we instead want to find the velocity V^B_{0i} of a frame $\mathscr{F}_i$ attached to the center of gravity of link i we get the following Jacobian:

The body Jacobian J_i^B for a frame $\mathscr{F}_i$ attached to the center of gravity of link i is given by

$$J_i^B(q) = \left[\mathrm{Ad}_{g_{\bar{1}i}^{-1}} X_1^1 \quad \mathrm{Ad}_{g_{\bar{2}i}^{-1}} X_2^2 \quad \mathrm{Ad}_{g_{\bar{3}i}^{-1}} X_3^3 \quad \cdots \quad \mathrm{Ad}_{g_{\bar{i}i}^{-1}} X_i^i \quad 0_{(n-i)\times 6}\right]. \tag{4.117}$$

The mapping from the joint velocities to the velocity of frame $\mathscr{F}_i$ attached to the center of gravity of link i is now given by

$$V_{0i}^B = J_i^B(q)\dot{q}. \tag{4.118}$$

4.4.4 Analytical Jacobian

As we have seen, it is common to use a velocity vector $\dot{\eta} = [\dot{x}_{0b}\ \dot{y}_{0b}\ \dot{z}_{0b}\ \dot{\phi}_{0b}\ \dot{\theta}_{0b}\ \dot{\psi}_{0b}]$ to denote the velocity state of a single rigid body. Similarly we can use the representation $\dot{\eta}_{0e} = [\dot{x}_{0e}\ \dot{y}_{0e}\ \dot{z}_{0e}\ \dot{\phi}_{0e}\ \dot{\theta}_{0e}\ \dot{\psi}_{0e}]$ to describe the velocity state of the end effector in the inertial frame. In this case the Jacobian maps a set of joint velocities $\dot{q}$ to the velocity of the end effector $\dot{\eta}_{0e}$ by the mapping

$$\dot{\eta}_{0e} = \frac{\partial f(q)}{\partial q}\dot{q} \tag{4.119}$$

where $f : \mathbb{R}^n \to \mathbb{R}^6$ is the forward kinematics mapping from the joint positions to η_{0e}. The analytical Jacobian is then given by the mapping $J_{m,a} = \frac{\partial f}{\partial q}$. It is not trivial to compute the partial derivatives of the forward mapping $f(q)$ because the rotational part in general requires us to compute the elements of the rotation matrix. Nor do the columns of the analytical Jacobian have a nice geometric interpretation as the joint twists of the manipulator, as they did for the geometric Jacobian.

It is important to note that the local parameterization of $SE(3)$ using $\eta_{0e} = [x_{0e}\ y_{0e}\ z_{0e}\ \phi_{0e}\ \theta_{0e}\ \psi_{0e}]$ leads to singularities. These singularities are important to identify as they may cause us to draw wrong conclusions about the ability of the robot to move in a certain direction. These singularities are a result of choosing a local parameterization and not due to the manipulator structure itself.

In the same way that we found a mapping from the body to the spatial Jacobian, we can find a mapping between the analytical and the geometric Jacobians. Let $T_{0e}(\eta_{0e,2})$ map the body angular velocities into the time derivative of the Euler angles, i.e., $\dot{\eta}_2 = T_{0e}(\eta_{0e,2})\omega_{0e}^B$ and let $R_{0e}(\eta_{0e,2})$ map the body linear velocities into the time derivative of the position vector, i.e., $\dot{\eta}_{0e,1} = R_{0e}(\eta_{0e,2})v_{0e}^B$. We then have

$$\begin{bmatrix} \dot{\eta}_{0e,1} \\ \dot{\eta}_{0e,2} \end{bmatrix} = \begin{bmatrix} R_{0e}(\eta_{0e,2}) & 0 \\ 0 & T_{0e}(\eta_{0e,2}) \end{bmatrix} \begin{bmatrix} v_{0e}^B \\ \omega_{0e}^B \end{bmatrix} \tag{4.120}$$

$$\begin{aligned} \begin{bmatrix} \dot{\eta}_{0e,1} \\ \dot{\eta}_{0e,2} \end{bmatrix} &= \begin{bmatrix} R_{0e}(\eta_{0e,2}) & 0 \\ 0 & T_{0e}(\eta_{0e,2}) \end{bmatrix} \begin{bmatrix} R_{0e}^{\mathsf{T}}(\eta_{0e,2}) & -R_{0e}^{\mathsf{T}}(\eta_{0e,2})\hat{p}_{0e} \\ 0 & R_{0e}^{\mathsf{T}}(\eta_{0e,2}) \end{bmatrix} \begin{bmatrix} v_{0e}^S \\ \omega_{0e}^S \end{bmatrix} \\ &= \begin{bmatrix} I & -\hat{p}_{0e} \\ 0 & T_{0e}(\eta_{0e,2}) R_{0e}^{\mathsf{T}}(\eta_{0e,2}) \end{bmatrix} \begin{bmatrix} v_{0e}^S \\ \omega_{0e}^S \end{bmatrix}. \end{aligned} \tag{4.121}$$

The relation between the spatial linear velocity and the inertial linear velocity is easily verified by comparing Eqs. (4.58) and (4.59).

We have now found the velocity transformation matrices of the robot end effector that give the relations $\dot{\eta}_{0e} = J_e(\eta_{0e,2}) V_{0e}^e$ and $\dot{\eta}_{0e} = J_e^S(\eta_{0e,2}) V_{0e}^S$. We can then write

$$\dot{\eta}_{0e} = J_e V_{0e}^B = J_e J_{m,g} \dot{q}, \tag{4.122}$$

$$\dot{\eta}_{0e} = J_e^S V_{0e}^S = J_e^S J_{m,g}^S \dot{q}. \tag{4.123}$$

The analytic Jacobians are then found as

$$J_{m,a} = J_e J_{m,g}, \tag{4.124}$$

$$J_{m,a}^S = J_e^S J_{m,g}^S. \tag{4.125}$$

Note that $J_{m,a}$ and $J_{m,a}^S$ represent the same transformation from the joint velocity space to $\dot{\eta}_{0e}$ and therefore presents us with two different ways of obtaining the same matrix.

4.5 Configuration States

In the previous section we wrote the kinematics in terms of a configuration variable Q so that the configuration of the rigid body was given by $g(Q)$. We can also write the configuration variables of a robotic manipulator on a fixed base using this framework. The joints of the robot are all restricted to a 1-DoF "matrix" which allows us to write the configuration of each joint as $Q_i = q_i \in \mathbb{R}$ for both linear and angular motion. We will write transformations due to the motion of a prismatic joint in the following way:

Theorem 4.4 *The configuration space of a prismatic joint can be expressed in the form of Definition* 3.3 *as a matrix* $Q_i \in \mathbb{R}$ *where*

$$Q_i = q_i \in \mathbb{R} \tag{4.126}$$

and the transformation of link i is given by

$$g_i(Q_i) = \begin{bmatrix} I & p_i q_i \\ 0 & 1 \end{bmatrix} \tag{4.127}$$

where p_i *is the direction of the prismatic joint represented in* $\bar{\mathscr{F}}_i$.

The velocity variable can also be written in the form of Definition 3.3 *as* $v_i \in \mathbb{R}$ *where*

$$v_i = \dot{q}_i \in \mathbb{R} \tag{4.128}$$

and the twist representing the velocity of $\bar{\mathscr{F}}_i$ *with respect to* $\mathscr{F}_0$ *in spatial coordinates is given by*

$$X_i = \begin{bmatrix} p_i \\ 0 \end{bmatrix} \tag{4.129}$$

where p_i *is represented in* $\mathscr{F}_0$.

X_i gives us the direction of the motion of the joint and we can write the velocity due to joint i as $X_i \dot{q}_i$.

For a linear motion in the direction of the x-axis we will for example write $p_i = [1\ 0\ 0]^\mathsf{T}$. We see that we in this way restrict the motion of the joints to a 1-DoF configuration space represented by a matrix Q_i. At the same time the configuration of the links in $SE(3)$ are given by a homogeneous transformation matrix so that we can apply the results from the previous chapters.

Also the configuration space of revolute joints can be written using the framework above:

Theorem 4.5 *The configuration space of a revolute joint can be expressed in the form of Definition* 3.3 *as a matrix* $Q_i \in \mathbb{R}$ *where*

$$Q_i = q_i \in \mathbb{R} \tag{4.130}$$

and the transformation of link i is given by

$$g_i(Q_i) = \begin{bmatrix} R(p_i, q_i) & 0 \\ 0 & 1 \end{bmatrix} \tag{4.131}$$

where $R(p_i, q_i)$ is the rotation matrix representation of a rotation q_i around the axis p_i, which for rotations around the coordinate axes are given by (3.10)–(3.12).

The velocity variable can also be written in the form of Definition 3.3 *as $v_i \in \mathbb{R}$ where*

$$v_i = \dot{q}_i \in \mathbb{R} \tag{4.132}$$

and the twist representing the velocity of $\bar{\mathscr{F}}_i$ with respect to $\mathscr{F}_0$ in spatial coordinates is given by

$$X_i = \begin{bmatrix} -p_i \times \bar{l}_i \\ p_i \end{bmatrix}. \tag{4.133}$$

where $\bar{l}_i$ is an arbitrary point on p_i (represented in $\mathscr{F}_0$).

X_i is the direction of the motion of the link and we can write the velocity of the link in spatial coordinates as $X_i \dot{q}_i$. Once the joint variables of each joint are found in this way we can find the configuration space of a robotic manipulator in a very simple form:

Theorem 4.6 *A robotic manipulator with n 1-DoF prismatic or revolute joints is completely characterized by the configuration states $Q_i \in \mathbb{R}$ and $v_i \in \mathbb{R}$ for $i = 1, \dots, n$ or as a vector $q \in \mathbb{R}^n$ representing the joint positions and a vector $v \in \mathbb{R}^n$ representing the velocities.*

This result differs somewhat from what we obtained in the previous sections because the transformations are Euclidean. This allows us to write the state space in a very simple form as vectors in $\mathbb{R}^n$. The reason that we adapt this framework and write the configuration space as matrices Q_i representing the configuration state of each link is that we will need this formulation in the next chapters when we study vehicle-manipulator systems which, as we have seen, are not Euclidean transformations. We thus adopt this formalism in order to obtain a single framework for both vehicles and multibody systems at the expense of a slightly more troublesome notation.

4.5.1 Local Coordinates

The position and velocity variables of a robotic manipulator can be written in terms of local position and velocity variables φ and $\dot{\varphi}$ in the same way as for a single rigid body in the previous chapter. We have already seen that the transformation from local to global velocity variables can be found in a very simple form for Euclidean transformations. We can write the local position variables of a revolute or prismatic joint as $\varphi_i = q_i$. Because the velocity variables can be found as $v_i = \dot{q}_i$ the we get $\dot{\varphi} = v_i$ and the velocity transformation for the robotic manipulator is written as

$$v = S(Q, \varphi)\dot{\varphi} \tag{4.134}$$

where $S(Q, \varphi) = I$.

References

Angeles, J. (1982). *Spatial kinematic chains: analysis, synthesis, optimization*. Berlin: Springer.
Brockett, R. W. (1984). Robotic manipulators and the product of exponentials formula. In *Proceedings of mathematical theory of networks and systems*, Beer Sheva, Israel (pp. 120–129).
Duindam, V. (2006). *Port-based modeling and control for efficient bipedal walking robots*. Ph.D. thesis, University of Twente.
Hervé, J. M., & Sparacino, F. (1991). Structural synthesis of parallel robots generating spatial translation. In *IEEE international conference on robotics and automation*, Sacramento, California (pp. 808–813).
Karger, A., & Novák, J. (1985). *Space kinematics and Lie groups*. New York: Gordon & Breach.
Murray, R. M., Li, Z., & Sastry, S. S. (1994). *A mathematical introduction to robotic manipulation*. Boca Raton: CRC Press.
Park, F. (1994). Computational aspects of the product-of-exponentials formula for robot kinematics. *IEEE Transactions on Automatic Control*, *39*(3), 643–647.
Selig, J. M. (2000). *Geometric fundamentals of robotics*. New York: Springer.
Siciliano, B., Sciavicco, L., Villani, L., & Oriolo, G. (2011). *Advanced textbooks in control and signal processing*. *Robotics: modelling, planning and control*. Berlin: Springer.
Spong, M., & Vidyasagar, M. (2008). *Robot dynamics and control*. India: Wiley.
Stramigioli, S. (2001) *Lecture notes*. *Modeling and IPC control of interactive mechanical systems: a coordinate-free approach* (p. 265). London: Springer.

Chapter 5
Kinematics of Vehicle-Manipulator Systems

Vehicle-manipulator systems consist of two parts: the first part is the vehicle which we can model using the framework presented in Chap. 3, and the second part is the robot manipulator which can be modeled using the framework presented in Chap. 4. In this chapter we will derive the kinematic relations of vehicle-manipulator systems, which is quite different from the kinematics of single rigid bodies and manipulators mounted on a fixed base. The main challenge is that we have a combination of Euclidean and non-Euclidean transformations and therefore the standard framework used for modeling rigid bodies or fixed-base manipulators cannot be applied directly (Duindam and Stramigioli 2008; Duindam 2006).

We start by choosing the reference frames of the system. Differently from fixed-base manipulators, the inertial frame cannot be chosen at the base of the manipulator, but must be chosen at a fixed location. We will attach a reference frame $\mathscr{F}_0$ to this fixed point and a reference frame $\mathscr{F}_b$ to the vehicle. $\mathscr{F}_b$ also defines the base of the robot and is no longer inertial. We denote the configuration of the vehicle with respect to the inertial frame by g_{0b}. The manipulator kinematics is given with respect to the base frame $\mathscr{F}_b$ (the vehicle), so that we can use the formulation of the manipulator kinematics that we derived in the previous chapter. The configuration of the end effector of the robot is identified with the reference frame $\mathscr{F}_e$ and is defined with respect to the *base frame* by the homogeneous transformation matrix g_{be}. Finally the configuration of the end effector with respect to the *inertial frame* is given by $g_{0e} = g_{0b}g_{be}$. The reference frames are illustrated in Fig. 5.1.

5.1 Configuration Space

Consider the setup of Fig. 5.1 describing a general n-link robotic manipulator arm attached to a vehicle. We will choose an inertial coordinate frame $\mathscr{F}_0$, a frame $\mathscr{F}_b$ rigidly attached to the vehicle, and n frames $\bar{\mathscr{F}}_i$ or $\mathscr{F}_i$ (not shown) attached to each link so that the origins coincide with the translational or rotational axes of the joints or the center of gravity, respectively. Finally, we choose a matrix g_{0b} that describes

P.J. From et al., *Vehicle-Manipulator Systems*, Advances in Industrial Control,
DOI 10.1007/978-1-4471-5463-1_5, © Springer-Verlag London 2014

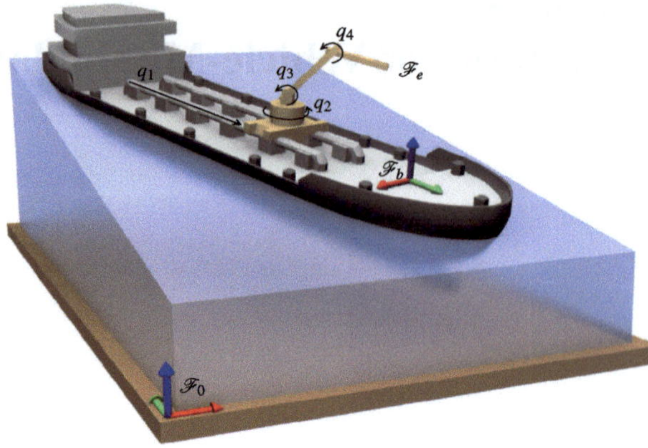

Fig. 5.1 Model setup for a four-link robot attached to a non-inertial base with coordinate frame $\mathscr{F}_b$. Frame $\mathscr{F}_0$ denotes the inertial reference frame and $\mathscr{F}_e$ the end-effector frame. Courtesy IEEE

the configuration of the vehicle, and a vector $q \in \mathbb{R}^n$ that describes the configuration of the n joints.

We will adopt the same framework as in the previous chapters and let Q_i describe the configuration of each rigid body in the system and v_i describe the velocity states. More specifically, the configuration space of the vehicle is denoted by a matrix Q_b, as we did in Chap. 3, where Q_b is a matrix Lie group (the special Euclidean group or one of its ten subgroups). An alternative way to write the configuration of the vehicle is as a vector η which gives $Q_b = \eta \in \mathbb{R}^6$. We will adopt the notation presented in Duindam (2006) and From et al. (2009, 2010).

The velocity variables of the vehicle can be described using body or spatial twists, or as the time derivative of the position variables η. In any case, the velocity vector is an element of $\mathbb{R}^6$ if the configuration space is the special Euclidean group, or a vector in $\mathbb{R}^m$ if the configuration space is an m-dimensional subgroup of $SE(3)$. In this chapter we will look at the advantages and disadvantages of the different formulations in more detail.

We will assume that the robotic manipulator consists only of 1-DoF joints with configuration space $\mathbb{R}$ or $\mathbb{S}^1$. We have seen that the configuration space of both prismatic and revolute joints can be described by the configuration states $Q_i \in \mathbb{R}$ for each joint $i = 1, \dots, n$. The configuration space of a robotic manipulator can therefore be written in terms of the joint positions as $Q_i = q_i \in \mathbb{R}$ with velocity variables $v_i = \dot{q}_i \in \mathbb{R}$ for both prismatic and revolute joints.

Using this approach the configuration space is written in terms of the configuration states Q_i which for a vehicle-manipulator system can be written as $Q = \{g_{0b}, q\}$ where $g_{0b} = g_{0b}(Q_b)$ is the vehicle position and orientation and $q \in \mathbb{R}^n$ collects the joint positions. The velocity variables are written as $v = \{v_b, \dot{q}\}$ where v_b is either the body or spatial twists or the time derivative of the position variables of the vehicle and $\dot{q} \in \mathbb{R}^n$ collects the joint velocities (From 2012a, 2012b).

Using the framework introduced in Chaps. 3 and 4, we can describe the pose of each frame $\mathscr{F}_i$ relative to $\mathscr{F}_0$ as a homogeneous transformation matrix $g_{0i} \in SE(3)$

of the form

$$g_{0i} = \begin{bmatrix} R_{0i} & p_{0i} \\ 0 & 1 \end{bmatrix} \in \mathbb{R}^{4\times 4} \tag{5.1}$$

with rotation matrix $R_{0i} \in SO(3)$ and translation vector $p_{0i} \in \mathbb{R}^3$. This pose can also be described using the vector of joint coordinates q and the vehicle pose Q_b as

$$g_{0i} = g_{0b} g_{bi} = g_{0b}(Q_b) g_{bi}(q). \tag{5.2}$$

The vehicle pose Q_b and the joint positions q thus fully determine the configuration state of the robot. Even though we use g_{0b} (6 DoF) to represent the vehicle configuration, the actual configuration space of the vehicle may be a subspace of $SE(3)$ of dimension $m < 6$. In this case the configuration space of the vehicle is represented by a matrix Q_b with m degrees of freedom, and we select the allowed transformations by selecting $g_{0b} = g_{0b}(Q_b)$ in the appropriate way. This was discussed in detail in Chap. 3.

Following the approach in Sect. 3.3 we can write the configuration of a vehicle-manipulator system as a vector

$$\xi = \begin{bmatrix} \eta \\ q \end{bmatrix} \in \mathbb{R}^{6+n} \tag{5.3}$$

where $\eta = [x_{0b}\ y_{0b}\ z_{0b}\ \phi_{0b}\ \theta_{0b}\ \psi_{0b}]^{\mathsf{T}}$ and $q = [q_1\ q_2 \cdots q_n]^{\mathsf{T}}$. In the case that the vehicle's configuration space is a subspace of the special Euclidean group we can write the vehicle configuration as a vector in $\mathbb{R}^m$ and the configuration space of the vehicle-manipulator system as a vector in $\mathbb{R}^{m+n}$.

The velocities of the VM system can be given in the inertial frame as

$$\dot{\xi} = \begin{bmatrix} \dot{\eta} \\ \dot{q} \end{bmatrix} \in \mathbb{R}^{6+n}. \tag{5.4}$$

This representation of the velocity variables has the apparent advantage that it can be integrated to find the position variables. It does not, however, give a geometrically meaningful representation of the vehicle velocity in the same way as the velocity twists. We will therefore use the twist representation of the velocity variables extensively also for vehicle-manipulator systems to represent the velocities of the vehicle, the robotic links, and the end effector.

Recall that the velocity vector depends on the reference frame in which it is observed. Depending on what reference frame the velocity is expressed in, we can write the velocity in terms of the spatial and body twists as

$$\zeta^S = \begin{bmatrix} V_{0b}^S \\ \dot{q} \end{bmatrix} \in \mathbb{R}^{6+n}, \qquad \zeta = \begin{bmatrix} V_{0b}^B \\ \dot{q} \end{bmatrix} \in \mathbb{R}^{6+n}. \tag{5.5}$$

The accelerations can be written in the same way as

$$\dot{\zeta}^S = \begin{bmatrix} \dot{V}_{0b}^S \\ \ddot{q} \end{bmatrix} \in \mathbb{R}^{6+n}, \qquad \dot{\zeta} = \begin{bmatrix} \dot{V}_{0b}^B \\ \ddot{q} \end{bmatrix} \in \mathbb{R}^{6+n}. \tag{5.6}$$

When the configuration space of the vehicle is a subspace of the special Euclidean group we can use the formulation in Chap. 3 and write the velocity variables as

$$\tilde{\zeta}^S = \begin{bmatrix} \tilde{V}_{0b}^S \\ \dot{q} \end{bmatrix} \in \mathbb{R}^{m+n}, \qquad \tilde{\zeta} = \begin{bmatrix} \tilde{V}_{0b}^B \\ \dot{q} \end{bmatrix} \in \mathbb{R}^{m+n}, \tag{5.7}$$

where $\tilde{V} \in \mathbb{R}^m$ represents the admissible velocities of a subspace of $SE(3)$. In this case when $m < 6$, we define a selection matrix $H \in \mathbb{R}^{6\times m}$ such that the velocity vector of the vehicle is given by

$$V_{0b}^B = H\tilde{V}_{0b}^B, \tag{5.8}$$

where $\tilde{V}_{0b}^B \in \mathbb{R}^m$ determines the velocity state of the vehicle by selecting elements of V_{0b}^B that are not trivially zero. This was studied in detail in Chap. 3.

We will start by looking at the case when Q_b is a vector in $\mathbb{R}^6$ or $\mathbb{R}^m$. In this case we need to find the velocity transformation matrices in the same way that we did for single rigid bodies and the fixed-base manipulator. This is treated in detail in the next section. In Sect. 5.3 we derive the vehicle-manipulator kinematics when the vehicle configuration space is written as a matrix Lie group $Q_b \in SE(3)$ or one of the subgroups of $SE(3)$.

5.2 Velocity Transformation Matrices

We have found the state variables of the vehicle-manipulator system using different formulations. In the following we will study the relation between the time derivative of the position variables and the velocity variables for these systems. This relation is given by the Jacobian of the vehicle-manipulator system, also called the velocity transformation matrix. We start with one of the most important relations, which is the mapping from the velocity variable in twists to the time derivative of the position variable ξ.

5.2.1 Twist and Position Variables

The kinematic relation between the velocity variables in twists ζ and the time derivative of the vehicle-manipulator system pose ξ can be written as

$$\dot{\xi} = J_a(\eta_2)\zeta, \tag{5.9}$$

where $\xi = [\eta^\mathsf{T}\ q^\mathsf{T}]^\mathsf{T}$ and $\zeta = [(V_{0b}^B)^\mathsf{T}\ \dot{q}^\mathsf{T}]^\mathsf{T}$ are given as above. This relation is found by combining (3.64)–(3.65) and the observation that the joint transformations are Euclidean and can be written as (From et al. 2010; Fossen 2002)

$$\begin{bmatrix} \dot{\eta}_1 \\ \dot{\eta}_2 \\ \dot{q} \end{bmatrix} = \begin{bmatrix} R_{0b}(\eta_2) & 0 & 0 \\ 0 & T_{0b}(\eta_2) & 0 \\ 0 & 0 & I \end{bmatrix} \begin{bmatrix} v_{0b}^b \\ \omega_{0b}^b \\ \dot{q} \end{bmatrix} \tag{5.10}$$

where I (no subscript) denotes the $n \times n$ identity matrix.

The *body analytical Jacobian* of the vehicle-manipulator system is defined as

$$J_a(\eta_2) = \begin{bmatrix} J_b(\eta_2) & 0 \\ 0 & I \end{bmatrix} = \begin{bmatrix} R_{0b}(\eta_2) & 0 & 0 \\ 0 & T_{0b}(\eta_2) & 0 \\ 0 & 0 & I \end{bmatrix} \in \mathbb{R}^{(6+n)\times(6+n)}. \tag{5.11}$$

J_b can be found explicitly in (3.67). We have already seen that this transformation has a singularity at a given configuration. The body analytical Jacobian $J_a(\eta_2)$ of the vehicle-manipulator system therefore has a singularity at the same point in the vehicle configuration space.

Similarly the spatial analytical Jacobian $J_a^S(\eta_2)$ of the vehicle-manipulator system takes us from ζ^S to $\dot{\xi}$ and thus requires the mapping from V_{0b}^S to $\dot{\eta}$. The spatial analytical Jacobian is then found by replacing $J_b(\eta_2)$ in (5.11) with $J_b^S(\eta)$, i.e.,

$$J_a^S(\eta_2) = \begin{bmatrix} J_b^S(\eta_2) & 0 \\ 0 & I \end{bmatrix} \in \mathbb{R}^{(6+n)\times(6+n)} \tag{5.12}$$

where $J_b^S(\eta_2)$ is given in (3.69).

5.2.2 Link Velocities

Let the twist coordinates V_{0i}^S denote the linear and angular velocity of body i represented in spatial coordinates, and $J_{gi}^S(\xi) \in \mathbb{R}^{6\times(6+n)}$ be the velocity transformation matrix of link i that gives the relation

$$V_{0i}^S = J_{gi}^S(\xi)\zeta. \tag{5.13}$$

This mapping is important as it gives us the velocity of a link i in the robotic chain, given the vehicle velocity twist and the joint velocities. The velocity transformation matrix $J_{gi}^S(\xi)$ of link i is given by

$$J_{gi}^S(\xi) = \frac{\partial V_{0i}^S}{\partial \zeta}. \tag{5.14}$$

We can also find the velocity transformation matrix $J_{gi}^S(\xi)$ in terms of the Adjoint maps and the Jacobians that we found in the previous section. We first note that the velocity of a link i with respect to the inertial frame can be written as the sum of the velocity of the vehicle and the velocity of the robot links with respect to the vehicle. We see this if we write the configuration of link i with respect to the inertial frame $\mathscr{F}_0$ as $g_{0i} = g_{0b}g_{bi}$, i.e., we transform first to the frame attached to the vehicle and then from the vehicle to link i. The corresponding velocity $\widehat{V}_{0i}^S$ of link i in spatial coordinates is found as (Murray et al. 1994)

$$\begin{aligned}
\widehat{V}_{0i}^S &= \dot{g}_{0i} g_{0i}^{-1} \\
&= (\dot{g}_{0b} g_{bi} + g_{0b} \dot{g}_{bi})(g_{0b} g_{bi})^{-1} \\
&= (\dot{g}_{0b} g_{bi} + g_{0b} \dot{g}_{bi}) g_{bi}^{-1} g_{0b}^{-1} \\
&= \dot{g}_{0b} g_{bi} g_{bi}^{-1} g_{0b}^{-1} + g_{0b} \dot{g}_{bi} g_{bi}^{-1} g_{0b}^{-1} \\
&= \dot{g}_{0b} g_{0b}^{-1} + g_{0b} \big(\dot{g}_{bi} g_{bi}^{-1}\big) g_{0b}^{-1} \\
&= \widehat{V}_{0b}^S + g_{0b} \big(\widehat{V}_{bi}^S\big) g_{0b}^{-1} \\
&= \widehat{V}_{0b}^S + \mathrm{Ad}_{g_{0b}} \widehat{V}_{bi}^S.
\end{aligned} \tag{5.15}$$

Recall that a spatial twist is represented with respect to some reference frame: $\widehat{V}_{0b}^S$ is the spatial twist with respect to $\mathscr{F}_0$, i.e., the velocities of the possibly imaginary point in the origin of $\mathscr{F}_0$; whereas $\widehat{V}_{bi}^S$ is the velocities in the origin of $\mathscr{F}_b$. To add these velocities we therefore need to transform $\widehat{V}_{bi}^S$ so that it represents the velocities in $\mathscr{F}_0$ and not in $\mathscr{F}_b$, which is obtained by the Adjoint map $\mathrm{Ad}_{g_{0b}}$. In this way, both $\widehat{V}_{0b}^S$ and $\mathrm{Ad}_{g_{0b}} \widehat{V}_{bi}^S$ are the spatial velocities with respect to the inertial frame $\mathscr{F}_0$. The relation $\widehat{V}_{0i}^S = \widehat{V}_{0b}^S + \mathrm{Ad}_{g_{0b}} \widehat{V}_{bi}^S$ will greatly simplify the dynamic equations of vehicle manipulator systems. It is, however, also useful to write the velocities in terms of the body velocity twist, i.e., as viewed from reference frame $\mathscr{F}_b$. We can then write the same relation, but as observed from the body frame, as

$$
\begin{aligned}
\widehat{V}_{0i}^{B} &= \mathrm{Ad}_{g_{0i}}^{-1}\widehat{V}_{0i}^{S} \\
&= \mathrm{Ad}_{g_{0i}}^{-1}\big(\widehat{V}_{0b}^{S} + \mathrm{Ad}_{g_{0b}}\widehat{V}_{bi}^{S}\big) \\
&= \mathrm{Ad}_{g_{0i}}^{-1}\big(\mathrm{Ad}_{g_{0b}}\widehat{V}_{0b}^{B} + \mathrm{Ad}_{g_{0b}}\mathrm{Ad}_{g_{bi}}\widehat{V}_{bi}^{B}\big) \\
&= \mathrm{Ad}_{g_{bi}}^{-1}\mathrm{Ad}_{g_{0b}}^{-1}\mathrm{Ad}_{g_{0b}}\widehat{V}_{0b}^{B} + \mathrm{Ad}_{g_{0i}}^{-1}\mathrm{Ad}_{g_{0b}}\mathrm{Ad}_{g_{bi}}\widehat{V}_{bi}^{B} \\
&= \mathrm{Ad}_{g_{bi}}^{-1}\widehat{V}_{0b}^{B} + \widehat{V}_{bi}^{B}.
\end{aligned}
\tag{5.16}
$$

We can now conclude with the following important result:

Theorem 5.1 *Assume a vehicle-manipulator system with inertial frame $\mathscr{F}_0$, a non-inertial frame $\mathscr{F}_b$ attached to the vehicle, and coordinate frames $\mathscr{F}_i$ attached to the center of gravity of each link $i = 1, \dots, n$ of the manipulator. Then the velocity of frame $\mathscr{F}_i$ with respect to the inertial frame $\mathscr{F}_0$ is given by the spatial twist V_{0i}^S and satisfies the relation*

$$
V_{0i}^{S} = V_{0b}^{S} + \mathrm{Ad}_{g_{0b}} V_{bi}^{S}. \tag{5.17}
$$

The velocity can also be written as observed from frame $\mathscr{F}_i$ in body coordinates as the twist V_{0i}^B, in which case the following is true:

$$
V_{0i}^{B} = \mathrm{Ad}_{g_{bi}}^{-1} V_{0b}^{B} + V_{bi}^{B}. \tag{5.18}
$$

We can now find $J_{gi}^S(\xi)$ in (5.13) from (5.17) if we use that

$$
V_{0b}^{S} = \mathrm{Ad}_{g_{0b}} V_{0b}^{B} \tag{5.19}
$$

and

$$
V_{bi}^{S} = J_i(q)\dot{q} \tag{5.20}
$$

where $J_i(q)$ is the same as $J_{m,g}^S$ in Theorem 4.3 but without the influence of the last joints (recall that joints j for $i < j$ do not affect the motion of link i for a serial manipulator):

$$
J_i(q) = \big[\mathrm{Ad}_{g_{0\bar{1}}}(q)X_1^1 \quad \mathrm{Ad}_{g_{0\bar{2}}}(q)X_2^2 \quad \mathrm{Ad}_{g_{0\bar{3}}}(q)X_3^3 \quad \cdots \quad \mathrm{Ad}_{g_{0\bar{i}}}(q)X_i^i \quad 0_{(n-i)\times 6}\big]. \tag{5.21}
$$

We denote $J_i(q)$ the *spatial geometric Jacobian of link i* and can be found from Proposition 4.1. The following conclusions can now be drawn:

Theorem 5.2 *The spatial geometric Jacobian $J^S_{gi}(\xi)$ that gives the relation $V^S_{0i} = J^S_{gi}(\xi)\zeta$ is given by*

$$J^S_{gi}(\xi) = \begin{bmatrix} \mathrm{Ad}_{g_{0b}} & \mathrm{Ad}_{g_{0b}} J_i \end{bmatrix} \in \mathbb{R}^{6\times(6+n)}. \tag{5.22}$$

Proof This expression is obtained by simply substituting (5.19) and (5.20) into (5.17):

$$\begin{aligned} V^S_{0i} &= V^S_{0b} + \mathrm{Ad}_{g_{0b}} V^S_{bi} \\ &= \mathrm{Ad}_{g_{0b}} V^B_{0b} + \mathrm{Ad}_{g_{0b}} J_i \dot{q} \\ &= \begin{bmatrix} \mathrm{Ad}_{g_{0b}} & \mathrm{Ad}_{g_{0b}} J_i \end{bmatrix} \begin{bmatrix} V^B_{0b} \\ \dot{q} \end{bmatrix} \\ &= J^S_{gi}(\xi)\zeta. \end{aligned} \tag{5.23}$$

□

Theorem 5.3 *The body geometric Jacobian $J^B_{gi}(q)$ that gives the relation $V^B_{0i} = J^B_{gi}(q)\zeta$ is given by*

$$J^B_{gi}(q) = \begin{bmatrix} \mathrm{Ad}^{-1}_{g_{bi}} & \mathrm{Ad}^{-1}_{g_{bi}} J_i \end{bmatrix} \in \mathbb{R}^{6\times(6+n)}. \tag{5.24}$$

Proof We see this if we write the expression as

$$\begin{aligned} V^B_{0i} &= \mathrm{Ad}^{-1}_{g_{bi}} V^B_{0b} + V^B_{bi} \\ &= \mathrm{Ad}^{-1}_{g_{bi}} V^B_{0b} + \mathrm{Ad}^{-1}_{g_{bi}} V^S_{bi} \\ &= \mathrm{Ad}^{-1}_{g_{bi}} V^B_{0b} + \mathrm{Ad}^{-1}_{g_{bi}} J_i \dot{q} \\ &= \begin{bmatrix} \mathrm{Ad}^{-1}_{g_{bi}} & \mathrm{Ad}^{-1}_{g_{bi}} J_i \end{bmatrix} \begin{bmatrix} V^B_{0b} \\ \dot{q} \end{bmatrix} \\ &= J^B_{gi}(q)\zeta. \end{aligned} \tag{5.25}$$

□

Note that $J_{gi}^B(q)$ only depends on the joint variables q and not the vehicle configuration state g_{0b}.

When the configuration space of the vehicle is a subspace of the special Euclidean group we will write the velocity of the vehicle as a vector $\tilde{V}_{0b}^B$ as in Eq. (3.80) in Sect. 3.3.5. The twist of the vehicle is then given as $V_{0b}^B = H\tilde{V}_{0b}^B$ and the twist of link i is found as

$$V_{0i}^B = \mathrm{Ad}_{g_{bi}}^{-1} V_{0b}^B + V_{bi}^B = \mathrm{Ad}_{g_{bi}}^{-1}\big(H\tilde{V}_{0b}^B + J_i\dot{q}\big). \tag{5.26}$$

Note once again that the body geometric Jacobian $J_{gi}^B(q)$ does not depend on the configuration of the vehicle. Similarly, in spatial coordinates we get

$$V_{0i}^S = V_{0b}^S + \mathrm{Ad}_{g_{0b}} V_{bi}^S = \mathrm{Ad}_{g_{0b}}\big(H\tilde{V}_{0b}^S + J_i\dot{q}\big). \tag{5.27}$$

Recall from the previous chapter that we can reduce the computational efforts by finding the velocities of the frames $\bar{\mathscr{F}}_i$ located at the axes of rotation or translation. As for fixed-base manipulators the velocity in spatial coordinates is the same for $\mathscr{F}_i$ and $\bar{\mathscr{F}}_i$, while in body coordinates we can use the transformation from $\mathscr{F}_b$ to frame $\bar{\mathscr{F}}_i$, which reduces computations slightly compared to $\mathscr{F}_i$:

Theorem 5.4 *Assume a vehicle-manipulator system with inertial frame $\mathscr{F}_0$, a non-inertial frame $\mathscr{F}_b$ attached to the vehicle, and coordinate frames $\bar{\mathscr{F}}_i$ attached to the link axis of each link $i = 1, \dots, n$ of the manipulator. Then the velocity of frame $\bar{\mathscr{F}}_i$ with respect to the inertial frame $\mathscr{F}_0$ is given in spatial coordinates by the spatial twist $V_{0\bar{i}}^S$ and satisfies the relation*

$$V_{0\bar{i}}^S = V_{0b}^S + \mathrm{Ad}_{g_{0b}} V_{b\bar{i}}^S = V_{0b}^S + \mathrm{Ad}_{g_{0b}} V_{bi}^S = V_{0i}^S. \tag{5.28}$$

The velocity can also be written as observed from frame $\bar{\mathscr{F}}_i$ in body coordinates as the twist $V_{0\bar{i}}^B$, in which case the following is true:

$$V_{0\bar{i}}^B = \mathrm{Ad}_{g_{b\bar{i}}}^{-1} V_{0b}^B + V_{b\bar{i}}^B. \tag{5.29}$$

We can now rewrite Theorems 5.2 and 5.3 for the link frames $\bar{\mathscr{F}}_i$ attached to the joint axes:

Theorem 5.5 *The spatial geometric Jacobian $\bar{J}^S_{gi}(\xi)$ that gives the relation $V^S_{0\bar{i}} = \bar{J}^S_{gi}(\xi)\zeta$ is given by*

$$\bar{J}^S_{gi}(\xi) = \begin{bmatrix} \mathrm{Ad}_{g_{0b}} & \mathrm{Ad}_{g_{0b}} J_i \end{bmatrix} = J^S_{gi}(\xi) \in \mathbb{R}^{6\times(6+n)}. \tag{5.30}$$

Proof This follows directly from Proposition 4.1 and Theorem 5.2. □

Theorem 5.6 *The body geometric Jacobian $\bar{J}^B_{gi}(q)$ that gives the relation $V^B_{0\bar{i}} = \bar{J}^B_{gi}(q)\zeta$ is given by*

$$\bar{J}^B_{gi}(q) = \begin{bmatrix} \mathrm{Ad}^{-1}_{g_{b\bar{i}}} & \mathrm{Ad}^{-1}_{g_{b\bar{i}}} J_i \end{bmatrix} \in \mathbb{R}^{6\times(6+n)}. \tag{5.31}$$

Proof We see this if we write the expression as

$$\begin{aligned} V^B_{0\bar{i}} &= \mathrm{Ad}^{-1}_{g_{b\bar{i}}} V^B_{0b} + V^B_{b\bar{i}} \\ &= \mathrm{Ad}^{-1}_{g_{b\bar{i}}} V^B_{0b} + \mathrm{Ad}^{-1}_{g_{b\bar{i}}} V^S_{bi} \\ &= \mathrm{Ad}^{-1}_{g_{b\bar{i}}} V^B_{0b} + \mathrm{Ad}^{-1}_{g_{b\bar{i}}} J_i \dot{q} \\ &= \begin{bmatrix} \mathrm{Ad}^{-1}_{g_{b\bar{i}}} & \mathrm{Ad}^{-1}_{g_{b\bar{i}}} J_i \end{bmatrix} \begin{bmatrix} V^B_{0b} \\ \dot{q} \end{bmatrix} \\ &= \bar{J}^B_{gi}(q)\zeta. \end{aligned} \tag{5.32}$$

□

5.2.3 The Geometric Jacobian

Of particular interest when working with any kind of robotics is the location and velocity of the robot end effector. We can find the end-effector velocity in the same way as we did for the links in the previous section.

Theorem 5.7 *The spatial geometric Jacobian $J^S_{ge}(\xi)$ that gives the relation $V^S_{0e} = J^S_{ge}(\xi)\zeta$ is given by*

$$J^S_{ge}(\xi) = \begin{bmatrix} \mathrm{Ad}_{g_{0b}} & \mathrm{Ad}_{g_{0b}} J^S_{m,g} \end{bmatrix} \in \mathbb{R}^{6\times(6+n)}. \tag{5.33}$$

Proof This follows directly from Theorem 5.2 by setting $i = e$ and noting that $J_e = J^S_{m,g}$. □

Theorem 5.8 *The body geometric Jacobian $J^B_{ge}(q)$ that gives the relation $V^B_{0e} = J^B_{ge}(q)\zeta$ is given by*

$$J^B_{ge}(q) = \begin{bmatrix} \mathrm{Ad}^{-1}_{g_{be}} & \mathrm{Ad}^{-1}_{g_{be}} J^S_{m,g} \end{bmatrix} \in \mathbb{R}^{6\times(6+n)}. \tag{5.34}$$

Proof This follows directly from Theorem 5.3 by setting $i = e$ and noting that $J_e = J^S_{m,g}$. □

5.2.4 Workspace Jacobian

One of the most important pieces of information that we need when dealing with vehicle-manipulator systems is how the end-effector is located with respect to the inertial frame. This allows us to use the robot to perform operations in its workspace and it also allows us to characterize the robot's environment and workspace relative to the inertial frame. This is more convenient than to use the base frame as a reference because the base frame is non-inertial.

We have already seen that the end effector of the robot is located at the pose given by $g_{0e} = g_{0b}g_{be}$. To perform operations in this workspace we also need to specify the *end-effector velocities in the inertial frame*. We can write the configuration of the vehicle-manipulator system in terms of the vehicle configuration $\eta = [x_{0b}\ y_{0b}\ z_{0b}\ \phi_{0b}\ \theta_{0b}\ \psi_{0b}]$ and the end effector configuration vector as $\eta_{0e} = [x_{0e}\ y_{0e}\ z_{0e}\ \phi_{0e}\ \theta_{0e}\ \psi_{0e}]^\mathsf{T}$. The vector $\varsigma = [\eta^\mathsf{T}\ \eta^\mathsf{T}_{0e}]^\mathsf{T} \in \mathbb{R}^{12}$ then relates to the vehicle twist and joint velocities through the *workspace Jacobian* $J_w(\xi)$ by

$$\dot{\varsigma} = J_w(\xi)\zeta. \tag{5.35}$$

We have found the mapping $J_b(\eta_2)$ from V^B_{0b} to $\dot{\eta}$ in (3.67). This gives us the first line of (5.35).

To find $\dot{\eta}_{0e}$ we first notice that it depends on both V_{0b}^B and $\dot{q}$ and that

$$\dot{\eta}_{0e} = \begin{bmatrix} R_{0e}(\eta_{0e,2}) & 0 \\ 0 & T_{0e}(\eta_{0e,2}) \end{bmatrix} \begin{bmatrix} v_{0e}^B \\ \omega_{0e}^B \end{bmatrix}, \tag{5.36}$$

where we define

$$J_e(\eta_{0e,2}) = \begin{bmatrix} R_{0e}(\eta_{0e,2}) & 0 \\ 0 & T_{0e}(\eta_{0e,2}) \end{bmatrix}. \tag{5.37}$$

This relation is equivalent to the one we found in (3.64) but with η_{0e} instead of η. Now that we have $\dot{\eta}_{0e} = J_e(\eta_{0e,2})V_{0e}^B$ we can write V_{0e}^B as

$$V_{0e}^B = \mathrm{Ad}_{g_{be}}^{-1} V_{0b}^B + V_{be}^B. \tag{5.38}$$

The part of the end-effector velocity that is due to the motion of the joints is found as

$$V_{be}^B = J_{m,g}(q)\dot{q} \tag{5.39}$$

from (4.77) in Theorem 4.2. We finally use that $\dot{\eta}_{0e} = J_e(\eta_{0e,2})V_{0e}^B$ and get an expression for $\dot{\eta}_{0e}$ in terms of V_{0b}^B and $\dot{q}$:

$$\begin{aligned} \dot{\eta}_{0e} &= J_e(\eta_{0e,2})V_{0e}^B \\ &= J_e(\eta_{0e,2})\big(\mathrm{Ad}_{g_{be}}^{-1}(q)V_{0b}^B + V_{be}^B\big) \\ &= J_e(\eta_{0e,2})\mathrm{Ad}_{g_{be}}^{-1}(q)V_{0b}^B + J_e(\eta_{0e,2})J_{m,g}(q)\dot{q}. \end{aligned} \tag{5.40}$$

The required mapping is then given by

$$\begin{bmatrix} \dot{\eta} \\ \dot{\eta}_{0e} \end{bmatrix} = \begin{bmatrix} J_b(\eta_2) & 0 \\ J_e(\eta_{0e,2})\mathrm{Ad}_{g_{be}}^{-1}(q) & J_e(\eta_{0e,2})J_{m,g}(q) \end{bmatrix} \begin{bmatrix} V_{0b}^B \\ \dot{q} \end{bmatrix}. \tag{5.41}$$

We can now define

$$J_w(\xi) = \begin{bmatrix} J_b(\eta_2) & 0 \\ J_e(\eta_{0e,2})\mathrm{Ad}_{g_{be}}^{-1}(q) & J_e(\eta_{0e,2})J_{m,g}(q) \end{bmatrix} \in \mathbb{R}^{12\times(6+n)} \tag{5.42}$$

as the workspace Jacobian of the VM system.

As expected, the vehicle velocity $\dot{\eta}$ does not depend on how the manipulator moves. The end-effector velocity $\dot{\eta}_{0e}$, on the other hand, depends on both the vehicle and manipulator motion and the Jacobian is dependent on the vehicle attitude η_2 and the joint positions q.

5.3 Configuration States

Using the framework presented in the previous chapters the configuration space of a multibody system can be written in its most general form as a set $Q = \{Q_i\}$ of configuration states Q_i (not necessarily Euclidean), a vector v of velocity states $v_i \in \mathbb{R}^{n_i}$, and several mappings (see Sect. 5.3.1) that describe the local Euclidean structure of the configuration states and their relation to the velocity states. For a vehicle-manipulator system the state space is denoted in the following way:

Theorem 5.9 *The state space of a vehicle-manipulator system can be written in terms of configuration states Q_i for $i = b, 1, 2, \dots, n$ where*

- *Q_b is an element of $SE(3)$ or one of its subgroups, and*
- *Q_i for $i = 1, \dots, n$ are elements in $\mathbb{R}$.*

The velocity state of the vehicle-manipulator system is described in terms of velocity states v_i for $i = b, 1, 2, \dots, n$ where

- *v_b is an element of $se(3)$ or one of its subgroups, and*
- *v_i for $i = 1, \dots, n$ are elements in $\mathbb{R}$.*

We will write the state space of the vehicle-manipulator system as $Q = \{g_{0b}, \mathbb{R}^n\}$ and $v = \{V_{0b}^B, \mathbb{R}^n\}$ if the configuration space is $SE(3)$ or as $Q = \{Q_b, \mathbb{R}^n\}$ and $v = \{\tilde{V}_{0b}^B, \mathbb{R}^n\}$ if the configuration space is a subspace of $SE(3)$.

The pose Q_b and velocity variable V_{0b}^B or $\tilde{V}_{0b}^B$ of the vehicle are typically taken from Table 3.1 and the configurations and velocities of the robotic links are simply chosen as the joint positions and velocities as in Chap. 4 so that $Q_i = q_i$ and $v_i = \dot{q}_i$. In addition we need to constrain the motion of the joint to the correct manifold, i.e., each joint corresponds to a homogeneous transformation $g_i(Q_i)$.

We use the notation v_i to describe a general velocity instead of restricting ourselves to velocities written in the specific form $\zeta = [(V_{0b}^B)^{\mathsf{T}}\ \dot{q}^{\mathsf{T}}]^{\mathsf{T}} \in \mathbb{R}^{6+n}$. We will often use $v = \zeta$, but in general we will allow for other representations. This will for example allow us to write the velocity of the vehicle as a matrix representation of the Lie algebra, i.e., an element of $se(3)$ or one of its subgroups.

5.3.1 Local Coordinates

Vehicle-manipulator systems can be described in terms of local coordinates in the same way as we did for single rigid bodies and fixed-base manipulators. We then choose six variables $\varphi_V = [\varphi_1\ \varphi_2 \cdots \varphi_6]^{\mathsf{T}}$ to represent the configuration of the vehicle (m variables for a subspace of $SE(3)$), and n variables $\varphi_q = [\varphi_7\ \varphi_8 \cdots \varphi_{6+n}]^{\mathsf{T}}$ for the manipulator. We need to find the relation between the local coordinates and the global velocity variables $v = [(V_{0b}^B)^\mathsf{T}\ \dot{q}^\mathsf{T}]^\mathsf{T}$. We can use the results from the previous chapters and write this in a block-diagonal form as (Rossmann 2002; Duindam and Stramigioli 2007)

$$S(Q,\varphi) = \begin{bmatrix} (I - \frac{1}{2}\operatorname{ad}_{\varphi_V} + \frac{1}{6}\operatorname{ad}^2_{\varphi_V} - \cdots) & 0 \\ 0 & I \end{bmatrix} \in \mathbb{R}^{(6+n)\times(6+n)} \tag{5.43}$$

when the configuration space of the vehicle is $SE(3)$ or as

$$S(Q,\varphi) = \begin{bmatrix} (I - \frac{1}{2}\operatorname{ad}_{\varphi_{\tilde{V}}} + \frac{1}{6}\operatorname{ad}^2_{\varphi_{\tilde{V}}} - \cdots) & 0 \\ 0 & I \end{bmatrix} \in \mathbb{R}^{(m+n)\times(m+n)} \tag{5.44}$$

for a subspace with m degrees of freedom in the vehicle motion. Here $\operatorname{ad}_{\varphi_{\tilde{V}}}$ is the adjoint map defining the Lie bracket of a subgroup of $SE(3)$ (see Sect. 2.6.6.6).

The velocity transformation matrix $S(q,\varphi)$ gives us an important kinematic relation of the vehicle-manipulator system: The mapping from local to global velocity variables can now be written as

$$\begin{bmatrix} V_{0b}^B \\ \dot{q} \end{bmatrix} = \begin{bmatrix} (I - \frac{1}{2}\operatorname{ad}_{\varphi_V} + \frac{1}{6}\operatorname{ad}^2_{\varphi_V} - \cdots) & 0 \\ 0 & I \end{bmatrix} \begin{bmatrix} \dot{\varphi}_V \\ \dot{\varphi}_q \end{bmatrix} \tag{5.45}$$

for a vehicle configuration space $SE(3)$, and

$$\begin{bmatrix} \tilde{V}_{0b}^B \\ \dot{q} \end{bmatrix} = \begin{bmatrix} (I - \frac{1}{2}\operatorname{ad}_{\varphi_{\tilde{V}}} + \frac{1}{6}\operatorname{ad}^2_{\varphi_{\tilde{V}}} - \cdots) & 0 \\ 0 & I \end{bmatrix} \begin{bmatrix} \dot{\varphi}_{\tilde{V}} \\ \dot{\varphi}_q \end{bmatrix} \tag{5.46}$$

for a Lie subgroup.

There are several important properties of these equations that give us insight into the kinematics of vehicle-manipulator systems. These properties will also be used in the chapters to come to simplify the dynamic equations. We first note that the velocity transformation matrix depends only on the local coordinates and that the global variables Q_b and Q_i do not appear in the expression. We also note that the velocity transformation matrix depends only on the local variables φ_V of the vehicle, and not the manipulator variables φ_q. The reason for this, of course, is that the transformation of the vehicle is non-Euclidean while we have assumed that the robot consists only of Euclidean joints, which gives $\dot{\varphi}_q = \dot{q}$.

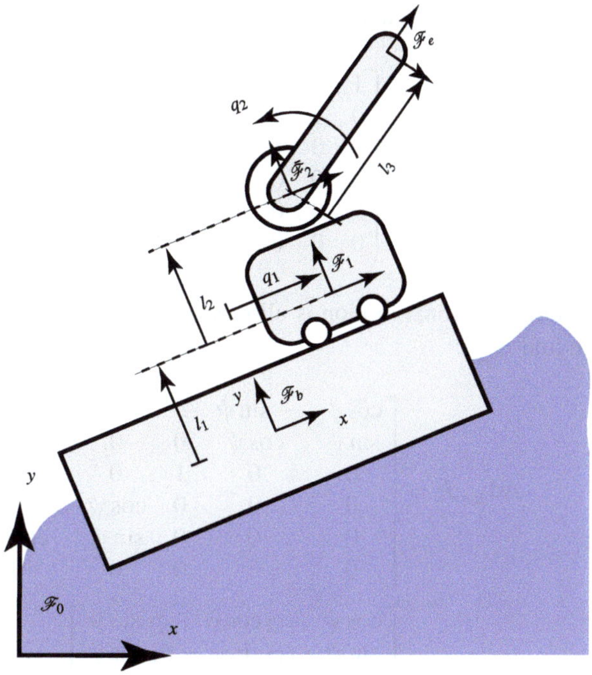

Fig. 5.2 Two-link robot with one prismatic and one revolute joint attached to a vehicle

5.4 Some Simple Examples

Example 5.1 We will now show how to find the spatial geometric Jacobian of the second joint for a simple robotic system. Consider the vehicle-manipulator system in Fig. 5.2 with planar motion only and where the vehicle only allows for rotational motion. The robotic manipulator has one prismatic and one revolute joint and we choose the home position $q_1 = q_2 = 0$ when the second link points in the direction of the y-axis. To simplify the computations we choose the inertial frame so that its origin coincides with the origin of the base frame, i.e., $\mathcal{O}_0 = \mathcal{O}_b$. The spatial geometric Jacobian of link i is given by

$$J_{gi}^S(\xi) = \bar{J}_{gi}^S(\xi) = \begin{bmatrix} \mathrm{Ad}_{g_{0b}} & \mathrm{Ad}_{g_{0b}} J_i \end{bmatrix}. \tag{5.47}$$

The manipulator Jacobian of joint 2 is given as

$$J_2 = \begin{bmatrix} X_1' & X_2' \end{bmatrix} = \begin{bmatrix} X_1^1 & \mathrm{Ad}_{g_{b\bar{2}}} X_2^2 \end{bmatrix} = \begin{bmatrix} 1 & l_{12} \\ 0 & -q_1 \\ 0 & 0 \\ 0 & 0 \\ 0 & 0 \\ 0 & 1 \end{bmatrix} \tag{5.48}$$

where $l_{12} = l_1 + l_2$ and we have used that

$$\mathrm{Ad}_{g_{b\bar{2}}} X_2^2 = \begin{bmatrix} 1 & 0 & 0 & 0 & 0 & l_{12} \\ 0 & 1 & 0 & 0 & 0 & -q_1 \\ 0 & 0 & 1 & -l_{12} & q_1 & 0 \\ 0 & 0 & 0 & 1 & 0 & 0 \\ 0 & 0 & 0 & 0 & 1 & 0 \\ 0 & 0 & 0 & 0 & 0 & 1 \end{bmatrix} \begin{bmatrix} 0 \\ 0 \\ 0 \\ 0 \\ 0 \\ 1 \end{bmatrix} = \begin{bmatrix} l_{12} \\ -q_1 \\ 0 \\ 0 \\ 0 \\ 1 \end{bmatrix}. \tag{5.49}$$

As only rotational motion is allowed for the vehicle and we assume that $\mathscr{O}_0 = \mathscr{O}_b$ we find

$$\mathrm{Ad}_{g_{0b}} J_2 = \begin{bmatrix} \cos\psi & -\sin\psi & 0 & 0 & 0 & 0 \\ \sin\psi & \cos\psi & 0 & 0 & 0 & 0 \\ 0 & 0 & 1 & 0 & 0 & 0 \\ 0 & 0 & 0 & \cos\psi & -\sin\psi & 0 \\ 0 & 0 & 0 & \sin\psi & \cos\psi & 0 \\ 0 & 0 & 0 & 0 & 0 & 1 \end{bmatrix} \begin{bmatrix} 1 & l_{12} \\ 0 & -q_1 \\ 0 & 0 \\ 0 & 0 \\ 0 & 0 \\ 0 & 1 \end{bmatrix}$$

$$= \begin{bmatrix} \cos\psi & l_{12}\cos\psi + q_1 \sin\psi \\ \sin\psi & l_{12}\sin\psi - q_1\cos\psi \\ 0 & 0 \\ 0 & 0 \\ 0 & 0 \\ 0 & 1 \end{bmatrix}. \tag{5.50}$$

J_2 is the manipulator geometric Jacobian and $J_2\dot{q}$ gives us the velocities in spatial coordinates, i.e., the velocities of an imaginary point attached to the second link as it travels through the point $\mathscr{O}_0 = \mathscr{O}_b$, represented in (the non-inertial) frame $\mathscr{F}_b$. $\mathrm{Ad}_{g_{0b}} J_i \dot{q}$ represents the same velocities but also takes into account the difference in orientation between $\mathscr{F}_0$ and $\mathscr{F}_b$ and represents the velocities in the inertial frame $\mathscr{F}_0$. The spatial geometric Jacobian is given by

$$J_{g2}^S(\xi) = \begin{bmatrix} \mathrm{Ad}_{g_{0b}} & \mathrm{Ad}_{g_{0b}} J_2 \end{bmatrix}$$

$$= \begin{bmatrix} \cos\psi & -\sin\psi & 0 & 0 & 0 & 0 & \cos\psi & l_{12}\cos\psi + q_1\sin\psi \\ \sin\psi & \cos\psi & 0 & 0 & 0 & 0 & \sin\psi & l_{12}\sin\psi - q_1\cos\psi \\ 0 & 0 & 1 & 0 & 0 & 0 & 0 & 0 \\ 0 & 0 & 0 & \cos\psi & -\sin\psi & 0 & 0 & 0 \\ 0 & 0 & 0 & \sin\psi & \cos\psi & 0 & 0 & 0 \\ 0 & 0 & 0 & 0 & 0 & 1 & 0 & 1 \end{bmatrix}. \tag{5.51}$$

Because only angular velocities are allowed for the vehicle we can write the vehicle velocities as $V_{0b}^B = [0\,0\,0\,0\,0\,r]^\mathsf{T}$ and the spatial velocities are given as

$$V_{02}^S = J_{g2}^S(\xi)v = \begin{bmatrix}\mathrm{Ad}_{g_{0b}} & \mathrm{Ad}_{g_{0b}}J_2\end{bmatrix}\begin{bmatrix}V_{0b}^b \\ \dot{q}\end{bmatrix}$$

$$= \begin{bmatrix} \cos\psi & -\sin\psi & 0 & 0 & 0 & 0 & \cos\psi & l_{12}\cos\psi + q_1\sin\psi \\ \sin\psi & \cos\psi & 0 & 0 & 0 & 0 & \sin\psi & l_{12}\sin\psi - q_1\cos\psi \\ 0 & 0 & 1 & 0 & 0 & 0 & 0 & 0 \\ 0 & 0 & 0 & \cos\psi & -\sin\psi & 0 & 0 & 0 \\ 0 & 0 & 0 & \sin\psi & \cos\psi & 0 & 0 & 0 \\ 0 & 0 & 0 & 0 & 0 & 1 & 0 & 1 \end{bmatrix}\begin{bmatrix}0\\0\\0\\0\\0\\r\\\dot{q}_1\\\dot{q}_2\end{bmatrix}$$

$$= \begin{bmatrix} \cos\psi\dot{q}_1 + (l_{12}\cos\psi + q_1\sin\psi)\dot{q}_2 \\ \sin\psi\dot{q}_1 + (l_{12}\sin\psi - q_1\cos\psi)\dot{q}_2 \\ 0 \\ 0 \\ 0 \\ r + \dot{q}_2 \end{bmatrix}. \tag{5.52}$$

Example 5.2 Before we find the body geometric Jacobian for the system in Example 5.1, illustrated in Fig. 5.2 we will look at a simplified example: we will start by finding the velocity of the second link as seen from the frame $\bar{\mathscr{F}}_2^0$, i.e., the frame attached to the second joint when $q_2 = 0$. This will make it easier to verify the final expressions. In the next example we will find the configuration dependent body velocity twist $V_{0\bar{2}}^{\bar{2}}$ as observed from $\bar{\mathscr{F}}_2$. Again we choose the inertial frame so that its origin coincides with the origin of the base frame, i.e., $\mathscr{O}_0 = \mathscr{O}_b$. The body geometric Jacobian is given by

$$J_{gi}^{\bar{2}}(\xi) = \begin{bmatrix}\mathrm{Ad}_{g_{b\bar{i}}}^{-1} & \mathrm{Ad}_{g_{b\bar{i}}}^{-1}J_i\end{bmatrix} \tag{5.53}$$

where we use the subscript to illustrate that we represent the velocity in frame $\bar{\mathscr{F}}_2^0$, and not in body coordinates (frame $\bar{\mathscr{F}}_2$). The manipulator Jacobian of joint 2 was found in (5.48). We now find

$$\mathrm{Ad}_{g_{b\bar{2}}}^{-1}J_2 = \begin{bmatrix} 1 & 0 & 0 & 0 & 0 & -l_{12} \\ 0 & 1 & 0 & 0 & 0 & q_1 \\ 0 & 0 & 1 & l_{12} & -q_1 & 0 \\ 0 & 0 & 0 & 1 & 0 & 0 \\ 0 & 0 & 0 & 0 & 1 & 0 \\ 0 & 0 & 0 & 0 & 0 & 1 \end{bmatrix}\begin{bmatrix} 1 & l_{12} \\ 0 & -q_1 \\ 0 & 0 \\ 0 & 0 \\ 0 & 0 \\ 0 & 1 \end{bmatrix} = \begin{bmatrix} 1 & 0 \\ 0 & 0 \\ 0 & 0 \\ 0 & 0 \\ 0 & 0 \\ 0 & 1 \end{bmatrix}. \tag{5.54}$$

We see that for this system the manipulator geometric Jacobian J_2 takes a particularly simple form when it is transformed to the body frame ($\mathrm{Ad}_{g_{b\bar{2}}}^{-1}J_2$). The body geometric Jacobian is given by

$$\bar{J}^{\bar{2}}_{g2}(\xi) = \begin{bmatrix} \mathrm{Ad}^{-1}_{g_{b\bar{2}}} & \mathrm{Ad}^{-1}_{g_{b\bar{2}}} J_2 \end{bmatrix}$$

$$= \begin{bmatrix} 1 & 0 & 0 & 0 & 0 & -l_{12} & 1 & 0 \\ 0 & 1 & 0 & 0 & 0 & q_1 & 0 & 0 \\ 0 & 0 & 1 & l_{12} & -q_1 & 0 & 0 & 0 \\ 0 & 0 & 0 & 1 & 0 & 0 & 0 & 0 \\ 0 & 0 & 0 & 0 & 1 & 0 & 0 & 0 \\ 0 & 0 & 0 & 0 & 0 & 1 & 0 & 1 \end{bmatrix}. \tag{5.55}$$

Here we have used $R_{0\bar{2}} = I$ because we want to find the transformation as represented in the frame $\bar{\mathscr{F}}^0_2$. Normally when computing the body link Jacobian we would also include the rotation around the z-axis to obtain the body velocities represented in the body frame $\bar{\mathscr{F}}_2$. Again we write the vehicle velocities as $V^B_{0b} = [0\,0\,0\,0\,0\,r]^\mathsf{T}$ and the velocities are given as

$$V^{\bar{2}}_{0\bar{2}} = \bar{J}^{\bar{2}}_{g2}(\xi)v = \begin{bmatrix} \mathrm{Ad}^{-1}_{g_{b\bar{2}}} & \mathrm{Ad}^{-1}_{g_{b\bar{2}}} J_2 \end{bmatrix} \begin{bmatrix} V^b_{0b} \\ \dot{q} \end{bmatrix} = \begin{bmatrix} -l_{12}r + \dot{q}_1 \\ q_1 r \\ 0 \\ 0 \\ 0 \\ r + \dot{q}_2 \end{bmatrix}. \tag{5.56}$$

We see that $V^{\bar{2}}_{0\bar{2}}$ gives us the velocity of a point $\mathscr{O}_2$ as seen from the frame $\mathscr{F}^0_2$, which can easily be verified.

Example 5.3 We will now find the body geometric Jacobian for the same system as in Example 5.1, illustrated in Fig. 5.2. Again we choose the inertial frame so that its origin coincides with the origin of the base frame, i.e., $\mathscr{O}_0 = \mathscr{O}_b$. The body geometric Jacobian is given by

$$\bar{J}^B_{gi}(\xi) = \begin{bmatrix} \mathrm{Ad}^{-1}_{g_{b\bar{i}}} & \mathrm{Ad}^{-1}_{g_{b\bar{i}}} J_i \end{bmatrix}. \tag{5.57}$$

The manipulator Jacobian of joint 2 was found in (5.48). We now find

$$\mathrm{Ad}^{-1}_{g_{b\bar{2}}} J_2 = \begin{bmatrix} cq_2 & sq_2 & 0 & 0 & 0 & 0 & 0 & q_1 sq_2 - l_{12}cq_2 \\ -sq_2 & cq_2 & 0 & 0 & 0 & 0 & 0 & q_1 cq_2 + l_{12}sq_2 \\ 0 & 0 & 1 & 0 & 0 & l_{12} & 0 & 0 \\ 0 & 0 & 0 & cq_2 & sq_2 & 0 & 0 & 0 \\ 0 & 0 & 0 & -sq_2 & cq_2 & 0 & 0 & 0 \\ 0 & 0 & 0 & 0 & 0 & 1 & 0 & 1 \end{bmatrix} \begin{bmatrix} 1 & l_{12} \\ 0 & -q_1 \\ 0 & 0 \\ 0 & 0 \\ 0 & 0 \\ 0 & 1 \end{bmatrix}$$

$$
= \begin{bmatrix} \cos q_2 & 0 \\ -\sin q_2 & 0 \\ 0 & 0 \\ 0 & 0 \\ 0 & 0 \\ 0 & 1 \end{bmatrix}. \tag{5.58}
$$

The body geometric Jacobian of link 2 is given by

$$
\begin{aligned}
J^B_{g\bar{2}}(\xi) &= \begin{bmatrix} \mathrm{Ad}^{-1}_{g_{b\bar{2}}} & \mathrm{Ad}^{-1}_{g_{b\bar{2}}} J_2 \end{bmatrix} \\
&= \begin{bmatrix} \cos q_2 & \sin q_2 & 0 & 0 & 0 & q_1 \sin q_2 - l_{12} \cos q_2 & \cos q_2 & 0 \\ -\sin q_2 & \cos q_2 & 0 & 0 & 0 & q_1 \cos q_2 + l_{12} \sin q_2 & -\sin q_2 & 0 \\ 0 & 0 & 1 & l_{12} & 0 & 0 & 0 & 0 \\ 0 & 0 & 0 & \cos q_2 & \sin q_2 & 0 & 0 & 0 \\ 0 & 0 & 0 & -\sin q_2 & \cos q_2 & 0 & 0 & 0 \\ 0 & 0 & 0 & 0 & 0 & 1 & 0 & 1 \end{bmatrix}.
\end{aligned} \tag{5.59}
$$

Again we write the vehicle velocities as $V^B_{0b} = \begin{bmatrix} 0 & 0 & 0 & 0 & 0 & r \end{bmatrix}^{\mathrm{T}}$ and the body velocities are given as

$$
\begin{aligned}
V^B_{0\bar{2}} = J^B_{g\bar{2}}(\xi) v &= \begin{bmatrix} \mathrm{Ad}^{-1}_{g_{b\bar{2}}} & \mathrm{Ad}^{-1}_{g_{b\bar{2}}} J_2 \end{bmatrix} \begin{bmatrix} V^b_{0b} \\ \dot{q} \end{bmatrix} \\
&= \begin{bmatrix} (q_1 \sin q_2 - l_{12} \cos q_2) r + \cos q_2 \dot{q}_1 \\ (q_1 \cos q_2 + l_{12} \sin q_2) r - \sin q_2 \dot{q}_1 \\ 0 \\ 0 \\ 0 \\ r + \dot{q}_2 \end{bmatrix}.
\end{aligned} \tag{5.60}
$$

Example 5.4 We can also use the framework above to find the end-effector velocity. If we follow the approach in Example 5.3 we can represent the body velocity in the end-effector frame. Let $\bar{l}_3 = \begin{bmatrix} 0 & l_3 & 0 \end{bmatrix}^{\mathrm{T}}$ denote the distance from $\mathscr{F}_2$ to $\mathscr{F}_e$. We can now write the body geometric Jacobian as

$$
J^B_{ge}(\xi) = \begin{bmatrix} \mathrm{Ad}^{-1}_{g_{be}} & \mathrm{Ad}^{-1}_{g_{be}} J_2 \end{bmatrix} \tag{5.61}
$$

where we use J_2 because we only have two joints. The manipulator Jacobian is the same as in the previous examples, see (5.48). We now find

$$\mathrm{Ad}_{g_{be}}^{-1} J_2 = \begin{bmatrix} cq_2 & sq_2 & 0 & 0 & 0 & 0 & 0 & q_1 sq_2 - l_{12} cq_2 - l_3 \\ -sq_2 & cq_2 & 0 & 0 & 0 & 0 & 0 & q_1 cq_2 + l_{12} sq_2 \\ 0 & 0 & 1 & 0 & 0 & l_{12} + l_3 cq_2 & l_3 sq_2 & 0 \\ 0 & 0 & 0 & cq_2 & sq_2 & 0 & 0 & 0 \\ 0 & 0 & 0 & -sq_2 & cq_2 & 0 & 0 & 0 \\ 0 & 0 & 0 & 0 & 0 & 1 & 0 & 1 \end{bmatrix}$$

$$\times \begin{bmatrix} 1 & l_{12} \\ 0 & -q_1 \\ 0 & 0 \\ 0 & 0 \\ 0 & 0 \\ 0 & 1 \end{bmatrix}$$

$$= \begin{bmatrix} \cos q_2 & -l_3 \\ -\sin q_2 & 0 \\ 0 & 0 \\ 0 & 0 \\ 0 & 0 \\ 0 & 1 \end{bmatrix}. \tag{5.62}$$

The body geometric Jacobian is given by

$$J_{ge}^B(\xi) = \left[\mathrm{Ad}_{g_{be}}^{-1} \quad \mathrm{Ad}_{g_{be}}^{-1} J_2\right]$$

$$= \left[\begin{matrix} \cos q_2 & \sin q_2 & 0 & 0 & 0 \\ -\sin q_2 & \cos q_2 & 0 & 0 & 0 \\ 0 & 0 & 1 & l_{12} + l_3 \cos q_2 & l_3 \sin q_2 \\ 0 & 0 & 0 & \cos q_2 & \sin q_2 \\ 0 & 0 & 0 & -\sin q_2 & \cos q_2 \\ 0 & 0 & 0 & 0 & 0 \end{matrix}\right.$$

$$\left.\begin{matrix} q_1 \sin q_2 - l_{12} \cos q_2 - l_3 & \cos q_2 & -l_3 \\ q_1 \cos q_2 + l_{12} \sin q_2 & -\sin q_2 & 0 \\ 0 & 0 & 0 \\ 0 & 0 & 0 \\ 0 & 0 & 0 \\ 1 & 0 & 1 \end{matrix}\right]. \tag{5.63}$$

Again we write the vehicle velocities as $V_{0b}^B = [0\,0\,0\,0\,0\,r]^\mathsf{T}$ and the body velocities are given as

$$V_{0e}^B = J_{ge}^B(\xi) v = \left[\mathrm{Ad}_{g_{be}}^{-1} \quad \mathrm{Ad}_{g_{be}}^{-1} J_2\right] \begin{bmatrix} V_{0b}^b \\ \dot{q} \end{bmatrix}$$

$$= \begin{bmatrix} (q_1 \sin q_2 - l_{12} \cos q_2 - l_3)r + \cos q_2 \dot{q}_1 - l_3 \dot{q}_2 \\ (q_1 \cos q_2 + l_{12} \sin q_2)r - \sin q_2 \dot{q}_1 \\ 0 \\ 0 \\ 0 \\ r + \dot{q}_2 \end{bmatrix}. \quad (5.64)$$

References

Duindam, V. (2006). *Port-based modeling and control for efficient bipedal walking robots*. Ph.D. thesis, University of Twente.

Duindam, V., & Stramigioli, S. (2007). Lagrangian dynamics of open multibody systems with generalized holonomic and nonholonomic joints. In *Proceedings of the IEEE/RSJ international conference on intelligent robots and systems*, San Diego, CA, USA (pp. 3342–3347).

Duindam, V., & Stramigioli, S. (2008). Singularity-free dynamic equations of open-chain mechanisms with general holonomic and nonholonomic joints. *IEEE Transactions on Robotics*, *24*(3), 517–526.

Fossen, T. I. (2002). *Marine control systems*. Trondheim: Marine Cybernetics AS. 3rd printing.

From, P. J. (2012a). An explicit formulation of singularity-free dynamic equations of mechanical systems in Lagrangian form—part one: single rigid bodies. *Modeling, Identification and Control*, *33*(2), 45–60.

From, P. J. (2012b). An explicit formulation of singularity-free dynamic equations of mechanical systems in Lagrangian form—part two: multibody systems. *Modeling, Identification and Control*, *33*(2), 61–68.

From, P. J., Duindam, V., Gravdahl, J. T., & Sastry, S. (2009). Modeling and motion planning for mechanisms on a non-inertial base. In *Proceedings of international conference of robotics and automation*, Kobe, Japan (pp. 3320–3326).

From, P. J., Duindam, V., Pettersen, K. Y., Gravdahl, J. T., & Sastry, S. (2010). Singularity-free dynamic equations of vehicle-manipulator systems. *Simulation Modelling Practice and Theory*, *18*(6), 712–731.

Murray, R. M., Li, Z., & Sastry, S. S. (1994). *A mathematical introduction to robotic manipulation*. Boca Raton: CRC Press.

Rossmann, W. (2002). *Lie groups—an introduction through linear algebra*. Oxford: Oxford Science Publications.

Chapter 6
Rigid Body Dynamics

There is a wide variety of approaches that serve as a starting point to derive the dynamics of mechanical systems. We need to choose the appropriate approach considering the dynamical properties that are of interest to our problem. For example, when we derive mathematical models for simulation and control we want to extract the components of the forces and state variables that affect the motion of the system. Constraint forces and other internal forces are important when it comes to mechanical design, durability analysis and so on, but in the setting of simulation and control we will generally try to eliminate these forces in order to keep the equations as simple as possible. One way to do this is to apply d'Alembert's principle for which the equations of motion are projected onto the directions associated with the admissible velocities. In this way the forces and torques of the constraints are eliminated from the equations. In other words, only the forces and moments that affect the motion of the rigid body are explicitly present in the final equations.

There are many different forces present in a mechanical system, but from a simulation and control point of view we are mainly interested in how the actuator and external forces affect the accelerations of the rigid bodies. Other forces, such as constraint forces between the links or other internal forces are not of interest when it comes to modeling for simulation and control. The principle of virtual work allows us to eliminate the constraint forces without actually deriving them explicitly, which can be quite hard. The starting point of d'Alembert's principle is that the constraint forces are perpendicular to the surface. For a virtual displacement δr we can thus write the virtual work δW as

$$\delta W = \sum_i (F_i - ma)\delta r_i = 0 \tag{6.1}$$

for any particle i. By introducing generalized coordinates x to represent r_i as $r_i = r_i(x_1, x_2, \dots, x_n, t)$ it is fairly straight forward (see Egeland and Gravdahl 2003; Marsden and Ratiu 1999) to show that this can be written as

$$\sum_j \left(\left[\frac{\mathrm{d}}{\mathrm{d}t} \left(\frac{\partial \mathcal{K}}{\partial \dot{x}_j} \right) - \frac{\partial \mathcal{K}}{\partial x_j} \right] - \tau_j \right) \delta x_j = 0 \tag{6.2}$$

P.J. From et al., *Vehicle-Manipulator Systems*, Advances in Industrial Control,
DOI 10.1007/978-1-4471-5463-1_6,

where $\mathscr{K} = \frac{1}{2}\sum_i m_i v_i^2$ is the kinetic energy and τ are the external forces. We will not prove this here, but we will show that this is true for a single particle in Sect. 6.1.3. The dynamic equations can now be written in terms of generalized coordinates as

$$\frac{\mathrm{d}x_j}{\mathrm{d}t} = \dot{x}_j \tag{6.3}$$

$$\frac{\mathrm{d}}{\mathrm{d}t}\left(\frac{\partial \mathscr{K}}{\partial \dot{x}_j}\right) - \frac{\partial \mathscr{K}}{\partial x_j} = \tau_j. \tag{6.4}$$

which are the Lagrange equations for a system with kinetic energy only. Note the required form of the velocity variables as the time derivative of the position variables. If potential energy is present the Lagrange equations become

$$\frac{\mathrm{d}}{\mathrm{d}t}\left(\frac{\partial L}{\partial \dot{x}_j}\right) - \frac{\partial L}{\partial x_j} = \tau_j \tag{6.5}$$

where the Lagrangian is the kinetic minus potential energy $L = \mathscr{K} - \mathscr{U}$. Lagrange's equations can also be written in vector form as

$$\frac{\mathrm{d}}{\mathrm{d}t}\left(\frac{\partial L}{\partial \dot{x}}\right) - \frac{\partial L}{\partial x} = \tau. \tag{6.6}$$

These equations are of vital importance in robotics because they can be derived directly from the system energy expressed in generalized coordinates. As we have seen, the configuration space of standard robotic manipulators is most naturally derived in terms of generalized coordinates because we do not need to represent the joint constraints explicitly. Taking the characteristics of a robotic manipulator into account we understand why the Lagrange formalism is often the method of choice when modeling standard robotic manipulators with 1-DoF joints (see also Zefran and Bullo 2004).

When it comes to modeling single rigid bodies, however, the Lagrangian formulation is not directly applicable because the configuration space cannot be described in terms of generalized coordinates and velocities. From the derivation above, we see that it is implicit in the derivation of Lagrange's equation that the configuration space of the system can be parameterized by a set $\mathbb{R}^m$ where m is the degree of freedom of the system (Murray et al. 1994). The configuration space of a single rigid body is $SE(3)$ or one of its subspaces. Because of this, Lagrange's equations cannot be applied in their standard form to derive the equations of motion of such systems. To derive the dynamics of a rigid body with a configuration space different from $\mathbb{R}^m$ using the Lagrangian approach, we must therefore choose a local parameterization for the configuration space in order to force the state space into the form of (6.3). From the previous chapters we have learned that choosing such local parameterizations, such as the Euler angles, leads to singularities and other artifacts in the representation.

Much work has, however, been done on finding a coordinate-free representation of the dynamics. One approach to this problem is the Newton–Euler equations of motion which give us a globally valid set of equations in the velocity space. The Newton–Euler equations are discussed briefly below. Another way to do deal with the problem of singularities is to apply quasi-coordinates. Kane et al. (1983) and Kane and Levinson (1985) present a detailed treatment of this topic and their approach for eliminating non-contributing forces provides physical insight, is control oriented, and allows for fast simulations. Kane's equations can be implemented using quasi-coordinates, which we will discuss later. Kane's equations are for example well suited to incorporate nonholonomic forces (Tanner and Kyriakopoulos 2001; Lesser 1992).

Formulations using quasi-coordinates have also been used to derive control laws. In particular, quasi-coordinates allow us to write the control laws in a very simple form. For more information on this topic, see for example Herman and Kozlowski (2006) and Kozlowski and Herman (2008).

Lie groups and algebras have been applied as a mathematical basis for the derivation of the dynamics of mechanical systems to obtain a singularity free formulation of dynamics (Selig 2000; Park et al. 1995). For these approaches to be valid globally the total configuration space needs to be covered by an atlas of local exponential coordinate patches. The appropriate equations must then be chosen according to the current configuration. The geometric approach presented in Bullo and Lewis (2000) can then be used to obtain a globally valid set of dynamic equations on a single Lie group, such as an AUV or spacecraft with no robotic manipulator attached. This approach is also used in Marsden and Ratiu (1999).

In our approach we choose the coordinates generated by the Lie algebra as local Euclidean coordinates which allows us to describe the dynamics locally. Lie groups are manifolds and thus also locally Euclidean. This means that locally we can write the vector of velocity variables as the derivative of the position vector. This is an important property that we can use to write the dynamics of a single rigid body in a singularity free manner.

6.1 Lagrangian Mechanics

The Euler–Lagrange equations of motion were first presented in Joseph-Louis Lagrange's *Mechanique Analytique* (Lagrange 1788). The equations presented by Lagrange are really just a coordinate invariant formulation of Newton's famous formula for motion $F = ma$:

We find the dynamic equations by the Lagrangian approach by following three steps:

1. Define a set of generalized coordinates x and corresponding generalized velocities $\dot{x}$;
2. find a Lagrangian $L(x, \dot{x})$; and

3. write the Euler–Lagrange equations as

$$\frac{\mathrm{d}}{\mathrm{d}t}\left(\frac{\partial L}{\partial \dot{x}}\right) - \frac{\partial L}{\partial x} = \tau. \tag{6.7}$$

These equations give the dynamics of a mechanical system when written in generalized coordinates and generalized velocities. Lagrange derived the dynamics for a fairly general Lagrangian L, but in the setting of this book we will choose the Lagrangian either as the kinetic energy of the mechanism or as the difference between the kinetic and potential energy.

It is, however, also possible to derive the dynamics of a single rigid body with *kinetic energy only* from d'Alembert's principle. The reason we can do this is that we neglect the position coordinates (and all properties that depend on these) and derive the equations in velocities only. The equations are then given in a similar form to Lagrange's equations, but in terms of the quasi-velocities. For example, for a pure rotational motion in $SO(3)$ the equations become

$$\frac{\mathrm{d}}{\mathrm{d}t}\left(\frac{\partial L}{\partial \omega_{0b}^{B}}\right) - \left(\widehat{\omega}_{0b}^{B}\right)^{\mathsf{T}} \frac{\partial L}{\partial \omega_{0b}^{B}} = \tau. \tag{6.8}$$

The Euler–Lagrange equations were later derived for a general Lie algebra in Poincaré (1901). The Poincaré equations, also referred to as the Euler–Poincaré equations, are given by

$$\frac{\mathrm{d}}{\mathrm{d}t}\left(\frac{\partial L}{\partial X}\right) - \mathrm{ad}_X^{*} \frac{\partial L}{\partial X} = \tau \tag{6.9}$$

for a general Lie algebra X where ad_X^{*} is found from $\langle \alpha, \mathrm{ad}_X Y\rangle = \langle \mathrm{ad}_X^{*}\alpha, Y\rangle$ where $\langle \cdot, \cdot \rangle$ denotes the pairing between the Lie algebra $\mathfrak{g}$ and its dual $\mathfrak{g}^{*}$.

The Poincaré equations are closely related to the Euler–Lagrange equations for rigid body motion. As an example, let ω_{0b}^{B} describe the angular body velocity of a rigid body, I_b its constant inertia matrix, and $\Pi = I_b \omega_{0b}^{B}$ the angular momentum in the body frame. Then the Euler equations for rigid body motion (angular motion only) are given by $\dot{\Pi} = \Pi \times \omega_{0b}^{B}$.

We will write the Lagrangian as the kinetic minus potential energy, i.e., $L = \mathscr{K} - \mathscr{U}$, or as the kinetic energy $L = \mathscr{K}$. The quantities that need to be derived in order to write the dynamics in terms of the Lagrangian are thus the kinetic and potential energy of the system. In the subsequent sections we study the energy and Lagrangians of a system written in terms of generalized coordinates and velocities. Then, in Sect. 6.4 we reformulate the Lagrangian so that we can write the dynamics in terms of generalized coordinates and quasi-velocities which in turn leads to the singularity-free formulation of the dynamics of a single rigid body, which is presented in Sect. 6.4.2.

6.1.1 Kinetic Energy

The first and most important quantity that we need to derive in order to obtain the dynamic equations of a rigid body is the kinetic energy. The kinetic energy $\mathscr{K}$ is given by the mass and the velocity of the body. For a particle with mass m and position $x(t) \in \mathbb{R}^3$, we can write the kinetic energy as

$$\mathscr{K} = \frac{1}{2}\dot{x}^{\mathsf{T}} m \dot{x}. \tag{6.10}$$

More generally, for a rigid body in $SE(3)$ we write the kinetic energy in terms of the inertia matrix.

Assume that the coordinate frame of the rigid body chosen so that the origin is located in the center of mass of the rigid body and the axes are aligned with the principle directions of the inertia, then the constant inertia matrix can be written as a diagonal matrix

$$M = I_b = \begin{bmatrix} m & 0 & 0 & 0 & 0 & 0 \\ 0 & m & 0 & 0 & 0 & 0 \\ 0 & 0 & m & 0 & 0 & 0 \\ 0 & 0 & 0 & I_x & 0 & 0 \\ 0 & 0 & 0 & 0 & I_y & 0 \\ 0 & 0 & 0 & 0 & 0 & I_z \end{bmatrix}. \tag{6.11}$$

If the coordinate frame attached to the rigid body is defined in this way we will denote this frame $\mathscr{F}_b$ (without bar) and the origin as $\mathscr{O}_b$. These are in general different from $\bar{\mathscr{F}}_b$ and $\bar{\mathscr{O}}_b$ which were chosen at the axis of rotation or translation in order to simplify the kinematic equations. The different coordinate frames are shown in Fig. 6.1. The total mass of the rigid body is given by m and I_x, I_y, and I_z are the moments of inertia around the x-, y-, and z-axes of $\mathscr{F}_b$. Most textbooks on mechanics and robotics show how to find the inertia matrix in more detail, see Murray et al. (1994) or Rao (2006).

The inertia matrix of a single rigid body, as shown above, is always constant when it is represented in the body frame $\mathscr{F}_b$. The inertia matrix can, however, be expressed in an arbitrary reference frame, but it will then become configuration dependent. We will denote the constant inertia matrix expressed in the body frame the *generalized inertia matrix*. For a single rigid body the generalized inertia matrix I_b is identical to the mass matrix M when both are expressed in the body frame $\mathscr{F}_b$. This is however not a general result and does not hold when the mass matrix M is expressed in another coordinate frame or for multibody systems.

Furthermore, if we write the velocity as a twist expressed in the body frame $\mathscr{F}_b$ the kinetic energy of the rigid body is given by

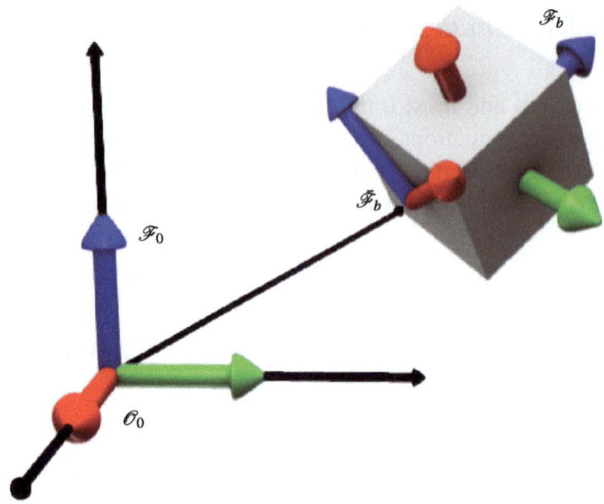

Fig. 6.1 The coordinate frame $\mathscr{F}_b$ is chosen so that it is aligned with the principal axes of inertia. In this case the generalized inertia matrix becomes diagonal. Note that $\mathscr{F}_b$ is chosen differently from $\bar{\mathscr{F}}_b$

$$\mathscr{K} = \frac{1}{2}(V_{0b}^B)^{\mathsf{T}} M V_{0b}^B. \tag{6.12}$$

It will also be useful to write the kinetic energy in terms of the time derivative of the position variables η. We then use Eq. (3.64) and get

$$\mathscr{K} = \frac{1}{2}\dot{\eta}^{\mathsf{T}} J_b^{-\mathsf{T}}(\eta_2) M J_b^{-1}(\eta_2)\dot{\eta}. \tag{6.13}$$

If the configuration space is a subspace of $SE(3)$ we get a simplified expression. For pure rotational motion in $SO(3)$ for example, the inertia matrix becomes

$$M = I_b = \begin{bmatrix} I_x & 0 & 0 \\ 0 & I_y & 0 \\ 0 & 0 & I_z \end{bmatrix} \tag{6.14}$$

with kinetic energy

$$\mathscr{K} = \frac{1}{2}(\omega_{0b}^B)^{\mathsf{T}} M \omega_{0b}^B. \tag{6.15}$$

We can obtain similar expressions for the other subspaces of $SE(3)$.

6.1.2 Potential Energy

While the kinetic energy is a very universal quantity that can be described in the same way for any kind of rigid body, potential energy enters the dynamic equations differently depending on the setting: for a land vehicle the potential energy is given by the gravitational forces and possibly spring forces in the suspension, for a ship it is the gravitational and buoyancy forces, and for a spacecraft or underwater vehicle the potential forces can in many cases be ignored due to the free-fall or weightless environment. Because of these differences, we will discuss the potential energy only briefly in this chapter and return to this topic in Chaps. 10–12 where we treat the different types of vehicle-manipulator systems.

Assume that we have a potential field $V(x)$. If this potential field represents the gravitational field we can write

$$V(x) = mgh(x) = mgh \tag{6.16}$$

where m is the weight, $g = 9.81\ \mathrm{m/s^2}$ is the gravitational force, and h is the height above some reference. In this case $V(x)$ gives us the potential energy at a configuration x and if this is the only potential field present we will write the potential energy as $\mathscr{U} = V$. We know that the forces that act on the rigid body in a gravitational field is given by $F = -mg$ in the direction of the gravitational forces in the inertial frame. More generally we can write the forces that act on the rigid body due to the potential energy $\mathscr{U}$ as

$$F = -\frac{\partial \mathscr{U}}{\partial x}. \tag{6.17}$$

6.1.3 The Euler–Lagrange Equations of Motion

Robot dynamics is the study of how actuator and external forces affect the motion of a rigid body. The fundamental laws of force balance were presented by Newton and gives the relation between the acceleration of the body and the forces that act on it. Newton's law for a particle with mass m is given by

$$F = ma \tag{6.18}$$

where F is the force and $a = \frac{\mathrm{d}^2 x}{\mathrm{d}t^2}$ is the acceleration.

Lagrange showed that Newton's second law can be cast into a set of second order differential equations in the following form:

$$\frac{\mathrm{d}}{\mathrm{d}t}\left(\frac{\partial L}{\partial \dot{x}}\right) - \frac{\partial L}{\partial x} = 0. \tag{6.19}$$

The Lagrangian $L(x, \dot{x})$ is normally taken as kinetic minus potential energy.

We can show that the dynamic equations in (6.19) and Newton's second law in (6.18) are equivalent. For a particle with mass m that moves in a potential field $\mathcal{U}$ the Lagrangian can be defined as

$$L(x, \dot{x}) = \frac{1}{2}\dot{x}^{\mathsf{T}} m \dot{x} - \mathcal{U}(x). \tag{6.20}$$

We now write

$$\frac{\partial L}{\partial \dot{x}} = m\dot{x} \tag{6.21}$$

$$\frac{\mathrm{d}}{\mathrm{d}t}\left(\frac{\partial L}{\partial \dot{x}}\right) = m\ddot{x} \tag{6.22}$$

$$\frac{\partial L}{\partial x} = -\frac{\partial \mathcal{U}}{\partial x}. \tag{6.23}$$

The dynamics in (6.19) now becomes

$$m\ddot{x} = -\frac{\partial \mathcal{U}}{\partial x} \tag{6.24}$$

which is essentially Newton's second law $F = ma$ for the motion of a particle in a potential field $\mathcal{U}$, see Eq. (6.17) and Marsden and Ratiu (1999).

The dynamic equations in this form are not restricted to a point mass, but can be generalized to rigid body motion on $SE(3)$:

Theorem 6.1 *Consider a mechanical system with generalized coordinates $x \in \mathbb{R}^m$ describing the position and a set of generalized velocities $\dot{x} = \frac{\mathrm{d}x}{\mathrm{d}t} \in \mathbb{R}^m$ describing the velocity state of the system. Furthermore, define the* Lagrangian

$$L(x, \dot{x}) = \frac{1}{2}\dot{x}^{\mathsf{T}} M(x)\dot{x} - \mathcal{U}(x) \tag{6.25}$$

as the kinetic minus the potential energy. Then the dynamics of the system can be written as a set of second order differential equations by the Euler–Lagrange equations *as*

$$\frac{\mathrm{d}}{\mathrm{d}t}\left(\frac{\partial L}{\partial \dot{x}}\right) - \frac{\partial L}{\partial x} = F \tag{6.26}$$

where F is the external force and is collocated with $\dot{x}$.

We have already seen that this expression makes sense for a point mass. We will not prove the general Lagrange equations as given above, but the proof can be found in most text books on dynamical systems, for example in Marsden and Ratiu (1999).

6.1.4 The Dynamic Equations in Matrix Form

It is very common to write the dynamic equations of single rigid bodies, standard robotic manipulators, and combinations of these in terms of vectors and matrices. The matrix form of the dynamic equations is very convenient for system and stability analysis and for model-based control design. By investigating the Euler–Lagrange equations (6.26) we will find that we can specify the system in terms of configuration- and velocity-dependent matrices. The most important matrices that are always present in robotic systems are the inertia matrix M and the Coriolis matrix C. The inertia matrix defines the inertial properties of the rigid body, that is the total mass of the body and the moments of inertia. Simply put, the inertia matrix gives us the relation between the forces that act on the system and the resulting acceleration, just like Newton's law.

However, for a system of rigid bodies, there are also other effects present, for example due to the non-inertial coordinate frames in which the different quantities are represented. These effects are added by the Coriolis matrix which includes the Coriolis and centrifugal forces. If we let $x \in \mathbb{R}^m$ denote the generalized coordinates, $\dot{x} \in \mathbb{R}^m$ the generalized velocities, and F the external forces collocated with $\dot{x}$, then the dynamic equations can be written in terms of the inertia and Coriolis matrices as

$$M(x)\ddot{x} + C(x,\dot{x})\dot{x} = F. \tag{6.27}$$

We will see that we can identify several important properties with the system matrices when the dynamics is written in this way and we will therefore always rewrite the Euler–Lagrange equations in (6.26) into this form.

If other forces are present in the system we write these so that they are collocated with $\dot{x}$ and add these to the dynamics in (6.27). We can for example add damping forces $D(\dot{x})\dot{x}$ (for floating or submerged vehicles) and gravitational forces $N(x)$ to obtain the dynamics including potential forces as

$$M(x)\ddot{x} + C(x,\dot{x})\dot{x} + D(\dot{x})\dot{x} + N(x) = F. \tag{6.28}$$

We thus see that the potential forces can be added in the dynamics either by including them in the Lagrangian as in (6.20), or as above by collocating them with the state variables.

6.2 Euler–Lagrange Equations for Rigid Body Motion

Leonhard Euler derived the dynamic equations of rigid body motion on $SO(3)$ in terms of the angular momentum vector Π which allowed him to present the dynamics in a very simple form. Euler's equations are easily derived from the conservation of the angular momentum vector Π^S relative to $\mathscr{F}_0$, which we will see. Alternatively we can find Euler's equations from the Lagrangian by applying Hamilton's extended principle. We can then go on to find a more general formulation of the equations on an arbitrary Lie algebra $\mathfrak{g}$.

We will first find Euler's equations from the conservation of angular momentum following the train of thought found in Arnold (1989). Let M be the inertia matrix of the rigid body, ω_{0b}^B the body angular velocities and $\Pi = M\omega_{0b}^B$ the angular momentum. Similarly we can write the velocities and momentum as observed from the inertial frame $\mathscr{F}_0$ as ω_{0b}^S and Π^S. If we think of Π and Π^S as vectors we write the time derivative of the momentum in the inertial frame in terms of the velocities and momenta in the body frame as

$$\begin{aligned}\dot{\Pi}^S &= R_{0b}\dot{\Pi} + \omega_{0b}^S \times \Pi^S \\ &= R_{0b}\dot{\Pi} + R_{0b}R_{0b}^{\mathsf{T}}\omega_{0b}^S \times R_{0b}R_{0b}^{\mathsf{T}}\Pi^S \\ &= R_{0b}\dot{\Pi} + R_{0b}\big(\omega_{0b}^B \times \Pi\big) \\ &= R_{0b}\big(\dot{\Pi} + \omega_{0b}^B \times \Pi\big). \end{aligned} \tag{6.29}$$

The angular momentum vector relative to $\mathscr{F}_0$ is constant for a free motion. Thus, if we use that $\dot{\Pi}^S = 0$ we get what is known as Euler's equations of motion on $SO(3)$ as

$$\begin{aligned}\dot{\Pi} + \omega_{0b}^B \times \Pi = 0 \\ \dot{\Pi} = \Pi \times \omega_{0b}^B\end{aligned} \tag{6.30}$$

which we can also write as

$$M\dot{\omega}_{0b}^B = M\omega_{0b}^B \times \omega_{0b}^B \tag{6.31}$$

since M is constant. We can further allow external forces to act on the rigid body. Let τ denote the external forces so that they are collocated with ω_{0b}^B. The equations can now be written explicitly as

$$I_x\dot{\omega}_x - (I_y - I_z)\omega_y\omega_z = \tau_1 \tag{6.32}$$

$$I_y\dot{\omega}_y - (I_z - I_x)\omega_z\omega_x = \tau_2 \tag{6.33}$$

$$I_z\dot{\omega}_z - (I_x - I_y)\omega_x\omega_y = \tau_3 \tag{6.34}$$

where $\omega_{0b}^B = [\omega_x\ \omega_y\ \omega_z]^{\mathsf{T}}$ and $M = I_b = \mathrm{diag}(I_x, I_y, I_z)$.

We can also find the Euler–Poincaré equations from Hamilton's extended principle, which states that for two fixed end points, we can write (Egeland and Gravdahl 2003)

$$\int_{t_1}^{t_2} (\delta L + W_\delta)\mathrm{d}t = 0. \tag{6.35}$$

Here, δL denotes the *variation* of the Lagrangian L and is to be considered a mathematical tool that reflects an infinitesimal change in L without any change in the physical variable L. Assume that only kinetic energy is present so $L = \mathscr{K} = \frac{1}{2}(V_{0b}^B)^\mathsf{T} M V_{0b}^B$, where $\mathscr{K}$ denotes the kinetic energy and $V_{0b}^B = [(v_{0b}^B)^\mathsf{T}\ (\omega_{0b}^B)^\mathsf{T}]^\mathsf{T}$ is the velocity twist. W_δ is the virtual work of the active generalized forces and is given by

$$W_\delta = \sum_{j=1}^{m} \tau_j \delta q_j. \tag{6.36}$$

From (6.35) we can derive what is known as the central equation (Bremer 1988). The central equation serves as a staring point to derive the dynamics in terms of Lagrange's equations, Boltzmann–Hamel equations and the Euler–Poincaré equations of motion, among other formulations. A slight modification of the central equation leads to

$$\frac{\mathrm{d}}{\mathrm{d}t}\left(\frac{\partial \mathscr{K}}{\partial v}\right)\sigma - \frac{\partial \mathscr{K}}{\partial v}(\delta v - \dot{\sigma}) - \tau^\mathsf{T}\sigma = 0 \tag{6.37}$$

for some velocity v and a variation vector σ which can be chosen freely for free motion. We can now derive the attitude dynamics of a rigid body on $SO(3)$ with only kinetic energy. For $SO(3)$ the variation of the angular velocities in body coordinates is given by $\delta\omega_{0b}^B = \dot{\sigma} + \hat{\omega}_{0b}^B\sigma$, and Hamilton's extended principle gives (Egeland and Gravdahl 2003)

$$\begin{aligned}
\frac{\mathrm{d}}{\mathrm{d}t}\left(\frac{\partial \mathscr{K}}{\partial \omega_{0b}^B}\right)\sigma - \frac{\partial \mathscr{K}}{\partial \omega_{0b}^B}\left(\delta\omega_{0b}^B - \dot{\sigma}\right) - \tau_\omega^\mathsf{T}\sigma &= 0 \\
\frac{\mathrm{d}}{\mathrm{d}t}\left(\frac{\partial \mathscr{K}}{\partial \omega_{0b}^B}\right)\sigma - \frac{\partial \mathscr{K}}{\partial \omega_{0b}^B}\hat{\omega}_{0b}^B\sigma - \tau_\omega^\mathsf{T}\sigma &= 0 \\
\left(\frac{\mathrm{d}}{\mathrm{d}t}\left(\frac{\partial \mathscr{K}}{\partial \omega_{0b}^B}\right) - \frac{\partial \mathscr{K}}{\partial \omega_{0b}^B}\hat{\omega}_{0b}^B - \tau_\omega^\mathsf{T}\right)\sigma &= 0.
\end{aligned} \tag{6.38}$$

The Euler–Poincaré equations of motion on $SO(3)$ can therefore be written as

$$\frac{\mathrm{d}}{\mathrm{d}t}\left(\frac{\partial \mathscr{K}}{\partial \omega_{0b}^B}\right) + (\widehat{\omega}_{0b}^B)^\mathsf{T}\left(\frac{\partial \mathscr{K}}{\partial \omega_{0b}^B}\right) = \tau_\omega. \tag{6.39}$$

The partial derivative of the kinetic energy $\mathscr{K} = \frac{1}{2}(\omega_{0b}^B)^\mathsf{T} M \omega_{0b}^B$ is given by $\frac{\partial \mathscr{K}}{\partial \omega_{0b}^B} = M\omega_{0b}^B$ and we get the Euler equations of rigid body motion

$$M\dot{\omega}_{0b}^B + \omega_{0b}^B \times \left(M\omega_{0b}^B\right) = \tau_\omega. \tag{6.40}$$

If we write $M = \mathrm{diag}(I_x, I_y, I_z)$ and $\omega_{0b}^B = [\omega_x\ \omega_y\ \omega_z]^\mathsf{T}$ we get (6.32)–(6.34).

The Euler equations of motion on $SE(3)$ can be found in the same way and are written as

$$\frac{\mathrm{d}}{\mathrm{d}t}\left(\frac{\partial \mathscr{K}}{\partial V_{0b}^B}\right) - \mathrm{ad}_V^\mathsf{T}\left(\frac{\partial \mathscr{K}}{\partial V_{0b}^B}\right) = \tau \tag{6.41}$$

where

$$\mathrm{ad}_V = \begin{bmatrix} \widehat{\omega}_{0b}^B & \widehat{v}_{0b}^B \\ 0 & \widehat{\omega}_{0b}^B \end{bmatrix}. \tag{6.42}$$

If we write $\tau = [\tau_v^\mathsf{T}\ \tau_\omega^\mathsf{T}]^\mathsf{T}$ we get the well known equations

$$\frac{\mathrm{d}}{\mathrm{d}t}\left(\frac{\partial \mathscr{K}}{\partial v_{0b}^B}\right) + \widehat{\omega}_{0b}^B\left(\frac{\partial \mathscr{K}}{\partial v_{0b}^B}\right) = \tau_v \tag{6.43}$$

$$\frac{\mathrm{d}}{\mathrm{d}t}\left(\frac{\partial \mathscr{K}}{\partial \omega_{0b}^B}\right) + \widehat{v}_{0b}^B\left(\frac{\partial \mathscr{K}}{\partial v_{0b}^B}\right) + \widehat{\omega}_{0b}^B\left(\frac{\partial \mathscr{K}}{\partial \omega_{0b}^B}\right) = \tau_\omega \tag{6.44}$$

which are also known as Kirchhoff's equations (Fossen 2002). This formulation of the dynamics is widely used to model ships and underwater marine vessels as well as spacecraft. It is interesting to note that the Euler–Poincaré equations of motion on both $SO(3)$ and $SE(3)$ assume kinetic energy only and do not depend on the position variables. We thus obtain a singularity free formulation at the expense of position variables and all the terms that depend on the position variables.

In the more general case where the center of gravity does not coincide with the origin of the body frame, i.e., $\mathscr{F}_{b1} \neq \mathscr{F}_b$, the inertia matrix will take the form (Fossen 2002)

$$\begin{aligned} M &= \begin{bmatrix} M_v & M_{\omega v}^\mathsf{T} \\ M_{\omega v} & M_\omega \end{bmatrix} = \begin{bmatrix} mI & -m\widehat{r}_{bg} \\ m\widehat{r}_{bg} & I_b \end{bmatrix} \\ &= \begin{bmatrix} m & 0 & 0 & 0 & mz_{bg} & -my_{bg} \\ 0 & m & 0 & -mz_{bg} & 0 & mx_{bg} \\ 0 & 0 & m & my_{bg} & -mx_{bg} & 0 \\ 0 & -mz_{bg} & my_{bg} & I_x & -I_{xy} & -I_{xz} \\ mz_{bg} & 0 & -mx_{bg} & -I_{yx} & I_y & -I_{yz} \\ -my_{bg} & mx_{bg} & 0 & -I_{zx} & -I_{zy} & I_z \end{bmatrix} \end{aligned} \tag{6.45}$$

where $r_{bg} = [x_{bg}\ y_{bg}\ z_{bg}]^{\mathsf{T}}$ is the vector from the origin of the body frame $\mathscr{F}_{b1}$ to the center of gravity $\mathscr{O}_b$. In this case the kinetic energy becomes

$$\mathscr{K} = \frac{1}{2}(v_{0b}^B)^{\mathsf{T}} M_v v_{0b}^B + \frac{1}{2}(v_{0b}^B)^{\mathsf{T}} M_{\omega v}^{\mathsf{T}} \omega_{0b}^B + \frac{1}{2}(\omega_{0b}^B)^{\mathsf{T}} M_{\omega v} v_{0b}^B + \frac{1}{2}(\omega_{0b}^B)^{\mathsf{T}} M_\omega \omega_{0b}^B \tag{6.46}$$

and the terms in (6.43)–(6.44) are written as

$$\frac{\partial \mathscr{K}}{\partial v_{0b}^B} = M_v v_{0b}^B + M_{\omega v}^{\mathsf{T}} \omega_{0b}^B, \qquad \frac{\mathrm{d}}{\mathrm{d}t}\left(\frac{\partial \mathscr{K}}{\partial v_{0b}^B}\right) = M_v \dot{v}_{0b}^B + M_{\omega v}^{\mathsf{T}} \dot{\omega}_{0b}^B, \tag{6.47}$$

$$\frac{\partial \mathscr{K}}{\partial \omega_{0b}^B} = M_\omega \omega_{0b}^B + M_{\omega v} v_{0b}^B, \qquad \frac{\mathrm{d}}{\mathrm{d}t}\left(\frac{\partial \mathscr{K}}{\partial \omega_{0b}^B}\right) = M_\omega \dot{\omega}_{0b}^B + M_{\omega v} \dot{v}_{0b}^B. \tag{6.48}$$

6.2.1 Euler–Lagrange Equations in Matrix Form

Because we are free to choose our reference frame as we like, we will normally simplify the dynamics of the single rigid body by choosing the reference frame $\mathscr{F}_b$ at the center of gravity of the rigid body. As we have seen, in this case the inertia matrix can be written as a diagonal matrix as in (6.11). To see what the Euler–Lagrange equations look like in matrix form in this case we first write the inertia matrix as a block diagonal matrix $M = \mathrm{diag}(mI, I_b) = \mathrm{diag}(m, m, m, I_x, I_y, I_z)$. The kinetic energy is then given by

$$\mathscr{K} = \frac{1}{2} m (v_{0b}^B)^{\mathsf{T}} v_{0b}^B + \frac{1}{2}(\omega_{0b}^B)^{\mathsf{T}} I_b \omega_{0b}^B \tag{6.49}$$

and the partial derivatives as

$$\frac{\partial \mathscr{K}}{\partial v_{0b}^B} = m v_{0b}^B, \qquad \frac{\mathrm{d}}{\mathrm{d}t}\left(\frac{\partial \mathscr{K}}{\partial v_{0b}^B}\right) = m \dot{v}_{0b}^B, \tag{6.50}$$

$$\frac{\partial \mathscr{K}}{\partial \omega_{0b}^B} = I_b \omega_{0b}^B, \qquad \frac{\mathrm{d}}{\mathrm{d}t}\left(\frac{\partial \mathscr{K}}{\partial \omega_{0b}^B}\right) = I_b \dot{\omega}_{0b}^B. \tag{6.51}$$

The rigid body dynamics on $SE(3)$ is therefore given by Eqs. (6.43)–(6.44) as

$$m \dot{v}_{0b}^B + \widehat{\omega}_{0b}^B m v_{0b}^B = \tau_v \tag{6.52}$$

$$I_b \dot{\omega}_{0b}^B + \widehat{v}_{0b}^B m v_{0b}^B + \widehat{\omega}_{0b}^B I_b \omega_{0b}^B = \tau_\omega. \tag{6.53}$$

We can write this in matrix form as

$$M \dot{V}_{0b}^B + C(V_{0b}^B) V_{0b}^B = \tau, \tag{6.54}$$

or more explicitly as

$$\begin{bmatrix} mI & 0 \\ 0 & I_b \end{bmatrix} \begin{bmatrix} \dot{v}_{0b}^B \\ \dot{\omega}_{0b}^B \end{bmatrix} - \begin{bmatrix} -m\widehat{\omega}_{0b}^B & 0 \\ 0 & \widehat{I_b \omega_{0b}^B} \end{bmatrix} \begin{bmatrix} v_{0b}^B \\ \omega_{0b}^B \end{bmatrix} = \begin{bmatrix} \tau_v \\ \tau_\omega \end{bmatrix} \tag{6.55}$$

where we have used that $\widehat{v}_{0b}^B v_{0b}^B = 0$. These are the Newton–Euler equations of motion for the choice of coordinate frames described above and are normally written as

$$\begin{aligned} m\dot{v}_{0b}^B + m\widehat{\omega}_{0b}^B v_{0b}^B &= \tau_v \\ I_b \dot{\omega}_{0b}^B + \widehat{\omega}_{0b}^B I_b \omega_{0b}^B &= \tau_\omega . \end{aligned} \tag{6.56}$$

We can also remove the assumption that $\mathscr{F}_{b1} = \mathscr{F}_b$ and write the dynamics in (6.47)–(6.48) in matrix form for which the Coriolis matrix becomes

$$C(V_{0b}^B) = -\begin{bmatrix} 0 & \widehat{M_v v_{0b}^B + M_{\omega v}^\mathsf{T} \omega_{0b}^B} \\ \widehat{M_v v_{0b}^B + M_{\omega v}^\mathsf{T} \omega_{0b}^B} & \widehat{M_\omega \omega_{0b}^B + M_{\omega v} v_{0b}^B} \end{bmatrix} . \tag{6.57}$$

This is the special case of Euler's equations when the inertia matrix is of the form shown in (6.45). The general form when only kinetic energy is present is given by

$$C(V_{0b}^B) = -\begin{bmatrix} 0 & \widehat{\frac{\partial \mathscr{K}}{\partial v_{0b}^B}} \\ \widehat{\frac{\partial \mathscr{K}}{\partial v_{0b}^B}} & \widehat{\frac{\partial \mathscr{K}}{\partial \omega_{0b}^B}} \end{bmatrix} . \tag{6.58}$$

Because the Coriolis matrix depends on the velocity variables and at the same time is multiplied with the same velocity variable—the vector appears in the dynamic equations as $C(V_{0b}^B)V_{0b}^B$—we have some freedom in choosing the Coriolis matrix, i.e., there are several different representations of $C(V_{0b}^B)$ that result in the same vector $C(V_{0b}^B)V_{0b}^B$. For other representations of the Coriolis matrix we refer to Sagatun and Fossen (1992) and Fossen and Fjellstad (1995).

The reason that we can obtain the singularity free representation shown above is basically that we are dealing with single rigid bodies and that we neglect gravitational and other potential forces. As a result the inertia matrix is constant and the Lagrangian is given by the kinetic energy $L(V_{0b}^B) = \mathscr{K}(V_{0b}^B) = \frac{1}{2}(V_{0b}^B)^\mathsf{T} M V_{0b}^B$. For multibody systems the inertia matrix depends on the configuration of the system, and so does forces such as gravity which are also configuration dependent. To be able to include these effects we therefore need the position vector to be explicit in the equations. In the next section we therefore present the dynamics in terms of the velocity variables $\dot{\eta}$ and the position variables η.

6.3 Full State Space Dynamics in Vector Form

To find the dynamic equations of a single rigid body with configuration space $SE(3)$ we first need to find a set of coordinates that describe the configuration of the system. We can then use the position variables $\eta = [x_{0b}\ y_{0b}\ z_{0b}\ \phi_{0b}\ \theta_{0b}\ \psi_{0b}]^{\mathsf{T}}$. The corresponding generalized velocities in this case is the set given by the vector $\dot{\eta} = \frac{\mathrm{d}\eta}{\mathrm{d}t} = [\dot{x}_{0b}\ \dot{y}_{0b}\ \dot{z}_{0b}\ \dot{\phi}_{0b}\ \dot{\theta}_{0b}\ \dot{\psi}_{0b}]^{\mathsf{T}}$. This choice of variables leads to a singularity-prone formulation of the kinematics, but it has the advantage that the use of generalized coordinates and generalized velocities allows us to formulate the Euler–Lagrange equations in the normal way. The dynamics then becomes

$$\frac{\mathrm{d}}{\mathrm{d}t}\left(\frac{\partial L}{\partial \dot{\eta}}\right) - \frac{\partial L}{\partial \eta} = F. \tag{6.59}$$

Following the approach in Fossen (2002) we can find the dynamics in terms of η and $\dot{\eta}$ by rewriting the dynamics in (6.54) instead of finding the partial derivatives in (6.59) directly. We first write

$$\dot{\eta} = J_b(\eta) V_{0b}^B \quad \Longleftrightarrow \quad V_{0b}^B = J_b^{-1}(\eta)\dot{\eta}, \tag{6.60}$$

$$\ddot{\eta} = J_b(\eta)\dot{V}_{0b}^B + \dot{J}_b(\eta) V_{0b}^B \quad \Longleftrightarrow \quad \dot{V}_{0b}^B = J_b^{-1}(\eta)\big(\ddot{\eta} - \dot{J}_b(\eta) J_b^{-1}(\eta)\dot{\eta}\big). \tag{6.61}$$

Now we can rewrite the dynamics in (6.54) as

$$M\dot{V}_{0b}^B + C\big(V_{0b}^B\big)V_{0b}^B = \tau \tag{6.62}$$

$$MJ_b^{-1}(\eta)\big(\ddot{\eta} - \dot{J}_b(\eta)J_b^{-1}(\eta)\dot{\eta}\big) + C\big(V_{0b}^B\big)J_b^{-1}(\eta)\dot{\eta} = \tau \tag{6.63}$$

$$MJ_b^{-1}(\eta)\ddot{\eta} - MJ_b^{-1}(\eta)\dot{J}_b(\eta)J_b^{-1}(\eta)\dot{\eta} + C(\eta,\dot{\eta})J_b^{-1}(\eta)\dot{\eta} = \tau. \tag{6.64}$$

We want the dynamics written so that the torques are collocated with the velocity variables. We obtain this by requiring that the work $W = (V_{0b}^B)^{\mathsf{T}}\tau$ generated by applying the external forces τ is the same regardless of the formulation used. The work is given by

$$W = \big(V_{0b}^B\big)^{\mathsf{T}}\tau = \big(J_b^{-1}(\eta)\dot{\eta}\big)^{\mathsf{T}}\tau = \dot{\eta}^{\mathsf{T}} J_b^{-\mathsf{T}}(\eta)\tau. \tag{6.65}$$

We see that we obtain the required external force if we pre-multiply (6.64) with $J_b^{-\mathsf{T}}(\eta)$ and write the dynamics as

$$J_b^{-\mathsf{T}}MJ_b^{-1}\ddot{\eta} - J_b^{-\mathsf{T}}MJ_b^{-1}\dot{J}_bJ_b^{-1}\dot{\eta} + J_b^{-\mathsf{T}}CJ_b^{-1}\dot{\eta} = J_b^{-\mathsf{T}}\tau \tag{6.66}$$

$$J_b^{-\mathsf{T}}MJ_b^{-1}\ddot{\eta} + J_b^{-\mathsf{T}}\big(CJ_b^{-1} - MJ_b^{-1}\dot{J}_bJ_b^{-1}\big)\dot{\eta} = J_b^{-\mathsf{T}}\tau \tag{6.67}$$

where we suppress the dependence of η and $\dot{\eta}$ to reduce notation.

In the previous section we found the dynamics in terms of the velocity variables only. We can now use these equations to obtain the full state space dynamics by including the kinematics:

The dynamics of a single rigid body can be written in terms of the position variables η and velocity variables V_{0b}^B as

$$\dot{\eta} = J_b(\eta) V_{0b}^B, \tag{6.68}$$

$$M\dot{V}_{0b}^B + C\big(V_{0b}^B\big)V_{0b}^B = \tau. \tag{6.69}$$

However, by Eqs. (6.60) and (6.61) we can eliminate the velocity variables V_{0b}^B to obtain the dynamics in the form of (6.67):

The dynamics of a single rigid body can be written in terms of the position variables η and velocity variables $\dot{\eta}$ as

$$\dot{\eta} = \frac{\mathrm{d}\eta}{\mathrm{d}t} \tag{6.70}$$

$$\tilde{M}(\eta)\ddot{\eta} + \tilde{C}(\eta, \dot{\eta})\dot{\eta} = \tilde{\tau} \tag{6.71}$$

where

$$\tilde{M}(\eta) = J_b^{-\mathsf{T}}(\eta) M J_b^{-1}(\eta) \tag{6.72}$$

$$\tilde{C}(\eta, \dot{\eta}) = J_b^{-\mathsf{T}}(\eta)\big(C(\eta, \dot{\eta})J_b^{-1}(\eta) - MJ_b^{-1}(\eta)\dot{J}_b(\eta)J_b^{-1}(\eta)\big) \tag{6.73}$$

$$\tilde{\tau}(\eta) = J_b^{-\mathsf{T}}(\eta)\tau. \tag{6.74}$$

A few comments regarding these equations are appropriate at this stage. The dynamics as written in (6.54) depends only on the velocity variables V_{0b}^B and no position variables appear in the equations. This allows us to formulate the dynamics as shown in (6.55) without singularities and the equations are well defined. This is also true for all the other formulations presented in the previous section. However, because the position variables of the rigid body do not appear in the dynamic equations we cannot directly include effects such as gravity and buoyancy. We obtain the full state space dynamics either by the equations in (6.68)–(6.69) or the equations in (6.70)–(6.74). It is important to note that we obtain the position variables through the velocity transformation matrix $J_b(\eta)$ which again introduces singularities to the system and the equations are only valid for $\theta \neq \pm\frac{\pi}{2}$.

In the following we will derive the dynamics in terms of generalized coordinates and quasi-velocities. We will in turn use this to express the dynamics without the presence of singularities. Finally we show that the potential forces can be expressed in terms of the generalized coordinates and therefore included in the dynamic equations in the standard way.

6.4 Lagrange Equations of Motion in Quasi-coordinates

In this section we will re-write the dynamics so that the equations are written in terms of the position and velocity variables x and v. Recall that using the standard formulation of Lagrange's equations requires the use of generalized coordinates $x \in \mathbb{R}^m$ and generalized velocities $\dot{x} = \frac{dx}{dt} \in \mathbb{R}^m$ to describe the state space of the rigid body. One example of such state variables is η for position and $\dot{\eta}$ for velocity which gives the dynamics in the form of (6.70). In this section we will reformulate the Lagrangian in terms of generalized coordinates x and quasi-velocities $v = S(x)\dot{x}$. The velocity variable v is called a quasi-velocity because v itself has a clear geometric interpretation as velocity but its integral $\int v$ has no physical interpretation. Quasi-velocities were discussed in great detail in Sect. 2.4.

We have seen that the Lagrangian can be written in terms of the generalized coordinates x and the body velocity twist V_{0b}^B as

$$L\left(x, V_{0b}^B\right) = \frac{1}{2}\left(V_{0b}^B\right)^\mathsf{T} M V_{0b}^B - \mathscr{U}(x) \tag{6.75}$$

for $SE(3)$ or more generally for a velocity v, also in the body frame $\mathscr{F}_b$, as

$$L(x, v) = \frac{1}{2} v^\mathsf{T} M v - \mathscr{U}(x). \tag{6.76}$$

Here x is a vector representation of the position of the rigid body, for example η for $SE(3)$, and v is a representation of the velocity variable, not necessarily given by the time derivative of the position. We will use v to illustrate that we do not need to restrict ourselves to twists such as V_{0b}^B. The partial derivatives of the Lagrangian in (6.76) will be used frequently in the following so we express these explicitly:

$$\frac{\partial L}{\partial v} = M v, \tag{6.77}$$

$$\frac{d}{dt}\left(\frac{\partial L}{\partial v}\right) = M \dot{v}, \tag{6.78}$$

$$\frac{\partial L}{\partial x} = -\frac{\partial \mathscr{U}(x)}{\partial x}. \tag{6.79}$$

Here we have used that M is constant, which is true for single rigid bodies with our choice of reference frame $\mathscr{F}_b$, but not a property of mechanical systems in general.

In Sect. 2.4 we saw that we can write the velocity v in terms of the time derivative of the generalized coordinates as $v = S(x)\dot{x}$. This allows us to write the Lagrangian as a function of generalized coordinates and generalized velocities as

$$\bar{L}(x, \dot{x}) = \frac{1}{2}\dot{x}^\mathsf{T} S(x)^\mathsf{T} M S(x)\dot{x} - \mathscr{U}(x) \tag{6.80}$$

which can be substituted into Lagrange's equations

$$\frac{\mathrm{d}}{\mathrm{d}t}\left(\frac{\partial \bar{L}}{\partial \dot{x}}\right) - \frac{\partial \bar{L}}{\partial x} = B(x)\tau \tag{6.81}$$

for some $B(x)$ yet to be determined. The reason that we need to include $B(x)$ in the equations is that we want to define τ so that it is collocated with v. In (6.81) we therefore need to find a $B(x)$ so that $B(x)\tau$ is collocated with the new velocity variable $\dot{x}$. We find the partial derivatives as (From 2012a, 2012b)

$$\frac{\partial \bar{L}}{\partial \dot{x}} = S^{\mathsf{T}}(x)MS(x)\dot{x} = S^{\mathsf{T}}(x)\underbrace{Mv}_{\frac{\partial L}{\partial v}} = S^{\mathsf{T}}(x)\frac{\partial L}{\partial v} \tag{6.82}$$

$$\frac{\mathrm{d}}{\mathrm{d}t}\left(\frac{\partial \bar{L}}{\partial \dot{x}}\right) = \dot{S}^{\mathsf{T}}(x)\frac{\partial L}{\partial v} + S^{\mathsf{T}}(x)\frac{\mathrm{d}}{\mathrm{d}t}\left(\frac{\partial L}{\partial v}\right) \tag{6.83}$$

$$\begin{aligned}\frac{\partial \bar{L}}{\partial x} &= \frac{\partial^{\mathsf{T}}(S(x)\dot{x})}{\partial x}MS(x)\dot{x} - \frac{\partial \mathcal{U}}{\partial x}\\ &= \frac{\partial^{\mathsf{T}}(S(x)\dot{x})}{\partial x}\underbrace{Mv}_{\frac{\partial L}{\partial v}}\underbrace{-\frac{\partial \mathcal{U}}{\partial x}}_{\frac{\partial L}{\partial x}}\\ &= \frac{\partial L}{\partial x} + \frac{\partial^{\mathsf{T}}(S(x)\dot{x})}{\partial x}\frac{\partial L}{\partial v}.\end{aligned} \tag{6.84}$$

From this we can conclude the following important relations:

Proposition 6.1 *The partial derivative of a Lagrangian in the form*

$$\bar{L}(x,\dot{x}) = \frac{1}{2}\dot{x}^{\mathsf{T}}S(x)^{\mathsf{T}}MS(x)\dot{x} - \mathcal{U}(x) \tag{6.85}$$

can be expressed in terms of the Lagrangian

$$L(x,v) = \frac{1}{2}v^{\mathsf{T}}Mv - \mathcal{U}(x), \tag{6.86}$$

where $v = S(x)\dot{x}$, as

$$\frac{\mathrm{d}}{\mathrm{d}t}\left(\frac{\partial \bar{L}}{\partial \dot{x}}\right) = S^{\mathsf{T}}(x)\frac{\mathrm{d}}{\mathrm{d}t}\left(\frac{\partial L}{\partial v}\right) + \dot{S}^{\mathsf{T}}(x)\frac{\partial L}{\partial v} \tag{6.87}$$

$$\frac{\partial \bar{L}}{\partial x} = \frac{\partial L}{\partial x} + \frac{\partial^{\mathsf{T}}(S(x)\dot{x})}{\partial x}\frac{\partial L}{\partial v}. \tag{6.88}$$

Proof The proof follows directly from Eqs. (6.82)–(6.84). □

The Euler–Lagrange equations are found by the partial derivatives of the Lagrangian $\bar{L}(x, \dot{x})$ as

$$\frac{\mathrm{d}}{\mathrm{d}t}\left(\frac{\partial \bar{L}}{\partial \dot{x}}\right) - \frac{\partial \bar{L}}{\partial x} = B(x)\tau \tag{6.89}$$

$$S^{\mathsf{T}}(x)\frac{\mathrm{d}}{\mathrm{d}t}\left(\frac{\partial L}{\partial v}\right) + \dot{S}^{\mathsf{T}}(x)\frac{\partial L}{\partial v} - \frac{\partial L}{\partial x} - \frac{\partial^{\mathsf{T}}(S(x)\dot{x})}{\partial x}\frac{\partial L}{\partial v} = B(x)\tau. \tag{6.90}$$

The torques τ are defined so that they are collocated with v (represented in the body frame). Following the general idea of (6.65) we find

$$W = v^{\mathsf{T}}\tau = \left(S(x)\dot{x}\right)^{\mathsf{T}}\tau = \dot{x}^{\mathsf{T}}S^{\mathsf{T}}(x)\tau \tag{6.91}$$

which gives $B(x) = S^{\mathsf{T}}(x)$ as expected. We then pre-multiply (6.90) with $S^{-\mathsf{T}}(x)$ to get the dynamics in the right form:

Proposition 6.2 *For a mechanical system with Lagrangian*

$$\bar{L}(x, \dot{x}) = \frac{1}{2}\dot{x}^{\mathsf{T}}S(x)^{\mathsf{T}}MS(x)\dot{x} - \mathscr{U}(x) \tag{6.92}$$

we can find the dynamic equations in terms of the Lagrangian $L(x, v)$ in (6.86) with $v = S(x)\dot{x}$ as

$$\frac{\mathrm{d}}{\mathrm{d}t}\left(\frac{\partial L}{\partial v}\right) - S^{-\mathsf{T}}(x)\frac{\partial L}{\partial x} + S^{-\mathsf{T}}(x)\left(\dot{S}^{\mathsf{T}}(x) - \frac{\partial^{\mathsf{T}}(S(x)\dot{x})}{\partial x}\right)\frac{\partial L}{\partial v} = \tau. \tag{6.93}$$

Proof The proof follows directly by substituting the expressions in Proposition 6.1 into (6.89) and pre-multiplying with $S^{-\mathsf{T}}(x)$. □

We would like to write the equations in terms of x and $v = S(x)\dot{x}$, but not with $\dot{x}$ explicitly present in the equations. We therefore need to take a closer look at the expressions involving the time derivative of the generalized coordinates. Following the notation in Duindam and Stramigioli (2007, 2008) we will write the dynamics as

$$\frac{\mathrm{d}}{\mathrm{d}t}\left(\frac{\partial L}{\partial v}\right) - S^{-\mathsf{T}}\frac{\partial L}{\partial x} + \left(\sum_k \gamma_k v_k\right)\frac{\partial L}{\partial v} = \tau \tag{6.94}$$

so it only remains to find an expression for the matrix γ_k.

Comparing (6.93) with (6.94) we can write

$$\sum_k \gamma_k v_k = \sum_k (\gamma_{k1} v_k + \gamma_{k2} v_k) \tag{6.95}$$

where

$$\sum_k \gamma_{k1} v_k = S^{-\mathsf{T}}(x)\dot{S}^{\mathsf{T}}(x) \tag{6.96}$$

$$\sum_k \gamma_{k2} v_k = -S^{-\mathsf{T}}(x)\frac{\partial^{\mathsf{T}}(S(x)\dot{x})}{\partial x}. \tag{6.97}$$

We will first look at $\sum_k \gamma_{k1} v_k = S^{-\mathsf{T}}(x)\dot{S}^{\mathsf{T}}(x)$. Note that $S^{-\mathsf{T}}(x)$ and $\dot{S}^{\mathsf{T}}(x)$ can be written as

$$S^{-\mathsf{T}}(x) = \begin{bmatrix} S_{11}^{-1} & S_{21}^{-1} & \cdots & S_{n1}^{-1} \\ S_{12}^{-1} & S_{22}^{-1} & \cdots & S_{n2}^{-1} \\ \vdots & \vdots & \ddots & \vdots \\ S_{1n}^{-1} & S_{2n}^{-1} & \cdots & S_{nn}^{-1} \end{bmatrix} \tag{6.98}$$

$$\dot{S}^{\mathsf{T}}(x) = \begin{bmatrix} \sum_m \frac{\partial S_{11}}{\partial x_m}\dot{x}_m & \sum_m \frac{\partial S_{21}}{\partial x_m}\dot{x}_m & \cdots & \sum_m \frac{\partial S_{n1}}{\partial x_m}\dot{x}_m \\ \sum_m \frac{\partial S_{12}}{\partial x_m}\dot{x}_m & \sum_m \frac{\partial S_{22}}{\partial x_m}\dot{x}_m & \cdots & \sum_m \frac{\partial S_{n2}}{\partial x_m}\dot{x}_m \\ \vdots & \vdots & \ddots & \vdots \\ \sum_m \frac{\partial S_{1n}}{\partial x_m}\dot{x}_m & \sum_m \frac{\partial S_{2n}}{\partial x_m}\dot{x}_m & \cdots & \sum_m \frac{\partial S_{nn}}{\partial x_m}\dot{x}_m \end{bmatrix} \tag{6.99}$$

where S_{ij}^{-1} is to be interpreted as element (i, j) of the matrix S^{-1} (not the inverse of the element S_{ij}). We now rewrite $\dot{x} = S^{-1}(x)v$ as $\dot{x}_m = \sum_k S_{mk}^{-1}(x)v_k$ and write the first part of $\gamma_k v_k$, i.e., $\sum_k \gamma_{k1} v_k = S^{-\mathsf{T}}(x)\dot{S}^{\mathsf{T}}(x)$ as

$$\begin{aligned} \sum_k (\gamma_{k1})_{ij} v_k &= \sum_l S_{li}^{-1} \sum_m \frac{\partial S_{jl}}{\partial x_m}\dot{x}_m \\ &= \sum_{l,m} S_{li}^{-1} \frac{\partial S_{jl}}{\partial x_m}\dot{x}_m \\ &= \sum_{l,m} S_{li}^{-1} \frac{\partial S_{jl}}{\partial x_m} \sum_k S_{mk}^{-1} v_k \\ &= \sum_k \underbrace{\left(\sum_{l,m} S_{li}^{-1} \frac{\partial S_{jl}}{\partial x_m} S_{mk}^{-1} \right)}_{(\gamma_k)_{ij}} v_k \end{aligned} \tag{6.100}$$

which eliminates $\dot{x}$ from the first part of γ_k.

Similarly for the second part of Eq. (6.95) given by $\sum_k \gamma_{k2} v_k = -S^{-\mathsf{T}}(x)\frac{\partial^{\mathsf{T}}(S(x)\dot{x})}{\partial x}$ we first write

$$\frac{\partial(S(x)\dot{x})}{\partial x} = \begin{bmatrix} \frac{\partial(S\dot{x})_1}{\partial x_1} & \frac{\partial(S\dot{x})_1}{\partial x_2} & \cdots & \frac{\partial(S\dot{x})_1}{\partial x_n} \\ \frac{\partial(S\dot{x})_2}{\partial x_1} & \frac{\partial(S\dot{x})_2}{\partial x_2} & \cdots & \frac{\partial(S\dot{x})_2}{\partial x_n} \\ \vdots & \vdots & \ddots & \vdots \\ \frac{\partial(S\dot{x})_n}{\partial x_1} & \frac{\partial(S\dot{x})_n}{\partial x_2} & \cdots & \frac{\partial(S\dot{x})_n}{\partial x_n} \end{bmatrix} \tag{6.101}$$

and then by the relation $(S(x)\dot{x})_l = \sum_m S_{lm}\dot{x}_m$ we can write

$$\frac{\partial^{\mathsf{T}}(S(x)\dot{x})}{\partial x} = \begin{bmatrix} \sum_m \frac{\partial S_{1m}}{\partial x_1}\dot{x}_m & \sum_m \frac{\partial S_{2m}}{\partial x_1}\dot{x}_m & \cdots & \sum_m \frac{\partial S_{nm}}{\partial x_1}\dot{x}_m \\ \sum_m \frac{\partial S_{1m}}{\partial x_2}\dot{x}_m & \sum_m \frac{\partial S_{2m}}{\partial x_2}\dot{x}_m & \cdots & \sum_m \frac{\partial S_{nm}}{\partial x_2}\dot{x}_m \\ \vdots & \vdots & \ddots & \vdots \\ \sum_m \frac{\partial S_{1m}}{\partial x_n}\dot{x}_m & \sum_m \frac{\partial S_{2m}}{\partial x_n}\dot{x}_m & \cdots & \sum_m \frac{\partial S_{nm}}{\partial x_n}\dot{x}_m \end{bmatrix}. \tag{6.102}$$

We find the expression for $\sum_k \gamma_{k2} v_k = -S^{-\mathsf{T}}(x)\frac{\partial^{\mathsf{T}}(S(x)\dot{x})}{\partial x}$ by multiplying (6.98) and (6.102) to be

$$\begin{aligned} \sum_k (\gamma_{k2})_{ij} v_k &= -\sum_l S_{li}^{-1} \sum_m \frac{\partial S_{jm}}{\partial x_l}\dot{x}_m \\ &= -\sum_{l,m} S_{li}^{-1} \frac{\partial S_{jm}}{\partial x_l} \sum_k S_{mk}^{-1} v_k \\ &= -\sum_k \underbrace{\left(\sum_{l,m} S_{li}^{-1} \frac{\partial S_{jm}}{\partial x_l} S_{mk}^{-1}\right)}_{(\gamma_k)_{ij}} v_k. \end{aligned} \tag{6.103}$$

We can now find $\gamma_k = \gamma_{k1} + \gamma_{k2}$ as

$$(\gamma_k)_{ij}(x) = \sum_{l,m} S_{li}^{-1}\left(\frac{\partial S_{jl}}{\partial x_m} - \frac{\partial S_{jm}}{\partial x_l}\right) S_{mk}^{-1}. \tag{6.104}$$

This gives us the dynamic equations in terms of generalized coordinates x and quasi-velocities v. We get the following important result:

Proposition 6.3 *The dynamic equations in Proposition* 6.2 *can be written in terms of position variables* x *and velocity variables* $v = S(x)\dot{x}$ *as*

$$\frac{\mathrm{d}}{\mathrm{d}t}\left(\frac{\partial L}{\partial v}\right) - S^{-\mathsf{T}}\frac{\partial L}{\partial x} + \left(\sum_k \gamma_k v_k\right)\frac{\partial L}{\partial v} = \tau \tag{6.105}$$

with

$$(\gamma_k)_{ij}(x) = \sum_{l,m} S_{li}^{-1}\left(\frac{\partial S_{jl}}{\partial x_m} - \frac{\partial S_{jm}}{\partial x_l}\right) S_{mk}^{-1} \tag{6.106}$$

and is thus independent of the generalized velocity variables $\dot{x}$.

Proof Proposition 6.3 follows directly from the derivation above. □

We have now derived the dynamic equations from the Lagrangian in the normal way, but in terms of generalized coordinates x and quasi-velocities v. These equations were presented in the form above in Duindam and Stramigioli (2008) where they are referred to as the Boltzmann–Hamel equations of motion (Hamel 1949). The Boltzmann–Hamel equations are also used in Cameron and Book (1997) and Jarzbowska (2008). In Maruskin and Bloch (2007) the same equations are used to solve the optimal control problem for nonholonomic systems. Nonholonomic constraints are easily included in the dynamics using the Boltzmann–Hamel equations, and formulations using quasi-coordinates in general, which explains to a certain extent why these formulations have become so popular. In Chap. 12 we will show how to write the dynamics given by (6.105) and (6.106) for nonhonolomic systems.

There are many ways to derive the dynamic equations in terms of generalized coordinates and quasi-velocities by using mappings such as the one above. Common for all these methods is, however, that the state space of the vehicle and robot is described as $x \in \mathbb{R}^m$. This is not a problem when dealing with 1-DoF revolute or prismatic joints but for more complicated joints such as ball-joints or free-floating joints this will lead to singularities in the formulation, as illustrated by (3.67). Joints with more than one degree of freedom are sometimes modeled as compound kinematic joints (Kwatny and Blankenship 2000), i.e., a combination of 1-DoF simple kinematic joints. For joints that use the Euler angles to represent the generalized coordinates also this solution leads to singularities in the representation. This approach is discussed in Kwatny and Blankenship (2000) and is also treated in more detail in Sect. 6.4.4.

6.4.1 The Euler–Lagrange Equations in Matrix Form

We have now found the Euler–Lagrange equations in terms of x and v as required. We will as always strive to obtain the equations in matrix form as in (6.27) in

Sect. 6.1.4. We use the property that the inertia matrix is constant for single rigid bodies and obtain the dynamic equations in matrix form as follows:

Proposition 6.4 *Consider a rigid body with generalized position coordinates* $x \in \mathbb{R}^m$, *quasi-velocity coordinates* $v \in \mathbb{R}^m$ *where* $v = S(x)\dot{x}$, *a constant inertia matrix* M, *a potential field* $\mathscr{U}(x)$, *and Lagrangian*

$$L(x, v) = \frac{1}{2}v^{\mathsf{T}}Mv - \mathscr{U}(x). \tag{6.107}$$

Then the Euler–Lagrange equations of motion can be written in matrix form as

$$M\dot{v} + C(x, v)v + N(x) = \tau \tag{6.108}$$

where M *is found in the normal way and*

$$C(x, v) = \Big(\sum_k \gamma_k v_k\Big)M, \qquad (\gamma_k)_{ij}(x) = \sum_{l,m} S_{li}^{-1}\left(\frac{\partial S_{jl}}{\partial x_m} - \frac{\partial S_{jm}}{\partial x_l}\right)S_{mk}^{-1} \tag{6.109}$$

$$N(x) = S^{-\mathsf{T}}\frac{\partial \mathscr{U}(x)}{\partial x}. \tag{6.110}$$

Proof Using the Lagrangian in (6.107) we substitute the partial derivatives obtained in (6.77)–(6.79) into (6.105) and find

$$\frac{\mathrm{d}}{\mathrm{d}t}\left(\frac{\partial L}{\partial v}\right) - S^{-\mathsf{T}}\frac{\partial L}{\partial x} + \Big(\sum_k \gamma_k v_k\Big)\frac{\partial L}{\partial v} = \tau \tag{6.111}$$

$$M\dot{v} + \underbrace{S^{-\mathsf{T}}\frac{\partial \mathscr{U}(x)}{\partial x}}_{N(x)} + \underbrace{\Big(\sum_k \gamma_k v_k\Big)M}_{C(x,v)}\, v = \tau \tag{6.112}$$

which defines the matrices in (6.109)–(6.110). □

6.4.2 Local Parameterization

In the previous section we showed how to derive the dynamics from the Lagrangian so that the equations were expressed in coordinates $Q = x \in \mathbb{R}^m$ and $v \in \mathbb{R}^m$ where in general $v \neq \dot{x}$. In order to write the dynamic equations of the full state space, including both position and velocity, we always need to know the relation between the position coordinates x and velocity coordinates v. In the previous section we solved this problem by introducing quasi-velocities through the transformation $v = S(x)\dot{x}$. This relation, which we found explicitly for $SE(3)$ in (3.67), does however contain singularities. In this section we therefore address the problem of singularities in the representation.

Recall that we can define a set of local coordinates $\varphi \in \mathbb{R}^m$ and $\dot{\varphi} \in \mathbb{R}^m$ to describe the position and velocity state locally around a configuration Q. In Sect. 2.9 we found a mapping from the local velocities $\dot{\varphi}$ to the velocity twist V_{0b}^B, or the body velocity $\tilde{V}_{0b}^B$ when the configuration space is a subspace of $SE(3)$. For a general velocity v, this mapping was in the form $v = S(Q, \varphi)\dot{\varphi}$. Because the mapping $S(Q, \varphi)$ is the identity matrix for Euclidean transformations and can be written in terms of the adjoint map ad_X for non-Euclidean transformations, we obtain a well-defined mapping from the local velocity state $\dot{\varphi}$ to the global velocity state v.

In order to obtain a set of globally defined dynamic equations we want to write the final set of equations in terms of Q and v, and not φ and $\dot{\varphi}$. We will do this following the approach in Duindam and Stramigioli (2008). We will go through how we do this in detail in this section, but for now we present the general idea: The trick that we will use is to first consider Q a parameter, even though it strictly speaking is a state variable, and denote it $\bar{Q}$. We then think of the local coordinate φ as the position variable and the global coordinates v as the velocity variables in the normal way. The Lagrangian is then written in terms of v for velocity and $\Phi(\bar{Q}, \varphi)$ for position, in the same way that we did in Sect. 2.8. In the previous section we wanted the dynamics in terms of x and v, but we differentiated the Lagrangian with respect to x and $\dot{x}$. In this section we would like to obtain the final equations in terms of Q and v but we choose to differentiate with respect to the velocity variable $\dot{\varphi}$ and the position variable φ, not $\bar{Q}$ which we for now consider a parameter. Finally, if we recall that $\Phi_i(\bar{Q}, 0) = \bar{Q}$, we see that evaluating the expressions at $\varphi = 0$ takes us back to the position variable Q. In other words, we follow the same approach as in the previous section, but in variables φ and $\dot{\varphi}$, and in addition we evaluate the expressions at $\varphi = 0$ to obtain a set of dynamic equations expressed in Q and v.

We note that even though local coordinates φ appear in the derivations of the equations, the final equations are all evaluated at $\varphi = 0$ and hence these final equations do not depend on local coordinates. After taking partial derivatives of the Lagrangian we evaluate the results at $\varphi = 0$ (i.e., at configuration Q) to obtain the dynamics expressed in global coordinates Q and v as desired.

> The global coordinates Q and v are the only dynamic state variables and the equations are valid globally without the need for coordinate transitions between various areas of the configuration space, as is required in methods that use an atlas of local coordinate patches.

This approach has two main advantages. Firstly, the dynamics do not depend on the local coordinates as these are eliminated from the equations and the global position and velocity coordinates are the only state variables. This makes the equations valid globally. Secondly, evaluating the equations at $\varphi = 0$ greatly simplifies the dynamics and makes the equations suited for implementation in simulation software. This is clear from Eq. (2.208) where, after differentiating and setting $\varphi = 0$, all the higher order terms vanish and only the second term in the parenthesis is non-zero. We will see several examples of how we can use this to simplify the expressions of the Coriolis matrices for different types of configuration spaces. Note also that the approach is well suited for model-based control as the equations are explicit and without constraints. Even though the expressions in the derivation of the dynamics are somewhat complex we will see that the final expressions can be written in a very simple form.

We will first re-write the Lagrangian using the new variables φ and v:

$$L(\varphi, v) = \frac{1}{2} v^{\mathsf{T}} M v - \mathscr{U}\big(\Phi(Q, \varphi)\big). \tag{6.113}$$

The partial derivatives of the Lagrangian then become

$$\frac{\mathrm{d}}{\mathrm{d}t}\left(\frac{\partial L}{\partial v}\right) = M\dot{v} \tag{6.114}$$

$$\frac{\partial L}{\partial \varphi} = -\frac{\partial \mathscr{U}(\varphi)}{\partial \varphi}. \tag{6.115}$$

In order to find the Euler–Lagrange equations we need the partial derivatives of the Lagrangian $\bar{L}_\varphi(\varphi, \dot{\varphi})$ expressed in terms of the local coordinates φ and $\dot{\varphi}$ (Duindam and Stramigioli 2008; From et al. 2012; From 2012a, 2012b):

Proposition 6.5 *The partial derivatives of the Lagrangian*

$$\bar{L}_\varphi(\varphi, \dot{\varphi}) = \frac{1}{2} \dot{\varphi}^{\mathsf{T}} S^{\mathsf{T}}(Q, \varphi) M S(Q, \varphi) \dot{\varphi} - \mathscr{U}\big(\Phi(Q, \varphi)\big), \tag{6.116}$$

with respect to φ and $\dot{\varphi}$, can be written in terms of the Lagrangian

$$L(\varphi, v) = \frac{1}{2} v^{\mathsf{T}} M v - \mathscr{U}\big(\Phi(Q, \varphi)\big) \tag{6.117}$$

as

$$\frac{\mathrm{d}}{\mathrm{d}t}\left(\frac{\partial \bar{L}_\varphi}{\partial \dot{\varphi}}\right) = \dot{S}^{\mathsf{T}}(Q, \varphi)\frac{\partial L}{\partial v} + S^{\mathsf{T}}(Q, \varphi)\frac{\mathrm{d}}{\mathrm{d}t}\left(\frac{\partial L}{\partial v}\right) \tag{6.118}$$

$$\frac{\partial \bar{L}_\varphi}{\partial \varphi} = \frac{\partial L}{\partial \varphi} + \frac{\partial^{\mathsf{T}}(S(Q, \varphi)\dot{\varphi})}{\partial \varphi}\frac{\partial L}{\partial v}. \tag{6.119}$$

Proof Proposition 6.5 follows directly from Proposition 6.1. □

Substituting this into Lagrange's equations will give us the Euler–Lagrange equations of motion in terms of local position and velocity variables φ and $\dot{\varphi}$. These equations are valid only in the vicinity of a configuration $\bar{Q}$. However, we want the dynamics expressed in terms of global position variables Q and velocity variables v. We can eliminate the local velocity variables $\dot{\varphi}$ by using the relation $v = S(Q,\varphi)\dot{\varphi}$ and eliminate the local position variables φ by representing the position as $\Phi(\bar{Q},\varphi)$ and then, after differentiating we let $\varphi = 0$ to obtain the global position variable through the relation $\Phi(\bar{Q},0) = \bar{Q}$. Note that the velocity transformation matrix $S(Q,\varphi)$ relating $\dot{\varphi}$ and v is well defined, as opposed to $S(x)$ which gave us the mapping from $\dot{x}$ to v. This is an important difference that we will make much use of in this chapter and the chapters to come. We can state this as the following important result:

Proposition 6.6 *Consider a single rigid body with local position and velocity coordinates φ and $\dot{\varphi}$ and global position and velocity coordinates Q and v. Write the kinetic energy as $\mathscr{K}(v) = \frac{1}{2}v^{\mathsf{T}}Mv$ with the inertia matrix M. The dynamics of this system then satisfies*

$$M\dot{v} + C(v)v = \tau \tag{6.120}$$

where M is found in the normal way, with τ the vector of external and control wrenches (collocated with v), and the matrix describing the Coriolis and centrifugal forces given by

$$C_{ij}(\varphi,v) = \sum_{l,m,s} S_{li}^{-1}\left(\frac{\partial S_{sl}}{\partial \varphi_m} - \frac{\partial S_{sj}}{\partial \varphi_l}\right) S_{mj}^{-1}\bigg|_{\varphi=0} (Mv)_s. \tag{6.121}$$

To compute the matrix $C(v)$ for a single rigid body with configuration space $SE(3)$ or one of its subgroups, we can use (2.209) to simplify $C(v)$ slightly to

$$C_{ij}(\varphi,v) = \sum_k \left(\frac{\partial S_{ki}}{\partial \varphi_j} - \frac{\partial S_{kj}}{\partial \varphi_i}\right)\bigg|_{\varphi=0} (Mv)_k. \tag{6.122}$$

Proof From Eqs. (6.106) and (6.105) and using φ as the position variable we find

$$\begin{aligned} C_{ij}(\varphi,v) &= \sum_s \left(\sum_k \gamma_k v_k\right)_{is} M_{sj}\bigg|_{\varphi=0} \\ &= \sum_s \sum_{l,m,k} S_{li}^{-1}\left(\frac{\partial S_{sl}}{\partial \varphi_m} - \frac{\partial S_{sm}}{\partial \varphi_l}\right) S_{mk}^{-1}\bigg|_{\varphi=0} v_k M_{sj}. \end{aligned} \tag{6.123}$$

We know that there are many different ways to write the Coriolis matrix which all result in the same vector $C(\varphi, v)v$. We denote this vector $C(\varphi, v)v = [c_1\ c_2 \cdots c_n]^{\mathsf{T}}$ and write

$$\begin{aligned}
c_i(\varphi, v) &= \sum_p C_{ip} v_p \\
&= \sum_p \Big(\sum_s \Big(\sum_k \gamma_k v_k \Big)_{is} M_{sp} \Big) v_p \\
&= \sum_p \sum_s \Big(\sum_k \gamma_k v_k \Big)_{is} M_{sp} v_p \\
&= \Big(\sum_s \sum_k \gamma_k v_k \Big)_{is} \sum_p M_{sp} v_p \\
&= \Big(\sum_s \sum_k \gamma_k v_k \Big)_{is} (Mv)_s \\
&= \Big(\sum_s \sum_{l,m,k} S_{li}^{-1} \Big(\frac{\partial S_{sl}}{\partial \varphi_m} - \frac{\partial S_{sm}}{\partial \varphi_l} \Big) S_{mk}^{-1} \Big|_{\varphi=0} v_k \Big) (Mv)_s \\
&= \sum_k \underbrace{\Big(\sum_{l,m,s} S_{li}^{-1} \Big(\frac{\partial S_{sl}}{\partial \varphi_m} - \frac{\partial S_{sm}}{\partial \varphi_l} \Big) S_{mk}^{-1} \Big|_{\varphi=0} (Mv)_s \Big)}_{C_{ik}} v_k. \qquad (6.124)
\end{aligned}$$

Here we have used the observation that M does not depend on φ so there is no need to evaluate M at $\varphi = 0$. Because $c_i = C_{ij} v_j$, a trivial change in the variable name gives the Coriolis matrix

$$C_{ij}(v) = \sum_{l,m,s} S_{li}^{-1} \Big(\frac{\partial S_{sl}}{\partial \varphi_m} - \frac{\partial S_{sm}}{\partial \varphi_l} \Big) S_{mj}^{-1} \Big|_{\varphi=0} (Mv)_s. \qquad (6.125)$$

Now, recall that the velocity transformation matrix can be written as

$$S(\varphi) = \Big(I - \frac{1}{2} \mathrm{ad}_\varphi + \frac{1}{6} \mathrm{ad}_\varphi^2 - \cdots \Big) \in \mathbb{R}^{m \times m} \qquad (6.126)$$

where ad_X is the adjoint map for a general Lie algebra X of dimension m. Because the expression in (6.125) is to be evaluated at $\varphi = 0$ the expression in (6.125) is non-zero only for $l = i$ and $m = j$. The final expressions then become

$$C_{ij}(v) = \sum_{l,m,s} S_{li}^{-1} \Big(\frac{\partial S_{sl}}{\partial \varphi_m} - \frac{\partial S_{sm}}{\partial \varphi_l} \Big) S_{mj}^{-1} \Big|_{\varphi=0} (Mv)_s$$

$$= \sum_s S_{ii}^{-1} \left(\frac{\partial S_{si}}{\partial \varphi_j} - \frac{\partial S_{sj}}{\partial \varphi_i} \right) S_{jj}^{-1} \bigg|_{\varphi=0} (Mv)_s$$

$$= \sum_s \left(\frac{\partial S_{si}}{\partial \varphi_j} - \frac{\partial S_{sj}}{\partial \varphi_i} \right) \bigg|_{\varphi=0} (Mv)_s \tag{6.127}$$

which after a change in the index name becomes

$$C_{ij}(v) = \sum_k \left(\frac{\partial S_{ki}}{\partial \varphi_j} - \frac{\partial S_{kj}}{\partial \varphi_i} \right) \bigg|_{\varphi=0} (Mv)_k. \tag{6.128}$$

□

We see that we have found a formulation of the dynamic equations in terms of the variables φ and v. However, when we evaluate the expressions at $\varphi = 0$ the position variables φ vanish from the equations and the position is written in terms of the configuration state matrix Q. Admittedly, the formulation is rather complex, but we will see in the next section that when we use the expressions for the velocity transformation matrices $S(\varphi)$ for the different configuration spaces the final expressions take a very simple form.

6.4.3 The Most Important Configuration Spaces

To apply this general result to single rigid bodies, we write the configuration space as a matrix Lie group $g_{0b}(Q)$, where Q is the Lie group $SE(3)$ or one of its subgroups, and the velocity state v as a twist V_{0b}^B or $\tilde{V}_{0b}^B$, i.e., the Lie algebra of g_{0b}.

The local Euclidean structure for the state g_{0b} is given by exponential coordinates. Mathematically, we can express configurations g_{0b} around a fixed state $\bar{g}_{0b}$ as

$$g_{0b} = \bar{g}_{0b} e^{\hat{\varphi}}, \tag{6.129}$$

for a local position variable φ. For $se(3)$ we have $m = 6$ and know that $e^{\widehat{\varphi}} \in SE(3)$ from Sect. 2.8. Similarly, when $m < 6$ we set $\varphi_i = 0$ for all the $6 - m$ entries that are trivially zero, corresponding to the all-zero rows of the selection matrix (Definition 3.4). The position and velocity variables φ and $\dot{\varphi}$ then become vectors in $\mathbb{R}^m$.

We can now derive an expression for the total kinetic energy. Let $I_b \in \mathbb{R}^{6\times 6}$ denote the constant positive definite diagonal inertia matrix of the rigid body. The kinetic energy $\mathscr{K}_b$ then follows as

$$\mathscr{K}_b = \frac{1}{2} \left(V_{0b}^B\right)^{\mathsf{T}} I_b V_{0b}^B$$

$$= \frac{1}{2} \left(H \tilde{V}_{0b}^B\right)^{\mathsf{T}} I_b \left(H \tilde{V}_{0b}^B\right)$$

$$
\begin{aligned}
&= \frac{1}{2}\big(\tilde{V}_{0b}^B\big)^{\mathsf{T}} H^{\mathsf{T}} I_b H \tilde{V}_{0b}^B \\
&= \frac{1}{2}\big(\tilde{V}_{0b}^B\big)^{\mathsf{T}} M \tilde{V}_{0b}^B \\
&= \frac{1}{2} v^{\mathsf{T}} M v.
\end{aligned} \tag{6.130}
$$

Here, H^{T} is the transpose of H which works fine when dealing with the Lie groups treated here, so we will use this notation throughout this book. We see that the selection matrix $H \in \mathbb{R}^{6\times m}$ guarantees that the inertia matrix M has the right dimension. Note that neither $\mathscr{K}(v)$ nor M depend on the pose g_{0b} nor the choice of inertial reference frame $\mathscr{F}_0$.

The reason that the inertia matrix is constant is that we represent the velocities in the body frame $\mathscr{F}_b$. If we choose to represent the velocities in a different frame, for example the inertial frame, we get a configuration dependent inertia matrix in the form

$$
\mathscr{K}_b = \frac{1}{2}\big(V_{0b}^S\big)^{\mathsf{T}} M(g_{0b}) V_{0b}^S = \frac{1}{2}\big(V_{0b}^S\big)^{\mathsf{T}} \mathrm{Ad}_{g_{b0}}^{\mathsf{T}} I_b \mathrm{Ad}_{g_{b0}} V_{0b}^S \tag{6.131}
$$

where $M(g_{0b}) = \mathrm{Ad}_{g_{b0}}^{\mathsf{T}} I_b \mathrm{Ad}_{g_{b0}}$ is the configuration dependent inertia matrix.

6.4.3.1 Rigid Bodies with Configuration Space $SO(3)$

For a rigid body with configuration space $SO(3)$ the selection matrix becomes

$$
H = \begin{bmatrix} 0 & 0 & 0 \\ 0 & 0 & 0 \\ 0 & 0 & 0 \\ 1 & 0 & 0 \\ 0 & 1 & 0 \\ 0 & 0 & 1 \end{bmatrix}. \tag{6.132}
$$

There are two ways to see this: either as a matrix that gives us the twist V_{0b}^B from the velocity vector $\tilde{V}_{0b}^B = \omega_{0b}^B$, i.e., $V_{0b}^B = H\tilde{V}_{0b}^B$, or as the matrix that transforms the generalized inertia matrix $I_b \in \mathbb{R}^{6\times 6}$ into the inertia matrix $M \in \mathbb{R}^{m\times m}$, i.e., $M = H^{\mathsf{T}} I_b H$. For $SO(3)$ the constant inertia matrix then becomes

$$
M = H^{\mathsf{T}} I_b H = \begin{bmatrix} I_x & 0 & 0 \\ 0 & I_y & 0 \\ 0 & 0 & I_z \end{bmatrix} \tag{6.133}
$$

as usual.

In Sect. 3.3.6.2 we found that the matrix relating the local velocities $\dot{\varphi}$ and global body velocities ω_{0b}^B is given by Rossmann (2002)

$$S(\varphi) = \left(I - \frac{1}{2}\widehat{\varphi}_V + \frac{1}{6}\widehat{\varphi}_V^2 - \cdots \right) \in \mathbb{R}^{3\times 3}. \tag{6.134}$$

We will first show that when differentiating with respect to φ and substituting $\varphi = 0$ in (6.122) the matrices $\sum_k \frac{\partial S_{ki}}{\partial \varphi_j}$ and $\sum_k \frac{\partial S_{kj}}{\partial \varphi_i}$ can be written in terms of very simple expressions. All constant terms will disappear when we differentiate with respect to φ and the terms that include higher order terms of φ will disappear when we evaluate the expressions at $\varphi = 0$. We disregard the terms that we already know will equal zero and denote this matrix S_r (the *reduced velocity transformation matrix*) which then becomes

$$S_r(\varphi) = -\frac{1}{2}\widehat{\varphi}_V = -\frac{1}{2}\begin{bmatrix} 0 & -\varphi_3 & \varphi_2 \\ \varphi_3 & 0 & -\varphi_1 \\ -\varphi_2 & \varphi_1 & 0 \end{bmatrix}. \tag{6.135}$$

We start with $\sum_k \frac{\partial S_{ki}}{\partial \varphi_j}$ in (6.122) which after multiplying with $(Mv)_k$ becomes:

$$\begin{aligned}
&\left\{ \sum_k \left(\frac{\partial S_{ki}}{\partial \varphi_j} \right)\bigg|_{\varphi=0} (Mv)_k \right\}_{ij} \\
&= -\frac{1}{2}\begin{bmatrix} \sum_k \frac{\partial \widehat{\varphi}_{k1}}{\partial \varphi_1}(Mv)_k & \sum_k \frac{\partial \widehat{\varphi}_{k1}}{\partial \varphi_2}(Mv)_k & \sum_k \frac{\partial \widehat{\varphi}_{k1}}{\partial \varphi_3}(Mv)_k \\ \sum_k \frac{\partial \widehat{\varphi}_{k2}}{\partial \varphi_1}(Mv)_k & \sum_k \frac{\partial \widehat{\varphi}_{k2}}{\partial \varphi_2}(Mv)_k & \sum_k \frac{\partial \widehat{\varphi}_{k2}}{\partial \varphi_3}(Mv)_k \\ \sum_k \frac{\partial \widehat{\varphi}_{k3}}{\partial \varphi_1}(Mv)_k & \sum_k \frac{\partial \widehat{\varphi}_{k3}}{\partial \varphi_2}(Mv)_k & \sum_k \frac{\partial \widehat{\varphi}_{k3}}{\partial \varphi_3}(Mv)_k \end{bmatrix} \\
&= -\frac{1}{2}\begin{bmatrix} 0 & -(Mv)_3 & (Mv)_2 \\ (Mv)_3 & 0 & -(Mv)_1 \\ -(Mv)_2 & (Mv)_1 & 0 \end{bmatrix} \\
&= -\frac{1}{2}\widehat{(Mv)}.
\end{aligned} \tag{6.136}$$

Similarly, $\sum_k (\frac{\partial S_{kj}}{\partial \varphi_i})|_{\varphi=0}(Mv)_k$ becomes

$$\begin{aligned}
&\left\{ \sum_k \left(\frac{\partial S_{kj}}{\partial \varphi_i} \right)\bigg|_{\varphi=0} (Mv)_k \right\}_{ij} \\
&= -\frac{1}{2}\begin{bmatrix} \sum_k \frac{\partial \widehat{\varphi}_{k1}}{\partial \varphi_1}(Mv)_k & \sum_k \frac{\partial \widehat{\varphi}_{k2}}{\partial \varphi_1}(Mv)_k & \sum_k \frac{\partial \widehat{\varphi}_{k3}}{\partial \varphi_1}(Mv)_k \\ \sum_k \frac{\partial \widehat{\varphi}_{k1}}{\partial \varphi_2}(Mv)_k & \sum_k \frac{\partial \widehat{\varphi}_{k2}}{\partial \varphi_2}(Mv)_k & \sum_k \frac{\partial \widehat{\varphi}_{k3}}{\partial \varphi_2}(Mv)_k \\ \sum_k \frac{\partial \widehat{\varphi}_{k1}}{\partial \varphi_3}(Mv)_k & \sum_k \frac{\partial \widehat{\varphi}_{k2}}{\partial \varphi_3}(Mv)_k & \sum_k \frac{\partial \widehat{\varphi}_{k3}}{\partial \varphi_3}(Mv)_k \end{bmatrix}
\end{aligned}$$

$$
\begin{aligned}
&= -\frac{1}{2}\begin{bmatrix} 0 & (Mv)_3 & -(Mv)_2 \\ -(Mv)_3 & 0 & (Mv)_1 \\ (Mv)_2 & -(Mv)_1 & 0 \end{bmatrix} \\
&= \frac{1}{2}\widehat{(Mv)}. \qquad (6.137)
\end{aligned}
$$

For a rigid body with configuration space $SO(3)$ the rather complicated expression for the Coriolis matrix in (6.122) becomes

$$
C(v) = -\widehat{(Mv)}. \qquad (6.138)
$$

The dynamic equations can now be written in matrix form as

$$
\begin{aligned}
M\dot{v} + C(v)v &= \tau \\
M\dot{v} - \widehat{(Mv)}v &= \tau.
\end{aligned} \qquad (6.139)
$$

If we use the relation $\widehat{(Mv)}v = -\widehat{v}Mv$ and denote the velocity as $v = \omega_{0b}^{B}$ we get Euler's equation of motion:

$$
M\dot{\omega}_{0b}^{B} + \widehat{\omega}_{0b}^{B} M \omega_{0b}^{B} = \tau \qquad (6.140)
$$

$$
M\dot{\omega}_{0b}^{B} - M\omega_{0b}^{B} \times \omega_{0b}^{B} = \tau \qquad (6.141)
$$

which in this case is identical to the equations that we found in (6.31).

6.4.3.2 Rigid Bodies with Configuration Space *SE*(3)

For a rigid body with configuration space $SE(3)$ the matrix relating the local and global velocities is given by Rossmann (2002)

$$
S(\varphi) = \left(I - \frac{1}{2}\mathrm{ad}_{\varphi} + \frac{1}{6}\mathrm{ad}_{\varphi}^{2} - \cdots \right) \in \mathbb{R}^{6\times 6}. \qquad (6.142)
$$

The precise computational details of the partial derivatives follow the same steps as for $SO(3)$ in the previous section. If we write $S = (I - \frac{1}{2}\mathrm{ad}_{\varphi} + \frac{1}{6}\mathrm{ad}_{\varphi}^{2} - \cdots)$ we note that after differentiating and evaluating at $\varphi = 0$ the matrices $\sum \frac{\partial S_{ki}}{\partial \varphi_j}$ are equal to $\frac{1}{2}\widetilde{\mathrm{ad}}_{e_k}$ where e_k is a 6-vector with 1 in the kth entry and zeros elsewhere and $\widetilde{\mathrm{ad}}_p$ for $p \in \mathbb{R}^6$ is defined as

$$
\widetilde{\mathrm{ad}}_p = \begin{bmatrix} 0 & 0 & 0 & 0 & p_3 & -p_2 \\ 0 & 0 & 0 & -p_3 & 0 & p_1 \\ 0 & 0 & 0 & p_2 & -p_1 & 0 \\ 0 & p_3 & -p_2 & 0 & p_6 & -p_5 \\ -p_3 & 0 & p_1 & -p_6 & 0 & p_4 \\ p_2 & -p_1 & 0 & p_5 & -p_4 & 0 \end{bmatrix}. \qquad (6.143)
$$

Similarly, $\sum \frac{\partial S_{kj}}{\partial \varphi_i}$ is equal to $-\frac{1}{2}\widetilde{\text{ad}}_{e_k}$. This is then multiplied by the kth element of Mv when differentiating with respect to φ_k. The C-matrix is thus given by

$$C(v) = \widetilde{\text{ad}}_{(Mv)} \tag{6.144}$$

where

$$\widetilde{\text{ad}}_{(Mv)} = \begin{bmatrix} 0 & 0 & 0 & 0 & (Mv)_3 & -(Mv)_2 \\ 0 & 0 & 0 & -(Mv)_3 & 0 & (Mv)_1 \\ 0 & 0 & 0 & (Mv)_2 & -(Mv)_1 & 0 \\ 0 & (Mv)_3 & -(Mv)_2 & 0 & (Mv)_6 & -(Mv)_5 \\ -(Mv)_3 & 0 & (Mv)_1 & -(Mv)_6 & 0 & (Mv)_4 \\ (Mv)_2 & -(Mv)_1 & 0 & (Mv)_5 & -(Mv)_4 & 0 \end{bmatrix}. \tag{6.145}$$

The dynamic equations are written in matrix form as

$$\begin{aligned} M\dot{v} + C(v)v &= \tau \\ M\dot{v} + \widetilde{\text{ad}}_{(Mv)}v &= \tau \end{aligned} \tag{6.146}$$

which in this case is identical to the Euler–Lagrange equations of rigid body motion. We see this if we use that $\widetilde{\text{ad}}_{(Mv)}v = -\text{ad}_V^{\mathsf{T}} Mv$ and we obtain the Newton–Euler equations

$$M\dot{V}_{0b}^B - \text{ad}_V^{\mathsf{T}} M V_{0b}^B = \tau \tag{6.147}$$

which we found in (6.41). For the special case when M is a constant diagonal matrix we can reproduce the equations that we found in (6.55) and write out the dynamics of a single rigid body on $SE(3)$ as

$$\begin{aligned} m\dot{u} - mrv + mqw &= \tau_u \\ m\dot{v} + mru - mpw &= \tau_v \\ m\dot{w} - mqu + mpv &= \tau_w \\ I_x\dot{p} - I_y rq + I_z qr &= \tau_p \\ I_y\dot{q} + I_x rp - I_z pr &= \tau_q \\ I_z\dot{r} - I_x qp + I_y pq &= \tau_r. \end{aligned}$$

6.4.3.3 Rigid Bodies with Configuration Space $SE(2)$

We can find a similar expression for the planar case. The adjoint map of a Lie algebra $V \in se(2)$ is given by

$$\text{ad}_V = \begin{bmatrix} 0 & -r & v \\ r & 0 & -u \\ 0 & 0 & 0 \end{bmatrix}, \quad \text{where} \quad \hat{V} = \begin{bmatrix} 0 & -r & u \\ r & 0 & v \\ 0 & 0 & 0 \end{bmatrix} \in se(2). \tag{6.148}$$

We thus write

$$S_r(\varphi) = \frac{1}{2}\begin{bmatrix} 0 & -\varphi_3 & \varphi_2 \\ \varphi_3 & 0 & -\varphi_1 \\ 0 & 0 & 0 \end{bmatrix} \tag{6.149}$$

and find the partial derivatives as

$$\begin{aligned}
&\left\{\sum_k \left(\frac{\partial S_{ki}}{\partial \varphi_j}\right)\Bigg|_{\varphi=0} (Mv)_k\right\}_{ij} \\
&= -\frac{1}{2}\begin{bmatrix} \sum_k \frac{\partial \widehat{\varphi}_{k1}}{\partial \varphi_1}(Mv)_k & \sum_k \frac{\partial \widehat{\varphi}_{k1}}{\partial \varphi_2}(Mv)_k & \sum_k \frac{\partial \widehat{\varphi}_{k1}}{\partial \varphi_3}(Mv)_k \\ \sum_k \frac{\partial \widehat{\varphi}_{k2}}{\partial \varphi_1}(Mv)_k & \sum_k \frac{\partial \widehat{\varphi}_{k2}}{\partial \varphi_2}(Mv)_k & \sum_k \frac{\partial \widehat{\varphi}_{k2}}{\partial \varphi_3}(Mv)_k \\ \sum_k \frac{\partial \widehat{\varphi}_{k3}}{\partial \varphi_1}(Mv)_k & \sum_k \frac{\partial \widehat{\varphi}_{k3}}{\partial \varphi_2}(Mv)_k & \sum_k \frac{\partial \widehat{\varphi}_{k3}}{\partial \varphi_3}(Mv)_k \end{bmatrix} \\
&= -\frac{1}{2}\begin{bmatrix} 0 & 0 & (Mv)_2 \\ 0 & 0 & -(Mv)_1 \\ -(Mv)_2 & (Mv)_1 & 0 \end{bmatrix} \\
&= \frac{1}{2}\widetilde{\text{ad}}_{(Mv)}.
\end{aligned} \tag{6.150}$$

Similarly, $\sum_k (\frac{\partial S_{kj}}{\partial \varphi_i})|_{\varphi=0}(Mv)_k$ becomes

$$\begin{aligned}
&\left\{\sum_k \left(\frac{\partial S_{kj}}{\partial \varphi_i}\right)\Bigg|_{\varphi=0} (Mv)_k\right\}_{ij} \\
&= -\frac{1}{2}\begin{bmatrix} \sum_k \frac{\partial \widehat{\varphi}_{k1}}{\partial \varphi_1}(Mv)_k & \sum_k \frac{\partial \widehat{\varphi}_{k2}}{\partial \varphi_1}(Mv)_k & \sum_k \frac{\partial \widehat{\varphi}_{k3}}{\partial \varphi_1}(Mv)_k \\ \sum_k \frac{\partial \widehat{\varphi}_{k1}}{\partial \varphi_2}(Mv)_k & \sum_k \frac{\partial \widehat{\varphi}_{k2}}{\partial \varphi_2}(Mv)_k & \sum_k \frac{\partial \widehat{\varphi}_{k3}}{\partial \varphi_2}(Mv)_k \\ \sum_k \frac{\partial \widehat{\varphi}_{k1}}{\partial \varphi_3}(Mv)_k & \sum_k \frac{\partial \widehat{\varphi}_{k2}}{\partial \varphi_3}(Mv)_k & \sum_k \frac{\partial \widehat{\varphi}_{k3}}{\partial \varphi_3}(Mv)_k \end{bmatrix} \\
&= -\frac{1}{2}\begin{bmatrix} 0 & 0 & -(Mv)_2 \\ 0 & 0 & (Mv)_1 \\ (Mv)_2 & -(Mv)_1 & 0 \end{bmatrix} \\
&= -\frac{1}{2}\widetilde{\text{ad}}_{(Mv)}.
\end{aligned} \tag{6.151}$$

For a rigid body with configuration space $SE(2)$ the Coriolis matrix in (6.122) becomes

$$C(v) = \widetilde{\text{ad}}_{(Mv)}. \tag{6.152}$$

The dynamic equations can now be written in matrix form as

$$\begin{aligned} M\dot{v} + C(v)v &= \tau \\ M\dot{v} + \widetilde{\text{ad}}_{(Mv)}v &= \tau \end{aligned} \tag{6.153}$$

If we use the relation $\widetilde{\text{ad}}_{(Mv)}v = -\text{ad}_v^\mathsf{T} M v$ we get the dynamics:

$$M\dot{v}_{0b}^B - \text{ad}_v^\mathsf{T} M v_{0b}^B = \tau. \tag{6.154}$$

If we assume that $\mathscr{F}_b$ is at the center of gravity we get the generalized inertia matrix $M = \text{diag}(m, m, I_z)$ and we can write the dynamics explicitly as

$$\begin{bmatrix} m & 0 & 0 \\ 0 & m & 0 \\ 0 & 0 & I_z \end{bmatrix} \begin{bmatrix} \dot{u} \\ \dot{v} \\ \dot{r} \end{bmatrix} - \begin{bmatrix} 0 & r & 0 \\ -r & 0 & 0 \\ v & -u & 0 \end{bmatrix} \begin{bmatrix} mu \\ mv \\ I_z r \end{bmatrix} = \begin{bmatrix} \tau_u \\ \tau_v \\ \tau_r \end{bmatrix} \tag{6.155}$$

where we have denoted the velocity as $v = [u\ v\ r]^\mathsf{T}$. The dynamics now becomes

$$\begin{aligned} m\dot{u} - mrv &= \tau_u \\ m\dot{v} + mru &= \tau_v \\ I_z\dot{r} &= \tau_r. \end{aligned} \tag{6.156}$$

We see that the cross terms from the Coriolis matrix $c_3 = -mvu + muv$ disappear from the equations when the reference frame is chosen so that $\mathscr{F}_{b1} = \mathscr{F}_b$.

6.4.3.4 The Schönflies Group $\mathscr{X}$

For the Shönflies group we write the reduced velocity transformation matrix as

$$S_r(\varphi) = \frac{1}{2} \begin{bmatrix} 0 & -\varphi_4 & 0 & \varphi_2 \\ \varphi_4 & 0 & 0 & -\varphi_1 \\ 0 & 0 & 0 & 0 \\ 0 & 0 & 0 & 0 \end{bmatrix} \tag{6.157}$$

and find the partial derivatives as

$$\begin{aligned} \left\{ \sum_k \left(\frac{\partial S_{ki}}{\partial \varphi_j} \right) \bigg|_{\varphi=0} (Mv)_k \right\}_{ij} &= -\frac{1}{2} \begin{bmatrix} 0 & 0 & 0 & (Mv)_2 \\ 0 & 0 & 0 & -(Mv)_1 \\ 0 & 0 & 0 & 0 \\ -(Mv)_2 & (Mv)_1 & 0 & 0 \end{bmatrix} \\ &= \frac{1}{2} \widetilde{\text{ad}}_{(Mv)}. \end{aligned} \tag{6.158}$$

Similarly, $\sum_k (\frac{\partial S_{kj}}{\partial \varphi_i})|_{\varphi=0}(Mv)_k$ becomes

$$\left\{ \sum_k \left(\frac{\partial S_{kj}}{\partial \varphi_i} \right) \bigg|_{\varphi=0} (Mv)_k \right\}_{ij} = -\frac{1}{2} \begin{bmatrix} 0 & 0 & 0 & -(Mv)_2 \\ 0 & 0 & 0 & (Mv)_1 \\ 0 & 0 & 0 & 0 \\ (Mv)_2 & -(Mv)_1 & 0 & 0 \end{bmatrix}$$

Table 6.1 The Coriolis matrix for different Lie subgroups of $SE(3)$ for velocities in body coordinates

Lie Group	S	C	Eq.
$SE(3)$	$I - \frac{1}{2}\text{ad}_\varphi + \frac{1}{6}\text{ad}_\varphi^2 - \cdots$	$\tilde{\text{ad}}_{(Mv)}$	(6.144)
$\mathscr{X}(z)$	$I - \frac{1}{2}\text{ad}_\varphi + \frac{1}{6}\text{ad}_\varphi^2 - \cdots$	$\tilde{\text{ad}}_{(Mv)}$	(6.160)
$\mathbb{R}^3$	$I_{3\times 3}$	0	
$SE(2)$	$I - \frac{1}{2}\text{ad}_\varphi + \frac{1}{6}\text{ad}_\varphi^2 - \cdots$	$\tilde{\text{ad}}_{(Mv)}$	(6.152)
$SO(3)$	$I - \frac{1}{2}\widehat{\varphi} + \frac{1}{6}\widehat{\varphi}^2 - \cdots$	$-\widehat{(Mv)}$	(6.138)
$\mathbb{R}^2$	$I_{2\times 2}$	0	
$\mathbb{R}, H, SO(2)$	$I_{1\times 1}$	0	

$$= -\frac{1}{2}\widetilde{\text{ad}}_{(Mv)}. \tag{6.159}$$

This gives us the Coriolis matrix

$$C(v) = \widetilde{\text{ad}}_{(Mv)}. \tag{6.160}$$

The dynamic equations can now be written in matrix form as

$$M\dot{v} + \widetilde{\text{ad}}_{(Mv)}v = \tau. \tag{6.161}$$

Again, we can use that $\widetilde{\text{ad}}_{(Mv)}v = -\text{ad}_v^{\mathsf{T}} Mv$ and write

$$M\dot{v} - \text{ad}_v^{\mathsf{T}} Mv = \tau \tag{6.162}$$

or, for a diagonal M, explicitly as

$$\begin{aligned} m\dot{u} - mrv &= \tau_x \\ m\dot{v} + mru &= \tau_y \\ m\dot{w} &= \tau_z \\ I_z\dot{r} &= \tau_r. \end{aligned} \tag{6.163}$$

6.4.3.5 Summary

Table 6.1 shows the mappings from local to global velocity coordinates and the corresponding C-matrices for different Lie Groups. We note that when the Lie bracket vanish, which is the case for Abelian Lie algebras, the Coriolis and centrifugal forces are not present in the dynamics, as expected.

6.4.4 Other Formulations in Quasi-coordinates

There are many ways to write the dynamics in terms of quasi-coordinates. The Gibbs-Appel equations (Gibbs 1879; Lewis 1996) are obtained by differentiation of a function of kinematic variables. In this way the Gibbs-Appel equations resemble Lagrange's equations and the Euler–Poincaré equations. On the other hand, Kane's equations (Kane et al. 1983; Kane and Levinson 1985) use the virtual work principle as a starting point and can be considered a generalization of the approaches that are derived from the virtual work approach.

There are also some more recent work. Quasi-coordinates are used in Kwatny and Blankenship (2000) where the Poincaré equations of motion are derived in terms of quasi-velocities. Let X be defined in terms of the Lie bracket as

$$X_j = \big[[t_j, t_1][t_j, t_2] \cdots [t_j, t_m]\big] \tag{6.164}$$

where $T = [t_1\ t_2 \cdots t_m]$ is the inverse of the velocity transformation matrix, i.e., $T(x) = S^{-1}(x)$ and the Lie bracket is defined as $[t_1, t_2] = \hat{t}_1\hat{t}_2 - \hat{t}_2\hat{t}_1$. The Poincaré equations of motion as presented in Kwatny and Blankenship (2000) are then given by

$$v = S(x)\dot{x}, \quad \dot{x} = T(x)v \tag{6.165}$$

$$\frac{\mathrm{d}}{\mathrm{d}t}\left(\frac{\partial L}{\partial v}\right) - T^{\mathsf{T}}\frac{\partial L}{\partial x} - \sum_{j=1}^{m} X_j^{\mathsf{T}} S^{\mathsf{T}} \frac{\partial L}{\partial v} v_j = T^{\mathsf{T}}\tau \tag{6.166}$$

for the Lagrangian $L(x, v)$. If we recall that $T(x) = S^{-1}(x)$ we see the clear resemblance with the formulation in (6.105). There are two differences between this approach and the one presented in the previous section. Firstly, in (6.166) the Lie bracket is used for differentiation, while in (6.105) we use the adjoint map ad_X. This is, of course, just a matter of notation and has no real significance. Secondly and more importantly, the difference lies in how the velocity transformation matrix is handled. In Kwatny and Blankenship (2000) the velocity transformation matrix $S(x)$ is calculated as in (3.67) and the formulation is thus singularity prone. However, using the same ideas as in the previous section, these singularities can be removed and we obtain a formulation similar to the one in (6.105).

References

Arnold, V. I. (1989). *Mathematical methods of classical mechanics*. Berlin: Springer.

Bremer, H. (1988). *Über eine zentralgleichung in der Dynamik* (Vol. 68, pp. 307–311).

Bullo, F., & Lewis, A. D. (2000). *Geometric control of mechanical systems: modeling, analysis, and design for simple mechanical control systems*. New York: Springer.

Cameron, J. M., & Book, W. J. (1997). Modeling mechanisms with nonholonomic joints using the Boltzmann-Hamel equations. *The International Journal of Robotics Research*, *16*(1), 47–59.

Duindam, V., & Stramigioli, S. (2007). Lagrangian dynamics of open multibody systems with generalized holonomic and nonholonomic joints. In *Proceedings of the IEEE/RSJ international conference on intelligent robots and systems*, San Diego, CA, USA (pp. 3342–3347).

Duindam, V., & Stramigioli, S. (2008). Singularity-free dynamic equations of open-chain mechanisms with general holonomic and nonholonomic joints. *IEEE Transactions on Robotics*, *24*(3), 517–526.

Egeland, O., & Gravdahl, J. T. (2003). *Modeling and simulation for automatic control*. Trondheim: Marine Cybernetics AS.

Fossen, T. I. (2002). *Marine control systems*. Trondheim: Marine Cybernetics AS. 3rd printing.

Fossen, T. I., & Fjellstad, O. E. (1995). Nonlinear modelling of marine vehicles in 6 degrees of freedom. *International Journal of Mathematical Modelling Systems*, *1*(1), 17–28.

From, P. J. (2012a). An explicit formulation of singularity-free dynamic equations of mechanical systems in Lagrangian form—part one: single rigid bodies. *Modeling, Identification and Control*, *33*(2), 45–60.

From, P. J. (2012b). An explicit formulation of singularity-free dynamic equations of mechanical systems in Lagrangian form—part two: multibody systems. *Modeling, Identification and Control*, *33*(2), 61–68.

From, P. J., Duindam, V., & Stramigioli, S. (2012). Corrections to "Singularity-free dynamic equations of open-chain mechanisms with general holonomic and nonholonomic joints". *IEEE Transactions on Robotics*, *28*(6), 1431–1432.

Gibbs, J. W. (1879). On the fundamental formulae of dynamics. *American Journal of Mathematics*, *2*(1), 49–64.

Hamel, G. (1949). *Theoretische Mechanik*. Berlin: Springer.

Herman, P., & Kozlowski, K. (2006). A survey of equations of motion in terms of inertial quasi-velocities for serial manipulators. *Archive of Applied Mechanics*, *76*(9–10), 579–614.

Jarzbowska, E. (2008). Quasi-coordinates based dynamics modeling and control design for non-holonomic systems. *Nonlinear Analysis*, *71*(12), 118–131.

Kane, T. R., & Levinson, D. A. (1985). *Dynamics: theory and applications*. New York: McGraw Hill.

Kane, T. R., Likins, P. W., & Levinson, D. A. (1983). *Spacecraft dynamics*. New York: McGraw Hill.

Kozlowski, K., & Herman, P. (2008). Control of robot manipulators in terms of quasi-velocities. *Journal of Intelligent & Robotic Systems*, *53*(3), 205–221.

Kwatny, H. G., & Blankenship, G. (2000). *Nonlinear control and analytical mechanics a computational approach*. Boston: Birkhäuser.

Lagrange, J.-L. (1788). *Mécanique analytique*. Chez la Veuve Desaint.

Lesser, M. (1992). A geometrical interpretation of Kanes equations. *Journal of Mathematical and Physical Sciences*, *436*(1896), 69–87.

Lewis, A. D. (1996). The geometry of the Gibbs-Appel equations and Gauss's principle of least constraint. *Reports on Mathematical Physics*, *38*(1), 11–28.

Marsden, J. E., & Ratiu, T. S. (1999). *Texts in applied mathematics. Introduction to mechanics and symmetry* (2nd ed.). New York: Springer.

Maruskin, J. M., & Bloch, A. M. (2007). The Boltzmann-Hamel equations for optimal control. In *IEEE conference on decision and control*, San Diego, CA, USA (pp. 554–559).

Murray, R. M., Li, Z., & Sastry, S. S. (1994). *A mathematical introduction to robotic manipulation*. Boca Raton: CRC Press.

Park, F. C., Bobrow, J. E., & Ploen, S. R. (1995). A Lie group formulation of robot dynamics. *The International Journal of Robotics Research*, *14*(6), 609–618.

Poincaré, H. (1901). Sur une forme nouvelle des équations de la mécanique. *Bull Astron.*

Rao, A. (2006). *Dynamics of particles and rigid bodies—a systematic approach*. Cambridge: Cambridge University Press.

Rossmann, W. (2002). *Lie groups—an introduction through linear algebra*. Oxford: Oxford science publications.

Sagatun, S. I., & Fossen, T. I. (1992). Lagrangian formulation of underwater vehicles. In *Conference of systems, man and cybernetics*, Charlottesville, VA, USA (pp. 1029–1034).

Selig, J. M. (2000). *Geometric fundamentals of robotics*. New York: Springer.

Tanner, H. G., & Kyriakopoulos, K. J. (2001). Mobile manipulator modeling with Kane's approach. *Robotica*, *19*, 675–690.

Zefran, M., & Bullo, F. (2004). *Lagrangian dynamics, robotics and automation handbook*. Boca Raton: CRC Press.

Chapter 7
Dynamics of Manipulators on a Fixed Base

Fixed-base manipulators with conventional rotational and prismatic joints are most efficiently modeled using a Lagrangian framework (Lagrange 1788). From the robot kinematics that we studied in the previous chapters we learned that the end-effector velocities and the workspace of the robot are determined by the allowed motions between any two consecutive links of the robot: the joints define the mechanical constrains of the robot and therefore also the allowed motion of the links and the end effector. With this as a starting point it is natural to model robotic manipulators using a Lagrangian framework because in this way the constraints are included in the dynamics by choosing the set of generalized coordinates as the joint variables.

7.1 Kinetic and Potential Energy in Multibody Systems

The total energy in a multibody mechanical system is found by adding the kinetic and potential energy of each rigid body with respect to some common inertial reference. An important difference compared to single rigid bodies is that the inertia matrix depends on the configuration of the multibody system, i.e., the position of the rigid bodies relative to each other. Because of this the partial derivatives of the inertia matrix will not vanish like it did for single rigid bodies.

Recall from Sect. 4.4.3 that the velocity of a link i of a robotic manipulator can be written with respect to the inertial base frame $\mathscr{F}_0$ in spatial coordinates as

$$V_{0i}^S = \begin{bmatrix} v_{0i}^S \\ \omega_{0i}^S \end{bmatrix} = J_i(q)\dot{q} \tag{7.1}$$

where v_{0i}^S and ω_{0i}^S are the linear and angular velocities, respectively, of link i in spatial coordinates, and $J_i(q) \in \mathbb{R}^{6\times n}$ is the geometric Jacobian of link i relative to $\mathscr{F}_0$. The velocity state is thus fully determined given the joint velocities $\dot{q}$ and the joint positions q.

P.J. From et al., *Vehicle-Manipulator Systems*, Advances in Industrial Control, DOI 10.1007/978-1-4471-5463-1_7,

The kinetic energy $\mathscr{K}_i$ of link i then follows as

$$\begin{aligned}\mathscr{K}_i &= \frac{1}{2}\left(V_{0i}^B\right)^\mathsf{T} I_i V_{0i}^B \\ &= \frac{1}{2}\left(\mathrm{Ad}_{g_{i0}} V_{0i}^S\right)^\mathsf{T} I_i \mathrm{Ad}_{g_{i0}} V_{0i}^S \\ &= \frac{1}{2}\left(J_i(q)\dot{q}_i\right)^\mathsf{T} \mathrm{Ad}_{g_{i0}}^\mathsf{T} I_i \mathrm{Ad}_{g_{i0}} J_i(q)\dot{q}_i \ , \\ &= \frac{1}{2}\dot{q}_i^\mathsf{T} J_i(q)^\mathsf{T} \mathrm{Ad}_{g_{i0}}^\mathsf{T} I_i \mathrm{Ad}_{g_{i0}} J_i(q)\dot{q}_i \\ &= \frac{1}{2}\dot{q}_i^\mathsf{T} M_i(q)\dot{q}_i. \end{aligned} \tag{7.2}$$

Because we only consider Euclidean transformations we can set $v = \dot{q}$ and we can write the kinetic energy as

$$\mathscr{K}_i = \frac{1}{2} v_i^\mathsf{T} M_i(q) v_i \tag{7.3}$$

which defines the inertia matrix of link i as

$$M_i(q) = J_i^\mathsf{T} \mathrm{Ad}_{g_{i0}}^\mathsf{T} I_i \mathrm{Ad}_{g_{i0}} J_i \in \mathbb{R}^{n \times n}. \tag{7.4}$$

Using our definition, I_i is the generalized inertia matrix expressed in the reference frame $\mathscr{F}_i$ attached to the rigid body and is therefore constant. The inertia matrix can also be represented in a different coordinate frame by the Adjoint map. More specifically, the inertia matrix represented in the inertial frame $\mathscr{F}_0$ is given by

$$I_i^S = \mathrm{Ad}_{g_{i0}}^\mathsf{T} I_i \mathrm{Ad}_{g_{i0}} \tag{7.5}$$

and is therefore configuration dependent. However, if we choose the coordinate frames $\mathscr{F}_i$ at the center of gravity and so that the principal axes are aligned with the principal axes of inertia, then the generalized inertia matrix I_i of each rigid body becomes constant and diagonal. We will therefore always choose the coordinate frame $\mathscr{F}_i$ so that I_i is the generalized inertia matrix.

The total kinetic energy of the mechanism is given by the sum of the kinetic energies of the mechanism links, that is,

$$\mathcal{K}(q, v) = \frac{1}{2} v^{\mathsf{T}} \underbrace{\left(\sum_{i=1}^{n} M_i(q)\right)}_{\text{inertia matrix } M(q)} v. \tag{7.6}$$

The inertia matrix $M(q)$ of the total system is then given by

$$M(q) = \sum_{i=1}^{n} M_i(q). \tag{7.7}$$

The potential energy is found in the same way by first expressing the potential energy $\mathcal{U}_i$ of each link i with respect to some common inertial reference $\mathcal{F}_0$ and then find the total potential energy of the system by

$$\mathcal{U} = \sum_{i=1}^{n} \mathcal{U}_i. \tag{7.8}$$

Potential forces are discussed in more detail in Chaps. 10–12.

Example 7.1 As an example we will study the two joint robot with rotational joints in Fig. 7.1. We assume that both joints rotate around the same axis which makes this a two-link planar manipulator. Further set $l_0 = 0$, i.e., we place $\mathcal{F}_0$ at $\bar{\mathcal{F}}_1$ (recall that the bar notation refers to the coordinate frame attached to the axis of rotation) and associate the zero pose position $q = 0$ with the joint positions when the robot is stretched out in the direction of the y-axis. The body twist of each joint is given by (4.27) as

$$X_1^1 = \begin{bmatrix} 0 \\ p_1^1 \end{bmatrix} = \begin{bmatrix} 0 \\ 0 \\ 0 \\ 0 \\ 0 \\ 1 \end{bmatrix}, \qquad X_2^2 = \begin{bmatrix} 0 \\ p_2^2 \end{bmatrix} = \begin{bmatrix} 0 \\ 0 \\ 0 \\ 0 \\ 0 \\ 1 \end{bmatrix}, \tag{7.9}$$

and the link velocities are given as

$$V_{01}^S = J_1(q)\dot{q} = \begin{bmatrix} X_1^1 & 0 \end{bmatrix} \begin{bmatrix} \dot{q}_1 \\ \dot{q}_2 \end{bmatrix} \tag{7.10}$$

$$V_{02}^S = J_2(q)\dot{q} = \begin{bmatrix} X_1^1 & \mathrm{Ad}_{g_{0\bar{2}}} X_2^2 \end{bmatrix} \begin{bmatrix} \dot{q}_1 \\ \dot{q}_2 \end{bmatrix}. \tag{7.11}$$

$\mathrm{Ad}_{g_{0\bar{2}}}$ is the transformation from $\mathcal{F}_0$ to $\bar{\mathcal{F}}_2^0$ which does not depend on q_2 and is

Fig. 7.1 Two-link robot with rotational joints

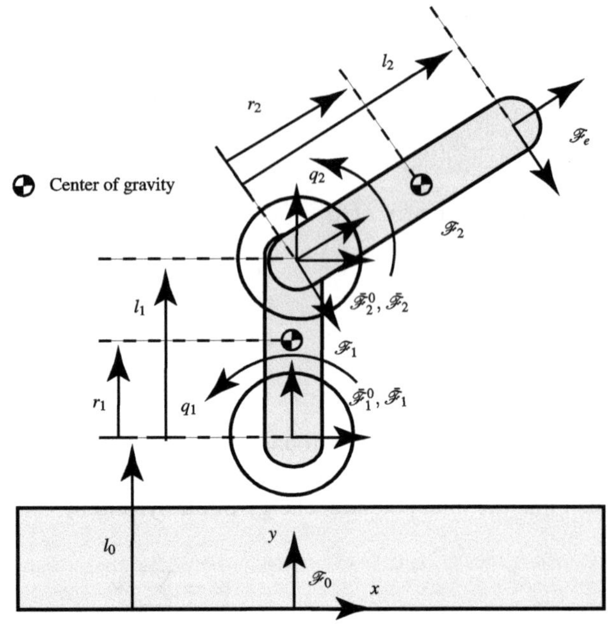

given by

$$\mathrm{Ad}_{g_{0\bar{2}}} = \begin{bmatrix} R_{0\bar{2}} & \hat{p}_{0\bar{2}} R_{0\bar{2}} \\ 0 & R_{0\bar{2}} \end{bmatrix} = \begin{bmatrix} \cos q_1 & -\sin q_1 & 0 & 0 & 0 & l_1 \cos q_1 \\ \sin q_1 & \cos q_1 & 0 & 0 & 0 & l_1 \sin q_1 \\ 0 & 0 & 1 & -l_1 & 0 & 0 \\ 0 & 0 & 0 & \cos q_1 & -\sin q_1 & 0 \\ 0 & 0 & 0 & \sin q_1 & \cos q_1 & 0 \\ 0 & 0 & 0 & 0 & 0 & 1 \end{bmatrix} \tag{7.12}$$

Note that we use $\mathrm{Ad}_{g_{0\bar{2}}}$ instead of $\mathrm{Ad}_{g_{02}}$ which reduces the computations somewhat. We can do this because the velocity of $\mathscr{F}_2$ and $\bar{\mathscr{F}}_2$ is the same in spatial coordinates. We now find

$$X_2 = \mathrm{Ad}_{g_{0\bar{2}}} X_2^2 = \begin{bmatrix} l_1 \cos q_1 & l_1 \sin q_1 & 0 & 0 & 0 & 1 \end{bmatrix}^{\mathsf{T}} \tag{7.13}$$

which gives the following Jacobians:

$$J_1(q) = \begin{bmatrix} X_1 & 0 \end{bmatrix} = \begin{bmatrix} 0 & 0 \\ 0 & 0 \\ 0 & 0 \\ 0 & 0 \\ 0 & 0 \\ 1 & 0 \end{bmatrix}, \qquad J_2(q) = \begin{bmatrix} X_1 & X_2 \end{bmatrix} = \begin{bmatrix} 0 & l_1 \cos q_1 \\ 0 & l_1 \sin q_1 \\ 0 & 0 \\ 0 & 0 \\ 0 & 0 \\ 1 & 1 \end{bmatrix}. \tag{7.14}$$

The spatial velocities in (7.10) and (7.11) can now be written as

$$V_{01}^S = \begin{bmatrix} 0 & 0 & 0 & 0 & 0 & \dot{q}_1 \end{bmatrix}^\mathsf{T} \tag{7.15}$$

$$V_{02}^S = \begin{bmatrix} l_1 \cos q_1 \dot{q}_2 & l_1 \sin q_1 \dot{q}_2 & 0 & 0 & 0 & \dot{q}_1 + \dot{q}_2 \end{bmatrix}^\mathsf{T}. \tag{7.16}$$

The inertia matrix is found from (7.4) and (7.7) as

$$M_i(q) = J_1^\mathsf{T} \mathrm{Ad}_{g_{10}}^\mathsf{T} I_1 \mathrm{Ad}_{g_{10}} J_1 + J_2^\mathsf{T} \mathrm{Ad}_{g_{20}}^\mathsf{T} I_2 \mathrm{Ad}_{g_{20}} J_2 \tag{7.17}$$

where we use $\mathrm{Ad}_{g_{i0}}$ (and not $\mathrm{Ad}_{g_{\bar{i}0}}$) because the inertia tensor is defined in $\mathscr{F}_i$ (not $\bar{\mathscr{F}}_i$). Now, if we first write

$$g_{01} = \begin{bmatrix} \cos q_1 & -\sin q_1 & 0 & -r_1 \sin q_1 \\ \sin q_1 & \cos q_1 & 0 & r_1 \cos q_1 \\ 0 & 0 & 1 & 0 \\ 0 & 0 & 0 & 1 \end{bmatrix} \tag{7.18}$$

$$g_{02} = \begin{bmatrix} \cos q_{12} & -\sin q_{12} & 0 & -l_1 \sin q_1 - r_2 \sin q_{12} \\ \sin q_{12} & \cos q_{12} & 0 & l_1 \cos q_1 + r_2 \cos q_{12} \\ 0 & 0 & 1 & 0 \\ 0 & 0 & 0 & 1 \end{bmatrix} \tag{7.19}$$

where q_{12} means $q_1 + q_2$, we can use Property 2.22 to write the Adjoint maps as

$$\mathrm{Ad}_{g_{01}}^{-1} = \begin{bmatrix} R_{01}^\mathsf{T} & -R_{01}^\mathsf{T} \hat{p}_{01} \\ 0 & R_{01}^\mathsf{T} \end{bmatrix} = \begin{bmatrix} \cos q_1 & \sin q_1 & 0 & 0 & 0 & -r_1 \\ -\sin q_1 & \cos q_1 & 0 & 0 & 0 & 0 \\ 0 & 0 & 1 & r_1 \cos q_1 & r_1 \sin q_1 & 0 \\ 0 & 0 & 0 & \cos q_1 & \sin q_1 & 0 \\ 0 & 0 & 0 & -\sin q_1 & \cos q_1 & 0 \\ 0 & 0 & 0 & 0 & 0 & 1 \end{bmatrix} \tag{7.20}$$

$$\mathrm{Ad}_{g_{02}}^{-1} = \begin{bmatrix} \cos q_{12} & \sin q_{12} & 0 & 0 & 0 & -l_1 \cos q_2 - r_2 \\ -\sin q_{12} & \cos q_{12} & 0 & 0 & 0 & l_1 \sin q_2 \\ 0 & 0 & 1 & l_1 \cos q_1 + r_2 \cos q_{12} & l_1 \sin q_1 + r_2 \sin q_{12} & 0 \\ 0 & 0 & 0 & \cos q_{12} & \sin q_{12} & 0 \\ 0 & 0 & 0 & -\sin q_{12} & \cos q_{12} & 0 \\ 0 & 0 & 0 & 0 & 0 & 1 \end{bmatrix} \tag{7.21}$$

and we find

$$\mathrm{Ad}_{g_{10}} J_1 = \begin{bmatrix} -r_1 & 0 \\ 0 & 0 \\ 0 & 0 \\ 0 & 0 \\ 0 & 0 \\ 1 & 0 \end{bmatrix}, \qquad \mathrm{Ad}_{g_{20}} J_2 = \begin{bmatrix} -l_1 \cos q_2 - r_2 & -r_2 \\ l_1 \sin q_2 & 0 \\ 0 & 0 \\ 0 & 0 \\ 0 & 0 \\ 1 & 1 \end{bmatrix}. \tag{7.22}$$

This can now be substituted directly into (7.17) to find the inertia matrix as

$$\begin{aligned} M_i(q) &= J_1^{\mathrm{T}} \mathrm{Ad}_{g_{10}}^{\mathrm{T}} I_1 \mathrm{Ad}_{g_{10}} J_1 + J_2^{\mathrm{T}} \mathrm{Ad}_{g_{20}}^{\mathrm{T}} I_2 \mathrm{Ad}_{g_{20}} J_2 \\ &= \begin{bmatrix} m_1 r_1^2 + I_{1,z} & 0 \\ 0 & 0 \end{bmatrix} \\ &\quad + \begin{bmatrix} m_2 l_1^2 + 2 m_2 r_2 l_1 \cos q_2 + m_2 r_2^2 + I_{2,z} & m_2 l_1 r_2 \cos q_2 + m_2 r_2^2 + I_{2,z} \\ m_2 l_1 r_2 \cos q_2 + m_2 r_2^2 + I_{2,z} & m_2 r_2^2 + I_{2,z} \end{bmatrix} \\ &= \begin{bmatrix} \alpha + 2\beta \cos q_2 & \delta + \beta \cos q_2 \\ \delta + \beta \cos q_2 & \delta \end{bmatrix} \end{aligned} \tag{7.23}$$

where

$$\begin{aligned} \alpha &= m_1 r_1^2 + m_2 r_2^2 + m_2 l_1^2 + I_{z,1} + I_{z,2}, \\ \beta &= m_2 r_2 l_1, \\ \delta &= m_2 r_2^2 + I_{z,2}. \end{aligned}$$

Example 7.2 Note that it is also possible to write the inertia matrix in terms of the body link velocities V_{0i}^B directly. We then write

$$V_{01}^B = J_1^B(q)\dot{q} = \begin{bmatrix} \mathrm{Ad}_{g_{\bar{1}1}^{-1}} X_1^1 & 0 \end{bmatrix} \begin{bmatrix} \dot{q}_1 \\ \dot{q}_2 \end{bmatrix} \tag{7.24}$$

$$V_{02}^B = J_2^B(q)\dot{q} = \begin{bmatrix} \mathrm{Ad}_{g_{\bar{1}2}^{-1}} X_1^1 & \mathrm{Ad}_{g_{\bar{2}2}^{-1}} X_2^2 \end{bmatrix} \begin{bmatrix} \dot{q}_1 \\ \dot{q}_2 \end{bmatrix} \tag{7.25}$$

where $g_{\bar{j}i}^{-1}$ gives us the transformation from frame $\bar{\mathscr{F}}_j$ (at the joint axis in which the twist is defined) to frame $\mathscr{F}_i$ (at the center of gravity of link i) so that

$$\mathrm{Ad}_{g_{\bar{1}1}}^{-1} = \begin{bmatrix} 1 & 0 & 0 & 0 & 0 & -r_1 \\ 0 & 1 & 0 & 0 & 0 & 0 \\ 0 & 0 & 1 & r_1 & 0 & 0 \\ 0 & 0 & 0 & 1 & 0 & 0 \\ 0 & 0 & 0 & 0 & 1 & 0 \\ 0 & 0 & 0 & 0 & 0 & 1 \end{bmatrix}, \qquad \mathrm{Ad}_{g_{\bar{2}2}}^{-1} = \begin{bmatrix} 1 & 0 & 0 & 0 & 0 & -r_2 \\ 0 & 1 & 0 & 0 & 0 & 0 \\ 0 & 0 & 1 & r_2 & 0 & 0 \\ 0 & 0 & 0 & 1 & 0 & 0 \\ 0 & 0 & 0 & 0 & 1 & 0 \\ 0 & 0 & 0 & 0 & 0 & 1 \end{bmatrix}$$

$$\mathrm{Ad}_{g_{\bar{1}2}}^{-1} = \begin{bmatrix} \cos q_2 & \sin q_2 & 0 & 0 & 0 & -l_1\cos q_2 - r_2 \\ -\sin q_2 & \cos q_2 & 0 & 0 & 0 & l_1 \sin q_2 \\ 0 & 0 & 1 & l_1 + r_2\cos q_2 & r_2 \sin q_2 & 0 \\ 0 & 0 & 0 & \cos q_2 & \sin q_2 & 0 \\ 0 & 0 & 0 & -\sin q_2 & \cos q_2 & 0 \\ 0 & 0 & 0 & 0 & 0 & 1 \end{bmatrix}.$$

We can now write the body link Jacobians as

$$J_1^B(q) = \begin{bmatrix} \mathrm{Ad}_{g_{\bar{1}1}^{-1}} X_1^1 & 0 \end{bmatrix} = \begin{bmatrix} -r_1 & 0 \\ 0 & 0 \\ 0 & 0 \\ 0 & 0 \\ 0 & 0 \\ 1 & 0 \end{bmatrix} \tag{7.26}$$

$$J_2^B(q) = \begin{bmatrix} \mathrm{Ad}_{g_{\bar{1}2}^{-1}} X_1^1 & \mathrm{Ad}_{g_{\bar{2}2}^{-1}} X_2^2 \end{bmatrix} = \begin{bmatrix} -l_1\cos q_2 - r_2 & -r_2 \\ l_1 \sin q_2 & 0 \\ 0 & 0 \\ 0 & 0 \\ 0 & 0 \\ 1 & 1 \end{bmatrix}. \tag{7.27}$$

The inertia matrix is given in terms of the body Jacobians as

$$M_i(q) = (J_1^B)^\top I_1 J_1^B + (J_2^B)^\top I_2 J_2^B \tag{7.28}$$

and the inertia matrix is found as in (7.23) in Example 7.1.

7.2 Lagrangian Dynamics

We can find the dynamic equations of a manipulator on a fixed base from the Lagrangian (Lagrange 1788; Arnold 1989; Marsden and Ratiu 1999; Zefran and Bullo 2004). Now that we have found the kinetic and potential energy of the system we write the Lagrangian in the normal way as $L(q,\dot{q}) = \mathscr{K}(q,\dot{q}) - \mathscr{U}(q)$ or, following the train of thought in Murray et al. (1994), as a sum

$$L(q,\dot{q}) = \frac{1}{2}\sum_{i,j=1}^{n} M_{ij}(q)\dot{q}_i\dot{q}_j - \mathscr{U}(q). \tag{7.29}$$

The dynamics is then found by the Lagrange equations in component form as

$$\frac{\mathrm{d}}{\mathrm{d}t}\frac{\partial L}{\partial \dot{q}_i} - \frac{\partial L}{\partial q_i} = \tau_i \tag{7.30}$$

where τ_i are the joint torques and other generalized forces collocated with $\dot{q}_i$. The partial derivatives of the Lagrangian are found as

$$\frac{\mathrm{d}}{\mathrm{d}t}\frac{\partial L}{\partial \dot{q}_i} = \frac{\mathrm{d}}{\mathrm{d}t}\left(\sum_{j=1}^{n} M_{ij}(q)\dot{q}_j\right) = \sum_{j=1}^{n}\big(\dot{M}_{ij}(q)\dot{q}_j + M_{ij}(q)\ddot{q}_j\big), \tag{7.31}$$

$$\frac{\partial L}{\partial q_i} = \frac{1}{2}\sum_{j,k=1}^{n}\frac{\partial M_{jk}(q)}{\partial q_i}\dot{q}_k\dot{q}_j - \frac{\partial \mathscr{U}}{\partial q_i}. \tag{7.32}$$

Substituting this into the Lagrange equations in the form of (7.30) and using $\dot{M}_{ij} = \sum_k \frac{\partial M_{ij}}{\partial q_k}\dot{q}_k$ gives

$$\begin{aligned}
\sum_{j=1}^{n} M_{ij}\ddot{q}_j + \sum_{j=1}^{n}\dot{M}_{ij}\dot{q}_j - \frac{1}{2}\sum_{j,k=1}^{n}\frac{\partial M_{jk}}{\partial q_i}\dot{q}_k\dot{q}_j + \frac{\partial \mathscr{U}}{\partial q_i} &= \tau_i \\
\sum_{j=1}^{n} M_{ij}\ddot{q}_j + \sum_{j=1}^{n}\sum_{k=1}^{n}\frac{\partial M_{ij}}{\partial q_k}\dot{q}_k\dot{q}_j - \frac{1}{2}\sum_{j,k=1}^{n}\frac{\partial M_{jk}}{\partial q_i}\dot{q}_k\dot{q}_j + \frac{\partial \mathscr{U}}{\partial q_i} &= \tau_i \\
\sum_{j=1}^{n} M_{ij}\ddot{q}_j + \sum_{j,k=1}^{n}\left(\frac{\partial M_{ij}}{\partial q_k}\dot{q}_k\dot{q}_j - \frac{1}{2}\frac{\partial M_{jk}}{\partial q_i}\dot{q}_k\dot{q}_j\right) + \frac{\partial \mathscr{U}}{\partial q_i} &= \tau_i \\
\sum_{j=1}^{n} M_{ij}\ddot{q}_j + \frac{1}{2}\sum_{j,k=1}^{n}\left(\frac{\partial M_{ij}}{\partial q_k}\dot{q}_k\dot{q}_j + \frac{\partial M_{ik}}{\partial q_j}\dot{q}_k\dot{q}_j - \frac{\partial M_{jk}}{\partial q_i}\dot{q}_k\dot{q}_j\right) + \frac{\partial \mathscr{U}}{\partial q_i} &= \tau_i
\end{aligned} \tag{7.33}$$

for $i = 1, \ldots, n$ where we have used that $\sum_{j,k}\frac{\partial M_{ij}}{\partial q_k}\dot{q}_k\dot{q}_j = \sum_{k,j}\frac{\partial M_{ik}}{\partial q_j}\dot{q}_j\dot{q}_k$. This gives us the dynamic equations of a robotic manipulator in terms of each joint variable.

7.2.1 Robot Dynamics in Matrix Form

As always we would like to write the dynamics in matrix form. The dynamics of a serial manipulator can be written in terms of the generalized coordinates q which are chosen as the joint positions, and the generalized velocities $\dot{q}$ which are the joint velocities. We can choose these variables for any manipulator which consists of 1-DoF revolute or prismatic joints. The aim is then to write the manipulator dynamics in the same form as we did for single rigid bodies in the previous chapter, i.e., in the form

$$M(q)\ddot{q} + C(q,\dot{q})\dot{q} + N(q) = \tau. \tag{7.34}$$

Here $M(q)$ is the manipulator inertia matrix, which is defined in (7.6). The Coriolis matrix $C(q,\dot{q})$ is normally written in terms of the Christoffel symbols Γ as

$$C_{ij}(q,\dot{q}) = \left\{ \sum_{k=1}^{n} \Gamma_{ijk}\dot{q}_k \right\} \tag{7.35}$$

where Γ is found directly from (7.33) as

$$\Gamma_{ijk} = \frac{1}{2}\left(\frac{\partial M_{ij}}{\partial q_k} + \frac{\partial M_{ik}}{\partial q_j} - \frac{\partial M_{jk}}{\partial q_i} \right). \tag{7.36}$$

There are many ways to define the Coriolis matrix for robotic manipulators, but this particular representation in terms of the Christoffel symbols Γ_{ijk} is particularly useful because it allows us to identify certain properties with the Coriolis matrix written in this form. This is discussed in more detail in Chap. 9.

If gravitational forces are present we also need to include the potential forces $N(q)$. Let the wrench associated with the gravitational force of link i with respect to coordinate frame $\mathscr{F}_0$ be given by Duindam (2006)

$$F_g^i = \begin{bmatrix} f_i \\ \widehat{r}_i^g f_i \end{bmatrix} = -m_i g \begin{bmatrix} R_{0i} e_z \\ \widehat{r}_i^g R_{0i} e_z \end{bmatrix} \tag{7.37}$$

where $e_z = [0\ 0\ 1]^\mathsf{T}$ and r_i^g is the center of gravity of link i expressed in $\mathscr{F}_i$. $f_i \in \mathbb{R}^3$ represents the forces that act on link i at the point r_i^g. In our case $\mathscr{F}_i$ is chosen so that r_i^g coincides with the origin of $\mathscr{F}_i$ so we have $r_i^g = 0$. The coordinate frames needed to compute the gravitational forces are illustrated in Fig. 7.2. The joint torque associated with link i is given by

$$\tau_i^g = J_i^\mathsf{T}(q)\mathrm{Ad}_{g_{0i}}^\mathsf{T}(q) F_g^i(q) \tag{7.38}$$

and the total effect of the gravitational forces is found by summing the effect of each link, which is given by

$$N(q) = \sum_{i=1}^{n} \tau_i^g \tag{7.39}$$

and thus enters (7.34) in the same way as τ.

This gives us the dynamics of the robot in a gravitational field. Other forces can also be included in the dynamics, but we will not discuss these in more detail in this chapter but return to this topic in Chaps. 10–12. For now we write the robot dynamics as follows:

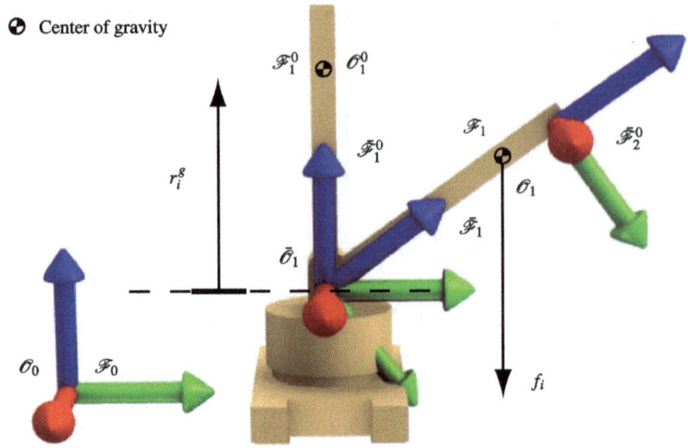

Fig. 7.2 The coordinate frames of a robotic link. The coordinate frame $\mathscr{F}_1$ is chosen so that its origin $\mathscr{O}_1$ is fixed for the joint motion while the coordinate frame $\bar{\mathscr{F}}_1$ is chosen so that it is aligned with the principal axes of inertia and the origin $\bar{\mathscr{O}}_1$ is in the center of gravity (represented by *black circles*)

Theorem 7.1 *The dynamic equations of a robotic manipulator with joint positions in generalized coordinates q and joint velocities in generalized velocities $\dot{q}$ can be written as*

$$M(q)\ddot{q} + C(q,\dot{q})\dot{q} + N(q) = \tau \tag{7.40}$$

where $M(q)$ is the manipulator inertia matrix found in (7.6), *$N(q)$ represents the potential forces, and $C(q,\dot{q})$ is the Coriolis matrix given by*

$$C_{ij}(q,\dot{q}) = \left\{ \sum_{k=1}^{n} \Gamma_{ijk}\dot{q}_k \right\}, \quad \textit{where} \quad \Gamma_{ijk} = \frac{1}{2}\left(\frac{\partial M_{ij}}{\partial q_k} + \frac{\partial M_{ik}}{\partial q_j} - \frac{\partial M_{jk}}{\partial q_i} \right). \tag{7.41}$$

7.2.2 Operational Space Approach

Recall that the analytical Jacobian gives the mapping from the joint velocities to the time derivative of the end-effector position η_{0e} through the relation $\dot{\eta}_{0e} = J_{m,a}(q)\dot{q}$. We can use this to reformulate the dynamics in Theorem 7.1 so that the equations are written in terms of the end-effector position and velocity instead of the joint positions and velocities (Siciliano et al. 2011). This is often useful since we normally want to specify the end-effector configuration directly in the operational space. Following the same approach as in Sect. 6.3 we first write

$$\dot{\eta}_{0e} = J_{m,a}(q)\dot{q}, \quad J_{m,a}(q) = \frac{\partial f}{\partial q} \tag{7.42}$$

where f is the mapping $f : q \to \eta_{0e}$. If the mapping is smooth and invertible we can write

$$\dot{q} = J_{m,a}^{-1}(q)\dot{\eta}_{0e}, \tag{7.43}$$

$$\ddot{q} = J_{m,a}^{-1}(q)\ddot{\eta}_{0e} + \dot{J}_{m,a}^{-1}(q)\dot{\eta}_{0e}. \tag{7.44}$$

We can now substitute this into (7.40) and pre-multiply by $J_{m,a}^{-\mathsf{T}}$ which gives (Murray et al. 1994; Fossen 2002)

$$J_{m,a}^{-\mathsf{T}} M J_{m,a}^{-1} \ddot{\eta}_{0e} + \big(J_{m,a}^{-\mathsf{T}} C J_{m,a}^{-1} + J_{m,a}^{-\mathsf{T}} M \dot{J}_{m,a}^{-1}\big)\dot{\eta}_{0e} + J_{m,a}^{-\mathsf{T}} N = J_{m,a}^{-\mathsf{T}} \tau. \tag{7.45}$$

The dynamic equations of a robotic manipulator can now be written in terms of the operational space variables as

$$\bar{M}(q)\ddot{\eta}_{0e} + \bar{C}(q,\dot{q})\dot{\eta}_{0e} + \bar{N}(q) = \bar{\tau} \tag{7.46}$$

where

$$\bar{M}(q) = J_{m,a}^{-\mathsf{T}}(q) M(q) J_{m,a}^{-1}(q), \tag{7.47}$$

$$\bar{C}(q) = J_{m,a}^{-\mathsf{T}}(q)\big(C(q,\dot{q}) J_{m,a}^{-1}(q) + M \dot{J}_{m,a}^{-1}(q)\big), \tag{7.48}$$

$$\bar{N}(q) = J_{m,a}^{-\mathsf{T}}(q) N(q), \tag{7.49}$$

$$\bar{\tau}(q) = J_{m,a}^{-\mathsf{T}}(q)\tau. \tag{7.50}$$

Again we see that we need to pre-multiply the torques with $J_{m,a}^{-\mathsf{T}}(q)$ so that $\bar{\tau}(q)$ is collocated with $\dot{\eta}_{0e}$. Note also the resemblance with the equations for a single rigid body formulated in position variables η as presented in (6.71)–(6.74).

There are, however, a few important differences between the operational space approach of robotic manipulators presented here and the formulation for single rigid bodies. First of all, caution needs to be taken if the robot has more degrees of freedom than the operational space. This is the case with kinematically redundant robots for which the manipulator Jacobian $J_{m,a}(q) \in \mathbb{R}^{r\times n}$ is not a square matrix but $n > r$, where r is the dimension of the operational space. For $SE(3)$, for example, we have $r = 6$. The Jacobian matrix of kinematically redundant robots have more columns than rows and the inverse kinematics problem has infinitely many solutions. This means that there are infinitely many joint velocities that result in the same end-effector velocity.

For non-redundant robots when $r = n$, attention must be given to the possibilities of kinematic singularities. Kinematic singularities occur at configurations for which the mobility is reduced and the end-effector cannot move in all directions. Classic examples of this is when the arm is totally stretched, which is a boundary singularity, or when two or more axes of motion are aligned, which are known as internal singularities. At these positions the inverse kinematics map may have infinitely many solutions and the manipulator Jacobian is not invertible.

It is important to note, however, that the problem that we encountered in the previous chapter when the Jacobian was singular due to the choice of representation (representation singularities) is not a problem for fixed-base serial manipulators because in order to find the manipulator Jacobian we can always choose the position variables as the joint positions for which the Jacobian is well defined, as we learned in Sect. 4.4. Kinematic singularities are fundamentally different from representation singularities because they occur as a result of the manipulator design and physical properties of the robot. These can only be avoided by changing the manipulator design.

7.3 Configuration States

In the previous chapter we derived Lagrange's equations of motion for a single rigid body in the form

$$\frac{\mathrm{d}}{\mathrm{d}t}\left(\frac{\partial L}{\partial v}\right) - S^{-\mathrm{T}}\frac{\partial L}{\partial q} + \left(\sum_k \gamma_k v_k\right)\frac{\partial L}{\partial v} = \tau. \tag{7.51}$$

We can also use these equations to describe multi-body systems. We will write the state variables in the normal way with position variables q by stacking the position q_i of each joint and the velocity variables $v = \dot{q}$ by stacking the velocities of the joints. The state space is thus written by the vectors

$$q = \begin{bmatrix} q_1 \\ q_2 \\ \vdots \\ q_n \end{bmatrix}, \qquad \dot{q} = \begin{bmatrix} \dot{q}_1 \\ \dot{q}_2 \\ \vdots \\ \dot{q}_n \end{bmatrix}. \tag{7.52}$$

For single rigid bodies this representation lead to singularities due to the (generally) non-Euclidean configuration space. As we will see shortly, this is not the case with robotic manipulators with conventional 1-DoF joints because $v = \frac{\mathrm{d}q}{\mathrm{d}t}$. One of the principal objectives of this book is to derive a singularity-free formulation of vehicle-manipulator systems, which is addressed in the next chapter. To this end, we will use the same framework when we derive the manipulator dynamics in this chapter that we did for single rigid bodies in the previous chapter. There exist other and more direct formulations of manipulator dynamics for robotic manipulators on a fixed base, for example the one in Sect. 7.2, but in order to obtain one uniform theory that can be applied to vehicle-manipulator systems we choose this approach also for fixed-base manipulators.

For a single rigid body we re-wrote these equations in terms of φ and $\dot{\varphi}$ and therefore needed to find the partial derivatives of the velocity transformation matrix $S(Q, \varphi)$. However, for a robotic manipulator with only 1-DoF joints, we have $v = \frac{\mathrm{d}q}{\mathrm{d}t} = \dot{q}$ and therefore $S(Q, \varphi) = I$. This simplifies the dynamics because all these terms are eliminated from the equations. We will therefore derive the dynamics in terms of the global state variables q and $\dot{q}$. We note, however, that because $q, \dot{q} \in \mathbb{R}^n$ we can with a trivial change of variables $(q, \dot{q}) \mapsto (\varphi, \dot{\varphi})$ obtain the equations in the desired variables. This is not necessary for the fixed-base manipulators discussed in this chapter, but will become useful in the next chapter when we model multibody systems with non-Euclidean configuration spaces.

7.4 The Euler–Lagrange Equations of Motion in Quasi-coordinates

Recall that while the inertia matrix of a single rigid body can be written as a constant matrix, the inertia matrix of multibody systems is configuration dependent, which can be seen from (7.4). We write the Lagrangian in the normal way in terms of q and v:

$$L(q, v) = \frac{1}{2} v^{\mathsf{T}} M(q) v - \mathscr{U}(q). \tag{7.53}$$

The partial derivatives of the Lagrangian for a multibody system (configuration dependent inertia matrix) with Euclidean transformations (constant velocity transformation matrix) is then given by Duindam and Stramigioli (2008), From et al. (2012)

$$\frac{\mathrm{d}}{\mathrm{d}t}\left(\frac{\partial L}{\partial v}\right) = M(q)\dot{v} + \dot{M}(q) v \tag{7.54}$$

$$\frac{\partial L}{\partial q} = \frac{1}{2} \frac{\partial^{\mathsf{T}}(M(q)v)}{\partial q} v - \frac{\partial \mathscr{U}(q)}{\partial q}. \tag{7.55}$$

We can also write the Lagrangian in terms of the variables $(q, \dot{q})$ (ignoring for now the fact that these are identical to the variables (q, v) and that $S = I$) as

$$\bar{L}(q, \dot{q}) = \frac{1}{2} \dot{q}^{\mathsf{T}} S^{\mathsf{T}} M S \dot{q} - \mathscr{U}(q) \tag{7.56}$$

and the partial derivatives with respect to q and $\dot{q}$ are given by From (2012a, 2012b)

$$\frac{\partial \bar{L}}{\partial \dot{q}} = S^{\mathsf{T}} M S \dot{q} = S^{\mathsf{T}} \underbrace{M v}_{\frac{\partial L}{\partial v}} = S^{\mathsf{T}} \frac{\partial L}{\partial v} \tag{7.57}$$

$$\frac{\mathrm{d}}{\mathrm{d}t}\left(\frac{\partial \bar{L}}{\partial \dot{q}}\right) = S^{\mathsf{T}} \frac{\mathrm{d}}{\mathrm{d}t}\left(\frac{\partial L}{\partial v}\right) + \dot{S}^{\mathsf{T}} \frac{\partial L}{\partial v} \tag{7.58}$$

$$\begin{aligned}
\frac{\partial \bar{L}}{\partial q} &= \frac{\partial^{\mathsf{T}} S\dot{q}}{\partial q} M S\dot{q} + \frac{1}{2}\frac{\partial^{\mathsf{T}}(Mv)}{\partial q} v - \frac{\partial \mathscr{U}}{\partial q} \\
&= \frac{\partial^{\mathsf{T}}(S\dot{q})}{\partial q} \underbrace{Mv}_{\frac{\partial L}{\partial v}} + \underbrace{\frac{1}{2}\frac{\partial^{\mathsf{T}}(Mv)}{\partial q} v - \frac{\partial \mathscr{U}}{\partial q}}_{\frac{\partial L}{\partial q}} \\
&= \frac{\partial L}{\partial q} + \frac{\partial^{\mathsf{T}}(S\dot{q})}{\partial q}\frac{\partial L}{\partial v}
\end{aligned} \tag{7.59}$$

where we have used the definitions in (7.54)–(7.55). Comparing this to Proposition 6.1, we note the following important result: we can write the partial derivatives of the Lagrangian $\bar{L}(q, \dot{q})$ in terms of the Lagrangian $L(q, v)$ for a fixed-base manipulator with configuration-dependent inertia matrix in the same way as we did for a single rigid body with constant inertia matrix. To verify that this also applies to manipulators with Euclidean joints is straight forward: for a manipulator with only Euclidean joints we can simplify the equations substantially by using $S = I$ and Proposition 6.1 which gives us

$$\begin{aligned}
\frac{\mathrm{d}}{\mathrm{d}t}\left(\frac{\partial \bar{L}}{\partial \dot{q}}\right) &= \frac{\mathrm{d}}{\mathrm{d}t}\left(\frac{\partial L}{\partial v}\right) \\
\frac{\partial \bar{L}}{\partial q} &= \frac{\partial L}{\partial q}
\end{aligned} \tag{7.60}$$

as expected. We can now write the dynamic equations as

$$\frac{\mathrm{d}}{\mathrm{d}t}\left(\frac{\partial \bar{L}}{\partial \dot{q}}\right) - \frac{\partial \bar{L}}{\partial q} = \tau \tag{7.61}$$

which is the standard formulation of the manipulator dynamics in Lagrangian form. For robotic manipulators with only Euclidean joints this can of course be written as

$$\frac{\mathrm{d}}{\mathrm{d}t}\left(\frac{\partial L}{\partial v}\right) - \frac{\partial L}{\partial q} = \tau \tag{7.62}$$

which is in the same form as in the dynamic equations for single rigid bodies presented in the previous section.

We could of course arrive at these equations directly as they are the exact definition of Lagrange's equations. It will, however, be important in the next chapters to know that we can write the robot dynamics in the same form as we did for a single rigid body with a non-Euclidean configuration space.

7.5 Robot Dynamics in Matrix Form

As always we will write the equations in matrix form. For standard robotic manipulators with Euclidean joints the formulation in the previous section will reduce to the formulation presented in Theorem 7.1:

Theorem 7.2 *Consider a robotic manipulator with Euclidean joints in a potential field* $\mathcal{U}(q)$*. The generalized coordinates* $q \in \mathbb{R}^n$ *are chosen as the joint positions and the generalized velocities as* $v = \dot{q} \in \mathbb{R}^n$*. The inertia matrix* M *is given by* (7.7)*, and the Lagrangian by*

$$L(q, \dot{q}) = \frac{1}{2}\dot{q}^{\mathsf{T}} M(q)\dot{q} - \mathcal{U}(q). \tag{7.63}$$

Then the Euler–Lagrange equations of motion can be written in matrix form as

$$M(q)\ddot{q} + C(q, \dot{q})\dot{q} + N(q) = \tau \tag{7.64}$$

where

$$C_{ij}(q, \dot{q}) = \left\{ \sum_{k=1}^{n} \Gamma_{ijk}\dot{q}_k \right\}, \quad \text{where} \quad \Gamma_{ijk} = \frac{1}{2}\left(\frac{\partial M_{ij}}{\partial q_k} + \frac{\partial M_{ik}}{\partial q_j} - \frac{\partial M_{jk}}{\partial q_i} \right) \tag{7.65}$$

$$N(q) = \frac{\partial \mathcal{U}(q)}{\partial q}. \tag{7.66}$$

Proof The proof follows directly from Theorem 7.1: for $S = I$, Lagrange's equations in (7.61) reduce to

$$\frac{\mathrm{d}}{\mathrm{d}t}\left(\frac{\partial L}{\partial \dot{q}} \right) - \frac{\partial L}{\partial q} = \tau \tag{7.67}$$

which allows us apply Theorem 7.1. □

7.6 Local Parameterization

To make the formulation of the dynamic equations presented in the previous section directly applicable to multibody systems with non-Euclidean configuration spaces, we need to express them in terms of state variables φ and $\dot{\varphi}$. We do this in the normal way by expressing the velocity as $v = \dot{\varphi}$ (recall that $\varphi = q$) and the position variables as $\Phi(Q, \varphi)$. The Lagrangian then becomes

$$\begin{aligned} L_\varphi(\varphi, \dot{\varphi}) &= \frac{1}{2}\dot{\varphi}^\mathsf{T} S^\mathsf{T}(Q, \varphi) M\big(\Phi(Q, \varphi)\big) S(Q, \varphi)\dot{\varphi} - \mathscr{U}\big(\Phi(Q, \varphi)\big) \\ &= \frac{1}{2}\dot{\varphi}^\mathsf{T} M\big(\Phi(Q, \varphi)\big)\dot{\varphi} - \mathscr{U}\big(\Phi(Q, \varphi)\big). \end{aligned} \quad (7.68)$$

For Euclidean configuration spaces we note that the exponential map given by $\Phi(Q, \varphi)$ is trivially $\Phi(\bar{q}, \varphi) = \bar{q} + \varphi$ where both $\bar{q}$ and φ are vectors in $\mathbb{R}^n$. The configuration of the robot can therefore be written in terms of the generalized coordinates $Q = q \in \mathbb{R}^n$ and the velocities as $v = \dot{q} \in \mathbb{R}^n$. This gives us the following important theorem:

Proposition 7.1 *The dynamics of a fixed-base manipulator with Euclidean joints can be written in terms of local position and velocity coordinates φ and $\dot{\varphi}$ and global position and velocity coordinates $Q = q \in \mathbb{R}^n$ and $v = \dot{q} \in \mathbb{R}^n$. Write the kinetic energy as $\mathscr{K}(Q, v) = \frac{1}{2}v^\mathsf{T} M(Q) v$ with the inertia matrix $M(Q)$. The dynamics of this system then satisfies*

$$M(Q)\dot{v} + C(Q, v)v = \tau \quad (7.69)$$

where $M(Q)$ is found in the normal way, with τ the vector of external and control wrenches (collocated with v), and $C(Q, v)$ the matrix describing the Coriolis and centrifugal forces given by

$$C_{ij}(Q, v) = \sum_{k,l} \left(\frac{\partial M_{ij}}{\partial \varphi_l} - \frac{1}{2} \frac{\partial M_{lj}}{\partial \varphi_i} \right)\bigg|_{\varphi=0} v_l. \quad (7.70)$$

Proof We can prove this by noting the similarity with the equations in (7.34) which give us the inertial and Coriolis matrix in (7.33) in q and $\dot{q}$. A change of variables into $(q, \dot{q}) \mapsto (\varphi, \dot{\varphi})$ and the fact that $S = I$ and from (7.33) the relation

$$\frac{1}{2} \sum_{j,k=1}^{n} \left(\frac{\partial M_{ij}}{\partial q_k} \dot{q}_k \dot{q}_j + \frac{\partial M_{ik}}{\partial q_j} \dot{q}_k \dot{q}_j - \frac{\partial M_{jk}}{\partial q_i} \dot{q}_k \dot{q}_j \right)$$

$$= \sum_{j,k=1}^{n} \left(\frac{\partial M_{ij}}{\partial q_k} \dot{q}_k \dot{q}_j - \frac{1}{2} \frac{\partial M_{jk}}{\partial q_i} \dot{q}_k \dot{q}_j \right) \tag{7.71}$$

gives us (7.70).

We can also substitute (7.54)–(7.55) into (7.62) and use Proposition 6.1 to get

$$M\dot{v} + \dot{M}(q)v - \frac{\partial^{\mathsf{T}}(M(\varphi)v)}{\partial \varphi} v + \frac{\partial \mathscr{U}(\varphi)}{\partial \varphi} = \tau. \tag{7.72}$$

These equations are the same as (7.30) because $v = \dot{q}$. Thus, if we follow the computations of (7.31)–(7.33) we arrive at (7.70). □

References

Arnold, V. I. (1989). *Mathematical methods of classical mechanics*. Berlin: Springer.

Duindam, V. (2006). *Port-based modeling and control for efficient bipedal walking robots*. Ph.D. thesis, University of Twente.

Duindam, V., & Stramigioli, S. (2008). Singularity-free dynamic equations of open-chain mechanisms with general holonomic and nonholonomic joints. *IEEE Transactions on Robotics*, *24*(3), 517–526.

Fossen, T. I. (2002). *Marine control systems*. Trondheim: Marine Cybernetics AS. 3rd printing.

From, P. J. (2012a). An explicit formulation of singularity-free dynamic equations of mechanical systems in Lagrangian form—part one: single rigid bodies. *Modeling, Identification and Control*, *33*(2), 45–60.

From, P. J. (2012b). An explicit formulation of singularity-free dynamic equations of mechanical systems in Lagrangian form—part two: multibody systems. *Modeling, Identification and Control*, *33*(2), 61–68.

From, P. J., Duindam, V., & Stramigioli, S. (2012). Corrections to "Singularity-free dynamic equations of open-chain mechanisms with general holonomic and nonholonomic joints". *IEEE Transactions on Robotics*, *28*(6), 1431–1432.

Lagrange, J.-L. (1788). In *Mécanique analytique*. Chez la Veuve Desaint.

Marsden, J. E., & Ratiu, T. S. (1999). *Texts in applied mathematics. Introduction to mechanics and symmetry* (2nd ed.). New York: Springer.

Murray, R. M., Li, Z., & Sastry, S. S. (1994). *A mathematical introduction to robotic manipulation*. Boca Raton: CRC Press.

Siciliano, B., Sciavicco, L., Villani, L., & Oriolo, G. (2011). *Advanced textbooks in control and signal processing. Robotics: modelling, planning and control*. Berlin: Springer.

Zefran, M., & Bullo, F. (2004). *Robotics and automation handbook. Lagrangian dynamics*. Boca Raton: CRC Press.

Chapter 8
Dynamics of Vehicle-Manipulator Systems

This chapter presents the main topic of the book, namely the dynamic equations of vehicle-manipulator systems. These systems need special consideration because of the combination of non-Euclidean configuration spaces and multibody systems. As a result, the formalisms normally used for single rigid bodies or multibody systems with Euclidean configuration spaces cannot be applied directly because, as we will see, this will not result in well-defined dynamic equations for the system.

We first present the dynamic equations in the form that they are normally presented in the space robotics community, for example spacecraft-manipulator systems, and in the underwater robotics literature, for example AUV-manipulator systems. We will see that these formulations are derived more or less directly from the equations presented in the previous chapters. The main problem with these formulations is that we cannot guarantee that the dynamics is not singularity prone when the configuration space of the vehicle (or any of the joints) is non-Euclidean.

A mechanical system consisting of joints and links can be modeled as a serial combination of several Lie groups, representing the freedom of each joint and the admissible velocities. These representations are not only mathematically correct, the singularities are also removed from the equations. Unfortunately, the formulation is more involved than other methods and even though combinations of Lie groups theoretically can be used to represent multibody systems, the formulation is very complex. These approaches are thus not suited for implementation in a simulation or control environment even though, mathematically speaking, they are correct.

The main topic of this chapter is to find a mathematically well defined formulation of the dynamic equations for vehicle-manipulator systems. To this end, we will derive the dynamic equations in terms of global configuration states Q and velocity variables υ using the geometric approach presented in the previous chapters. This will lead to a well-defined and singularity-free formulation of the dynamics of vehicle-manipulator systems which is well suited both for system analysis and for implementation in a controller or simulation environment. Vehicle-manipulator systems have been treated to some extent in literature, for example in From et al. (2010a, 2012a, 2012b). Underwater vehicle-manipulator systems have been studied in Antonelli (2006) and From et al. (2010b), and space systems were treated in

P.J. From et al., *Vehicle-Manipulator Systems*, Advances in Industrial Control,
DOI 10.1007/978-1-4471-5463-1_8,

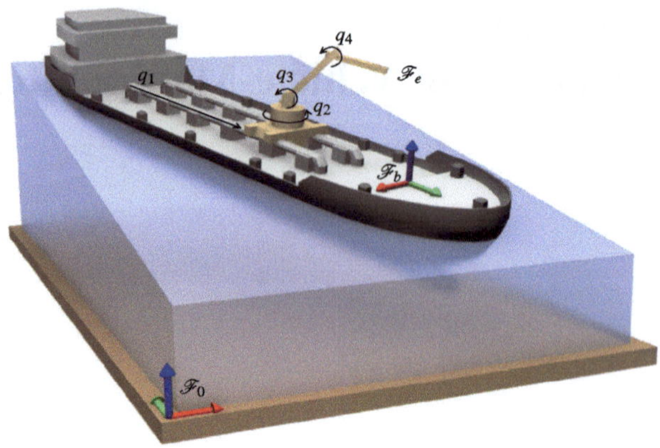

Fig. 8.1 Model setup for a four-link robot attached to a non-inertial base with coordinate frame $\mathscr{F}_b$. Frame $\mathscr{F}_0$ denotes the inertial reference frame and $\mathscr{F}_e$ the end-effector frame. Courtesy IEEE

Dubowsky and Papadopoulos (1993), Liang et al. (1998), From et al. (2011a, 2011b, 2010c) and From et al. (2012b).

8.1 The Dynamic Equations in Terms of Quasi-velocities

In this section we derive the dynamics of a robotic manipulator mounted on a free-floating base in terms of quasi-velocities. The approach is based on Egeland and Pettersen (1998), but we will provide some more details in the derivation to get a deeper understanding of vehicle-manipulator systems dynamics. When we say that the manipulator is mounted on a free-floating base we mean that the vehicle is free to move in space without any constraints on its configuration space.

Recall that for robotic manipulators with Euclidean joints, the quasi-velocities are identical to the time derivative of the joint positions, while for the vehicle the quasi-velocities are given by V_{0b}^B. The quasi-velocity vector of the whole system can therefore be written as

$$\zeta = \begin{bmatrix} V_{0b}^B \\ \dot{q} \end{bmatrix} = \begin{bmatrix} u \\ v \\ w \\ p \\ q \\ r \\ \dot{q}_1 \\ \vdots \\ \dot{q}_n \end{bmatrix}. \tag{8.1}$$

We will generally use body coordinates for the vehicle velocities because these are the state variables that are normally measured, for example by an inertial measurement unit (IMU) fixed to the vehicle. Now write the linear and angular velocities

V_{0i}^B of each link i with respect to the inertial frame in body coordinates as

$$V_{0i}^B = \begin{bmatrix} \upsilon_{0i}^B \\ \omega_{0i}^B \end{bmatrix} = \frac{\partial V_{0i}^B}{\partial \zeta} \zeta \tag{8.2}$$

where we recognize the matrix $\frac{\partial V_{0i}^B}{\partial \zeta} \in \mathbb{R}^{6\times(6+n)}$ as the link geometric Jacobian in Theorem 5.2.

The kinetic energy of a vehicle-manipulator system is given by summing the kinetic energy of each link, including the base. If we write the kinetic energy of link i as

$$\mathcal{K}_i = \frac{1}{2}\left(V_{0i}^B\right)^{\mathsf{T}} I_i V_{0i}^B \tag{8.3}$$

for $i = b, 1, \ldots, n$, we find the kinetic energy of the vehicle-manipulator system as

$$\begin{aligned} \mathcal{K} &= \sum_{i=b}^{n} \frac{1}{2}\left(V_{0i}^B\right)^{\mathsf{T}} I_i V_{0i}^B \\ &= \sum_{i=b}^{n} \frac{1}{2} \zeta^{\mathsf{T}} \underbrace{\frac{\partial V_{0i}^B}{\partial \zeta}^{\mathsf{T}} I_i \frac{\partial V_{0i}^B}{\partial \zeta}}_{M(q)} \zeta \end{aligned} \tag{8.4}$$

which also gives us the inertia matrix of the vehicle-manipulator system.

$I_i \in \mathbb{R}^{6\times 6}$ denotes the generalized inertia tensor of link i expressed in $\mathcal{F}_i$. Again, we assume that we choose the reference frame $\mathcal{F}_i$ so that I_i is diagonal.

Equation (6.41) gave us Euler's equation of motion on $SE(3)$ for a single rigid body as

$$\frac{\mathrm{d}}{\mathrm{d}t}\frac{\partial \mathcal{K}_i}{\partial V_{0i}^B} + \begin{bmatrix} \hat{\omega}_{0i}^B & 0 \\ \hat{\upsilon}_{0i}^B & \hat{\omega}_{0i}^B \end{bmatrix} \frac{\partial \mathcal{K}_i}{\partial V_{0i}^B} = \tau_i . \tag{8.5}$$

The corresponding dynamics for a multibody system is given by adding the effects of all the robotic links, as well as the vehicle, and map these to the state space of the vehicle-manipulator system: we then obtain the dynamics of the multibody system by mapping the dynamics of each link in V_{0i}^B and τ_i, into the dynamics in ζ and τ. This mapping is given by the Jacobian in Eq. (8.2). The Jacobian also guarantees that the mechanical constraints of the joints are preserved. We can now find the dynamic equations from the principal of virtual work given in (6.38) as

$$\sum_{i=b}^{6}\left\{\left[\frac{\mathrm{d}}{\mathrm{d}t}\frac{\partial^{\mathsf{T}}\mathscr{K}_i}{\partial V_{0i}^B}+\frac{\partial^{\mathsf{T}}\mathscr{K}_i}{\partial V_{0i}^B}\begin{bmatrix}\hat{\omega}_{0i}^B & 0\\ \hat{v}_{0i}^B & \hat{\omega}_{0i}^B\end{bmatrix}^{\mathsf{T}}-\tau_i^{\mathsf{T}}\right]\frac{\partial V_{0i}^B}{\partial\zeta}\right\}\sigma=0. \tag{8.6}$$

If we recall that σ is arbitrary we can write the dynamics of a multibody system as

$$\sum_{i=b}^{6}\left\{\frac{\partial V_{0i}^B}{\partial\zeta}^{\mathsf{T}}\left[\frac{\mathrm{d}}{\mathrm{d}t}\frac{\partial\mathscr{K}_i}{\partial V_{0i}^B}+\begin{bmatrix}\hat{\omega}_{0i}^B & 0\\ \hat{v}_{0i}^B & \hat{\omega}_{0i}^B\end{bmatrix}\frac{\partial\mathscr{K}_i}{\partial V_{0i}^B}\right]\right\}=\sum_{i=b}^{6}\frac{\partial V_{0i}^B}{\partial\zeta}^{\mathsf{T}}\tau_i \tag{8.7}$$

$$\sum_{i=b}^{6}\left\{\frac{\partial V_{0i}^B}{\partial\zeta}^{\mathsf{T}}\left[\frac{\mathrm{d}}{\mathrm{d}t}\frac{\partial\mathscr{K}_i}{\partial V_{0i}^B}+\begin{bmatrix}\hat{\omega}_{0i}^B & 0\\ \hat{v}_{0i}^B & \hat{\omega}_{0i}^B\end{bmatrix}\frac{\partial\mathscr{K}_i}{\partial V_{0i}^B}\right]\right\}=\tau \tag{8.8}$$

where

$$\tau=\begin{bmatrix}\tau_v\\ \tau_\omega\\ \tau_q\end{bmatrix}=\sum_{i=b}^{6}\left\{\frac{\partial V_{0i}^B}{\partial\zeta}^{\mathsf{T}}\tau_i\right\}. \tag{8.9}$$

This formulation of the dynamics was found in Egeland and Pettersen (1998) in the form presented above. We will now look at what these equations look like in matrix form, i.e., we will find the explicit expressions for the inertia and Coriolis matrices when we choose ζ as the state variable. First write, using (8.2) and (8.3),

$$\frac{\mathrm{d}}{\mathrm{d}t}\frac{\partial\mathscr{K}_i}{\partial V_{0i}^B}=\frac{\mathrm{d}}{\mathrm{d}t}(I_iV_{0i}^B)=I_i\dot{V}_{0i}^B=I_i\left(\frac{\partial V_{0i}^B}{\partial\zeta}\dot{\zeta}+\frac{\mathrm{d}}{\mathrm{d}t}\left(\frac{\partial V_{0i}^B}{\partial\zeta}\right)\zeta\right), \tag{8.10}$$

and

$$\begin{aligned}\begin{bmatrix}\hat{\omega}_{0i}^B & 0\\ \hat{v}_{0i}^B & \hat{\omega}_{0i}^B\end{bmatrix}\begin{bmatrix}\frac{\partial\mathscr{K}_i}{\partial v_{0i}^B}\\ \frac{\partial\mathscr{K}_i}{\partial\omega_{0i}^B}\end{bmatrix}&=\begin{bmatrix}\hat{\omega}_{0i}^B\frac{\partial\mathscr{K}_i}{\partial v_{0i}^B} & 0\\ \hat{v}_{0i}^B\frac{\partial\mathscr{K}_i}{\partial v_{0i}^B} & \hat{\omega}_{0i}^B\frac{\partial\mathscr{K}_i}{\partial\omega_{0i}^B}\end{bmatrix}\\ &=\begin{bmatrix}-\frac{\partial\mathscr{K}_i}{\partial v_{0i}^B}\times\omega_{0i}^B & 0\\ -\frac{\partial\mathscr{K}_i}{\partial v_{0i}^B}\times v_{0i}^B & -\frac{\partial\mathscr{K}_i}{\partial\omega_{0i}^B}\times\omega_{0i}^B\end{bmatrix}\\ &=-\begin{bmatrix}0 & \widehat{\frac{\partial\mathscr{K}_i}{\partial v_{0i}^B}}\\ \widehat{\frac{\partial\mathscr{K}_i}{\partial v_{0i}^B}} & \widehat{\frac{\partial\mathscr{K}_i}{\partial\omega_{0i}^B}}\end{bmatrix}\begin{bmatrix}v_{0i}^B\\ \omega_{0i}^B\end{bmatrix}.\end{aligned} \tag{8.11}$$

Substituting (8.10) and (8.11) into (8.8) we get

$$\sum_{i=b}^{6}\left\{\frac{\partial V_{0i}^{B}}{\partial\zeta}^{\mathsf{T}}\left[\frac{\mathrm{d}}{\mathrm{d}t}\frac{\partial\mathscr{K}_i}{\partial V_{0i}^{B}}+\begin{bmatrix}\hat{\omega}_{0i}^{B} & 0\\ \hat{v}_{0i}^{B} & \hat{\omega}_{0i}^{B}\end{bmatrix}\frac{\partial\mathscr{K}_i}{\partial V_{0i}^{B}}\right]\right\}=\tau$$

$$\sum_{i=b}^{6}\left\{\frac{\partial V_{0i}^{B}}{\partial\zeta}^{\mathsf{T}}\left[I_i\left(\frac{\partial V_{0i}^{B}}{\partial\zeta}\dot{\zeta}+\frac{\mathrm{d}}{\mathrm{d}t}\left(\frac{\partial V_{0i}^{B}}{\partial\zeta}\right)\zeta\right)-\begin{bmatrix}0 & \widehat{\frac{\partial\mathscr{K}_i}{\partial v_{0i}^{B}}}\\ \widehat{\frac{\partial\mathscr{K}_i}{\partial v_{0i}^{B}}} & \widehat{\frac{\partial\mathscr{K}_i}{\partial\omega_{0i}^{B}}}\end{bmatrix}\begin{bmatrix}v_{0i}^{B}\\ \omega_{0i}^{B}\end{bmatrix}\right]\right\}=\tau$$

$$\sum_{i=b}^{6}\left\{\frac{\partial V_{0i}^{B}}{\partial\zeta}^{\mathsf{T}}I_i\frac{\partial V_{0i}^{B}}{\partial\zeta}\dot{\zeta}+\frac{\partial V_{0i}^{B}}{\partial\zeta}^{\mathsf{T}}I_i\frac{\mathrm{d}}{\mathrm{d}t}\left(\frac{\partial V_{0i}^{B}}{\partial\zeta}\right)\zeta-\frac{\partial V_{0i}^{B}}{\partial\zeta}^{\mathsf{T}}\begin{bmatrix}0 & \widehat{\frac{\partial\mathscr{K}_i}{\partial v_{0i}^{B}}}\\ \widehat{\frac{\partial\mathscr{K}_i}{\partial v_{0i}^{B}}} & \widehat{\frac{\partial\mathscr{K}_i}{\partial\omega_{0i}^{B}}}\end{bmatrix}\begin{bmatrix}v_{0i}^{B}\\ \omega_{0i}^{B}\end{bmatrix}\right\}=\tau$$

$$\sum_{i=b}^{6}\left[\frac{\partial V_{0i}^{B}}{\partial\zeta}^{\mathsf{T}}I_i\frac{\partial V_{0i}^{B}}{\partial\zeta}\right]\dot{\zeta}$$

$$+\sum_{i=b}^{6}\left[\frac{\partial V_{0i}^{B}}{\partial\zeta}^{\mathsf{T}}I_i\frac{\mathrm{d}}{\mathrm{d}t}\left(\frac{\partial V_{0i}^{B}}{\partial\zeta}\right)-\frac{\partial V_{0i}^{B}}{\partial\zeta}^{\mathsf{T}}\begin{bmatrix}0 & \widehat{\frac{\partial\mathscr{K}_i}{\partial v_{0i}^{B}}}\\ \widehat{\frac{\partial\mathscr{K}_i}{\partial v_{0i}^{B}}} & \widehat{\frac{\partial\mathscr{K}_i}{\partial\omega_{0i}^{B}}}\end{bmatrix}\frac{\partial V_{0i}^{B}}{\partial\zeta}\right]\zeta=\tau \tag{8.12}$$

where we have used the relation in (8.2). We have now found the dynamic equations of a vehicle-manipulator system in matrix form. If we note that we can write the velocity transformation in (8.2) explicitly in terms of the geometric link Jacobian $J_{gi}^{B}(q)$ given in (5.24) we can formulate the dynamics as follows:

Theorem 8.1 *The dynamics of a vehicle-manipulator system with only kinetic energy and velocity variables* $\zeta=[(V_{0b}^{B})^{\mathsf{T}}\ \dot{q}]^{\mathsf{T}\mathsf{T}}$ *can be written as*

$$M(q)\dot{\zeta}+C(q,\zeta)\zeta=\tau \tag{8.13}$$

where the inertia matrix is given by

$$M(q)=\sum_{i=b}^{n}(J_{gi}^{B})^{\mathsf{T}}(q)I_iJ_{gi}^{B}(q) \tag{8.14}$$

where

$$J_{gi}^{B}(q)=\frac{\partial V_{0i}^{B}}{\partial\zeta}=\left[\mathrm{Ad}_{g_{ib}}\ \mathrm{Ad}_{g_{ib}}J_i\right] \tag{8.15}$$

and the Coriolis matrix is given by

$$C(q,\zeta)=\sum_{i=b}^{n}\left(\left(J_{gi}^{B}\right)^{\mathsf{T}}(q)I_i\dot{J}_{gi}^{B}(q)-\left(J_{gi}^{B}\right)^{\mathsf{T}}(q)W_i\left(V_{0i}^{B}\right)J_{gi}^{B}(q)\right) \tag{8.16}$$

where

$$W_i\left(V_{0i}^{B}\right)=\begin{bmatrix} 0 & \widehat{\frac{\partial\mathcal{K}_i}{\partial v_{0i}^{B}}} \\ \widehat{\frac{\partial\mathcal{K}_i}{\partial v_{0i}^{B}}} & \widehat{\frac{\partial\mathcal{K}_i}{\partial \omega_{0i}^{B}}} \end{bmatrix}. \tag{8.17}$$

Proof The proof follows directly from Eq. (8.12), the definition of the inertia- and Coriolis matrices and the definition of $J_{gi}^{B}(q)$ in Theorem 5.3. □

We note that we have derived the dynamics in terms of the velocity variables of the vehicle and manipulator, but only the joint positions and not the vehicle configuration appear in the equations. Hence, the equations are well defined as long as $J_{gi}^{B}(q)$ is well defined. This is always the case with 1-DoF revolute or prismatic joints, so in our case the dynamics is singularity free.

For many applications the position and orientation of the vehicle are needed explicitly in the equations. For the dynamics as presented in Theorem 8.1 it is not straight forward to include the vehicle attitude because the velocity vector is not integrable. We arrived at this formulation because we started off with the Lagrangian as the kinetic energy only. In the next section we present the dynamics of vehicle-manipulator systems where the complete state space, including the vehicle position and pose, is explicit in the equations.

8.2 The Dynamic Equations in Terms of Generalized Coordinates and Quasi-velocities

In this section we present the dynamic equations as they are normally presented in the underwater robotics literature, for example in Schjølberg (1996) and Antonelli (2006), and we will generalize this to vehicles with configuration spaces other than $SE(3)$. For ships and underwater vehicles we will need the pose of the vehicle to calculate effects such as buoyancy and damping. To include these in the dynamics in the normal way we will write the vehicle position as $\eta=[x_{0b}\ y_{0b}\ z_{0b}\ \phi_{0b}\ \theta_{0b}\ \psi_{0b}]^{\mathsf{T}}$. Using this vector representation for $SE(3)$ presented in Sect. 6.3, we can write the position of the vehicle-manipulator system by stacking the vehicle position and the robot position. We can do the same with the velocity, so the state space of the vehicle-manipulator system is written as

$$\xi = \begin{bmatrix} \eta \\ q \end{bmatrix} = \begin{bmatrix} x_{0b} \\ y_{0b} \\ z_{0b} \\ \phi_{0b} \\ \theta_{0b} \\ \psi_{0b} \\ q_1 \\ \vdots \\ q_n \end{bmatrix}, \qquad \zeta = \begin{bmatrix} V_{0b}^B \\ \dot{q} \end{bmatrix} = \begin{bmatrix} u \\ v \\ w \\ p \\ q \\ r \\ \dot{q}_1 \\ \vdots \\ \dot{q}_n \end{bmatrix}. \tag{8.18}$$

We would like to write the dynamics in the same way as we did for single rigid bodies in Sect. 6.3, i.e., in the form

$$\dot{\xi} = J_a(\xi)\zeta, \tag{8.19}$$

$$M(q)\dot{\zeta} + C(q,\zeta)\zeta + N(\xi) = \tau \tag{8.20}$$

where $\xi = [\eta^{\mathsf{T}}\ q^{\mathsf{T}}]^{\mathsf{T}}$ and $\zeta = [(V_{0b}^B)^{\mathsf{T}}\ \dot{q}^{\mathsf{T}}]^{\mathsf{T}}$ are as defined above. $M(q) \in \mathbb{R}^{(6+n)\times(6+n)}$ is the inertia matrix, $C(q,\zeta) \in \mathbb{R}^{(6+n)\times(6+n)}$ is the Coriolis and centripetal matrix, and $N(\xi)$ is the potential forces. The velocity transformation matrix is given by

$$J_a(\xi) = \begin{bmatrix} R_{0b}(\eta_2) & 0 & 0 \\ 0 & T_{0b}(\eta_2) & 0 \\ 0 & 0 & I \end{bmatrix} \in \mathbb{R}^{(6+n)\times(6+n)} \tag{8.21}$$

where I (no subscript) denotes the identity matrix. Already at this point we note that because $T_{0b}(\eta_2)$ appears in the kinematics, the equations are singularity prone. The reason that the singularity appears in the equations is of course that we choose the same vector representation for the vehicle state space as we did in Sect. 3.3.1. We also note the difference between the non-Euclidean transformation of the vehicle where the transformation is configuration dependent, and the Euclidean transformations of the manipulator where we have a linear and constant relation between the time derivative of the position variables and the velocity variables through the identity matrix.

In the more general case when the vehicle has a configuration space other than $SE(3)$, but rather one of its subgroups, we will write the dynamics in the same way as

$$\tilde{\xi} = \begin{bmatrix} \tilde{\eta} \\ q \end{bmatrix}, \qquad \tilde{\zeta} = \begin{bmatrix} \tilde{V}_{0b}^B \\ \dot{q} \end{bmatrix}. \tag{8.22}$$

The state space is then written in terms of the vectors $\tilde{\xi} \in \mathbb{R}^{m+n}$ and $\tilde{\zeta} \in \mathbb{R}^{m+n}$ where m is the dimension of the configuration space of the vehicle.

We will now find the dynamics of the vehicle-manipulator system following an approach similar to the one in the previous section, but with more emphasis on finding the explicit expressions, and we will also generalize the expressions to arbitrary configuration spaces. We first write the velocity of the vehicle with respect to the body frame $\mathscr{F}_b$ as V_{0b}^B and the link velocities as functions of V_{0b}^B and $\dot{q}$. For vehicle-manipulator systems the motion of the links are found by adding the motion of the vehicle and the robotic joints. We can write this mathematically as

$$g_{0i} = g_{0b} g_{bi} = g_{0b} g_{bi}(q). \tag{8.23}$$

Similarly the velocity of link i with respect to the base frame is given by Theorem 5.1 as

$$V_{0i}^B = \mathrm{Ad}_{g_{bi}}^{-1} V_{0b}^B + V_{bi}^B = \mathrm{Ad}_{g_{bi}}^{-1} V_{0b}^B + \mathrm{Ad}_{g_{bi}}^{-1} J_i \dot{q}. \tag{8.24}$$

The kinetic energy is found in terms of the link velocities represented in the link frame $\mathscr{F}_i$. We first need to find the link velocities in the body frame as above, i.e., by adding the vehicle velocity V_{0b}^B represented in the body frame $\mathscr{F}_b$ and the velocities of the links, also with respect to the body frame, which is given in spatial coordinates as $V_{bi}^S = J_i(q)\dot{q}$. The kinetic energy of each link is then given by From et al. (2009, 2010a)

$$\begin{aligned}
\mathscr{K}_i &= \frac{1}{2}\left(V_{0i}^B\right)^\mathsf{T} I_i V_{0i}^B \\
&= \frac{1}{2}\left(V_{0b}^B + J_i(q)\dot{q}\right)^\mathsf{T} \mathrm{Ad}_{g_{bi}}^{-\mathsf{T}} I_i \mathrm{Ad}_{g_{bi}}^{-1}\left(V_{0b}^B + J_i(q)\dot{q}\right) \\
&= \frac{1}{2}\left(V_{0b}^B + J_i(q)\dot{q}\right)^\mathsf{T} \mathrm{Ad}_{g_{ib}}^\mathsf{T} I_i \mathrm{Ad}_{g_{ib}}\left(V_{0b}^B + J_i(q)\dot{q}\right) \\
&= \frac{1}{2}\begin{bmatrix}\left(V_{0b}^B\right)^\mathsf{T} & \dot{q}^\mathsf{T}\end{bmatrix} M_i(q) \begin{bmatrix} V_{0b}^B \\ \dot{q}\end{bmatrix} \\
&= \frac{1}{2} v^\mathsf{T} M_i(q) v
\end{aligned} \tag{8.25}$$

for $i = b, 1, \dots, n$. We see that the inertia matrix of the whole system can be written in terms of the Adjoint map $\mathrm{Ad}_{g_{ib}}(q)$ and the link Jacobian $J_i(q)$ as

$$M_i(q) = \begin{bmatrix} \mathrm{Ad}_{g_{ib}}^\mathsf{T} I_i \mathrm{Ad}_{g_{ib}} & \mathrm{Ad}_{g_{ib}}^\mathsf{T} I_i \mathrm{Ad}_{g_{ib}} J_i \\ J_i^\mathsf{T} \mathrm{Ad}_{g_{ib}}^\mathsf{T} I_i \mathrm{Ad}_{g_{ib}} & J_i^\mathsf{T} \mathrm{Ad}_{g_{ib}}^\mathsf{T} I_i \mathrm{Ad}_{g_{ib}} J_i \end{bmatrix} \in \mathbb{R}^{(6+n)\times(6+n)}. \tag{8.26}$$

Using the definition of the body geometric Jacobian of link i that we found in Sect. 5.2.2, we can write this as

$$M_i(q) = \left(J_{gi}^B\right)^\mathsf{T} I_i J_{gi}^B \tag{8.27}$$

where $J_{gi}^B(q) = [\mathrm{Ad}_{g_{ib}} \; \mathrm{Ad}_{g_{ib}} J_i] \in \mathbb{R}^{6\times(6+n)}$.

For a vehicle with an m-dimensional configuration space we recall that $V_{0b}^B = H\tilde{V}_{0b}^B$ and the same computations give

$$
\begin{aligned}
\mathscr{K}_i &= \frac{1}{2}\left(V_{0i}^B\right)^\mathsf{T} I_i V_{0i}^B \\
&= \frac{1}{2}\left(H\tilde{V}_{0b}^B + J_i(q)\dot{q}\right)^\mathsf{T} \mathrm{Ad}_{g_{ib}}^\mathsf{T} I_i \mathrm{Ad}_{g_{ib}} \left(H\tilde{V}_{0b}^B + J_i(q)\dot{q}\right) \\
&= \frac{1}{2}\left(\left(\tilde{V}_{0b}^B\right)^\mathsf{T} H^\mathsf{T} + \dot{q}^\mathsf{T} J_i(q)^\mathsf{T}\right) \mathrm{Ad}_{g_{ib}}^\mathsf{T} I_i \mathrm{Ad}_{g_{ib}} \left(H\tilde{V}_{0b}^B + J_i(q)\dot{q}\right) \\
&= \frac{1}{2}\begin{bmatrix} \left(\tilde{V}_{0b}^B\right)^\mathsf{T} & \dot{q}^\mathsf{T} \end{bmatrix} M_i(q) \begin{bmatrix} \tilde{V}_{0b}^B \\ \dot{q} \end{bmatrix} \\
&= \frac{1}{2} v^\mathsf{T} M_i(q) v
\end{aligned}
\tag{8.28}
$$

with $M_b = \begin{bmatrix} I_b & 0 \\ 0 & 0 \end{bmatrix}$ for the vehicle and

$$
M_i(q) = \begin{bmatrix} H^\mathsf{T}\mathrm{Ad}_{g_{ib}}^\mathsf{T} I_i \mathrm{Ad}_{g_{ib}} H & H^\mathsf{T}\mathrm{Ad}_{g_{ib}}^\mathsf{T} I_i \mathrm{Ad}_{g_{ib}} J_i \\ J_i^\mathsf{T}\mathrm{Ad}_{g_{ib}}^\mathsf{T} I_i \mathrm{Ad}_{g_{ib}} H & J_i^\mathsf{T}\mathrm{Ad}_{g_{ib}}^\mathsf{T} I_i \mathrm{Ad}_{g_{ib}} J_i \end{bmatrix} \in \mathbb{R}^{(m+n)\times(m+n)} \tag{8.29}
$$

for the links.

The total kinetic energy of the VM system is given by the sum of the kinetic energies of the mechanism links and the vehicle, that is,

$$
\mathscr{K}\left(q, V_{0b}^S\right) = \frac{1}{2}\zeta^\mathsf{T}\left(\sum_{i=b}^{n} M_i(q)\right)\zeta = \frac{1}{2}\zeta^\mathsf{T}\underbrace{\left(\begin{bmatrix} I_b & 0 \\ 0 & 0 \end{bmatrix} + \sum_{i=1}^{n} M_i(q)\right)}_{\text{inertia matrix } M(q)}\zeta \tag{8.30}
$$

with $M(q)$ the inertia matrix of the total system.

We see that the inertia matrix of the vehicle is constant because we choose to represent the velocities in the vehicle frame $\mathscr{F}_b$. Note that neither $\mathscr{K}(q, V_{0b}^B)$ nor $M(q)$ depend on the pose g_{0b} nor the choice of inertial reference frame $\mathscr{F}_0$.

We can now write the inertia matrix in terms of the Adjoint map $\mathrm{Ad}_{g_{ib}}(q)$ and the spatial link Jacobian J_i:

Theorem 8.2 *The inertia matrix $M(q) \in \mathbb{R}^{(m+n)\times(m+n)}$ of a vehicle-manipulator system with configuration space $\tilde{\xi} = [\tilde{\eta}^{\mathsf{T}}\ q^{\mathsf{T}}]^{\mathsf{T}}$ where $\tilde{\eta} \in \mathbb{R}^m$ is the vehicle position and $q \in \mathbb{R}^n$ is the manipulator position, is given by*

$$\begin{aligned} M(q) &= \sum_{i=b}^{n} M_i(q) \\ &= \sum_{i=b}^{n} (J_{gi}^B)^{\mathsf{T}}(q) I_i J_{gi}^B(q) \\ &= \sum_{i=b}^{n} \begin{bmatrix} H^{\mathsf{T}}\mathrm{Ad}_{g_{ib}}^{\mathsf{T}} I_i \mathrm{Ad}_{g_{ib}} H & H^{\mathsf{T}}\mathrm{Ad}_{g_{ib}}^{\mathsf{T}} I_i \mathrm{Ad}_{g_{ib}} J_i \\ J_i^{\mathsf{T}}\mathrm{Ad}_{g_{ib}}^{\mathsf{T}} I_i \mathrm{Ad}_{g_{ib}} H & J_i^{\mathsf{T}}\mathrm{Ad}_{g_{ib}}^{\mathsf{T}} I_i \mathrm{Ad}_{g_{ib}} J_i \end{bmatrix} \end{aligned} \tag{8.31}$$

where $J_{gi}^B = [\mathrm{Ad}_{g_{ib}} H\ \mathrm{Ad}_{g_{ib}} J_i]$. The inertia matrix only depends on the joint positions of the robot, and not the position of the vehicle frame $\mathscr{F}_b$ with respect to the inertial reference frame $\mathscr{F}_0$.

Proof The proof follows directly from the expression in (8.26). □

We can obtain the dynamics of the system in the same way that we did in the previous section by substituting $J_{gi}^B = \frac{\partial V_{0i}^B}{\partial \zeta}$ and including the potential forces $\mathscr{U}$ as

$$\sum_{i=b}^{6} \left\{ (J_{gi}^B)^{\mathsf{T}} \left[\frac{\mathrm{d}}{\mathrm{d}t} \frac{\partial \mathscr{K}_i}{\partial V_{0i}^B} + \begin{bmatrix} \hat{\omega}_{0i}^B & 0 \\ \hat{v}_{0i}^B & \hat{\omega}_{0i}^B \end{bmatrix} \frac{\partial \mathscr{K}_i}{\partial V_{0i}^B} + \frac{\partial \mathscr{U}}{\partial \eta_{0i}} \right] \right\} = \tau$$

$$\sum_{i=b}^{6} [(J_{gi}^B)^{\mathsf{T}} I_i J_{gi}^B]\dot{\zeta} + \sum_{i=b}^{6} [(J_{gi}^B)^{\mathsf{T}} I_i \dot{J}_{gi} - (J_{gi}^B)^{\mathsf{T}} W_i (J_{gi}^B)]\zeta + (J_{gi}^B)^{\mathsf{T}} \frac{\partial \mathscr{U}}{\partial \eta_{0i}} = \tau. \tag{8.32}$$

This allows us to state the following important theorem:

Theorem 8.3 *The dynamics of a vehicle-manipulator system with position variables $\tilde{\xi} = [\tilde{\eta}^{\mathsf{T}}\ q^{\mathsf{T}}]^{\mathsf{T}}$ and velocity variables $\tilde{\zeta} = [(\tilde{V}_{0b}^B)^{\mathsf{T}}\ \dot{q}^{\mathsf{T}}]^{\mathsf{T}}$ can be written as*

$$\dot{\tilde{\xi}} = J_a(\tilde{\xi})\tilde{\zeta}, \tag{8.33}$$

$$M(q)\dot{\tilde{\zeta}} + C(q,\tilde{\zeta})\tilde{\zeta} + N(\tilde{\xi}) = \tau. \tag{8.34}$$

The inertia matrix is given by

$$M(q) = \sum_{i=b}^{n} (J_{gi}^B)^{\mathsf{T}}(q) I_i (J_{gi}^B)(q) \tag{8.35}$$

where

$$J_{gi}^B(q) = \begin{bmatrix} \mathrm{Ad}_{g_{ib}} H & \mathrm{Ad}_{g_{ib}} J_i \end{bmatrix}. \tag{8.36}$$

The Coriolis matrix is given by

$$C(q, \zeta) = \sum_{i=b}^{n} (J_{gi}^B)^{\mathsf{T}}(q) I_i (\dot{J}_{gi}^B)(q) - (J_{gi}^B)^{\mathsf{T}}(q) W_i (V_{0i}^B)(J_{gi}^B)(q) \tag{8.37}$$

where

$$W_i(V_{0i}^B) = \begin{bmatrix} 0 & \widehat{\frac{\partial \mathcal{K}_i}{\partial v_{0i}^B}} \\ \widehat{\frac{\partial \mathcal{K}_i}{\partial v_{0i}^B}} & \widehat{\frac{\partial \mathcal{K}_i}{\partial \omega_{0i}^B}} \end{bmatrix}, \tag{8.38}$$

and the potential forces are given by

$$N(\xi) = \sum_{i=b}^{n} (J_{gi}^B)^{\mathsf{T}}(q) \mathrm{Ad}_{g_{i0}}(g_{0b}, q) F_g^i(g_{0b}, q) \tag{8.39}$$

where $F_g^i(g_{0b}, q)$ represents the potential forces of link i represented in the inertial frame $\mathcal{F}_0$.

Proof We find the potential forces by ensuring that the work is the same regardless of the variables used, i.e.,

$$N^{\mathsf{T}}\zeta = \sum_{i=b}^{n} (F_g^i)^{\mathsf{T}} V_{0i}^S = \sum_{i=b}^{n} (F_g^i)^{\mathsf{T}} \mathrm{Ad}_{g_{0i}} V_{0i}^B = \sum_{i=b}^{n} (F_g^i)^{\mathsf{T}} \mathrm{Ad}_{g_{0i}} J_{gi}^B \zeta \tag{8.40}$$

which, if we compare the first and last expressions, gives

$$N(\xi) = \sum_{i=b}^{n} (J_{gi}^b(q))^{\mathsf{T}} \mathrm{Ad}_{g_{i0}}(g_{0b}, q) F_g^i. \tag{8.41}$$

The proof now follows more or less directly from Theorem 8.1. □

8.2.1 Full State Space Dynamics in Vector Form

We can eliminate the velocity variables ζ from the equations in (8.33)–(8.34) in the same way that we did in Sect. 6.3. We then write

$$\dot{\xi} = J_a(\xi)\zeta \quad \Longleftrightarrow \quad \zeta = J_a^{-1}(\xi)\dot{\xi}, \tag{8.42}$$

$$\ddot{\xi} = J_a(\xi)\dot{\zeta} + \dot{J}_a(\xi)\zeta \quad \Longleftrightarrow \quad \dot{\zeta} = J_a^{-1}(\xi)\big(\ddot{\xi} - \dot{J}_a(\xi)J_a^{-1}(\xi)\dot{\xi}\big). \tag{8.43}$$

The dynamics is now given by (Murray et al. 1994; Fossen 2002)

$$M(q)\dot{\zeta} + C(q,\zeta)\zeta = \tau \tag{8.44}$$

$$M(q)J_a^{-1}(\xi)\big(\ddot{\xi} - \dot{J}_a(\xi)J_a^{-1}(\xi)\dot{\xi}\big) + C\big(q, J_a^{-1}(\xi)\dot{\xi}\big)J_a^{-1}(\xi)\dot{\xi} = \tau \tag{8.45}$$

$$M(q)J_a^{-1}(\xi)\ddot{\xi} - M(q)J_a^{-1}(\xi)\dot{J}_a(\xi)J_a^{-1}(\xi)\dot{\xi} + C(q,\dot{\xi})J_a^{-1}(\xi)\dot{\xi} = \tau. \tag{8.46}$$

Once again, we will look at the work and make sure that this is invariant with respect to the coordinate frame in which the torques and velocities are observed. The work is given by

$$W = \zeta^{\mathsf{T}}\tau = \big(J_a^{-1}(\xi)\dot{\xi}\big)^{\mathsf{T}}\tau = \dot{\xi}^{\mathsf{T}}J_a^{-\mathsf{T}}(\xi)\tau. \tag{8.47}$$

The required external force is obtained by pre-multiplying (8.46) with $J_a^{-\mathsf{T}}(\xi)$ which gives us the dynamics as

$$J_a^{-\mathsf{T}}MJ_a^{-1}\ddot{\xi} + J_a^{-\mathsf{T}}\big(CJ_a^{-1} - MJ_a^{-1}\dot{J}_aJ_a^{-1}\big)\dot{\xi} = J_a^{-\mathsf{T}}\tau. \tag{8.48}$$

We have thus obtained the dynamic equations of a vehicle-manipulator system for a vehicle with configuration space $SE(3)$ and a robotic manipulator with n Euclidean joints:

The dynamic equations of a vehicle-manipulator system can be written in terms of the position variables ξ and velocity variables $\dot{\xi}$ as

$$\tilde{M}(\xi)\ddot{\xi} + \tilde{C}(\xi,\dot{\xi})\dot{\xi} = \tilde{\tau} \tag{8.49}$$

where

$$\tilde{M}(\xi) = J_a^{-\mathsf{T}}(\xi)M(q)J_a^{-1}(\xi) \tag{8.50}$$

$$\tilde{C}(\xi,\dot{\xi}) = J_a^{-\mathsf{T}}(\xi)\big(J_a^{-\mathsf{T}}(\xi)C(q,\dot{\xi})J_a^{-1}(\xi) - M(q)J_a^{-1}(\xi)\dot{J}_a(\xi)J_a^{-1}(\xi)\big) \tag{8.51}$$

$$\tilde{\tau}(\xi) = J_a^{-\mathsf{T}}(\xi)\tau. \tag{8.52}$$

Because J_a appears in the equations, we need to keep in mind the singularity at $\theta = \pm\frac{\pi}{2}$ whenever we use this formulation of the dynamics.

8.2.2 Operational Space Approach

As for fixed-base manipulators it is often more convenient to represent the dynamics in terms of the end-effector position and velocity than the joint variables. In this section we therefore reformulate the dynamics in terms of the vector $\varsigma = [\eta^{\mathsf{T}}\ (\eta_{0e})^{\mathsf{T}}]^{\mathsf{T}} \in \mathbb{R}^{12}$ where η is the position of the vehicle and η_{0e} is the end-effector position vector in the inertial frame. In Sect. 5.2.4 we found the relation between this operational space configuration vector and the vehicle twist and joint velocities through the workspace Jacobian $J_w(\xi)$ as

$$\dot{\varsigma} = J_w(\xi)\zeta \tag{8.53}$$

where

$$J_w(\xi) = \begin{bmatrix} J_b(\eta_2) & 0 \\ J_e(\eta_{0e,2})\mathrm{Ad}_{g_{eb}} & J_e(\eta_{0e,2})J_{m,g}(q) \end{bmatrix} \in \mathbb{R}^{12\times(6+n)}. \tag{8.54}$$

Following the same approach as in the previous section we can find the operational space dynamic equations as

$$\bar{M}(\xi)\ddot{\varsigma} + \bar{C}(\xi,\zeta)\dot{\varsigma} = \bar{\tau}, \tag{8.55}$$

where we define

$$\bar{M}(\xi) = J_w^{-\mathsf{T}}(\xi)M(q)J_w^{-1}(\xi) \tag{8.56}$$

$$\bar{C}(\xi,\dot{\xi}) = J_w^{-\mathsf{T}}(\xi)\big(J_w^{-\mathsf{T}}(\xi)C(q,\dot{\xi})J_w^{-1}(\xi) - M(q)J_w^{-1}(\xi)\dot{J}_w(\xi)J_w^{-1}(\xi)\big) \tag{8.57}$$

$$\bar{\tau}(\xi) = J_w^{-\mathsf{T}}(\xi)\tau. \tag{8.58}$$

We can now look at what these equations look like for our system. We will first write the dynamics as

$$\begin{bmatrix} M_{VV} & M_{qV}^{\mathsf{T}} \\ M_{qV} & M_q \end{bmatrix}\begin{bmatrix} \dot{V}_{0b}^B \\ \ddot{q} \end{bmatrix} + \begin{bmatrix} C_{VV} & C_{Vq} \\ C_{qV} & C_q \end{bmatrix}\begin{bmatrix} V_{0b}^B \\ \dot{q} \end{bmatrix} = \begin{bmatrix} \tau_V \\ \tau_q \end{bmatrix} \tag{8.59}$$

and we rewrite the workspace Jacobian $J_w(\xi)$ as

$$J_w(\xi) = \begin{bmatrix} J_b & 0 \\ J_1 & J_2 \end{bmatrix} \tag{8.60}$$

so that

$$\dot{\eta}_{0e} = J_1 V_{0b}^B + J_2\dot{q}. \tag{8.61}$$

The change of variables given by the workspace Jacobian in (8.54) can now be found as

$$\dot{\eta} = J_b V_{0b}^B \iff V_{0b}^B = J_b^{-1}\dot{\eta}, \tag{8.62}$$

$$\ddot{\eta} = \dot{J}_b V_{0b}^B + J_b \dot{V}_{0b}^B \iff \dot{V}_{0b}^B = J_b^{-1}\left(\ddot{\eta} - \dot{J}_b J_b^{-1}\dot{\eta}\right). \tag{8.63}$$

for the first line of (8.59). We can also rewrite the equations

$$\dot{\eta}_{0e} = J_1 V_{0b}^B + J_2\dot{q} \tag{8.64}$$

$$\ddot{\eta}_{0e} = \dot{J}_1 V_{0b}^B + J_1 \dot{V}_{0b}^B + \dot{J}_2\dot{q} + J_2\ddot{q} \tag{8.65}$$

as

$$\dot{q} = J_2^{-1}\left(\dot{\eta}_{0e} - J_1 J_b^{-1}\dot{\eta}\right), \tag{8.66}$$

$$\ddot{q} = J_2^{-1}\left(\ddot{\eta}_{0e} - \dot{J}_1 J_b^{-1}\dot{\eta} - J_1 J_b^{-1}\left(\ddot{\eta} - \dot{J}_b J_b^{-1}\dot{\eta}\right) - \dot{J}_2 J_2^{-1}\left(\dot{\eta}_{0e} - J_1 J_b^{-1}\dot{\eta}\right)\right) \tag{8.67}$$

which gives us the change of variables in the second line of (8.59). Substituting (8.62)–(8.63) and (8.66)–(8.67) into (8.59) gives us an idea of what the dynamics looks like. Alternatively we can find the dynamics directly by substituting J_w into (8.55)–(8.58).

The expressions in (8.62)–(8.63) and (8.66)–(8.67) give us valuable insight into how the Jacobian maps work for vehicle-manipulator systems: the Jacobians merely assure that the different velocity variables are correctly collocated. For example in Eqs. (8.62)–(8.63), J_b collocates V_{0b}^B with $\dot{\eta}$, and in Eqs. (8.66)–(8.67) we see that J_2^{-1} collocates $\dot{\eta}_{0e}$ with $\dot{q}$. Similarly, $\dot{J}_1 J_b^{-1}$ allows us to add $\dot{\eta}$ with $\ddot{\eta}_{0e}$ (see Eq. (8.67)), and finally all the terms, now collocated with $\ddot{\eta}_{0e}$, are collocated with $\ddot{q}$ through the Jacobian mapping of J_2^{-1}. Working with vehicle-manipulator systems requires several different variables to represent the positions and velocities. The Jacobians help us to make sure that the different variables can be added and subtracted in a mathematically meaningful manner.

Up until now we have assumed that the inverse of the Jacobian matrix always exists. However, we have already seen that the upper left part of the workspace Jacobian that gives the relation $\dot{\eta} = J_b V_{0b}^B$ is not well defined. Also J_1 and J_2 contain singularities at isolated points, and as for manipulators on a fixed base, caution must be taken with J_2 if the manipulator part of the vehicle-manipulator system is redundant. For redundant manipulators the inverse of $J_{m,g}(q)$ will in general not give us a one-to-one mapping and several solutions to the inverse kinematics problem exist.

8.3 Configuration States

In this section we will derive the dynamics of vehicle-manipulator systems in terms of a general configuration vector x and velocity vector v. We will write

$v = [(\tilde{V}_{0b}^B)^\mathsf{T}\ \dot{q}^\mathsf{T}]^\mathsf{T} \in \mathbb{R}^{m+n}$ for velocity and $x \in \mathbb{R}^{m+n}$ for position, which can be either ξ or $\tilde{\xi}$, or for that sake any other representation of the vehicle configuration. The mapping between the time derivative of the position variable $\dot{x}$ and the velocity variable v is given as earlier by $S(x)$. In this section we will use this relation to eliminate $\dot{x}$ from the equations. The Lagrangian of the vehicle-manipulator system written in terms of this general configuration space is given by

$$L(x, v) = \frac{1}{2} v^\mathsf{T} M(x) v - \mathcal{U}(x). \tag{8.68}$$

We will find the partial derivatives of the Lagrangian in the same way as we did in the previous chapters.

The derivatives of the Lagrangian in (8.68) for a configuration-dependent matrix $M(x)$ are found with respect to v and x as

$$\frac{\partial L}{\partial v} = M(x) v, \tag{8.69}$$

$$\frac{\mathrm{d}}{\mathrm{d}t}\left(\frac{\partial L}{\partial v}\right) = M(x)\dot{v} + \dot{M}(x) v, \tag{8.70}$$

$$\frac{\partial L}{\partial x} = \frac{1}{2}\frac{\partial^\mathsf{T} M(x) v}{\partial x} v - \frac{\partial \mathcal{U}(x)}{\partial x}. \tag{8.71}$$

Using the relation $v = S(x)\dot{x}$ we write the Lagrangian as a function of generalized coordinates and velocities as

$$\bar{L}(x, \dot{x}) = \frac{1}{2}\dot{x}^\mathsf{T} S(x)^\mathsf{T} M(x) S(x) \dot{x} - \mathcal{U}(x). \tag{8.72}$$

The dynamics is then found by Lagrange's equations as

$$\frac{\mathrm{d}}{\mathrm{d}t}\left(\frac{\partial \bar{L}}{\partial \dot{x}}\right) - \frac{\partial \bar{L}}{\partial x} = B(x)\tau \tag{8.73}$$

for some $B(x)$ yet to be determined. To find the explicit equations we need the time derivatives of the Lagrangian $\bar{L}$ which are different from the single rigid body because $M(x)$ is configuration dependent. We find these as (From 2012a, 2012b; From et al. 2012a)

$$\frac{\partial \bar{L}}{\partial \dot{x}} = S^\mathsf{T}(x) M(x) S(x) \dot{x} = S^\mathsf{T}(x) \underbrace{M(x) v}_{\frac{\partial L}{\partial v}} = S^\mathsf{T}(x) \frac{\partial L}{\partial v}, \tag{8.74}$$

$$
\begin{aligned}
\frac{\mathrm{d}}{\mathrm{d}t}\left(\frac{\partial \bar{L}}{\partial \dot{x}}\right) &= \dot{S}^{\mathsf{T}}(x)M(x)v + S^{\mathsf{T}}(x)\dot{M}(x)v + S^{\mathsf{T}}(x)M(x)\dot{v} \\
&= \dot{S}^{\mathsf{T}}(x)\underbrace{M(x)v}_{\frac{\partial L}{\partial v}} + S^{\mathsf{T}}(x)\underbrace{\big(\dot{M}(x)v + M(x)\dot{v}\big)}_{\frac{\mathrm{d}}{\mathrm{d}t}(\frac{\partial L}{\partial v})} \\
&= \dot{S}^{\mathsf{T}}(x)\frac{\partial L}{\partial v} + S^{\mathsf{T}}(x)\frac{\mathrm{d}}{\mathrm{d}t}\left(\frac{\partial L}{\partial v}\right),
\end{aligned} \tag{8.75}
$$

$$
\begin{aligned}
\frac{\partial \bar{L}}{\partial x} &= \frac{\partial^{\mathsf{T}}(S(x)\dot{x})}{\partial x}M(x)S(x)\dot{x} + \frac{1}{2}\frac{\partial^{\mathsf{T}}(M(x)v)}{\partial x}S(x)\dot{x} - \frac{\partial U(x)}{\partial x} \\
&= \frac{\partial^{\mathsf{T}}(S(x)\dot{x})}{\partial x}\underbrace{M(x)v}_{\frac{\partial L}{\partial v}} + \underbrace{\frac{1}{2}\frac{\partial^{\mathsf{T}}(M(x)v)}{\partial x}v - \frac{\partial U(x)}{\partial x}}_{\frac{\partial L}{\partial x}} \\
&= \frac{\partial L}{\partial x} + \frac{\partial^{\mathsf{T}}(S(x)\dot{x})}{\partial x}\frac{\partial L}{\partial v}.
\end{aligned} \tag{8.76}
$$

We will use this result frequently, so we write it as a proposition:

Proposition 8.1 *The partial derivative of a Lagrangian in the form*

$$\bar{L}(x,\dot{x}) = \frac{1}{2}\dot{x}^{\mathsf{T}}S(x)^{\mathsf{T}}M(x)S(x)\dot{x} - \mathscr{U}(x) \tag{8.77}$$

can be expressed in terms of the Lagrangian

$$L(x,v) = \frac{1}{2}v^{\mathsf{T}}M(x)v - \mathscr{U}(x) \tag{8.78}$$

as

$$\frac{\mathrm{d}}{\mathrm{d}t}\left(\frac{\partial \bar{L}}{\partial \dot{x}}\right) = \dot{S}^{\mathsf{T}}(x)\frac{\partial L}{\partial v} + S^{\mathsf{T}}(x)\frac{\mathrm{d}}{\mathrm{d}t}\left(\frac{\partial L}{\partial v}\right) \tag{8.79}$$

$$\frac{\partial \bar{L}}{\partial x} = \frac{\partial L}{\partial x} + \frac{\partial^{\mathsf{T}}(S(x)\dot{x})}{\partial x}\frac{\partial L}{\partial v}. \tag{8.80}$$

Proof The proof follows directly from Eqs. (8.74)–(8.76). □

The Euler–Lagrange equations are found by the partial derivatives of the Lagrangian $\bar{L}(x,\dot{x})$ as

$$
\begin{gathered}
\frac{\mathrm{d}}{\mathrm{d}t}\left(\frac{\partial \bar{L}}{\partial \dot{x}}\right) - \frac{\partial \bar{L}}{\partial x} = B(x)\tau \\
S^{\mathsf{T}}(x)\frac{\mathrm{d}}{\mathrm{d}t}\left(\frac{\partial L}{\partial v}\right) + \dot{S}^{\mathsf{T}}(x)\frac{\partial L}{\partial v} - \frac{\partial L}{\partial x} - \frac{\partial^{\mathsf{T}}(S(x)\dot{x})}{\partial x}\frac{\partial L}{\partial v} = B(x)\tau.
\end{gathered} \tag{8.81}
$$

The torques $\tau = [\tau_V^\mathsf{T}\ \tau_q^\mathsf{T}]^\mathsf{T}$ are defined in the usual way so that they are collocated with $v = [(V_{0b}^B)^\mathsf{T}\ \dot{q}^\mathsf{T}]^\mathsf{T}$. To find the corresponding external forces that are collocated with $\dot{x}$ we write

$$W = v^\mathsf{T}\tau = \big(S(x)\dot{x}\big)^\mathsf{T}\tau = \dot{x}^\mathsf{T} S^\mathsf{T}(x)\tau \tag{8.82}$$

and again we have $B(x) = S^\mathsf{T}(x)$ as expected. $S(x)$ will be of the form

$$S(x) = \begin{bmatrix} J^\mathsf{T} & 0 \\ 0 & I \end{bmatrix} \tag{8.83}$$

for some vehicle Jacobian $J(x)$ depending on the choice of position and velocity variables. We then pre-multiply (8.81) with $S^{-\mathsf{T}}(x)$ we thus get the dynamic equations.

The dynamic equations of a vehicle-manipulator system is given by

$$\frac{\mathrm{d}}{\mathrm{d}t}\left(\frac{\partial L}{\partial v}\right) - S^{-\mathsf{T}}(x)\frac{\partial L}{\partial x} + S^{-\mathsf{T}}(x)\left(\dot{S}^\mathsf{T}(x) - \frac{\partial^\mathsf{T}(S(x)\dot{x})}{\partial x}\right)\frac{\partial L}{\partial v} = \tau. \tag{8.84}$$

If we substitute the expressions in (8.69)–(8.71) into (8.84), the dynamics becomes

$$\underbrace{M(x)\dot{v} + \dot{M}(x)v - \frac{1}{2}S^{-\mathsf{T}}(x)\frac{\partial^\mathsf{T} M(x)v}{\partial x}v}_{\text{Multibody terms}}$$
$$+\left(\sum_k \gamma_k v_k\right)M(x)v + S^{-\mathsf{T}}(x)\frac{\partial \mathscr{U}(x)}{\partial x} = \tau \tag{8.85}$$

where we recognize

$$\left(\sum_k \gamma_k v_k\right) = S^{-\mathsf{T}}(x)\left(\dot{S}^\mathsf{T}(x) - \frac{\partial^\mathsf{T}(S(x)\dot{x})}{\partial x}\right) \tag{8.86}$$

from Chap. 6. Note the terms that arise due to the configuration dependent inertia matrix $M(x)$ which were not present for single rigid bodies with a constant inertia matrix.

Similar to what we did in the previous chapters we would like to write the equations in terms of x and $v = S(x)\dot{x}$, but not with $\dot{x}$ explicitly present in the equations. We can follow the same train of thought as we did for single rigid body systems, but in addition we need to take into account that the inertia matrix is not constant. The inertia matrix of vehicle-manipulator systems differs from single rigid bodies in that it is configuration dependent and if we compare these equations to the ones in (6.112) we also notice that two new multibody terms arise in the dynamics (8.85). From the derivation of γ_k we see, however, that this will not change the expression in (8.86) in any way, and we can use the expression of γ_k that we found in (6.106) also for vehicle-manipulator systems.

It now only remains to look at the part of the Coriolis matrix that arises as a result of a configuration-dependent inertia matrix. If we use that $\dot{x}_l = \sum_k S_{lk}^{-1} v_k$ we first note that $\dot{M}(x)$ can be written as

$$\begin{aligned} \dot{M}_{ij}(x) &= \sum_l \frac{\partial M_{ij}(x)}{\partial x_l} \dot{x}_l \\ &= \sum_{k,l} \frac{\partial M_{ij}(x)}{\partial x_l} S_{lk}^{-1} v_k. \end{aligned} \tag{8.87}$$

Further, we write the matrices $S^{-\mathsf{T}}(x)$ and $\frac{\partial^{\mathsf{T}} M(x) v}{\partial x}$ as

$$S^{-\mathsf{T}}(x) = \begin{bmatrix} S_{11}^{-1} & S_{21}^{-1} & \cdots & S_{(m+n)1}^{-1} \\ S_{12}^{-1} & S_{22}^{-1} & \cdots & S_{(m+n)2}^{-1} \\ \vdots & \vdots & \ddots & \vdots \\ S_{1(m+n)}^{-1} & S_{2(m+n)}^{-1} & \cdots & S_{(m+n)(m+n)}^{-1} \end{bmatrix} \tag{8.88}$$

$$\frac{\partial^{\mathsf{T}} M(x) v}{\partial x} = \begin{bmatrix} \frac{\partial (Mv)_1}{\partial x_1} & \frac{\partial (Mv)_2}{\partial x_1} & \cdots & \frac{\partial (Mv)_{m+n}}{\partial x_1} \\ \frac{\partial (Mv)_1}{\partial x_2} & \frac{\partial (Mv)_2}{\partial x_2} & \cdots & \frac{\partial (Mv)_{m+n}}{\partial x_2} \\ \vdots & \vdots & \ddots & \vdots \\ \frac{\partial (Mv)_1}{\partial x_{m+n}} & \frac{\partial (Mv)_2}{\partial x_{m+n}} & \cdots & \frac{\partial (Mv)_{m+n}}{\partial x_{m+n}} \end{bmatrix} \tag{8.89}$$

which gives

$$\begin{aligned} & S^{-\mathsf{T}}(x) \frac{\partial^{\mathsf{T}} M(x) v}{\partial x} \\ &= \begin{bmatrix} \sum_l S_{l1}^{-1} \frac{\partial (Mv)_1}{\partial x_l} & \sum_l S_{l1}^{-1} \frac{\partial (Mv)_2}{\partial x_l} & \cdots & \sum_l S_{l1}^{-1} \frac{\partial (Mv)_{m+n}}{\partial x_l} \\ \sum_l S_{l2}^{-1} \frac{\partial (Mv)_1}{\partial x_l} & \sum_l S_{l2}^{-1} \frac{\partial (Mv)_2}{\partial x_l} & \cdots & \sum_l S_{l2}^{-1} \frac{\partial (Mv)_{m+n}}{\partial x_l} \\ \vdots & \vdots & \ddots & \vdots \\ \sum_l S_{l(m+n)}^{-1} \frac{\partial (Mv)_1}{\partial x_l} & \sum_l S_{l(m+n)}^{-1} \frac{\partial (Mv)_2}{\partial x_l} & \cdots & \sum_l S_{l(m+n)}^{-1} \frac{\partial (Mv)_{m+n}}{\partial x_l} \end{bmatrix} \end{aligned} \tag{8.90}$$

and we obtain the required expression as

$$\left(S^{-\mathsf{T}}(x) \frac{\partial^{\mathsf{T}} M(x) v}{\partial x} \right)_{ij} = \sum_l S_{li}^{-1} \frac{\partial (Mv)_j}{\partial x_l} = \sum_{k,l} S_{li}^{-1} \frac{\partial M_{jk}}{\partial x_l} v_k. \tag{8.91}$$

We have found the dynamics of vehicle-manipulator systems without $\dot{x}$ explicit in the equations and we can conclude with the following important result:

Theorem 8.4 *The dynamic equations of a vehicle-manipulator system with generalized coordinates x and quasi-velocity coordinates $v = S(x)\dot{x}$ can be written as*

$$M(x)\dot{v} + C(x, v)v + N(x) = \tau \tag{8.92}$$

where the inertia matrix is given by

$$M(x) = \sum_{i=b}^{n} \begin{bmatrix} H^{\mathsf{T}}\mathrm{Ad}_{g_{ib}}^{\mathsf{T}} I_i \mathrm{Ad}_{g_{ib}} H & H^{\mathsf{T}}\mathrm{Ad}_{g_{ib}}^{\mathsf{T}} I_i \mathrm{Ad}_{g_{ib}} J_i \\ J_i^{\mathsf{T}}\mathrm{Ad}_{g_{ib}}^{\mathsf{T}} I_i \mathrm{Ad}_{g_{ib}} H & J_i^{\mathsf{T}}\mathrm{Ad}_{g_{ib}}^{\mathsf{T}} I_i \mathrm{Ad}_{g_{ib}} J_i \end{bmatrix}, \tag{8.93}$$

the Coriolis matrix is given by

$$C(x, v) = \sum_k \alpha_k v_k + \sum_k \beta_k v_k, \tag{8.94}$$

where

$$(\alpha_k)_{ij}(x) = \sum_{k,l} \left(\frac{\partial M_{ij}}{\partial x_l} S_{lk}^{-1} - \frac{1}{2} S_{li}^{-1} \frac{\partial M_{jk}}{\partial x_l} \right), \tag{8.95}$$

$$(\beta)_{ij}(x) = \sum_s \gamma_{is} M_{sj} = \sum_{l,m,s} \left(S_{li}^{-1} \left(\frac{\partial S_{sl}}{\partial x_m} - \frac{\partial S_{sm}}{\partial x_l} \right) S_{mk}^{-1} \right)_{is} M_{sj}, \tag{8.96}$$

and the potential forces are given by

$$N(x) = S^{-\mathsf{T}} \frac{\partial \mathscr{U}(x)}{\partial x}. \tag{8.97}$$

Proof We first multiply (8.86) with the inertia matrix:

$$\begin{aligned} \left(\left(\sum_k \gamma_k v_k \right) M \right)_{ij} &= \sum_s \left(\sum_k \gamma_k v_k \right)_{is} M_{sj} \\ &= \sum_s \left(\sum_k \sum_{l,m} S_{li}^{-1} \left(\frac{\partial S_{sl}}{\partial x_m} - \frac{\partial S_{sm}}{\partial x_l} \right) S_{mk}^{-1} v_k \right)_{is} M_{sj} \\ &= \sum_k \sum_{l,m,s} \left(S_{li}^{-1} \left(\frac{\partial S_{sl}}{\partial x_m} - \frac{\partial S_{sm}}{\partial x_l} \right) S_{mk}^{-1} \right)_{is} M_{sj} v_k \end{aligned} \tag{8.98}$$

where we have used the expression for γ_k that we found in (6.109), which we have already seen is the same for multibody systems.

We rewrite Eq. (8.85) by substituting Eqs. (8.87), (8.91) and (8.98) into (8.85) and get

$$\begin{aligned} M(x)\dot{v} + \dot{M}(x)v - \frac{1}{2}S^{-\mathsf{T}}(x)\frac{\partial^{\mathsf{T}} M(x)v}{\partial x}v \\ + \Big(\sum_k \gamma_k v_k\Big) M(x)v + S^{-\mathsf{T}}(x)\frac{\partial \mathscr{U}(x)}{\partial x} = \tau \\ M\dot{v} + \sum_{k,l} \frac{\partial M_{ij}}{\partial x_l} S_{lk}^{-1} v_k - \frac{1}{2}\sum_{k,l} S_{li}^{-1} \frac{\partial M_{jk}}{\partial x_l} v_k \\ + \sum_k \Big(\sum_{l,m,s} S_{li}^{-1}\Big(\frac{\partial S_{sl}}{\partial x_m} - \frac{\partial S_{sm}}{\partial x_l}\Big) S_{mk}^{-1}\Big)_{is} M_{sj} v_k + S^{-\mathsf{T}}\frac{\partial \mathscr{U}}{\partial x} = \tau \end{aligned} \tag{8.99}$$

which defines the matrices in the theorem. □

8.3.1 Local Parameterization

In Chap. 6 we found the dynamics of a single rigid body in terms of the state variables $Q = g_{0b}$ for position and $v = V_{0b}^B$ for velocity. In Chap. 7 we extended this formulation to multibody systems, but only with Euclidean joints. In this section we will use the same formulation but also allow for non-Euclidean transformations in multibody systems. This allows us to derive the dynamics of vehicle-manipulator systems in terms of the well-defined state variables $Q = \{g_{0b}, q\}$ for position and $v = \{V_{0b}^B, \dot{q}\}$ for velocity.

We see that the definition of the state variables adopted in this section is more general than in the previous sections because we do not force the state variables into a vector form. For position we write the state space in terms of the global configuration states $Q = \{Q_b, Q_1, Q_2, \dots\}$ where Q_i denotes the configuration of rigid body i. The vehicle configuration state Q_b is then a matrix representation of $SE(3)$ or one of its subgroups. For the links in the robotic manipulator we can stack the joint positions in a vector q in the normal way, i.e., $Q_i = q_i \in \mathbb{R}$.

We can also write the velocity state space as $v = \{v_b, v_1, v_2, \dots\}$ where v_i is the velocity variable of rigid body i. For the velocity state we can choose the twist representation for the vehicle velocity and the time derivative of the joint positions as the robot velocity. Because both can be written in vector form we can write the velocity state as $v = \zeta = [(V_{0b}^B)^{\mathsf{T}}\ \dot{q}^{\mathsf{T}}]^{\mathsf{T}}$.

The Lagrangian is given in terms of the velocity variable v and the configuration-dependent inertia matrix $M(Q)$ as

$$L(Q, v) = \frac{1}{2}v^{\mathsf{T}} M(Q)v - \mathscr{U}(Q). \tag{8.100}$$

In the previous section we found the dynamics of a very general multibody system with state variables x and v. We also eliminated the time derivative of the position variable $\dot{x}$ from the equations. We did this by introducing the velocity transformation matrix $S(x)$. However, as we have seen several times, these kinds of transformations are not well-defined if we use the Euler angles to describe the orientation of one or more of the transformations described by x. In this section we use local state variables to eliminate these singularities. More details on the general idea applied in this section can be found in Sect. 6.4.2.

Recall that locally the position variables can be written in terms of the exponential map as $\Phi(Q,\varphi)$ where φ is the local position variables in the vicinity of Q.

We can re-write the Lagrangian using the new variables φ and v as

$$L(\varphi, v) = \frac{1}{2} v^{\mathsf{T}} M\big(\Phi(Q,\varphi)\big)v - \mathcal{U}\big(\Phi(Q,\varphi)\big). \tag{8.101}$$

The partial derivatives of the Lagrangian then become

$$\frac{\mathrm{d}}{\mathrm{d}t}\left(\frac{\partial L}{\partial v}\right) = M\big(\Phi(Q,\varphi)\big)\dot{v} + \dot{M}\big(\Phi(Q,\varphi)\big)v \tag{8.102}$$

$$\frac{\partial L}{\partial \varphi} = \frac{1}{2}\frac{\partial^{\mathsf{T}}(M(\Phi(Q,\varphi))v)}{\partial \varphi}v - \frac{\partial \mathcal{U}(\varphi)}{\partial \varphi}. \tag{8.103}$$

We consider φ and $\dot{\varphi}$ as variables and would like to differentiate with respect to these, so we need the partial derivatives of the Lagrangian $\bar{L}_{\varphi}(\varphi,\dot{\varphi})$ expressed in terms of the local coordinates φ and $\dot{\varphi}$. $\bar{L}_{\varphi}(\varphi,\dot{\varphi})$ can be written as

$$\bar{L}_{\varphi}(\varphi,\dot{\varphi}) = \frac{1}{2}\dot{\varphi}^{\mathsf{T}} S^{\mathsf{T}}(Q,\varphi) M\big(\Phi(Q,\varphi)\big) S(Q,\varphi)\dot{\varphi} - \mathcal{U}\big(\Phi(Q,\varphi)\big). \tag{8.104}$$

From Proposition 8.1 we find the partial derivatives with respect to φ and $\dot{\varphi}$ as

$$\frac{\mathrm{d}}{\mathrm{d}t}\left(\frac{\partial \bar{L}_{\varphi}}{\partial \dot{\varphi}}\right) = \dot{S}^{\mathsf{T}}(Q,\varphi)\frac{\partial L}{\partial v} + S^{\mathsf{T}}(Q,\varphi)\frac{\mathrm{d}}{\mathrm{d}t}\left(\frac{\partial L}{\partial v}\right) \tag{8.105}$$

$$\frac{\partial \bar{L}_{\varphi}}{\partial \varphi} = \frac{\partial L}{\partial \varphi} + \frac{\partial (S(Q,\varphi)\dot{\varphi})}{\partial \varphi}\frac{\partial L}{\partial v}. \tag{8.106}$$

Substituting this into Lagrange's equations

$$\frac{\mathrm{d}}{\mathrm{d}t}\left(\frac{\partial \bar{L}_\varphi}{\partial \dot{\varphi}}\right) - \frac{\partial \bar{L}_\varphi}{\partial \varphi} = B(\varphi)\tau \tag{8.107}$$

gives us the equations of motion in terms of local position and velocity variables φ and $\dot{\varphi}$. To obtain the dynamics in terms of the global position and velocity variables Q and v we use the general idea presented in Sect. 6.4.2: we use the relation $v = S(Q, \varphi)\dot{\varphi}$ to eliminate the local velocity variables and eliminate the local position variables φ by representing the position as $\Phi(\bar{Q}, \varphi)$ which, after differentiating and evaluating at $\varphi = 0$, takes us back to Q through the relation $\Phi(\bar{Q}, 0) = \bar{Q}$. We then treat $\bar{Q}$ as a variable and get the dynamics in terms of the desired variables Q and v. This gives us the following important result:

Theorem 8.5 *Consider a general multibody system with local position and velocity coordinates φ and $\dot{\varphi}$ and global position and velocity coordinates Q and v. Write the kinetic energy as $\mathscr{K}(v) = \frac{1}{2}v^{\mathsf{T}}M(Q)v$ with the inertia matrix $M(Q)$. The dynamics of this system then satisfies*

$$M(Q)\dot{v} + C(Q, v)v + N(Q) = \tau \tag{8.108}$$

where M is found in the normal way, with τ the vector of external and control wrenches (collocated with v), the matrix describing the Coriolis and centrifugal forces given by

$$C_{ij}(x, v) = \sum_{l,k}\left(\frac{\partial M_{ij}}{\partial \varphi_l}S_{lk}^{-1} - \frac{1}{2}S_{li}^{-1}\frac{\partial M_{jk}}{\partial \varphi_l}\right)\bigg|_{\varphi=0} v_k$$
$$+ \sum_{l,m,k,s}\left(S_{li}^{-1}\left(\frac{\partial S_{sl}}{\partial \varphi_m} - \frac{\partial S_{sm}}{\partial \varphi_l}\right)S_{mk}^{-1}M_{sj}\right)\bigg|_{\varphi=0} v_k, \tag{8.109}$$

and the potential forces are given by

$$N(Q) = S^{-\mathsf{T}}\frac{\partial \mathscr{U}(\varphi)}{\partial \varphi}\bigg|_{\varphi=0}. \tag{8.110}$$

To compute the matrix $C(Q, v)$ for a single rigid body with configuration space $SE(3)$ or one of its subgroups, we can use (2.209) to simplify $C(Q, v)$ slightly to

$$C_{ij}(x, v) = \sum_{k}\left(\frac{\partial M_{ij}}{\partial \varphi_k} - \frac{1}{2}\frac{\partial M_{jk}}{\partial \varphi_i}\right)\bigg|_{\varphi=0} v_k$$
$$+\sum_{k,s}\left(\left(\frac{\partial S_{si}}{\partial \varphi_k} - \frac{\partial S_{sk}}{\partial \varphi_i}\right)M_{sj}\right)\bigg|_{\varphi=0} v_k. \tag{8.111}$$

Proof We will start by writing the first part of the Coriolis matrix in (8.94) as

$$\begin{aligned}\left(\sum_k \alpha_k v_k\right)_{ij}\bigg|_{\varphi=0} &= \sum_{k,l}\left(\frac{\partial M_{ij}}{\partial \varphi_l} S_{lk}^{-1} - \frac{1}{2} S_{li}^{-1}\frac{\partial M_{jk}}{\partial \varphi_l}\right) v_k\bigg|_{\varphi=0} \\ &= \sum_{k,l}\left(\frac{\partial M_{ij}}{\partial \varphi_l} S_{lk}^{-1} - \frac{1}{2} S_{li}^{-1}\frac{\partial M_{jk}}{\partial \varphi_l}\right)\bigg|_{\varphi=0} v_k. \end{aligned} \tag{8.112}$$

The second part of the Coriolis matrix in (8.94) can be written as

$$\begin{aligned}\sum_k \beta_k v_k &= \sum_s \left(\sum_k \gamma_k v_k\right)_{is} M_{sj}\bigg|_{\varphi=0} \\ &= \sum_s \sum_{l,m,k} S_{li}^{-1}\left(\frac{\partial S_{sl}}{\partial \varphi_m} - \frac{\partial S_{sm}}{\partial \varphi_l}\right) S_{mk}^{-1} v_k M_{sj}\bigg|_{\varphi=0} \\ &= \sum_k \sum_{l,m,s} S_{li}^{-1}\left(\frac{\partial S_{sl}}{\partial \varphi_m} - \frac{\partial S_{sm}}{\partial \varphi_l}\right) S_{mk}^{-1} M_{sj}\bigg|_{\varphi=0} v_k. \end{aligned}$$

If we also add the first and the second part of the Coriolis matrix in (8.94) we get the required expression for the Coriolis matrix of a general multibody mechanism as

$$\begin{aligned}C_{ij}(v) &= \sum_{k,l}\left(\frac{\partial M_{ij}}{\partial \varphi_l} S_{lk}^{-1} - \frac{1}{2} S_{li}^{-1}\frac{\partial M_{jk}}{\partial \varphi_l}\right)\bigg|_{\varphi=0} v_k \\ &\quad + \sum_{k,l,m,s} S_{li}^{-1}\left(\frac{\partial S_{sl}}{\partial \varphi_m} - \frac{\partial S_{sm}}{\partial \varphi_l}\right) S_{mk}^{-1} M_{sj}\bigg|_{\varphi=0} v_k. \end{aligned} \tag{8.113}$$

Now, recall that the velocity transformation matrix can be written as

$$S(\varphi) = \left(I - \frac{1}{2}\mathrm{ad}_\varphi + \frac{1}{6}\mathrm{ad}_\varphi^2 - \cdots\right) \in \mathbb{R}^{m\times m} \tag{8.114}$$

where ad_X is the adjoint map for a general Lie algebra X of dimension m. Because the expression in (8.113) is to be evaluated at $\varphi = 0$ this expression is non-zero only for the diagonal elements of S_{ij}, i.e., $i = j$. The final expressions then become

$$\begin{aligned}C_{ij}(v) &= \sum_{k}\left(\frac{\partial M_{ij}}{\partial \varphi_k} S_{kk}^{-1} - \frac{1}{2} S_{ii}^{-1}\frac{\partial M_{jk}}{\partial \varphi_i}\right)\bigg|_{\varphi=0} v_k \\ &\quad + \sum_{k,s} S_{ii}^{-1}\left(\frac{\partial S_{si}}{\partial \varphi_k} - \frac{\partial S_{sk}}{\partial \varphi_i}\right) S_{kk}^{-1} M_{sj}\bigg|_{\varphi=0} v_k \end{aligned}$$

$$= \sum_{k} \left(\frac{\partial M_{ij}}{\partial \varphi_k} - \frac{1}{2} \frac{\partial M_{jk}}{\partial \varphi_i} \right) \bigg|_{\varphi=0} v_k + \sum_{k,s} \left(\frac{\partial S_{si}}{\partial \varphi_k} - \frac{\partial S_{sk}}{\partial \varphi_i} \right) M_{sj} \bigg|_{\varphi=0} v_k. \tag{8.115}$$

□

We can further reduce the number of summation indexes by one by following the mathematics in (6.124) and keeping in mind that the inertia matrix is not constant, which gives

$$C_{ij}(v) = \sum_{k=1}^{m+n} \left(\frac{\partial M_{ij}}{\partial \varphi_k} - \frac{1}{2} \frac{\partial M_{jk}}{\partial \varphi_i} \right) \bigg|_{\varphi=0} v_k + \sum_{k=1}^{m+n} \left(\frac{\partial S_{ki}}{\partial \varphi_j} - \frac{\partial S_{kj}}{\partial \varphi_i} \right) (Mv)_k \bigg|_{\varphi=0}. \tag{8.116}$$

8.3.2 Dynamic Structure of Vehicle-Manipulator Systems

The dynamic equations as presented in the previous section are in a very general form in the sense that we only require that there exist some local position and velocity variables together with a mapping to the global state variables. To apply this general result to vehicle-manipulator systems in the form of Fig. 8.1, we write $Q = \{g_{0b}, q\}$ as the set of configuration states where g_{0b} is the Lie group $SE(3)$ or one of its sub-groups, and $v = [(\tilde{V}_{0b}^B)^{\mathsf{T}} \ \dot{q}^{\mathsf{T}}]^{\mathsf{T}}$ as the vector of velocity states. The local Euclidean structure for the state g_{0b} is given by exponential coordinates (Murray et al. 1994), while the state q is itself globally Euclidean. Mathematically, we can express configurations (g_{0b}, q) around a fixed state $(\bar{g}_{0b}, \bar{q})$ as (6.129)

$$g_{0b} = \bar{g}_{0b} e^{\hat{\varphi}}, \qquad q_i = \bar{q}_i + \varphi_i \quad \forall i \in \{1, \dots, n\}, \tag{8.117}$$

where $\hat{\varphi}$ is the matrix representation of φ which, because the local position coordinates live on the same space as the Lie algebra elements, is the same as the matrix representation of a Lie algebra element on $se(3)$ or one of its subalgebras. When $m < 6$ we set $\varphi_i = 0$ for all the $6 - m$ entries that are trivially zero, corresponding to Eq. (5.8).

We can use this local structure to find the mapping from the local to the global velocity variables and in the next step to find the explicit dynamic equations of vehicle-manipulator systems. Particularly, we will find that for this kind of systems we are able to write the dynamics in a very simple form.

We will write the dynamics of vehicle-manipulator systems in block-form as

$$\begin{bmatrix} M_{VV} & M_{qV}^{\mathsf{T}} \\ M_{qV} & M_q \end{bmatrix} \begin{bmatrix} \dot{\tilde{V}}_{0b}^B \\ \ddot{q} \end{bmatrix} + \begin{bmatrix} C_{VV} & C_{Vq} \\ C_{qV} & C_q \end{bmatrix} \begin{bmatrix} \tilde{V}_{0b}^B \\ \dot{q} \end{bmatrix} = \begin{bmatrix} \tau_V \\ \tau_q \end{bmatrix} \tag{8.118}$$

and the corresponding kinematics as

$$\begin{bmatrix} \tilde{V}_{0b}^B \\ \dot{q} \end{bmatrix} = \begin{bmatrix} (I - \frac{1}{2}\hat{\varphi}_V + \frac{1}{6}\hat{\varphi}_V^2 - \cdots) & 0 \\ 0 & I \end{bmatrix} \begin{bmatrix} \dot{\varphi}_V \\ \dot{\varphi}_q \end{bmatrix}. \tag{8.119}$$

Here, τ_V is a wrench of control and external forces acting on the vehicle, expressed in coordinates $\mathscr{F}_b$ (such that it is collocated with $\tilde{V}_{0b}^B$) and τ_q is the robot control torque and external forces collocated with $\dot{q}$. As previously, the subscript V refers to the first m entries representing the vehicle and the subscript q to the last n entries of the state vector representing the manipulator arm.

When expressed as in the previous section, the expression for the Coriolis matrix $C(Q, v)$ is rather complex. We will identify some important properties and observations regarding dynamic systems in the form of (8.118) which will allow us to write the dynamics of these systems in a much simpler form:

Property 8.1 The inertia matrix $M(Q) = M(q)$ is independent of g_{0b}.

We learned in Sect. 8.1 that the inertia matrix does not depend on the position of the vehicle frame $\mathscr{F}_b$ with respect to the inertial frame $\mathscr{F}_0$. As a result, the partial derivative of $M(\varphi)$ with respect to φ_V is zero.

Property 8.2 The velocity transformation matrix $S(Q) = S(g_{0b})$ is independent of q.

This property is a result of the robotic joints being Euclidean. As a consequence, the partial derivatives of $S(Q, \varphi)$ with respect to φ_q are zero.

Property 8.3 $S(Q, 0) = I$.

This is easily observed from (8.119). We have already used this observation in the proof of Theorem 8.5 to simplify the expressions. Furthermore, we see that for our systems $S(Q, \varphi)$ does in fact only depend on the local position variables and not the global variables Q. For vehicle-manipulator systems in this form we can therefore write $S(\varphi)$.

8.3.3 The Most Important Configuration Spaces

Based on the structural properties of the dynamic equations that we found in the previous section we can find the explicit dynamic equations for vehicle-manipulator systems for vehicles with different configuration spaces. We have already seen that the inertia matrix can be written as

$$M(q) = \begin{bmatrix} M_V & M_{qV}^{\mathsf{T}} \\ M_{qV} & M_q \end{bmatrix} = \sum_{i=b}^{n} \begin{bmatrix} H^{\mathsf{T}}\mathrm{Ad}_{g_{ib}}^{\mathsf{T}} I_i \mathrm{Ad}_{g_{ib}} H & H^{\mathsf{T}}\mathrm{Ad}_{g_{ib}}^{\mathsf{T}} I_i \mathrm{Ad}_{g_{ib}} J_i \\ J_i^{\mathsf{T}}\mathrm{Ad}_{g_{ib}}^{\mathsf{T}} I_i \mathrm{Ad}_{g_{ib}} H & J_i^{\mathsf{T}}\mathrm{Ad}_{g_{ib}}^{\mathsf{T}} I_i \mathrm{Ad}_{g_{ib}} J_i \end{bmatrix} \tag{8.120}$$

where the selection matrix H guarantees that the inertia matrix has the right dimensions, i.e., $M(q) \in \mathbb{R}^{(m+n)\times(m+n)}$. This formulation of the inertia matrix is both well defined and easy to implement. The main topic of this section is to find a similarly simple and explicit expression for the Coriolis matrix

$$C_{ij}(v) = \sum_{k=1}^{m+n} \left(\frac{\partial M_{ij}}{\partial \varphi_k} - \frac{1}{2}\frac{\partial M_{jk}}{\partial \varphi_i} \right)\bigg|_{\varphi=0} v_k + \sum_{k=1}^{m+n} \left(\frac{\partial S_{ki}}{\partial \varphi_j} - \frac{\partial S_{kj}}{\partial \varphi_i} \right)(Mv)_k \bigg|_{\varphi=0}. \tag{8.121}$$

To simplify these expressions we will use the structural properties of vehicle-manipulator systems that we identified in the previous section. We will therefore look at the different vehicle configuration spaces independently in the following sections.

8.3.3.1 Vehicles with Configuration Space $SO(3)$

The configuration space of a free-floating vehicle with one fixed point can be described by the matrix Lie group $SO(3)$. The velocity state is thus fully determined by only three variables and we choose H so that

$$V_{0b}^{B} = H\tilde{V}_{0b}^{B} \tag{8.122}$$

with $\tilde{V}_{0b}^{B} = \omega_{0b}^{B}$ and

$$H = \begin{bmatrix} 0_{3\times3} \\ I_{3\times3} \end{bmatrix}. \tag{8.123}$$

As we have seen, the velocity transformation matrix for a vehicle with configuration space $SO(3)$ is given by

$$\tilde{V}_{0b}^{B} = \left(I - \frac{1}{2}\hat{\varphi}_V + \frac{1}{6}\hat{\varphi}_V^2 - \cdots \right)\dot{\varphi}_V. \tag{8.124}$$

The velocity transformation matrix of the whole system is found by collecting the matrices S_i in one block-diagonal matrix S given by

$$S(Q,\varphi)=\begin{bmatrix}\left(I-\frac{1}{2}\hat{\varphi}_V+\frac{1}{6}\hat{\varphi}_V^2-\cdots\right) & 0\\ 0 & I\end{bmatrix}\in\mathbb{R}^{(3+n)\times(3+n)}. \tag{8.125}$$

Differentiating this expression with respect to φ and substituting $\varphi=0$ will simplify this expression substantially. We will start by looking at the first part of the Coriolis matrix in (8.121) given by

$$C_{ij,1}(q,v)=\sum_{k=1}^{3+n}\left(\frac{\partial M_{ij}}{\partial\varphi_k}-\frac{1}{2}\frac{\partial M_{jk}}{\partial\varphi_i}\right)\bigg|_{\varphi=0}v_k. \tag{8.126}$$

From Property 8.1 we see that the first part of this expression is non-zero only for $k>3$. The second part is non-zero only for $i>3$ which means that the upper part of second matrix is zero. The first part of the Coriolis matrix can therefore be written as

$$C_{ij,1}(q,v)=\sum_{k=4}^{3+n}\begin{bmatrix}\frac{\partial M_{VV}}{\partial\varphi_k} & \frac{\partial M_{qV}^{\mathrm{T}}}{\partial\varphi_k}\\ \frac{\partial M_{qV}}{\partial\varphi_k} & \frac{\partial M_q}{\partial\varphi_k}\end{bmatrix}\Bigg|_{\varphi=0}v_k-\frac{1}{2}\sum_{k=1}^{3+n}\begin{bmatrix}0 & 0\\ \frac{\partial M_{jk}}{\partial\varphi_i} & \frac{\partial M_{jk}}{\partial\varphi_i}\end{bmatrix}\Bigg|_{\varphi=0}v_k. \tag{8.127}$$

The first part of (8.127) is only to be differentiated with respect to the local position variables $(\varphi_4,\varphi_5,\dots,\varphi_{(3+n)})$. Note that the variables $(\varphi_4,\varphi_5,\dots,\varphi_{(3+n)})$ correspond only to the joint variables, and not the vehicle, so we can use the observation that the joint variables q are Euclidean and substitute these directly (see Sect. 7.6). This will thus eliminate the local velocity variables from the expression as we can write

$$\sum_{k=4}^{3+n}\begin{bmatrix}\frac{\partial M_{VV}}{\partial\varphi_k} & \frac{\partial M_{qV}^{\mathrm{T}}}{\partial\varphi_k}\\ \frac{\partial M_{qV}}{\partial\varphi_k} & \frac{\partial M_q}{\partial\varphi_k}\end{bmatrix}\Bigg|_{\varphi=0}v_k=\sum_{k=1}^{n}\begin{bmatrix}\frac{\partial M_{VV}}{\partial q_k} & \frac{\partial M_{qV}^{\mathrm{T}}}{\partial q_k}\\ \frac{\partial M_{qV}}{\partial q_k} & \frac{\partial M_q}{\partial q_k}\end{bmatrix}\dot{q}_k \tag{8.128}$$

because the variable φ_{m+i} (the position variable of the VM system) corresponds to position variable q_i of the robot.

We will re-write the second matrix in (8.127) as

$$\sum_{k=1}^{3+n}\begin{bmatrix} 0 & 0 \\ \frac{\partial M_{jk}}{\partial \varphi_i} & \frac{\partial M_{jk}}{\partial \varphi_i}\end{bmatrix} v_k = \begin{bmatrix} 0 & 0 & \cdots & 0 \\ \vdots & \vdots & \ddots & \vdots \\ 0 & 0 & \cdots & 0 \\ \sum_k \frac{\partial M_{1k}}{\partial \varphi_4} v_k & \sum_k \frac{\partial M_{2k}}{\partial \varphi_4} v_k & \cdots & \sum_k \frac{\partial M_{(3+n)k}}{\partial \varphi_4} v_k \\ \sum_k \frac{\partial M_{1k}}{\partial \varphi_5} v_k & \sum_k \frac{\partial M_{2k}}{\partial \varphi_5} v_k & \cdots & \sum_k \frac{\partial M_{(3+n)k}}{\partial \varphi_5} v_k \\ \vdots & \vdots & \ddots & \vdots \\ \sum_k \frac{\partial M_{1k}}{\partial \varphi_{3+n}} v_k & \sum_k \frac{\partial M_{2k}}{\partial \varphi_{3+n}} v_k & \cdots & \sum_k \frac{\partial M_{(3+n)k}}{\partial \varphi_{3+n}} v_k \end{bmatrix}. \tag{8.129}$$

Also this expression is independent of the vehicle configuration, so we can substitute the global position variables q into the expressions directly:

$$\begin{aligned} \sum_{k=1}^{3+n}\begin{bmatrix} 0 & 0 \\ \frac{\partial M_{jk}}{\partial \varphi_i} & \frac{\partial M_{jk}}{\partial \varphi_i}\end{bmatrix}\Bigg|_{\varphi=0} & v_k \\ &= \begin{bmatrix} 0 & 0 \\ \sum_{k=1}^{3+n} \frac{\partial M_{jk}}{\partial q_i} v_k & \sum_{k=1}^{3+n} \frac{\partial M_{jk}}{\partial q_i} v_k \end{bmatrix} \\ &= \begin{bmatrix} 0 & 0 \\ \frac{\partial^{\mathsf{T}}}{\partial q}\left([M_{VV}\ M_{qV}^{\mathsf{T}}]\left[\begin{smallmatrix}\omega_{0b}^B \\ \dot{q}\end{smallmatrix}\right]\right) & \frac{\partial^{\mathsf{T}}}{\partial q}\left([M_{qV}\ M_{q}]\left[\begin{smallmatrix}\omega_{0b}^B \\ \dot{q}\end{smallmatrix}\right]\right) \end{bmatrix}. \end{aligned} \tag{8.130}$$

Similarly, the second part of the Coriolis matrix in (8.121) given by

$$C_{ij,2}(q,v) = \sum_{k=1}^{3+n}\left(\frac{\partial S_{ki}}{\partial \varphi_j} - \frac{\partial S_{kj}}{\partial \varphi_i}\right)(Mv)_k\Bigg|_{\varphi=0} \tag{8.131}$$

can be simplified by observing that $M(\varphi_q)$ is independent of φ_V. The inertia matrix is therefore written in terms of the position variables $\Phi(\bar{q},\varphi_q) = \bar{q} + \varphi_q$ where $\bar{q}$ is considered a parameter. However, after we differentiate and substitute $\varphi = 0$ we recover the global position variable q. Once again we have used the property that the joints are Euclidean. Thus, after evaluating at $\varphi = 0$ we get the inertia matrix $M(q)$ which does no longer depend on φ and we can write

$$C_{ij,2}(q,v) = \sum_{k=1}^{3+n}\left(\frac{\partial S_{ki}}{\partial \varphi_j} - \frac{\partial S_{kj}}{\partial \varphi_i}\right)\Bigg|_{\varphi=0}\big(M(q)v\big)_k. \tag{8.132}$$

From Property 8.2 and the fact that $S(Q,\varphi)$ is block diagonal, the partial derivative of $S(Q,\varphi)$ with respect to φ_q is zero and only the upper left part $S_{VV}(Q,\varphi)$ is non-zero when differentiated. More specifically, $\frac{\partial S_{ki}}{\partial \varphi_j}$ is non-zero only when $i,j,k \le 3$, which means that only the upper left part of the matrix is non-zero. The same is true for the second term $\frac{\partial S_{kj}}{\partial \varphi_i}$. The second part of the Coriolis matrix then becomes

$$C_{ij,2}(q,v) = \sum_{k=1}^{3} \begin{bmatrix} \frac{\partial S_{ki}}{\partial \varphi_j} & 0 \\ 0 & 0 \end{bmatrix} \Bigg|_{\varphi=0} \big(M(q)v\big)_k - \sum_{k=1}^{3} \begin{bmatrix} \frac{\partial S_{kj}}{\partial \varphi_i} & 0 \\ 0 & 0 \end{bmatrix} \Bigg|_{\varphi=0} \big(M(q)v\big)_k. \tag{8.133}$$

We first note from the summation indexes that the elements of (Mv) that correspond to the manipulator vanish, and only the first three elements are of importance. We denote the first three elements of this vector as $(Mv)_\omega$. We recognize this expression as the same that we found for single rigid bodes given by (6.138). We can therefore write

$$C_{VV,2}(q,v) = -\widehat{(Mv)}_\omega. \tag{8.134}$$

The final expression for the Coriolis matrix is now found by adding (8.128), (8.130), and (8.134) which gives

$$\begin{aligned} C(q,v) = {} & \sum_{k=1}^{n} \frac{\partial M}{\partial q_k} \dot{q}_k \\ & - \frac{1}{2} \begin{bmatrix} 2\widehat{(Mv)}_\omega & 0 \\ \frac{\partial^\mathsf{T}}{\partial q}\left([M_{VV}\ M_{qV}^\mathsf{T}] \left[\begin{smallmatrix} \omega_{0b}^B \\ \dot{q} \end{smallmatrix}\right] \right) & \frac{\partial^\mathsf{T}}{\partial q}\left([M_{qV}\ M_q] \left[\begin{smallmatrix} \omega_{0b}^B \\ \dot{q} \end{smallmatrix}\right] \right) \end{bmatrix}. \end{aligned} \tag{8.135}$$

We see that we have obtained the dynamic equations in terms of global state variables only. The local state variables are eliminated from the equations either by observing that the joint transformations are Euclidean or by writing the velocity transformation in terms of the adjoint map.

8.3.3.2 Vehicles with Configuration Space $SE(3)$

The dynamics of a vehicle-manipulator system for a vehicle with configuration space $SE(3)$ is derived in the same way. In this case we have the mapping

$$V_{0b}^B = \left(I - \frac{1}{2}\mathrm{ad}_{\varphi_V} + \frac{1}{6}\mathrm{ad}_{\varphi_V}^2 - \cdots \right) \dot{\varphi}_V \tag{8.136}$$

with $\mathrm{ad}_p = \left[\begin{smallmatrix} \hat{p}_{4\ldots6} & \hat{p}_{1\ldots3} \\ 0 & \hat{p}_{4\ldots6} \end{smallmatrix}\right] \in \mathbb{R}^{6\times6}$ for $p \in \mathbb{R}^6$ relating the local and global velocity variables, and $\tilde{V}_{0b}^B = V_{0b}^B$. The corresponding matrices S_i can be collected in one block-diagonal matrix S given by

$$S(Q,\varphi)=\begin{bmatrix}\left(I-\frac{1}{2}\mathrm{ad}_{\varphi_V}+\frac{1}{6}\mathrm{ad}^2_{\varphi_V}-\cdots\right) & 0\\ 0 & I\end{bmatrix}\in\mathbb{R}^{(6+n)\times(6+n)}. \tag{8.137}$$

The dynamics of a vehicle-manipulator system is given in the same form as in (8.135). Changing the summation indexes and using the results from Sect. 6.4.3.2 we can write the partial derivatives of the velocity transformation matrix as

$$C_{VV,2}(q,v)=-\widetilde{\mathrm{ad}}_{(Mv)_V}. \tag{8.138}$$

The final expression for the Coriolis matrix of a vehicle-manipulator system with a vehicle configuration space $SE(3)$ is given by

$$C(q,v)=\sum_{k=1}^{n}\frac{\partial M}{\partial q_k}\dot{q}_k - \frac{1}{2}\begin{bmatrix}2\widetilde{\mathrm{ad}}_{(Mv)_V} & 0\\ \frac{\partial^{\mathrm{T}}}{\partial q}\left([M_{VV}\ M_{qV}^{\mathrm{T}}]\left[\begin{smallmatrix}V_{0b}^B\\ \dot{q}\end{smallmatrix}\right]\right) & \frac{\partial^{\mathrm{T}}}{\partial q}\left([M_{qV}\ M_q]\left[\begin{smallmatrix}V_{0b}^B\\ \dot{q}\end{smallmatrix}\right]\right)\end{bmatrix}. \tag{8.139}$$

8.3.3.3 Vehicles with Configuration Space $\mathbb{R}^3$

We will now look at what the dynamics looks like when the vehicle configuration space is Euclidean. In this case we can consider the vehicle-manipulator system as one serial robot. For pure translational motion, for example, we model the first three joints that represent the vehicle as prismatic joints. The Coriolis matrix is then given by (7.70) and as long as the mathematical model is concerned, there is no difference between a serial manipulator and a vehicle-manipulator system of this kind. The Coriolis matrix in (8.121) can therefore be simplified to

$$C_{ij}(v)=\sum_{k=1}^{m+n}\left(\frac{\partial M_{ij}}{\partial q_k}-\frac{1}{2}\frac{\partial M_{jk}}{\partial q_i}\right)\dot{q}_k \tag{8.140}$$

which is the same that we found for fixed-base manipulators. There is one important difference between an $(m+n)$-dimensional serial manipulator and an n-dimensional manipulator mounted on a vehicle with m degrees of freedom. This is clear because changing the location g_{0b} of the vehicle in the inertial frame will not change the dynamic properties of the VM system (it does not affect the inertia matrix), but changing the position of any of the first joints (representing the vehicle motion) in a serial manipulator will. We solve this by collecting the mass of the vehicle in the last vehicle link (link m). For example, if we want to represent only translational motion by using three prismatic joints, we set the mass and inertia of the two first links to zero ($I_1 = I_2 = 0$) and the last vehicle link to $I_3 = I_b$. Here, I_b is the inertial properties of the vehicle.

We see that there are two apparent disadvantages with this approach. The first is that we need to adapt this somewhat artificial solution of eliminating the inertial properties of all the links representing the vehicle, but the last. This is necessary to make sure that changing any of the first joint variables $q_1, \dots, q_m$ will not affect the inertia matrix (recall that the inertia matrix is independent of the vehicle position and orientation). Secondly, modeling non-Euclidean configuration spaces by using 1-parameter subgroups may lead to singularities. This is for example the case if we represent $SO(3)$ by three revolute joints. All these drawbacks may suggest that we are better off modeling these systems using the framework for vehicle-manipulator systems that we have developed in this chapter.

We can consider the vehicle and the manipulator as two different systems, as we did for vehicles with non-Euclidean configuration spaces. The dynamics can then be written in a very simple form because the velocity transformation matrix S is identical to the identity matrix. Because the velocity transformation matrix is constant we get

$$C_{VV,2}(q, v) = 0. \tag{8.141}$$

The final expression for the Coriolis matrix of a vehicle-manipulator system with a vehicle configuration space $\mathbb{R}^3$ is given by

$$C(q, v) = \sum_{k=1}^{n} \frac{\partial M}{\partial q_k} \dot{q}_k - \frac{1}{2} \begin{bmatrix} 0 & 0 \\ \frac{\partial^{\mathsf{T}}}{\partial q}\left(\begin{bmatrix} M_{VV} & M_{qV}^{\mathsf{T}} \end{bmatrix} \begin{bmatrix} v_{0b}^{B} \\ \dot{q} \end{bmatrix}\right) & \frac{\partial^{\mathsf{T}}}{\partial q}\left(\begin{bmatrix} M_{qV} & M_{q} \end{bmatrix} \begin{bmatrix} v_{0b}^{B} \\ \dot{q} \end{bmatrix}\right) \end{bmatrix} \tag{8.142}$$

which is of course just a reformulation of (8.140).

8.3.3.4 Summary

We see that the dynamics of vehicle-manipulator systems has the same form regardless of the configuration space of the vehicle. The Coriolis matrix is given by

Table 8.1 The Coriolis matrices for different Lie subgroups of $SE(3)$ and $SE(3)$ itself

Lie Group	S_{VV}	C
$SE(3)$	$I - \frac{1}{2}\mathrm{ad}_{\varphi_V} + \frac{1}{6}\mathrm{ad}^2_{\varphi_V} - \cdots$	$\sum_{k=1}^n \frac{\partial M}{\partial q_k}\dot{q}_k - \frac{1}{2}\begin{bmatrix} 2\widetilde{\mathrm{ad}}_{(Mv)_V} & 0 \\ A & B \end{bmatrix}$
$\mathscr{X}(z)$	$I - \frac{1}{2}\mathrm{ad}_{\varphi_X} + \frac{1}{6}\mathrm{ad}^2_{\varphi_X} - \cdots$	$\sum_{k=1}^n \frac{\partial M}{\partial q_k}\dot{q}_k - \frac{1}{2}\begin{bmatrix} 2\widetilde{\mathrm{ad}}_{(Mv)_X} & 0 \\ A & B \end{bmatrix}$
$\mathbb{R}^3$	$I_{3\times 3}$	$\sum_{k=1}^n \frac{\partial M}{\partial q_k}\dot{q}_k - \frac{1}{2}\begin{bmatrix} 0 & 0 \\ A & B \end{bmatrix}$
$SE(2)$	$I - \frac{1}{2}\mathrm{ad}_{\varphi_V} + \frac{1}{6}\mathrm{ad}^2_{\varphi_V} - \cdots$	$\sum_{k=1}^n \frac{\partial M}{\partial q_k}\dot{q}_k - \frac{1}{2}\begin{bmatrix} 2\widetilde{\mathrm{ad}}_{(Mv)_V} & 0 \\ A & B \end{bmatrix}$
$SO(3)$	$I - \frac{1}{2}\hat{\varphi}_V + \frac{1}{6}\hat{\varphi}_V^2 - \cdots$	$\sum_{k=1}^n \frac{\partial M}{\partial q_k}\dot{q}_k - \frac{1}{2}\begin{bmatrix} -2\widehat{(Mv)}_\omega & 0 \\ A & B \end{bmatrix}$
$\mathbb{R}^2, C$	$I_{2\times 2}$	$\sum_{k=1}^n \frac{\partial M}{\partial q_k}\dot{q}_k - \frac{1}{2}\begin{bmatrix} 0 & 0 \\ A & B \end{bmatrix}$
$\mathbb{R}, H, SO(2)$	$I_{1\times 1}$	$\sum_{k=1}^n \frac{\partial M}{\partial q_k}\dot{q}_k - \frac{1}{2}\begin{bmatrix} 0 & 0 \\ A & B \end{bmatrix}$
	$A = \frac{\partial^\mathsf{T}}{\partial q}\left(\begin{bmatrix} M_{VV} & M_{qV}^\mathsf{T} \end{bmatrix}\begin{bmatrix} \tilde{V}_{0b}^B \\ \dot{q} \end{bmatrix}\right)$	$B = \frac{\partial^\mathsf{T}}{\partial q}\left(\begin{bmatrix} M_{qV} & M_q \end{bmatrix}\begin{bmatrix} \tilde{V}_{0b}^B \\ \dot{q} \end{bmatrix}\right)$

$$C(q,v) = \sum_{k=1}^{n} \frac{\partial M}{\partial q_k}\dot{q}_k - \frac{1}{2}\begin{bmatrix} 2\widetilde{\mathrm{ad}}_{(Mv)_{\tilde{V}}} & 0 \\ \frac{\partial^\mathsf{T}}{\partial q}\left(\begin{bmatrix} M_{VV} & M_{qV}^\mathsf{T} \end{bmatrix}\begin{bmatrix} \tilde{V}_{0b}^B \\ \dot{q} \end{bmatrix}\right) & \frac{\partial^\mathsf{T}}{\partial q}\left(\begin{bmatrix} M_{qV} & M_q \end{bmatrix}\begin{bmatrix} \tilde{V}_{0b}^B \\ \dot{q} \end{bmatrix}\right) \end{bmatrix} \tag{8.143}$$

where the matrix $\widetilde{\mathrm{ad}}_{(Mv)_{\tilde{V}}}$ is the $m \times m$ adjoint map of $\tilde{V}$ where $\tilde{V}$ can be an element of $SE(3)$ or one of the subgroups. The matrices $\widetilde{\mathrm{ad}}_{(Mv)_{\tilde{V}}}$ were found in Chap. 6 and are the same for VM systems as for single rigid bodies. Table 8.1 shows the mapping from local to global velocity coordinates and the corresponding C-matrices for different Lie Groups.

8.3.4 Examples

To get a better understanding of what the dynamic equations look like and how to compute these for different vehicle-manipulator systems we include a few examples.

Example 8.1 Consider the general structure of the equations for a mechanism with one joint with joint variable q_1 mounted on a vehicle with configuration space $SE(3)$ as shown in Fig. 8.2. The Jacobian J_i for the first (and only) joint is then given by

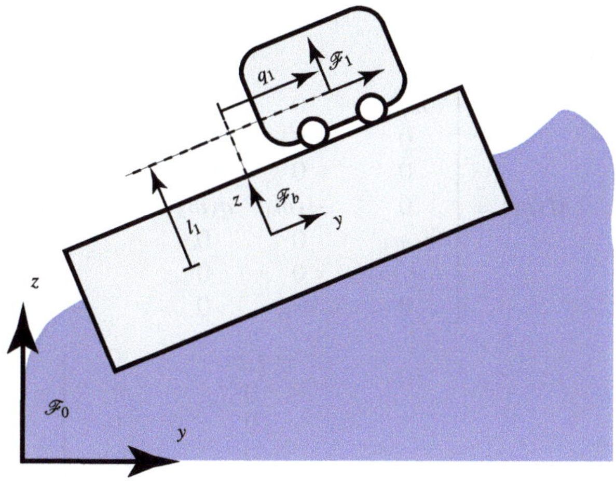

Fig. 8.2 One-link robot with one prismatic joint attached to a vehicle

the joint twist as

$$J_1 = X_1^1 = [0 \quad 1 \quad 0 \quad 0 \quad 0 \quad 0]^\mathsf{T}. \tag{8.144}$$

We can write the inertia matrix (8.120) as follows

$$M(q_1) = \begin{bmatrix} I_b + \mathrm{Ad}_{g_{1b}}^\mathsf{T} I_1 \mathrm{Ad}_{g_{1b}} & \mathrm{Ad}_{g_{1b}}^\mathsf{T} I_1 \mathrm{Ad}_{g_{1b}} X_1 \\ X_1^\mathsf{T} \mathrm{Ad}_{g_{1b}}^\mathsf{T} I_1 \mathrm{Ad}_{g_{1b}} & X_1^\mathsf{T} \mathrm{Ad}_{g_{1b}}^\mathsf{T} I_1 \mathrm{Ad}_{g_{1b}} X_1 \end{bmatrix}. \tag{8.145}$$

To find the explicit equations we write $I_b = \mathrm{diag}(m_b, m_b, m_b, I_{b,x}, I_{b,y}, I_{b,z})$ and $I_1 = \mathrm{diag}(m_1, m_1, m_1, I_{1,x}, I_{1,y}, I_{1,z})$, and the Adjoint map as

$$\mathrm{Ad}_{g_{1b}} = \begin{bmatrix} R_{1b} & \hat{p}_{1b} R_{1b} \\ 0 & R_{1b} \end{bmatrix} = \begin{bmatrix} 1 & 0 & 0 & 0 & l_1 & -q_1 \\ 0 & 1 & 0 & -l_1 & 0 & 0 \\ 0 & 0 & 1 & q_1 & 0 & 0 \\ 0 & 0 & 0 & 1 & 0 & 0 \\ 0 & 0 & 0 & 0 & 1 & 0 \\ 0 & 0 & 0 & 0 & 0 & 1 \end{bmatrix}. \tag{8.146}$$

We need to find the expression for the inertia matrix in the base frame

$$\mathrm{Ad}_{g_{1b}}^\mathsf{T} I_1 \mathrm{Ad}_{g_{1b}} = \begin{bmatrix} m_1 & 0 & 0 & 0 & l_1 m_1 & -q_1 m_1 \\ 0 & m_1 & 0 & -l_1 m_1 & 0 & 0 \\ 0 & 0 & m_1 & q_1 m_1 & 0 & 0 \\ 0 & -l_1 m_1 & q_1 m_1 & (l_1^2 + q_1^2) m_1 + I_{1,x} & 0 & 0 \\ l_1 m_1 & 0 & 0 & 0 & l_1^2 m_1 + I_{1,y} & -q_1 l_1 m_1 \\ -q_1 m_1 & 0 & 0 & 0 & -q_1 l_1 m_1 & q_1^2 m_1 + I_{1,z} \end{bmatrix}. \tag{8.147}$$

We can now write $\mathrm{Ad}_{g_{1b}}^{\mathsf{T}} I_1 \mathrm{Ad}_{g_{1b}} X_1 = [0\ m_1\ 0\ -l_1 m_1\ 0\ 0]^{\mathsf{T}}$ and find the inertia matrix as

$$M(q_1) = \begin{bmatrix} m_{b1} & 0 & 0 & 0 & l_1 m_1 & -q_1 m_1 & 0 \\ 0 & m_{b1} & 0 & -l_1 m_1 & 0 & 0 & m_1 \\ 0 & 0 & m_{b1} & q_1 m_1 & 0 & 0 & 0 \\ 0 & -l_1 m_1 & q_1 m_1 & (l_1^2 + q_1^2) m_1 + I_{1b,x} & 0 & 0 & -l_1 m_1 \\ l_1 m_1 & 0 & 0 & 0 & l_1^2 m_1 + I_{1b,y} & -q_1 l_1 m_1 & 0 \\ -q_1 m_1 & 0 & 0 & 0 & -q_1 l_1 m_1 & q_1^2 m_1 + I_{1b,z} & 0 \\ 0 & m_1 & 0 & -l_1 m_1 & 0 & 0 & m_1 \end{bmatrix} \tag{8.148}$$

where $m_{b1} = m_b + m_1$ and $I_{b1,x} = I_{b,x} + I_{1,x}$, etc.

Its partial derivative with respect to q is given as

$$\frac{\partial M(q_1)}{\partial q_1} = \begin{bmatrix} I \\ X_1^{\mathsf{T}} \end{bmatrix} \left[\frac{\partial^{\mathsf{T}} \mathrm{Ad}_{g_{1b}}}{\partial q_1} I_1 \mathrm{Ad}_{g_{1b}} + \mathrm{Ad}_{g_{1b}}^{\mathsf{T}} I_1 \frac{\partial \mathrm{Ad}_{g_{1b}}}{\partial q_1} \right] [I \quad X_1]. \tag{8.149}$$

We can calculate $\frac{\partial \mathrm{Ad}_{g_{1b}}}{\partial q_1}$ by

$$\frac{\partial g_{1b}}{\partial q_1} = -g_{1b} \hat{X}_1 g_{bb} = -g_{1b} \hat{X}_1 \tag{8.150}$$

$$= -\begin{bmatrix} 1 & 0 & 0 & 0 \\ 0 & 1 & 0 & -q_1 \\ 0 & 0 & 1 & -l_1 \\ 0 & 0 & 0 & 1 \end{bmatrix} \begin{bmatrix} 0 & 0 & 0 & 0 \\ 0 & 0 & 0 & 1 \\ 0 & 0 & 0 & 0 \\ 0 & 0 & 0 & 0 \end{bmatrix} = \begin{bmatrix} 0 & 0 & 0 & 0 \\ 0 & 0 & 0 & -1 \\ 0 & 0 & 0 & 0 \\ 0 & 0 & 0 & 0 \end{bmatrix} \tag{8.151}$$

which gives

$$\frac{\partial \mathrm{Ad}_{g_{1b}}}{\partial q_1} = \begin{bmatrix} 0 & 0 & 0 & 0 & 0 & -1 \\ 0 & 0 & 0 & 0 & 0 & 0 \\ 0 & 0 & 0 & 1 & 0 & 0 \\ 0 & 0 & 0 & 0 & 0 & 0 \\ 0 & 0 & 0 & 0 & 0 & 0 \\ 0 & 0 & 0 & 0 & 0 & 0 \end{bmatrix} \tag{8.152}$$

which we, for this simple example, can see directly from (8.146). The way that we compute $\frac{\partial g_{1b}}{\partial q_1}$ in (8.150) is very simple and effective when implementing this in a simulation software. Note that the Jacobian matrix is constant and hence no partial derivatives are taken.

The Coriolis terms are given by

$$C(q,v) = \frac{\partial M}{\partial q_1}\dot{q}_1 - \frac{1}{2}\begin{bmatrix} 2\widetilde{\mathrm{ad}}_{(Mv)_v} & 0 \\ \frac{\partial^{\mathsf{T}}}{\partial q_1}([M_{VV}\ M_{qV}^{\mathsf{T}}][\begin{smallmatrix} V_{0b}^B \\ \dot{q}_1 \end{smallmatrix}]) & \frac{\partial^{\mathsf{T}}}{\partial q_1}([M_{qV}\ M_q][\begin{smallmatrix} V_{0b}^B \\ \dot{q}_1 \end{smallmatrix}]) \end{bmatrix}. \tag{8.153}$$

We first write

$$(Mv) = \begin{bmatrix} m_{b1}u + l_1 m_1 q - q_1 m_1 r \\ m_{b1}v - l_1 m_1 p + m_1 \dot{q}_1 \\ m_{b1}w + q_1 m_1 p \\ -l_1 m_1 v + q_1 m_1 w + ((l_1^2 + q_1^2)m_1 + I_{1b,x})p - l_1 m_1 \dot{q}_1 \\ l_1 m_1 u + (l_1^2 m_1 + I_{1b,y})q - q_1 l_1 m_1 r \\ -q_1 m_1 u - q_1 l_1 m_1 q + (q_1^2 m_1 + I_{1b,z})r \\ m_1 v - l_1 m_1 p + m_1 \dot{q}_1 \end{bmatrix} \tag{8.154}$$

where $V_{0b}^B = [u\ v\ w\ p\ q\ r]^{\mathsf{T}}$. The partial derivatives now become

$$\frac{\partial}{\partial q_1}\left(M(q_1)\begin{bmatrix} V_{0b}^B \\ \dot{q}_1 \end{bmatrix}\right) = \begin{bmatrix} -m_1 r \\ 0 \\ m_1 p \\ m_1 w + 2q_1 m_1 p \\ -l_1 m_1 r \\ -m_1 u - l_1 m_1 q + 2q_1 m_1 r \\ 0 \end{bmatrix}. \tag{8.155}$$

Assume we are interested in the dynamics of the prismatic joint and that the motion of the base is given. The motion of the prismatic joint is given by the last row of the inertia and Coriolis matrix. The Coriolis matrix is given by adding the two matrices in (8.153) where the first part is zero (for the last row) and the second part is given by (8.155):

$$\begin{aligned} &C(q, V_{0b}^B, \dot{q}) \\ &= \begin{bmatrix} * & * & * & * & * & * & * \\ * & * & * & * & * & * & * \\ * & * & * & * & * & * & * \\ * & * & * & * & * & * & * \\ * & * & * & * & * & * & * \\ \frac{1}{2}m_1 r & 0 & -\frac{1}{2}m_1 p & -\frac{1}{2}m_1 w - q_1 m_1 p & \frac{1}{2}l_1 m_1 r & \frac{1}{2}(m_1 u + l_1 m_1 q) - q_1 m_1 r & 0 \end{bmatrix}. \end{aligned} \tag{8.156}$$

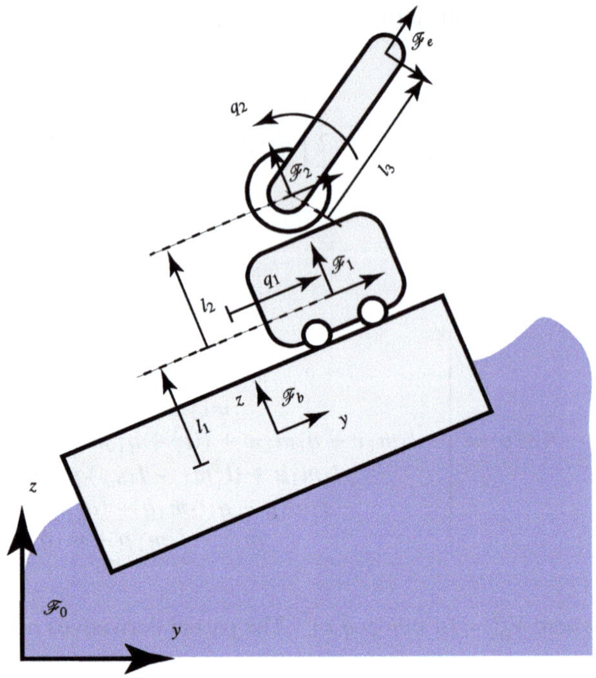

Fig. 8.3 Two-link robot with one prismatic and one revolute joint attached to a vehicle

Using these expressions, we can write the dynamics of the prismatic joint, assuming the motion of the vehicle known, as

$$\begin{aligned}
&[M_{qV} \quad M_q]\begin{bmatrix}\dot{V}_{0b}^B \\ \ddot{q}_1\end{bmatrix} + \begin{bmatrix}C_{qV} & C_q^{\mathrm{T}}\end{bmatrix}\begin{bmatrix}V_{0b}^B \\ \dot{q}_1\end{bmatrix} = \tau, \\
&m_1\ddot{q}_1 + m_1\dot{v} - l_1 m_1 \dot{p} + \frac{1}{2}m_1 ru - \frac{1}{2}m_1 pw - \frac{1}{2}m_1 wp \\
&\quad - q_1 m_1 p^2 + \frac{1}{2}l_1 m_1 rq + \frac{1}{2}(m_1 u + l_1 m_1 q)r - q_1 m_1 r^2 = \tau, \\
&\ddot{q}_1 + \dot{v} - l_1\dot{p} + ru + l_1 q_1 r - pw - q_1\left(p^2 + r^2\right) = \frac{\tau}{m_1}.
\end{aligned} \tag{8.157}$$

Example 8.2 We will now look at a two joint mechanism mounted on a vehicle that rotates around a fixed point, as illustrated in Fig. 8.3. This can for example be used to model a 1-DoF robotic arm mounted on a rail on a satellite, a concept similar to the one on the International Space Station. We will assume that the first joint is prismatic and that the second joint is rotational with twists

$$X_1^1 = [0 \quad 1 \quad 0 \quad 0 \quad 0 \quad 0]^{\mathrm{T}}, \tag{8.158}$$

$$X_2^2 = [0 \quad 0 \quad 0 \quad 1 \quad 0 \quad 0]^{\mathrm{T}}, \tag{8.159}$$

respectively. The Adjoint map $\mathrm{Ad}_{g_{b\bar{2}}}$ is given by

$$\mathrm{Ad}_{g_{b\bar{2}}} = \begin{bmatrix} R_{b\bar{2}} & \hat{p}_{b\bar{2}} R_{b\bar{2}} \\ 0 & R_{b\bar{2}} \end{bmatrix}$$

$$= \begin{bmatrix} 1 & 0 & 0 & 0 & -l_{12}\cos q_2 + q_1 \sin q_2 & l_{12}\sin q_2 + q_1 \cos q_2 \\ 0 & \cos q_2 & -\sin q_2 & l_{12} & 0 & 0 \\ 0 & \sin q_2 & \cos q_2 & -q_1 & 0 & 0 \\ 0 & 0 & 0 & 1 & 0 & 0 \\ 0 & 0 & 0 & 0 & \cos q_2 & -\sin q_2 \\ 0 & 0 & 0 & 0 & \sin q_2 & \cos q_2 \end{bmatrix} \tag{8.160}$$

which gives us the following Jacobians:

$$J_1 = \begin{bmatrix} X_1^1 & 0 \end{bmatrix} = \begin{bmatrix} 0 & 0 \\ 1 & 0 \\ 0 & 0 \\ 0 & 0 \\ 0 & 0 \\ 0 & 0 \end{bmatrix}, \qquad J_2 = \begin{bmatrix} X_1^1 & \mathrm{Ad}_{g_{0\bar{2}}} X_2^2 \end{bmatrix} = \begin{bmatrix} 0 & 0 \\ 1 & l_{12} \\ 0 & -q_1 \\ 0 & 1 \\ 0 & 0 \\ 0 & 0 \end{bmatrix}. \tag{8.161}$$

For a two-joint mechanism the inertia matrix (8.120) becomes

$$M(q_1, q_2) = \begin{bmatrix} I_b & 0 \\ 0 & 0 \end{bmatrix} + \begin{bmatrix} \mathrm{Ad}_{g_{1b}}^{\mathrm{T}} I_1 \mathrm{Ad}_{g_{1b}} & \mathrm{Ad}_{g_{1b}}^{\mathrm{T}} I_1 \mathrm{Ad}_{g_{1b}} J_1 \\ J_1^{\mathrm{T}} \mathrm{Ad}_{g_{1b}}^{\mathrm{T}} I_1 \mathrm{Ad}_{g_{1b}} & J_1^{\mathrm{T}} \mathrm{Ad}_{g_{1b}}^{\mathrm{T}} I_1 \mathrm{Ad}_{g_{1b}} J_1 \end{bmatrix}$$
$$+ \begin{bmatrix} \mathrm{Ad}_{g_{2b}}^{\mathrm{T}} I_2 \mathrm{Ad}_{g_{2b}} & \mathrm{Ad}_{g_{2b}}^{\mathrm{T}} I_2 \mathrm{Ad}_{g_{2b}} J_2 \\ J_2^{\mathrm{T}} \mathrm{Ad}_{g_{2b}}^{\mathrm{T}} I_2 \mathrm{Ad}_{g_{2b}} & J_2^{\mathrm{T}} \mathrm{Ad}_{g_{2b}}^{\mathrm{T}} I_2 \mathrm{Ad}_{g_{2b}} J_2 \end{bmatrix}. \tag{8.162}$$

The explicit expressions become rather lengthy, but are found in the same way as in the previous example.

References

Antonelli, G. (2006). *Underwater robots. Motion and force control of vehicle-manipulator systems*. Berlin: Springer.

Dubowsky, S., & Papadopoulos, E. (1993). The kinematics, dynamics and control of free-flying and free-floating space robotic systems. *IEEE Transactions on Robotics and Automation*, *9*(5), 531–541.

Egeland, O., & Pettersen, K. Y. (1998). Free-floating robotic systems. *Control Problems in Robotics and Automation*, *230*, 119–134.

Fossen, T. I. (2002). *Marine control systems*. Trondheim: Marine Cybernetics AS. 3rd printing.

From, P. J. (2012a). An explicit formulation of singularity-free dynamic equations of mechanical systems in Lagrangian form—part one: single rigid bodies. *Modeling, Identification and Control*, *33*(2), 45–60.

From, P. J. (2012b). An explicit formulation of singularity-free dynamic equations of mechanical systems in Lagrangian form—part two: multibody systems. *Modeling, Identification and Control*, *33*(2), 61–68.

From, P. J., Duindam, V., Gravdahl, J. T., & Sastry, S. (2009). Modeling and motion planning for mechanisms on a non-inertial base. In *Proceedings of international conference of robotics and automation*, Kobe, Japan (pp. 3320–3326).

From, P. J., Duindam, V., Pettersen, K. Y., Gravdahl, J. T., & Sastry, S. (2010a). Singularity-free dynamic equations of vehicle-manipulator systems. *Simulation Modelling Practice and Theory*, *18*(6), 712–731.

From, P. J., Duindam, V., & Stramigioli, S. (2012a). Corrections to "Singularity-free dynamic equations of open-chain mechanisms with general holonomic and nonholonomic joints". *IEEE Transactions on Robotics*, *28*(6), 1431–1432.

From, P. J., Pettersen, K. Y., & Gravdahl, J. T. (2010b). Singularity-free dynamic equations of AUV-manipulator systems. In *Proceeding of symposium on intelligent autonomous vehicles*, Lecce, Italy.

From, P. J., Pettersen, K. Y., & Gravdahl, J. T. (2011a). Singularity-free dynamic equations of spacecraft-manipulator systems. *Acta Astronautica*, *69*(11–12), 1057–1065.

From, P. J., Pettersen, K. Y., & Gravdahl, J. T. (2011b). Singularity-free formulation of the dynamically equivalent manipulator mapping for space manipulators. In *AIAA SPACE*, Long Beach, California, USA.

From, P. J., Schjølberg, I., Gravdahl, J. T., Pettersen, K. Y., & Fossen, T. I. (2010c). On the boundedness property of the inertia matrix and skew-symmetric property of the Coriolis matrix for vehicle-manipulator systems. In *Proceeding of symposium on intelligent autonomous vehicles*, Lecce, Italy.

From, P. J., Schjølberg, I., Gravdahl, J. T., Pettersen, K. Y., & Fossen, T. I. (2012b). On the boundedness and skew-symmetric properties of the inertia and Coriolis matrices for vehicle-manipulator systems. *Journal of Dynamic Systems, Measurement, and Control*, *134*(4), 2012.

Liang, B., Xu, Y., & Bergerman, M. (1998). Mapping a space manipulator to a dynamically equivalent manipulator. *ASME Journal of Dynamic Systems, Measurement, and Control*, *120*(1), 1–7.

Murray, R. M., Li, Z., & Sastry, S. S. (1994). *A mathematical introduction to robotic manipulation*. Boca Raton: CRC Press.

Schjølberg, I. (1996). *Modeling and control of underwater robotic systems*. Ph.D. thesis, Department of Engineering Cybernetics, Norwegian University of Science and Technology, Trondheim, Norway.

Chapter 9
Properties of the Dynamic Equations in Matrix Form

Most modern control laws take advantage of one or more structural properties of the system that is to be controlled. When the dynamics is written in matrix form, like they were in the previous chapters, we can define the properties of the dynamics as properties or restrictions imposed on the matrices. The structural properties on the dynamics is then cast into properties on the system matrices, such as skew-symmetry, boundedness, positive definiteness, and so on. Because control laws like model predictive control, robust control, and Lyapunov-based control all use the inertia and/or Coriolis matrix to compute the control signal and to prove stability, it is important to identify these properties to make sure that the mathematical model that we derive can be used for stability analysis and controller design.

The fist property that we will address is the boundedness of the inertia matrix $M(x)$. For a single rigid body the inertia matrix should always be bounded: an unbounded inertia matrix means that there are singularities present in the formulation. These are representation singularities that arise due to our choice of state variables and not the dynamic properties of the system. The inertia matrix can be looked upon as a mapping from the velocity space to the kinetic energy of the rigid body. This mapping should always be smooth, i.e., a small change in the velocity should result in a small change in the kinetic energy. If this is not the case and a small change in the velocity leads to a large change in the kinetic energy we know that this is due to a representation singularity and does not describe the actual behavior of the system. These singularities should therefore always be avoided.

For robotic manipulators on a fixed base or vehicle-manipulator systems there are two cases in which the inertia matrix is not bounded. The first is, as for single rigid bodies, when there are singularities in the representation. For a given robotic structure there may exist one mathematical representation for which the inertia matrix is bounded and another for which it is not. We will show that for the most commonly used mathematical representation of vehicle-manipulator systems this property does not hold. This is one of the main motivations behind the work that we did in the previous chapters on deriving the dynamics of these systems without the presence of representation singularities. We should always seek to represent the dynamics without this kind of singularities.

P.J. From et al., *Vehicle-Manipulator Systems*, Advances in Industrial Control,
DOI 10.1007/978-1-4471-5463-1_9, © Springer-Verlag London 2014

The second case is when the geometry of the robotic arm allows for singular configurations. In this case the problem cannot be solved by choosing a different set of variables because the singularities arise as a result of the physical properties of the manipulator. This is thus an intrinsic property of the robotic manipulator and these singularities can only be avoided if the geometric structure of the robot is modified. For robot manipulators the boundedness property is addressed in Ghorbel et al. (1998) where the class of robots for which the inertia matrix is bounded is characterized. The work of Ghorbel et al. (1998) addresses the geometric design of the manipulator. This is thus fundamentally different from the case when singularities arise as a result of the mathematical representation chosen. Singularities in VM systems is discussed in From et al. (2012, 2010a, 2010b, 2011).

The second property that we are concerned with is the parameterization of the Coriolis matrix $C(x, v)$: we want to find a parameterization so that the matrix $\dot{M} - 2C$ is skew-symmetric. This property is extensively used in Lyapunov-based control laws and is thus very important to identify. An important observation here is that this property is not intrinsic, i.e., it does not depend on the geometry of the system at hand, but on the choice of state variables. Such a parameterization is easy to find for fixed-base robots and for single rigid bodies, but not always for vehicle-manipulator systems or multibody systems with non-Euclidean joints.

9.1 Misconceptions in the Literature

Lyapunov-based controllers are based on several assumptions that make the controller design both more convenient and physically meaningful. These assumptions therefore need to reflect the physical properties of the system. This is important to keep in mind when choosing the mathematical formalism used to derive the dynamics. With the increasing popularity of Lyapunov design some of these properties are almost universally taken for granted, and the correct proofs are often left out. The reason for this is that both the boundedness property of the inertia matrix and the skew-symmetric property of the Coriolis matrix are relatively straight forward to prove for single rigid bodies and standard robotic manipulators with 1-DoF Euclidean joints; however, for vehicle-manipulator systems and general multibody systems with non-Euclidean joints, the proofs become more involved.

It seems that because these properties are fairly straight forward to show for some systems, they are simply taken for granted also for vehicle-manipulator systems: most work on this topic does not show these properties, and when referring to the proofs, references to the proofs for other systems like fixed-base manipulators are often used. As a result the otherwise rigorous stability proofs fall apart. To ensure that proofs are formally correct, it is important to prove the properties for the specific system at hand and for the specific parameterization chosen.

From et al. (2010c, 2012) stated a general concern that some frequently used properties of the inertia and Coriolis matrices for vehicle-manipulator systems are assumed true based on the proofs for other systems. It was shown that the proofs of these properties for fixed-base robot manipulators or single rigid bodies (vehicles) cannot be generalized to vehicle-manipulator systems directly. In fact, the most commonly used formulation of the dynamic equations for vehicle-manipulator systems do not possess both the boundedness and skew-symmetric properties. There is thus a need to clarify to what extent these properties are true and to find a rigorous mathematical representation of these systems for use in simulation and controller design.

In this section we will focus on two important classes of vehicle-manipulator systems—underwater manipulator systems and spacecraft-manipulator systems—but the results are general and also applicable to other vehicles. Underwater manipulator systems are extensively treated in Antonelli (2006), Schjølberg (1996), and Schjølberg and Fossen (1994). For the choice of state variables used in most of the literature, the boundedness property does not hold for the whole configuration space, i.e., there exist isolated points where the inertia matrix becomes singular. This can, however, be dealt with by introducing a non-minimal representation such as the unit quaternion. The quaternion representation is well suited for single rigid bodies, but for multibody systems the Euler angles are normally adopted. The reason for this is, as we have already seen, that the quaternions are not generalized coordinates and thus not suited for the Lagrangian formalism. The problems regarding the Euler angle singularities are pointed out in most books and papers when it comes to modeling, but is often left out when dealing with stability proofs. As a result of this the control laws presented are not valid for the whole configuration space.

Similarly, the skew-symmetric property of the Coriolis matrix is in general not treated correctly and is in most cases assumed true without any further proof. This is a strong weakness because this property depends on how we choose to represent the Coriolis matrix and it is not always trivial to find a representation for which this property holds. It is thus not sufficient to refer to an arbitrary proof of skew-symmetry: one must refer to a proof for the specific parameterization of the Coriolis matrix chosen. Most papers on the topic refer to Antonelli (2006), Fossen and Fjellstad (1995), de Wit et al. (1998) or Schjølberg and Fossen (1994) for this proof for VM systems. However, none of these references actually show the proof. Given the velocity state v, Schjølberg and Fossen (1994) state that $v^{\mathsf{T}}(\dot{M} - 2C)v = 0$, which is true, but a weaker result than skew-symmetry. This property is known as the *principle of conservation of energy* and is always true. This is often used to show skew-symmetry, which is not correct. Other commonly used references are taken from the fixed-base robotics literature, such as Murray et al. (1994) and Sciavicco and Siciliano (2005). None of these references show this proof—nor do they claim to do so—for vehicle-manipulator systems. The proof can be found in Schjølberg (1996), but only for systems where the boundedness property does not hold.

Spacecraft are normally modeled using quaternions and the inertia matrix is thus bounded for the whole configuration manifold (Wen and Kreutz-Delgado 1991). A Lie group formulation of the dynamics of a rigid body also leads to a well-defined set of equations and is studied in Bullo and Lewis (2000) and Marsden and Ratiu (1999). For spacecraft-manipulator systems, however, a Lagrangian approach is normally adopted and again the dynamics is not globally valid. Such systems are discussed in Hughes (2002), Moosavian and Papadopoulos (2007), Liang et al. (1997,

1998) and Vafa and Dubowsky (1987). As for the underwater systems, most papers concerned with modeling address the boundedness property, but it is often not noted in the stability proofs. Also for the skew-symmetric property the most commonly used references—such as in Murray et al. (1994), Sciavicco and Siciliano (2005) and Craig (1987)—only show this property for fixed-base manipulators. Again this cannot be taken as a general result that is also valid for vehicle-manipulator systems without further consideration.

It should be clear from the short literature review above that there are many misconceptions and misunderstandings regarding the boundedness and skew-symmetry properties for vehicle-manipulator systems: because these properties are so easy to prove for other systems they are taken as true also for vehicle-manipulator systems without any further proofs. Based on the work of From et al. (2010c, 2012) we will therefore study these properties in more detail in this chapter.

9.2 The Boundedness and Skew-Symmetric Properties in Control

In this section we illustrate how the boundedness property of the inertia matrix appears in control schemes such as robust control and how the skew-symmetric property of the Coriolis matrix appears in the stability proof of PD control laws, among other control laws. The next two sections are based on the control laws presented in Sciavicco and Siciliano (2005) and Murray et al. (1994), respectively. These properties do also appear in several other control laws for both robots and rigid bodies such as ships, satellites, and underwater vehicles.

9.2.1 Properties of the Dynamics in Matrix Form

Assume for now that we can write the dynamic equations of a mechanical system in the form

$$M(x)\ddot{x} + C(x, \dot{x})\dot{x} = \tau \tag{9.1}$$

where x is the state of the system, $M(x)$ is the inertia matrix and $C(x, \dot{x})$ is the Coriolis and centripetal matrix. The following properties can be associated with the inertia and Coriolis matrices:

Definition 9.1 (The boundedness property) The inertia matrix $M(x)$ is uniformly bounded in x if there exist constants d_1 and d_2, such that

$$0 < d_1 \le \|M(x)\| \le d_2 < \infty, \quad \forall x \in \mathbb{R}^n, \tag{9.2}$$

where $\|\cdot\|$ is the induced norm for matrices, i.e., a max-bound on the maximum singular value and a min-bound on the minimum singular value of the matrix.

Definition 9.1 is true only when there are no singularities present in the formulation. Thus, if the Euler angles are used to represent the attitude of the vehicle, as in Fossen (2002), Schjølberg (1996) and Børhaug (2008), this is not satisfied for a configuration-dependent inertia matrix. The existence of the boundaries d_1 and d_2 is the basis of gain controller design and global Lyapunov stability, and this property of the inertia matrix is used in several manipulator control laws such as robust control (Ghorbel et al. 1998; Sciavicco and Siciliano 2005). Given a computed estimate of the inertia matrix denoted $\hat{M}$ many controllers assume the property

$$\left\| M(x)^{-1}\hat{M}(x) - I \right\| \leq d < 1, \quad \forall x \in \mathbb{R}^n \tag{9.3}$$

which is automatically satisfied if the constant d is chosen as

$$d = \frac{d_2 - d_1}{d_2 + d_1} \tag{9.4}$$

with d_1 and d_2 as in Definition 9.1. Recall that Definition 9.1 guarantees that the constant d is bounded and is thus important in a large class of existing control laws. We will see one such example in Sect. 9.2.2.

The second property that we are concerned with is the skew-symmetric property of the Coriolis matrix:

Definition 9.2 (The skew-symmetric property) The matrix $(\dot{M}(x) - 2C(x, \dot{x}))$ is skew-symmetric if and only if

$$\left(\dot{M}(x) - 2C(x, \dot{x})\right)^{\mathrm{T}} = -\left(\dot{M}(x) - 2C(x, \dot{x})\right). \tag{9.5}$$

Definition 9.2 is true for a certain parameterization of the Coriolis matrix. Such a representation is well known for robotic manipulators on a fixed base (Murray et al. 1994; Sciavicco and Siciliano 2005) and for vehicles with no manipulator attached (Fossen 2002). One formulation for spacecraft-manipulator systems is found in Egeland and Pettersen (1998) where the boundedness and skew-symmetric properties are both true. The formulation uses quasi-velocities and the final equations resemble Kirchhoff's equations (Fossen 2002), but for multibody systems.

Finally, positive definite matrices frequently occur in the robot dynamics and can also simplify the proofs of control laws. This is thus an important property and is defined in the following way:

Definition 9.3 (Positive definiteness) A symmetric matrix $M \in \mathbb{R}^{n\times n}$ is positive definite if and only if

$$x^{\mathsf{T}} M x > 0, \quad \forall x \in \mathbb{R}^n. \tag{9.6}$$

We see from the definition that positive definiteness also implies that the matrix is symmetric. We will often write $M > 0$ which means that the matrix M is symmetric and positive definite.

9.2.2 Robust Control

Assume the dynamics of a robotic manipulator written in the form of (9.1) with the control

$$\tau = \hat{M}(x)y + \hat{C}(x, \dot{x})\dot{x} \tag{9.7}$$

where $\hat{M}$ and $\hat{C}$ represent the computed estimates of the dynamic model. This is a control law with an imperfect compensation due to the uncertainties, and the errors are given by

$$\tilde{M}(x) = \hat{M}(x) - M(x), \qquad \tilde{C}(x) = \hat{C}(x) - C(x). \tag{9.8}$$

We choose the control action as

$$y = \ddot{x}_d + K_D(\dot{x}_d - \dot{x}) + K_P(x_d - x) \tag{9.9}$$

where x_d is the desired trajectory. We further choose K_D and K_P as positive definite matrices. We can now combine (9.1) and (9.7) which gives

$$M(x)\ddot{x} + C(x, \dot{x})\dot{x} = \hat{M}(x)y + \hat{C}(x, \dot{x})\dot{x}. \tag{9.10}$$

If we assume that the inertia matrix of the manipulator is invertible we can rewrite this as

$$\ddot{x} = y + \big(M^{-1}(x)\hat{M}(x) - I\big)y + M^{-1}(x)\big(\hat{C}(x, \dot{x}) - C(x, \dot{x})\big)\dot{x}. \tag{9.11}$$

Thus, as we would like to have $\ddot{x} = y$, the uncertainty is represented by

$$\gamma = \big(I - M^{-1}(x)\hat{M}(x)\big)y - M^{-1}(x)\big(\hat{C}(x, \dot{x}) - C(x, \dot{x})\big)\dot{x}. \tag{9.12}$$

We can now write the error dynamics as

$$\ddot{e} + K_D\dot{e} + K_P e = \gamma \tag{9.13}$$

where $e = x_d - x$. Due to the uncertainties on the right hand side we cannot guarantee that the error converges to zero. We will therefore turn to the Lyapunov direct method to obtain a control law that is robust also when uncertainties are present. This is called a robust control law.

The right hand side of (9.13) represents the uncertainties of the system. We need to find a control law that guarantees asymptotic stability given that an estimate of the range of variation of the uncertainties is available. Based on (9.12) we can set up the following assumptions that will guarantee asymptotic stability (Sciavicco and Siciliano 2005):

$$\sup_{t\geq 0} \|\ddot{x}_d\| < d_M < \infty \qquad \forall \ddot{x}_d, \tag{9.14}$$

$$\left\| I - M^{-1}(x)\hat{M}(x) \right\| \leq d < 1 \qquad \forall x, \tag{9.15}$$

$$\left\| \hat{C}(x,\dot{x}) - C(x,\dot{x}) \right\| \leq d_\phi < \infty \quad \forall x, \dot{x}. \tag{9.16}$$

Assumption (9.14) is trivially satisfied as our planned trajectory should not require infinite accelerations. Assumption (9.16) depends only on x and $\dot{x}$, and is satisfied if we assume that the joint ranges are limited and that there exist maximum saturations on the velocities of the motors, which is the case in most mechanical systems and certainly for standard robotic manipulators.

Of special interest in the setting of this chapter is assumption (9.15). For this to be true, we need to guarantee that the inequality

$$0 < d_1 \leq \left\| M^{-1}(x) \right\| \leq d_2 < \infty, \qquad \forall x \in \mathbb{R}^n \tag{9.17}$$

holds. If $M(x)$ is bounded with lower and upper bounds, this inequality holds and we can always find a matrix $\hat{M}(x)$ that satisfies (9.15). For example, if we set

$$\hat{M} = \frac{2}{d_2 + d_1} I \tag{9.18}$$

where I is the identity matrix, we get

$$\left\| M(x)^{-1}\hat{M}(x) - I \right\| \leq \frac{d_2 - d_1}{d_2 + d_1} < 1, \quad \forall x \in \mathbb{R}^n \tag{9.19}$$

and (9.15) is satisfied. If all the properties in (9.14)–(9.16) hold we can find a controller that is robust with respect to the uncertainties in the system. We refer to Skogestad and Postlethwaite (2005) and Sciavicco and Siciliano (2005) for a more detailed treatment.

We see that the property that the inertia matrix is strictly positive definite and bounded is used explicitly in the stability proofs. This is therefore an important property to identify both for simulation and control of these systems.

9.2.3 PD Control Law

Stability in the sense of Lyapunov can be summarized in the following way: *Let $V(x,t)$ be a non-negative function with derivative $\dot{V}$ along the trajectories of the system. If $V(x,t)$ is locally positive definite and $\dot{V}(x,t) \le 0$ locally in x and for all t, then the origin of the system is locally stable.*

We will now see an example of how the skew-symmetric property of the Coriolis matrix plays an important role in Lyapunov-based stability proofs. Assume the system (9.1) and the augmented PD control law

$$\tau = M(x)\ddot{x}_d + C(x,\dot{x})\dot{x}_d - K_D\dot{e} - K_P e \tag{9.20}$$

where $e = x_d - x$. The closed loop system is then given by

$$M(x)\ddot{e} + C(x,\dot{x})\dot{e} + K_D\dot{e} + K_P e = 0. \tag{9.21}$$

To show stability we choose the Lyapunov function

$$V(e,\dot{e},t) = \frac{1}{2}\dot{e}^{\mathsf{T}} M(x)\dot{e} + \frac{1}{2} e^{\mathsf{T}} K_P e + \varepsilon e^{\mathsf{T}} M(x)\dot{e} \tag{9.22}$$

which is positive definite for sufficiently small ε. We now evaluate $\dot{V}$ along the trajectories of (9.21):

$$\dot{V} = \dot{e}^{\mathsf{T}} M\ddot{e} + \frac{1}{2}\dot{e}^{\mathsf{T}}\dot{M}\dot{e} + \dot{e}^{\mathsf{T}} K_P e + \varepsilon\dot{e}^{\mathsf{T}} M\dot{e} + \varepsilon e^{\mathsf{T}}(M\ddot{e} + \dot{M}\dot{e}) \tag{9.23}$$

$$= \dot{e}^{\mathsf{T}}(-K_P e - K_D\dot{e} - C\dot{e}) + \frac{1}{2}\dot{e}^{\mathsf{T}}\dot{M}\dot{e} + \dot{e}^{\mathsf{T}} K_P e + \varepsilon\dot{e}^{\mathsf{T}} M\dot{e} + \varepsilon e^{\mathsf{T}}(M\ddot{e} + \dot{M}\dot{e})$$

$$= -\dot{e}^{\mathsf{T}}(K_D - \varepsilon M)\dot{e} + \frac{1}{2}\dot{e}^{\mathsf{T}}(\dot{M} - 2C)\dot{e} + \varepsilon e^{\mathsf{T}}(-K_P e - K_D\dot{e} - C\dot{e} + \dot{M}\dot{e})$$

$$= -\dot{e}^{\mathsf{T}}(K_D - \varepsilon M)\dot{e} - \varepsilon e^{\mathsf{T}} K_P e - \varepsilon e^{\mathsf{T}}(K_D - \dot{M} + C)\dot{e} \tag{9.24}$$

which, for sufficiently small $\varepsilon > 0$ guarantees that $\dot{V}$ is negative definite and that the system is exponentially stable (Murray et al. 1994). The stability proof thus requires that the term with $(\dot{M} - 2C)$ vanishes.

We note that the stability proofs of neither the robust controller nor the PD controller are valid if Properties 9.1 and 9.2, respectively, are not true. With this as motivation we now investigate the validity of these properties for different formulations of the dynamic equations in matrix form for mechanical systems.

9.3 Single Rigid Bodies

For a single rigid body it is always possible to choose a parameterization of the inertia matrix so that it is bounded. The inertia matrix gives us a mapping from the velocity state to the kinetic energy, so a small change in the state variables should lead to a small change in the kinetic energy of the system. There is thus no point in the configuration space where a small change in the state variables should lead to a very large or infinite change in the kinetic energy because the physics of the system does not allow it. When singularities are present, however, our mathematical model allows for such effects. Because this comes as a result of our mathematical model and not the physics of the system, these singularities should be avoided.

For a well-defined choice of reference frames, the inertia matrix is constant so the problem of finding a skew-symmetric matrix $(\dot{M} - 2C(v))$ reduces to finding a skew-symmetric $C(v)$. Finding a skew-symmetric Coriolis matrix is fairly straight forward for single rigid bodies. Also when the inertia matrix is not constant, it is fairly straight forward to parameterize the inertia matrix so that $(\dot{M}(x) - 2C(x, v))$ becomes skew-symmetric.

9.3.1 The Boundedness Property

Many of the formulations of the inertia matrix that we found for single rigid bodies resulted in a constant inertia matrix. This is for example the case with Euler's equations of motion in (6.31)–(6.34), the Euler-Lagrange equations of motion in (6.54)–(6.55), and the general formulation of the inertia matrix in (6.130). Naturally, for these systems the inertia matrix is always bounded. In this sense it is advantageous to formulate the dynamics in the body frame.

On the other hand, it is sometimes more convenient to write the state variables in another reference frame and to reformulate the dynamics in terms of other state variables. In this case, the inertia matrix will no longer be constant. Consider for example the system (6.71)–(6.74) and recall that $J(\eta)$ is not defined for $\theta = \pm\pi/2$. This is the well known Euler angle singularity for the zyx-sequence. The inverse mappings $T_{\Theta}^{-1}(\Theta)$ and $J^{-1}(\eta)$ are defined for all $\theta \in \mathbb{R}$ but singular for $\theta = \pm\pi/2$. The boundedness of the inertia matrix as defined in Definition 9.1 is thus not true. We can only obtain a weaker result than the one found in Definition 9.1:

Definition 9.4 (The weak boundedness property) The inertia matrix $M(\eta)$ is uniformly bounded in η for θ separated from $\pm\pi/2$ if there exist constants d_1 and d_2 such that

$$0 < d_1 \leq \|M(\eta)\| \leq d_2 < \infty, \quad \forall \eta \in \mathbb{R}^6 \text{ where } \left| |\theta| - \frac{\pi}{2} \right| \geq \delta \tag{9.25}$$

for some small positive delta.

We note that the lower bound $d_1 > 0$ only exists if $|\theta|$ is separated from $\pi/2$ by some constant δ. If a different convention than the zyx-sequence is chosen the conditions on η need to be changed accordingly because the singularity is no longer at $\theta = \pm\frac{\pi}{2}$. As a direct consequence of the constraint given by (9.25) we see that for the singular configuration, in this case when $\theta = \pm\frac{\pi}{2}$, we have $d_1 = 0$ and the property of Definition (9.3) does not hold because $d = 1$.

When the dynamics is written as in (6.71), Definition 9.1 does not hold for the inertia matrix $\tilde{M}(\eta)$ because of the singularity. However, using Proposition 9.4 the inertia matrix $\tilde{M}(\eta)$ satisfies the weak boundedness property, which means that the boundedness property holds as long as we do not get too close to the singular configuration. It is important to understand that this is sufficient for many applications. For ship motion control, for example, we can simply choose the singular configuration at some point that the ship will never reach (laying on its side or upside down, for example). This means that any controller or stability proof that requires the boundedness of the inertia matrix is valid as long as we can guarantee that the current configuration is always at a certain distance from this point, which is always the case for ships but not necessarily for other systems such as a satellite.

It is important to note that it is possible to write the inertia matrix without the presence of singularities also when the inertia matrix is configuration dependent. The inertia matrix becomes configuration dependent whenever we choose to represent it in a frame other than the body frame $\mathscr{F}_b$. The formulation given in (6.131) is defined with respect to the inertial frame and is given by

$$M(g_{0b}) = \mathrm{Ad}_{g_{b0}}^{\mathsf{T}} I_b \mathrm{Ad}_{g_{b0}}. \tag{9.26}$$

This is an example of a well defined inertia matrix because there are no singularities in the Adjoint map $\mathrm{Ad}_{g_{0b}}$. In this case the inertia matrix is bounded and Definition 9.1 holds. We obtain this well-defined formulation because we use the state variables V_{0i}^B instead of for example $\dot{\eta}$.

9.3.2 The Skew-Symmetric Property

There are many ways to choose the Coriolis matrix so that $\dot{M} - 2C$ is skew-symmetric. We first note that if M is constant, and thus $\dot{M} = 0$, this is true if C is skew-symmetric. The parameterization chosen for the Coriolis matrices in (6.55), (6.57), and (6.58) are all skew symmetric and satisfy Definition 9.2. Several other representations for which the Coriolis matrix is skew-symmetric are found in Fossen and Fjellstad (1995).

If the inertia matrix is not constant and the dynamics is in the form of (6.71), we can also show this property. Consider the dynamics

$$\tilde{M}(x) = J^{-\mathsf{T}}(x) M(x) J^{-1}(x) \tag{9.27}$$

$$\tilde{C}(x, \dot{x}) = J^{-\mathsf{T}}(x)\big(C(x, \dot{x}) J^{-1}(x) - M(x) J^{-1}(x) \dot{J}(x) J^{-1}(x)\big). \tag{9.28}$$

Here we consider the more general formulation when also the inertia matrix $M(x)$ is configuration dependent, and not only $\dot{\tilde{M}}(x)$. We can also use the results below when the inertia matrix is constant and show the skew-symmetric property for the dynamics in the form of (6.71).

We first need to show that because $M(x) = M^{\mathsf{T}}(x)$ we have

$$\tilde{M}^{\mathsf{T}} = \left(J^{-\mathsf{T}} M J^{-1}\right)^{\mathsf{T}} = J^{-\mathsf{T}} M^{\mathsf{T}} J^{-1} = J^{-\mathsf{T}} M J^{-1} = \tilde{M} \tag{9.29}$$

which shows that $\tilde{M}^{\mathsf{T}} = \tilde{M}$. The time derivative of the inertia matrix can be written as (Fossen 1991)

$$\begin{aligned}\dot{\tilde{M}} &= \dot{J}^{-\mathsf{T}} M J^{-1} + J^{-\mathsf{T}} \dot{M} J^{-1} + J^{-\mathsf{T}} M \dot{J}^{-1} \\ &= -\left(J^{-1} \dot{J} J^{-1}\right)^{\mathsf{T}} M J^{-1} + J^{-\mathsf{T}} \dot{M} J^{-1} - J^{-\mathsf{T}} M J^{-1} \dot{J} J^{-1} \\ &= J^{-\mathsf{T}} \dot{M} J^{-1} - J^{-\mathsf{T}} \dot{J}^{\mathsf{T}} J^{-\mathsf{T}} M J^{-1} - J^{-\mathsf{T}} M J^{-1} \dot{J} J^{-1}.\end{aligned} \tag{9.30}$$

We can now write the required expression $\dot{\tilde{M}} - 2\tilde{C}$ as

$$\begin{aligned}(\dot{\tilde{M}} - 2\tilde{C}) &= J^{-\mathsf{T}} \dot{M} J^{-1} - J^{-\mathsf{T}} \dot{J}^{\mathsf{T}} J^{-\mathsf{T}} M J^{-1} - J^{-\mathsf{T}} M J^{-1} \dot{J} J^{-1} \\ &\quad - 2J^{-\mathsf{T}} C J^{-1} + 2J^{-\mathsf{T}} M J^{-1} \dot{J} J^{-1} \\ &= J^{-\mathsf{T}} (\dot{M} - 2C) J^{-1} - J^{-\mathsf{T}} \dot{J}^{\mathsf{T}} J^{-\mathsf{T}} M J^{-1} + J^{-\mathsf{T}} M J^{-1} \dot{J} J^{-1}.\end{aligned} \tag{9.31}$$

The quadratic form of the first matrix is zero because $x^{\mathsf{T}} J^{-\mathsf{T}} (\dot{M} - 2C) J^{-1} x = 0$ which follows from the fact that $(\dot{M} - 2C)$ is skew symmetric. Based on the property that $(\dot{M} - 2C)$ is skew-symmetric, we can show that the matrix $J^{-\mathsf{T}}(\dot{M} - 2C)J^{-1}$ is skew symmetric by direct calculation:

$$\left(J^{-\mathsf{T}} (\dot{M} - 2C) J^{-1}\right)^{\mathsf{T}} = J^{-\mathsf{T}} (\dot{M} - 2C)^{\mathsf{T}} J^{-1} = -J^{-\mathsf{T}} (\dot{M} - 2C) J^{-1} \tag{9.32}$$

where we have used that $(\dot{M} - 2C)^{\mathsf{T}} = -(\dot{M} - 2C)$.

For the second and third matrices we have

$$\begin{aligned}&\left(-J^{-\mathsf{T}} \dot{J}^{\mathsf{T}} J^{-\mathsf{T}} M J^{-1} + J^{-\mathsf{T}} M J^{-1} \dot{J} J^{-1}\right)^{\mathsf{T}} \\ &\quad = -\left(-J^{-\mathsf{T}} \dot{J}^{\mathsf{T}} J^{-\mathsf{T}} M J^{-1} + J^{-\mathsf{T}} M J^{-1} \dot{J} J^{-1}\right)\end{aligned} \tag{9.33}$$

which is also skew symmetric. We can therefore conclude the following:

Theorem 9.1 *Assume that* $(\dot{M} - 2C)$ *is skew-symmetric. Then the matrix* $(\dot{\tilde{M}} - 2\tilde{C})$ *where*

$$\tilde{M} = J^{-\mathsf{T}} M J^{-1} \tag{9.34}$$

$$\tilde{C} = J^{-\mathsf{T}} \left(C J^{-1} - M J^{-1} \dot{J} J^{-1}\right) \tag{9.35}$$

is also skew symmetric.

Proof The proof follows directly from the derivation in Eqs. (9.27) to (9.33). □

In Sect. 6.4 the dynamics were written as

$$\begin{aligned} M\dot{v} + C(v)v &= \tau \\ M\dot{v} + \widetilde{\mathrm{ad}}_{(Mv)}v &= \tau. \end{aligned} \tag{9.36}$$

We saw that the Coriolis matrix $C(v) = \widetilde{\mathrm{ad}}_{(Mv)}v$ is skew symmetric for all the subroups of $SE(3)$ that we discussed in Sect. 6.4.3. To show that this is in fact a general result we need to look at the expression for the Coriolis matrix in (6.128) given by

$$C_{ij}(v) = \sum_k \left(\frac{\partial S_{ki}}{\partial \varphi_j} - \frac{\partial S_{kj}}{\partial \varphi_i} \right) \bigg|_{\varphi=0} (Mv)_k. \tag{9.37}$$

We can write

$$C_{ij}(v) = \sum_k \left(\frac{\partial S_{ki}}{\partial \varphi_j} - \frac{\partial S_{kj}}{\partial \varphi_i} \right) = -\sum_k \left(\frac{\partial S_{kj}}{\partial \varphi_i} - \frac{\partial S_{ki}}{\partial \varphi_j} \right) = -C_{ji}(v) \tag{9.38}$$

which shows that $C_{ij}(v) = -C_{ij}^{\mathrm{T}}(v)$. We see this if we recall that the adjoint matrix $\widetilde{\mathrm{ad}}$ is in the form

$$\widetilde{\mathrm{ad}}_p = \begin{bmatrix} 0 & 0 & 0 & 0 & p_3 & -p_2 \\ 0 & 0 & 0 & -p_3 & 0 & p_1 \\ 0 & 0 & 0 & p_2 & -p_1 & 0 \\ 0 & p_3 & -p_2 & 0 & p_6 & -p_5 \\ -p_3 & 0 & p_1 & -p_6 & 0 & p_4 \\ p_2 & -p_1 & 0 & p_5 & -p_4 & 0 \end{bmatrix}. \tag{9.39}$$

Because we differentiate with respect to the elements of this matrix, the Coriolis matrix in (9.38) is skew symmetric. We can therefore conclude that we can take as a general result that the Coriolis matrix is skew-symmetric whenever it is of the form (9.37). This corresponds well with the explicit expressions for the Coriolis matrix that we found in Sect. 6.4.3.

9.4 Robotic Manipulators on a Fixed Base

A robotic manipulator is a multibody system and as a result the inertia matrix is configuration dependent because it depends on the position of the links with respect to each other. It is, however, fairly straight forward to find a set of state variables for which the inertia matrix is well defined and mathematical singularities are not present for robotic manipulators with Euclidean joints. There are, on the other hand, another group of singularities called kinematic singularities that arise due to the geometric structure of the manipulator. These singularities give us important information about the admissible velocities of the manipulator at a given configuration and are therefore also important to identify.

Because manipulators with Euclidean joints are easily described in terms of generalized coordinates and velocities, we can find the Coriolis matrix by the Christoffel symbols. We will show that when the Coriolis matrix is written in this form the matrix $(\dot{M} - 2C)$ is always skew-symmetric, also when $M(x)$ is configuration dependent.

9.4.1 The Boundedness Property

There are two main types of singularities that appear in the dynamics of fixed-base manipulators, namely mathematical singularities and kinematic singularities. Mathematical singularities, or representation singularities, will not appear in the manipulator dynamics when they are written in the standard way with the joint positions and velocities as state variables: the inertia matrix of a robotic manipulator is found from the constant generalized inertia matrices of each link together with the link Jacobians $J_i(q)$, which are both well-defined for robotic manipulators with Euclidean joints.

However, when the dynamics is written in terms of other state variables it is important to verify that the inertia matrix is still bounded. For the operational space formulation of the dynamics, for example, we need to keep in mind the singularities in the analytical Jacobian. When the dynamics is written in the operational space as in (7.46) the inertia matrix is given as

$$\bar{M}(q) = J_{m,a}^{-\mathsf{T}}(q) M(q) J_{m,a}^{-1}(q), \tag{9.40}$$

where $J_{m,a}(q)$ is the analytical Jacobian. We see that Definition 9.1 does not hold because this Jacobian is singular and we can only show the weak boundedness Property 9.4.

It is important to note that the fact that $J_{m,a}(q)$, or for that sake $J_{m,g}(q)$, are not always invertible is not always due to singularities in the mathematical representation. Depending on the structure of the manipulator there might exist configurations where the mappings from the joint velocities $\dot{q}$ to the operational space velocities $\dot{\eta}_{0e}$ or V_{0b}^B are not well defined. This is the case when the manipulator structure does not allow for a certain end-effector velocity for a given configuration. These singularities are called kinematic singularities. Kinematic singularities are configurations at which the mobility of the manipulator is reduced, i.e., it is not possible to impose an arbitrary motion to the end effector. This is thus a property of the manipulator design and not the singularities in the mathematical representation.

Ghorbel et al. (1998) identified all manipulators for which Definition 9.1 is satisfied. They find that for a large class of manipulators, including manipulators with only revolute or only prismatic joints, this property is always satisfied. They do not address the mathematical representation of the inertia matrix.

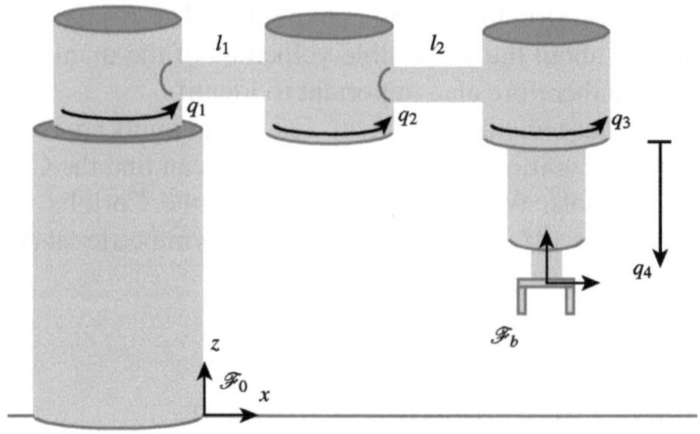

Fig. 9.1 The SCARA manipulator with a Schönflies configuration space in Example 9.1

As long as we can use generalized coordinates to represent the state of the robot we can do this without the presence of representation singularities. In this case, all manipulators characterized by Ghorbel et al. (1998) to satisfy Definition 9.1 from a design point of view, will also satisfy Definition 9.1 from mathematical point of view.

We see this from the derivation of the inertia matrix for fixed-base manipulators which was found as

$$M(q) = \sum_{i=1}^{n} M_i(q) \tag{9.41}$$

where

$$M_i(q) = J_i^{\mathsf{T}} \mathrm{Ad}_{g_{i0}}^{\mathsf{T}} I_i \mathrm{Ad}_{g_{i0}} J_i \in \mathbb{R}^{n\times n}. \tag{9.42}$$

$\mathrm{Ad}_{g_{i0}}$ is always well defined. Therefore, if the formulation is well defined from a design point of view (Ghorbel et al. 1998), i.e., J_i is well defined, then we can conclude that also M is well defined.

We see that when we write the dynamics in the form

$$M(q)\ddot{q} + C(q,\dot{q})\dot{q} = \tau \tag{9.43}$$

the inertia matrix is bounded. When we use the workspace representation

$$\bar{M}(q)\ddot{q} + \bar{C}(q,\dot{q})\dot{q} = \tau, \tag{9.44}$$

however, with $\bar{M}(q)$ as in (9.40) the inertia matrix is not bounded for all configurations. Definition 9.1 thus depends on the choice of state variables used.

We see that it is important to treat kinematic and representation singularities differently. Even though the Jacobian J_i in (9.42) is well defined in the mathematical sense, it is not always invertible. Whenever the manipulator is close the a kinematic singularity the Jacobian becomes ill posed and the robot loses one or more degrees of mobility for the end effector. We will look at a simple example to illustrate how kinematic singularities arise.

Example 9.1 As an example we look at the geometric Jacobian of the SCARA robot (Murray et al. 1994) which has three revolute joints and one prismatic joint normally chosen in the direction of the z-axis, shown in Fig. 9.1. The end effector of the SCARA manipulator can thus translate freely in $\mathbb{R}^3$ and rotate around the z-axis. This is known as the Schönflies motion. The Jacobian is given by

$$\begin{aligned} J^S_{m,g}(q) &= \begin{bmatrix} X'_1 & X'_2 & X'_3 & X'_4 \end{bmatrix} \\ &= \begin{bmatrix} 0 & l_1 \cos q_1 & l_1 \cos q_1 + l_2 \cos(q_1 + q_2) & 0 \\ 0 & l_1 \sin q_1 & l_1 \sin q_1 + l_2 \sin(q_1 + q_2) & 0 \\ 0 & 0 & 0 & -1 \\ 0 & 0 & 0 & 0 \\ 0 & 0 & 0 & 0 \\ 1 & 1 & 1 & 0 \end{bmatrix} \end{aligned} \tag{9.45}$$

where l_1 and l_2 are the lengths of links 1 and 2, respectively. The determinant of the 4×4-matrix of (9.45) representing the 4-DoF motion (cancelling rows 4 and 5) is given by

$$\begin{aligned} \det\big(J(q)\big) &= -l_1 l_2 \big(\cos q_1 \sin(q_1 + q_2) - \sin q_1 \cos(q_1 + q_2)\big) \\ &= -l_1 l_2 \sin q_2. \end{aligned} \tag{9.46}$$

We note that the determinant is zero only when $q_2 = \{0, \pm\pi\}$. This is the case when the arm is stretched out ($q_2 = 0$) and the manipulator loses mobility, in this case in the direction of the stretched out arm. This kinematic singularity is thus due to the geometry of the manipulator and not due to the mathematical representation. As we see, the kinematic singularity gives us valuable insight into the geometry and admissible velocities of the robotic arm.

We see from (9.45) that manipulators with only revolute and prismatic joints will not have representation singularities. It is, however, important to check for kinematic singularities whenever the boundedness Property 9.1 is required in simulation and control. Most industrial manipulators are therefore programmed to avoid the vicinity of all the kinematic singularities, which are normally known.

9.4.2 The Skew-Symmetric Property

For robotic manipulators represented in generalized coordinates the Coriolis matrix is normally obtained by the Christoffel symbols of the first kind as (Murray et al.

1994)

$$C(q,\dot{q}) = \{c_{ij}\} = \left\{ \sum_{k=1}^{n} c_{ijk}\dot{q}_k \right\}, \tag{9.47}$$

$$c_{ijk} = \frac{1}{2}\left(\frac{\partial M_{ij}}{\partial q_k} + \frac{\partial M_{ik}}{\partial q_j} - \frac{\partial M_{jk}}{\partial q_i} \right) \tag{9.48}$$

where $M(q) = \{M_{ij}\}$. Given this representation it is straight forward to show that the matrix $(\dot{M} - 2C)$ is skew-symmetric (Murray et al. 1994; Sciavicco and Siciliano 2005). We see this if we write out the components of $(\dot{M} - 2C)$:

$$\begin{aligned}(\dot{M} - 2C)_{ij} &= \dot{M}_{ij} - 2C_{ij} \\ &= \sum_{k=1}^{n} \left(\frac{\partial M_{ij}}{\partial q_k}\dot{q}_k - \frac{\partial M_{ij}}{\partial q_k}\dot{q}_k - \frac{\partial M_{ik}}{\partial q_j}\dot{q}_k + \frac{\partial M_{jk}}{\partial q_i}\dot{q}_k \right) \\ &= \sum_{k=1}^{n} \left(\frac{\partial M_{jk}}{\partial q_i}\dot{q}_k - \frac{\partial M_{ik}}{\partial q_j}\dot{q}_k \right). \end{aligned} \tag{9.49}$$

From $(\dot{M} - 2C)_{ij} = -(\dot{M} - 2C)_{ji}$ we conclude that $(\dot{M} - 2C)^{\mathsf{T}} = -(\dot{M} - 2C)$ and Definition 9.2 is satisfied. This is an important result:

> For a robotic manipulator with 1-DoF Euclidean joints written as generalized coordinates, the matrix $(\dot{M} - 2C)$ is always skew symmetric when the Coriolis matrix C is written in terms of the Christoffel symbols (9.47)–(9.48).

The skew-symmetric property of the formulation in operational space given by (7.46) can be verified using the same approach as in the previous section for single rigid bodies. We first write

$$\bar{M}(q)\ddot{\eta}_{0e} + \bar{C}(q,\dot{q})\dot{\eta}_{0e} = \bar{\tau} \tag{9.50}$$

where

$$\bar{M}(q) = J_{m,a}^{-\mathsf{T}}(q)M(q)J_{m,a}^{-1}(q), \tag{9.51}$$

$$\bar{C}(q) = J_{m,a}^{-\mathsf{T}}(q)\big(C(q,\dot{q})J_{m,a}^{-1}(q) + M\dot{J}_{m,a}^{-1}(q)\big), \tag{9.52}$$

and the skew symmetry is shown by writing $(\dot{\bar{M}} - 2\bar{C})$ as

$$(\dot{\bar{M}} - 2\bar{C}) = J^{-\mathsf{T}}(\dot{M} - 2C)J^{-1} - J^{-\mathsf{T}}\dot{J}^{\mathsf{T}}J^{-\mathsf{T}}MJ^{-1} + J^{-\mathsf{T}}MJ^{-1}\dot{J}J^{-1}. \tag{9.53}$$

Skew symmetry follows directly from the discussion in Theorem 9.1 in the previous section:

Assume that $(\dot{M} - 2C)$ is skew-symmetric. Then the matrix $(\dot{\bar{M}} - 2\bar{C})$ where

$$\bar{M} = J^{-\mathsf{T}} M J^{-1} \tag{9.54}$$

$$\bar{C} = J^{-\mathsf{T}}(CJ^{-1} - MJ^{-1}\dot{J}J^{-1}) \tag{9.55}$$

is also skew symmetric.

Because the formulations in Sects. 7.4–7.6 all reduce to the expressions above for manipulator arms with Euclidean joints, also these formulations satisfy the skew-symmetric Property 9.2.

9.5 Vehicle-Manipulator Systems

For vehicle-manipulator systems we can follow more or less the same argumentation as for single rigid bodies and fixed-base manipulators when we discuss the boundedness and skew-symmetric properties. As in the previous section the choice of state variables more or less determine whether the properties hold or not.

9.5.1 The Boundedness Property

The inertia matrix of a vehicle-manipulator system is given in its most general form as

$$M(x) = \sum_{i=b}^{6} \left[\frac{\partial V_{0i}^{B}}{\partial \zeta}^{\mathsf{T}} I_i \frac{\partial V_{0i}^{B}}{\partial \zeta} \right]. \tag{9.56}$$

This is well defined as long as the Jacobian

$$J_{gi}^{B}(q) = \frac{\partial V_{0i}^{B}}{\partial \zeta} \tag{9.57}$$

is well defined. This matrix is always well defined with respect to representation singularities for a suitable choice of state variables. One way to write the geometric Jacobian of link i is

$$J_{gi}^{B}(q) = \begin{bmatrix} \mathrm{Ad}_{g_{ib}} & \mathrm{Ad}_{g_{ib}} J_i \end{bmatrix} \tag{9.58}$$

from which it is clear that there are no representation singularities present. When the inertia matrix is written as in Sect. 8.2 as

$$M(x) = \sum_{i=b}^{6} \begin{bmatrix} H^{\mathsf{T}} \mathrm{Ad}_{g_{ib}}^{\mathsf{T}} I_i \mathrm{Ad}_{g_{ib}} H & H^{\mathsf{T}} \mathrm{Ad}_{g_{ib}}^{\mathsf{T}} I_i \mathrm{Ad}_{g_{ib}} J_i \\ J_i^{\mathsf{T}} \mathrm{Ad}_{g_{ib}}^{\mathsf{T}} I_i \mathrm{Ad}_{g_{ib}} H & J_i^{\mathsf{T}} \mathrm{Ad}_{g_{ib}}^{\mathsf{T}} I_i \mathrm{Ad}_{g_{ib}} J_i \end{bmatrix}, \tag{9.59}$$

i.e., by substituting (9.58) into (9.56), the inertia matrix is well defined.

In the same way as a change of state variables may lead to singularities for single rigid bodies and fixed-base manipulators, the dynamics written as

$$\tilde{M}(\xi)\ddot{\xi} + \tilde{C}(\xi, \dot{\xi})\dot{\xi} = \tilde{\tau} \tag{9.60}$$

with state variables $\xi = [\eta^{\mathsf{T}}\ q^{\mathsf{T}}]^{\mathsf{T}}$ and the inertia matrix

$$\tilde{M}(\xi) = J_a^{-\mathsf{T}}(\xi) M(q) J_a^{-1}(\xi) \tag{9.61}$$

is also singularity prone because J_a contains singular points. Thus, once again the operational space formulation in terms of the vehicle position η comes at the expense of an unbounded inertia matrix and only the weak boundedness property of Definition 9.4 holds for Eq. (9.60). This is the most commonly found formulation of vehicle-manipulator system dynamics in literature. The singularity is normally pointed out when deriving the models of these systems. It is, however, often left out when discussing the control problem which is a considerable weakness and often leads to control laws that cannot guarantee global stability properties.

If we choose to write the state space as

$$v = \begin{bmatrix} \tilde{V}_{0b}^{B} \\ \dot{q} \end{bmatrix} \tag{9.62}$$

the dynamic equations are written as

$$M(Q)\dot{v} + C(C, v)v = \tau \tag{9.63}$$

with inertia matrix

$$M(q) = \sum_{i=b}^{n} \begin{bmatrix} H^{\mathsf{T}} \mathrm{Ad}_{g_{ib}}^{\mathsf{T}} I_i \mathrm{Ad}_{g_{ib}} H & H^{\mathsf{T}} \mathrm{Ad}_{g_{ib}}^{\mathsf{T}} I_i \mathrm{Ad}_{g_{ib}} J_i \\ J_i^{\mathsf{T}} \mathrm{Ad}_{g_{ib}}^{\mathsf{T}} I_i \mathrm{Ad}_{g_{ib}} H & J_i^{\mathsf{T}} \mathrm{Ad}_{g_{ib}}^{\mathsf{T}} I_i \mathrm{Ad}_{g_{ib}} J_i \end{bmatrix} \tag{9.64}$$

which is well defined and Definition 9.1 holds also for the formulations presented in Sect. 8.3.

9.5.2 The Skew-Symmetric Property

In Sects. 8.1 and 8.2 the Coriolis matrix was given as

$$C(q, \zeta) = \sum_{i=b}^{n} \big(J_{gi}^{\mathsf{T}}(q) I_i \dot{J}_{gi}(q) - J_{gi}^{\mathsf{T}}(q) W_i \big(V_{0i}^{B} \big) J_{gi}(q) \big). \tag{9.65}$$

We can then write

$$
\begin{aligned}
(\dot{M}-2C) &= \frac{\mathrm{d}}{\mathrm{d}t}\left(\sum_{i=b}^{n} J_{gi}^{\mathsf{T}}(q)I_i J_{gi}(q)\right) \\
&\quad - 2\sum_{i=b}^{n}\left(J_{gi}^{\mathsf{T}}(q)I_i \dot{J}_{gi}(q) - J_{gi}^{\mathsf{T}}(q)W_i\left(V_{0i}^{B}\right)J_{gi}(q)\right) \\
&= \sum_{i=b}^{n}\left(\dot{J}_{gi}^{\mathsf{T}}(q)I_i J_{gi}(q) + J_{gi}^{\mathsf{T}}(q)I_i \dot{J}_{gi}(q)\right. \\
&\quad \left. - 2J_{gi}^{\mathsf{T}}(q)I_i \dot{J}_{gi}(q) + 2J_{gi}^{\mathsf{T}}(q)W_i\left(V_{0i}^{B}\right)J_{gi}(q)\right) \\
&= \sum_{i=b}^{n}\left(\dot{J}_{gi}^{\mathsf{T}}(q)I_i J_{gi}(q) - J_{gi}^{\mathsf{T}}(q)I_i \dot{J}_{gi}(q) + 2J_{gi}^{\mathsf{T}}(q)W_i\left(V_{0i}^{B}\right)J_{gi}(q)\right).
\end{aligned}
\tag{9.66}
$$

From the discussion in the previous sections we can conclude that $J_{gi}^{\mathsf{T}}(q)W_i(V_{0i}^{B})$ $J_{gi}(q)$ is skew symmetric because $W_i(V_{0i}^{B})$ is skew symmetric and we can further write (recall that I_i is diagonal and trivially $I_i = I_i^{\mathsf{T}}$)

$$
\left(\dot{J}_{gi}^{\mathsf{T}}(q)I_i J_{gi}(q) - J_{gi}^{\mathsf{T}}(q)I_i \dot{J}_{gi}(q)\right)^{\mathsf{T}} = -\left(\dot{J}_{gi}^{\mathsf{T}}(q)I_i J_{gi}(q) - J_{gi}^{\mathsf{T}}(q)I_i \dot{J}_{gi}(q)\right)
\tag{9.67}
$$

which is therefore also skew symmetric. We can now conclude the following:

Theorem 9.2 *Given the dynamic equations of a vehicle-manipulator system in the form*

$$
M(q)\dot{\zeta} + C(q,\zeta)\zeta = \tau \tag{9.68}
$$

where

$$
M(q) = \sum_{i=b}^{n}\left(J_{gi}^{\mathsf{T}}(q)I_i J_{gi}(q)\right) \tag{9.69}
$$

$$
C(q,\zeta) = \sum_{i=b}^{n}\left(J_{gi}^{\mathsf{T}}(q)I_i \dot{J}_{gi}(q) - J_{gi}^{\mathsf{T}}(q)W_i\left(V_{0i}^{B}\right)J_{gi}(q)\right) \tag{9.70}
$$

where $W_i(V_{0i}^{B})$ is skew symmetric. Then the matrix $(\dot{M}-2C)$ is skew symmetric.

Proof The proof follows directly from Eqs. (9.66) and (9.67) and the discussion above. □

We can use the results from the previous sections to show that when the dynamics is written in the form

$$\tilde{M}(\xi)\ddot{\xi} + \tilde{C}(\xi,\dot{\xi})\dot{\xi} = \tilde{\tau} \tag{9.71}$$

where

$$\tilde{M}(\xi) = J_a^{-\mathsf{T}}(\xi)M(q)J_a^{-1}(\xi) \tag{9.72}$$

$$\tilde{C}(\xi,\dot{\xi}) = J_a^{-\mathsf{T}}(\xi)\big(J_a^{-\mathsf{T}}(\xi)C(q,\dot{\xi})J_a^{-1}(\xi) - M(q)J_a^{-1}(\xi)\dot{J}_a(\xi)J_a^{-1}(\xi)\big), \tag{9.73}$$

the skew-symmetric Property 9.2 of $(\dot{\tilde{M}} - 2\tilde{C})$ in (8.49) follows directly from the skew symmetry of $(\dot{M} - 2C)$. This follows from Eq. (9.31) and the discussion thereafter. The same is the case for $(\dot{\bar{M}} - 2\bar{C})$ as defined in (8.55)–(8.58).

Finally we will look at the skew-symmetric property when the Coriolis matrix is in the form

$$C_{ij}(v) = \sum_k \left(\frac{\partial M_{ij}}{\partial \varphi_k} - \frac{1}{2}\frac{\partial M_{jk}}{\partial \varphi_i}\right)\bigg|_{\varphi=0} v_k + \sum_k \left(\frac{\partial S_{ki}}{\partial \varphi_j} - \frac{\partial S_{kj}}{\partial \varphi_i}\right)(Mv)_k\bigg|_{\varphi=0}. \tag{9.74}$$

The skew symmetry of the Coriolis matrix in this form follows directly from proofs of skew symmetry in the previous sections. The first part is identical to the formulation in Sect. 9.4.2 and is therefore skew symmetric by the same argumentation. Similarly, skew symmetry for the second part is shown in Sect. 9.3.2. The sum of two skew-symmetric matrices is also skew symmetric, so it follows that (9.74) is skew symmetric.

References

Antonelli, G. (2006). *Underwater robots. Motion and force control of vehicle-manipulator systems*. Berlin: Springer.

Bullo, F., & Lewis, A. D. (2000). *Geometric control of mechanical systems: modeling, analysis, and design for simple mechanical control systems*. New York: Springer.

Børhaug, E. (2008). *Nonlinear control and synchronization of mechanical systems*. Ph.D. thesis, Department of Engineering Cybernetics, Norwegian University of Science and Technology, Trondheim, Norway.

Craig, J. J. (1987). *Adaptive control of mechanical manipulators*. Boston: Addison-Wesley.

Egeland, O., & Pettersen, K. Y. (1998). Free-floating robotic systems. *Control Problems in Robotics and Automation*, *230*, 119–134.

Fossen, T. I. (1991). *Nonlinear modelling and control of underwater vehicles*. Ph.D. thesis, Department of Engineering Cybernetics, The Norwegian Institute of Technology.

Fossen, T. I. (2002). *Marine control systems*. Trondheim: Marine Cybernetics AS. 3rd printing.

Fossen, T. I., & Fjellstad, O. E. (1995). Nonlinear modelling of marine vehicles in 6 degrees of freedom. *International Journal of Mathematical Modelling Systems*, *1*(1), 17–28.

From, P. J., Duindam, V., Pettersen, K. Y., Gravdahl, J. T., & Sastry, S. (2010a). Singularity-free dynamic equations of vehicle-manipulator systems. *Simulation Modelling Practice and Theory*, *18*(6), 712–731.

From, P. J., Pettersen, K. Y., & Gravdahl, J. T. (2010b). Singularity-free dynamic equations of AUV-manipulator systems. In *Proceeding of symposium on intelligent autonomous vehicles*, Lecce, Italy.

From, P. J., Pettersen, K. Y., & Gravdahl, J. T. (2011). Singularity-free dynamic equations of spacecraft-manipulator systems. *Acta Astronautica*, *69*(11–12), 1057–1065.

From, P. J., Schjølberg, I., Gravdahl, J. T., Pettersen, K. Y., & Fossen, T. I. (2010c). On the boundedness property of the inertia matrix and skew-symmetric property of the Coriolis matrix for vehicle-manipulator systems. In *Proceeding of symposium on intelligent autonomous vehicles*, Lecce, Italy.

From, P. J., Schjølberg, I., Gravdahl, J. T., Pettersen, K. Y., & Fossen, T. I. (2012). On the boundedness and skew-symmetric properties of the inertia and Coriolis matrices for vehicle-manipulator systems. *Journal of Dynamic Systems, Measurement, and Control*, *134*(4), 2012.

Ghorbel, F., Srinivasan, B., & Spong, M. W. (1998). On the uniform boundedness of the inertia matrix of serial robot manipulators. *Journal of Robotic Systems*, *15*(1), 17–28.

Hughes, P. C. (2002). *Spacecraft attitude dynamics*. Mineola: Dover.

Liang, B., Xu, Y., & Bergerman, M. (1997). Dynamically equivalent manipulator for space manipulator system. 1. In *Proceedings of IEEE international conference on robotics and automation* (Vol. 4, pp. 2765–2770).

Liang, B., Xu, Y., & Bergerman, M. (1998). Mapping a space manipulator to a dynamically equivalent manipulator. *ASME Journal of Dynamic Systems, Measurement, and Control*, *120*(1), 1–7.

Marsden, J. E., & Ratiu, T. S. (1999). *Texts in applied mathematics. Introduction to mechanics and symmetry* (2nd ed.). New York: Springer.

Moosavian, S. A. A., & Papadopoulos, E. (2007). Free-flying robots in space: an overview of dynamics modeling, planning and control. *Robotica*, *25*(5), 537–547.

Murray, R. M., Li, Z., & Sastry, S. S. (1994). *A mathematical introduction to robotic manipulation*. Boca Raton: CRC Press.

Schjølberg, I. (1996). *Modeling and control of underwater robotic systems*. Ph.D. thesis, Department of Engineering Cybernetics, Norwegian University of Science and Technology, Trondheim, Norway.

Schjølberg, I., & Fossen, T. I. (1994). Modelling and control of underwater vehicle-manipulator systems. In *Conference on marine craft maneuvering and control*, Southampton, UK (pp. 45–57).

Sciavicco, L., & Siciliano, B. (2005). *Modelling and control of robot manipulators*. New York: Springer.

Skogestad, S., & Postlethwaite, I. (2005). *Multivariable feedback control: analysis and design*. New York: Wiley.

Vafa, Z., & Dubowsky, S. (1987). On the dynamics of manipulators in space using the virtual manipulator approach. In *Proceedings of IEEE international conference on robotics and automation*, North Carolina, USA (pp. 579–585).

Wen, J. T.-Y., & Kreutz-Delgado, K. (1991). The attitude control problem. *IEEE Transactions on Automatic Control*, *30*(10), 1148–1162.

de Wit, C. C., Diaz, E., & Perrier, M. (1998). Robust nonlinear control of an underwater vehicle/manipulator system with composite dynamics. In *Proceedings of IEEE international conference on robotics and automation*, Leuven, Belgium (pp. 452–457).

Chapter 10
Underwater Robotic Systems

One of the most promising application areas for vehicle-manipulator systems is underwater robotics. Robotic solutions will have huge impacts on operation and maintenance of underwater installations as well as exploration and research missions in distant underwater locations. In fact, underwater robotic systems are already being used for several complex subsea operations, many of which would be impossible to perform without the use of robots. This includes time-consuming operations such as inspection of the seabed, installation and operation of underwater fields at great depths and in areas with difficult access, and search and rescue missions.

10.1 Introduction

During the last decades, a lot of work and progress has been done on mobile marine robots for applications such as guided and autonomous inspection tasks, simple subsea intervention, and environmental monitoring, to name a few. These are all relatively mature fields and several commercially available products have been developed. This includes both Remotely Operated Underwater Vehicles (ROUVs, or simply ROVs) and vehicles with one or more manipulator arms. At the time being, attention is given to develop more advanced underwater systems for more complex manipulation and autonomous interaction tasks using Autonomous Underwater Vehicles (AUV). These systems are often termed intervention ROVs or I-ROVs for short, or AUV-manipulator systems if they can operate autonomously.

There are several success stories in the field of underwater robotics. Here we will mention two important happenings that recently have attracted a lot of attention in both the underwater community and the international press:

- One underwater operation of great impact was when the Remora 6000 ROV retrieved the black box of Air France flight 447 that crashed off the coast of Brazil in 2009 (Fig. 10.1). The Remora robot consists of a 900 kg underwater ROV with two manipulator arms and is controlled through a 6700 meter long fiber optic umbilical. The robot located the black box at 3900 meters depth. The operation led to great in-sight into the happenings just before flight 447 lost control.

P.J. From et al., *Vehicle-Manipulator Systems*, Advances in Industrial Control,
DOI 10.1007/978-1-4471-5463-1_10,

(a) The Remora ROV-manipulator system. Copyright 2003 Phoenix International Holdings, Inc. All Rights Reserved. Reprinted with express permission. Inquiries plehardy@phnx-international.com.

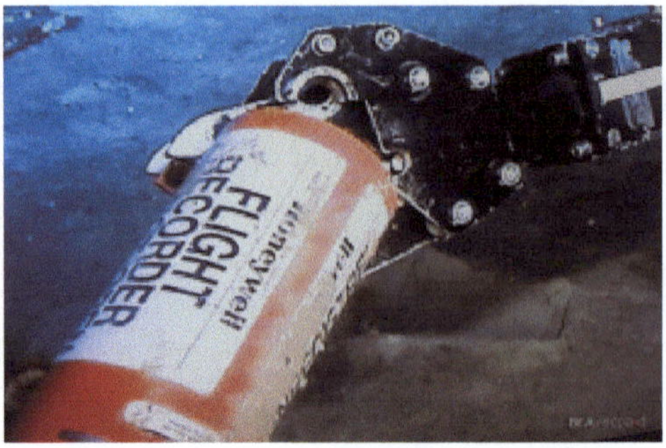

(b) The manipulator hand grasping the black box at 3900 meters depth

Fig. 10.1 The Remora 6000 ROV that retrieved that *black box* after Air France flight 447 crashed off the coast of Brazil in 2009

- Another area where underwater robots are frequently used is in the petroleum industry. Rig operators normally have several underwater robots available, both for inspection and assembly tasks. During the dramatic days after the explosion at the Deepwater Horizon platform in the Gulf of Mexico in 2010 where as much as 780.000 cubic meters of oil leaked into the sea, several ROVs with manipulator arms were deployed in the operation of stopping the leak. Underwater and surface vehicles were also used to monitor the oil spill in the following weeks and months.

Working in an underwater environment is different from traditional operation in many aspects. Even though several technical solutions that work on dry land can be applied also to subsea systems, modifications need to be made and there are many challenges that need to be overcome when we introduce these systems to a wet environment. Underwater robots are normally electrically and/or hydraulically driven ROVs or AUVs with one or more manipulator arms mounted on them. Most manipulator arms are hydraulically driven with very high work loads. There also exist electrically driven arms for lighter operation. A more detailed study of unmanned marine vehicles can be found in Sutton and Roberts (2006) and a treatment of underwater robotics in general can be found in Moore et al. (2009). A detailed treatment of underwater robots with focus an AUV-manipulator systems can be found in Antonelli (2006).

10.1.1 Operating Under Water

There are many remarkable characteristics of water, most of which are necessary for life on Earth and therefore we as humans depend on. However, some of these properties present us with challenging conditions when operating electrical and mechanical equipment submerged in water.

First of all, when an object is immersed in water, buoyancy forces tend to lift the object towards the surface. These forces work in the opposite direction of the gravitational forces. If the center of buoyancy and the center of gravity coincide, this is a linear force that either reduces the effects of the gravitational forces and slows down the falling motion of the object (if the mass of the water that would normally occupy the space of the object is less than that of the object), or it can lift the object towards the surface (if the density of the object is lower than the density of water). If the density of the object is equal to that of the water we say that the object is neutrally buoyant, which means that it acts as if no gravitational forces were present. Most underwater vehicles are made so that they are approximately neutrally buoyant so that as little energy as possible is used to keep the underwater robot hovering during operation. If the density of the underwater vehicle is slightly lower than water we also guarantee that the vehicle will float to the surface in the case of a system breakdown.

If the center of gravity does not coincide with the center of buoyancy, the buoyancy forces also add a rotational motion always forcing the center of buoyancy to lie directly above the center of gravity. This pendulum-like behavior is normally taken advantage of when constructing underwater vehicles by placing most of the mass at the bottom of the vehicle. This will then automatically keep the vehicle in the desired orientation without excessive use of control forces.

Secondly, the alignment of the atoms in the water molecules allows the water to absorb salt almost without increasing its volume. This has great implications when it comes to determining the mass of the water and the buoyancy forces that act on a submerged vehicle. The salt can also be blamed for another well known problem, namely corrosion. Salt water, and especially in combination with electric voltages, speed up the corrosion process which can change the kinematic and dynamic properties of the robot. If corrosion occurs over longer periods of time this can eventually cause the robot to break down mechanically or cause other types of failures. In addition, the salt makes the water into a good conductor (also fresh water is a better conductor than air) which increases the possibility of short-circuits. All electrical equipment should therefore be protected against water and water vapor.

Because liquid has a very large inertia compared to gases, acceleration under water requires more force than in air. The reason for this is that any motion of a submerged vehicle also requires acceleration of the surrounding water. The friction between the water and the underwater vehicle further increases the drag forces.

Finally, communication and localization present two challenging areas in underwater operation. The most common and useful tool for localization on the surface is the Global Positioning System (GPS). Unfortunately, GPS signals are not very powerful and do not transmit very well in water. As a result, the GPS requires a direct line-of-sight between the transmitter and the receiver and does not work in an underwater environment. There are, however, quite a few useful instruments for underwater navigation. Some examples are sonars, compasses, depth measuring systems, inertial measurement units, and so on.

As with the GPS signals, radio waves for communication do not travel very far in water and are therefore not very useful in communication with underwater robots. Most remotely controlled underwater systems therefore use umbilicals for communication. These are cables, often several kilometers long, that allow for fast and reliable communication as well as electricity supply to the remote vehicle.

Other complicating issues when dealing with both underwater and floating marine vehicles are waves and ocean currents. Waves affect the high-frequency motion of ships and platforms, but also underwater vehicles are affected by waves when operating in shallow waters and during lifting in launch and recovery operations. Ocean currents will almost always need to be compensated for when operating on the seabed. Currents are caused by the Earth's rotation, gravity, wind and tides, and can in some cases make large amount of water move at velocities up to several meters per second.

10.1.2 Underwater Vehicle-Manipulator Systems—A Brief Historical Overview

From the early 90s up until today a large number of underwater robots with manipulator arms have seen the light of day. The very first projects on underwater vehicle-manipulator systems were research projects developed mainly for demonstration purposes while we today have several suppliers of more or less advanced utility vehicles used for everyday operations of underwater fields and exploration of remotely located areas.

Some of the first robots developed for autonomous underwater manipulation studies were built by the University of Hawaii. The ODIN (Omni-Directional Intelligent Navigator) underwater robot was developed during the 1990s and consist of a fully actuated AUV with 6-DoF motion and a 1-DoF manipulator arm. ODIN can be operated either in autonomous mode or remotely controlled through a tether (Choi et al. 2003).

The underwater robot OTTER (Ocean Technology Testbed for Engineering Research) developed at Monterey Bay Aquarium Research Institute is similar to ODIN in that the vehicle has six degrees of freedom and a one degree of freedom removable robot arm is attached to the AUV. The robot is used as an experimental platform for various research areas, for example autonomous underwater operations. The OTTER project is described in Wang et al. (1996).

One of the first projects with focus on advanced underwater manipulation was the AMADEUS project (Advanced MAnipulation for DEep Underwater Sampling), see Lane et al. (1997), Casalino et al. (2001). The main objective of this project was to develop improved solutions for sampling and manipulation at great depths. During the first phase of the project a dexterous gripper for underwater use was developed and the second phase included, among other objectives, the realization of two Ansaldo 7-DoF robotic arms for cooperative underwater operation. The robot was controlled in tele-operaton and tele-assistance modes.

The SAUVIM AUV (Semi-Autonomous Underwater Vehicle for Intervention Missions) program was initiated in 1997 with the objective to develop the first autonomous underwater manipulation demonstration in an unstructured environment (Yuh et al. 1998). Over a 10 year period the researchers developed an AUV-manipulator system that was able to perform coordinated motion control of the vehicle and the manipulator and at the same time perform autonomous manipulation tasks. The SAUVIM was equipped with a 7-DoF manipulator arm, the MARIS 7080 arm manufactured by Ansaldo that can operate at 6000 meters depth. SAUVIM performed the first fully autonomous underwater manipulation task in an unstructured environment in 2010.

Another and more recent project on underwater manipulation is the TRIDENT project (Sanz et al. 2010). This large consortium of universities develop new solutions to autonomous intervention tasks. An intervention AUV is developed to perform underwater operations with the objective of reducing the cost of such operations and remove humans from the low-level control loop. It is also proposed to let an autonomous surface vehicle accompany the underwater vehicle. The project started in 2010 with a duration of 36 months.

In addition to the research projects mentioned above, the are also several underwater robots available on the market that perform operations on a daily basis around the world. Many of these are developed for the petroleum industry, but they can also be found elsewhere. The typical ROV-manipulator system available on the market today consists of one large ROV that typically weighs around 1000 kg, with one or two manipulator arms attached to it. Some of the main suppliers of underwater VMS are Cybernétix, Seebyte/Subsea7, and Oceaneering.

A big operator in underwater robotics is Cybernétix. They developed the SWIMMER concept which consists of a shuttle and an ROV where the shuttle operates as an AUV and autonomously takes the ROV to a pre-defined location on the subsea production site. The ROV is then released from the shuttle and can be operated remotely like a traditional ROV by utilizing the umbilical connecting the off-site control room and the subsea site for power and real-time transfer of information and control signals. Cybernétix has also developed the ALIVE concept which is equipped with two docking manipulators for safe docking and for keeping the AUV steady during manipulation, in addition to a 6-DOF intervention manipulator which can perform light intervention tasks (Evans et al. 2003). The AUV is commercially available and is used for example in subsea offshore installations.

Subsea 7 provides ROV-manipulator systems principally to the oil and gas industry. The robots are used for both observation and interaction tasks. The ROVs can be equipped with robotic arms from Schilling Robotics. In cooperation with Seebyte the company is also developing autonomous underwater robots for inspection and intervention tasks.

Oceaneering is another provider of underwater robots. Oceaneering has a large selection of ROVs of different sizes and have in cooperation with Schilling Robotics developed several robotic arms for underwater use. The larger work class ROVs are equipped with two hydraulically driven manipulator arms for complex manipulation tasks at great depths. They also provide smaller electrically driven ROVs for observation and lighter intervention tasks.

Even though many operating underwater vehicles are found in the petroleum industry, for example Cybernétix' Alive and Subsea7's Hercules, these vehicles also play an important role in other fields. In underwater scientific missions robots are frequently used and several of these have one or more robotic arms attached, for example the Hercules ROV built for the Institute For Exploration, the Jason and Medea ROVs designed at the Deep Submergence Laboratory at Woods Hole Oceanographic Institution, the Tiburon ROV designed at Monterey Bay Aquarium Research Institute, and the Remotely Operated Platform for Ocean Science (ROPOS). All of these are equipped with robotic arms and are mainly used for research purposes.

10.1.3 Underwater Manipulators

There are several underwater manipulator arms available on the market. They differ from standard robotic manipulators mainly in the way they are built in order to sustain the environmental challenges of operating in water, such as high pressure and the corrosive environments.

The vast majority of underwater robot arms are hydraulically driven. The hydraulic fluid is normally pressurized by an electric or combustion engine driving the hydraulic pump. The engine will also deliver energy to the other parts of the underwater system, such as the on-board computers, sensors, and the thrusters of the vehicle. The main advantage of hydraulically driven manipulator arms is the superior payload compared to electric systems. This payload is often needed in work class ROVs and is the main reason why most underwater arms are hydraulic. Another advantage is that hydraulic systems are inherently pressurized. Because the internal pressure is higher than the pressure outside, hydraulic systems are not as susceptible to the sea water as their electric counterparts. On the other hand, hydraulic systems require more maintenance than electric systems and are also more expensive to install. Smaller ROVs therefore use electric power both for propulsion and for the manipulator arm.

One of the very first advanced underwater robotic arms was developed by the AMADEUS project which developed the Ansaldo underwater arm in the 1990s (Casalino et al. 2001). The Ansaldo arm is an electrically driven 7-DoF robotic arm which allowed for dexterous manipulation and coordinated motion in an underwater environment. The Ansaldo arm was also used in the SAUVIM project (Yuh et al. 1998) for picking up objects from the sea bed.

Schilling Robotics is one of the main suppliers of underwater robotic arms with over 2000 arms in operation around the world. The manipulators are made of titanium, anodized aluminum and stainless steel in order to cope with the harsh conditions of an underwater environment. All Schilling manipulators are hydraulically driven. These robots can operate at depths of 3000–7000 msw (meters of seawater) and the payload range is 60–270 kg. The robots are remotely operated with a kinematically equivalent master device for position control or a rate controller for velocity control.

Kraft Telerobotics is another supplier of heavy duty arms that operate in an underwater environment. The Kraft Predator arm is constructed to operate both on land and in up to 6500 msw. It is made with anodized aluminum and stainless steel and weighs 80 kg in air and 50 kg in water. The maximum lift capacity is 227 kg. The Kraft Predator arm is operated remotely with position control and force feedback.

An electrically driven underwater arm has been developed by ECA Robotics. The electrically driven arms are lighter than their hydraulically driven counterparts, which comes at the expense of a lower payload. The ECA 5E arms weighs 27 kg and has a total lift of 25 kg at full reach. The fully electric arm is a zero oil system and does not require any hydraulic oils, which makes this a good alternative in environmentally sensitive areas. The depth rating of the 5E arm is 3000 meters.

10.2 Dynamics of Underwater Vehicles

The rigid body characteristics of underwater vehicles were discussed in Chaps. 3 and 6. There are also other important properties that need to be included to obtain a complete description of the dynamics of underwater vehicles. These additional terms in the dynamic equations arise mainly due to the hydrodynamic effects that occur when a rigid body is submerged in water. The main contributions of the hydrodynamic forces and moments are added mass, radiation-induced potential damping, and restoring forces (Faltinsen 1990).

The dynamic equations of a completely or partially submerged rigid body is written as Fossen (2002)

$$\dot{\eta} = J(\eta)\nu,$$

$$\underbrace{M_{RB}\dot{\nu} + C_{RB}(\nu)\nu}_{\text{Rigid body}} + \underbrace{M_A\dot{\nu} + C_A(\nu)\nu}_{\text{Added Mass}} + \underbrace{D(\nu)\nu}_{\substack{\text{Potential}\\ \text{damping}}} + \underbrace{N_g(\eta) + N_b(\eta)}_{\substack{\text{Restoring}\\ \text{forces}}} = \underbrace{Bu}_{\text{Actuation}} \tag{10.1}$$

where $\eta = [x_{0b}\ y_{0b}\ z_{0b}\ \phi\ \theta\ \psi]^{\mathsf{T}}$ is the position and orientation of the vessel given in the inertial frame and $\nu = V_{0b}^{B} = [u\ v\ w\ p\ q\ r]^{\mathsf{T}}$ is the linear and angular velocities given in the body frame. We choose to use the notation ν for velocity, which is normally used in the underwater robotics literature.

The first terms correspond to the rigid body dynamics that we find from Newton's second law. These are independent of the environment in which the rigid body operates and are found in the same way for all types of mechanical systems. We found the inertia matrix M_{RB} and the Coriolis matrix C_{RB} for single rigid bodies in Chaps. 3 and 6.

Added Mass The second term in Eq. (10.1) corresponds to the added mass of the rigid body. Any motion of an underwater vehicle will generate a flow in the surrounding fluid. This effect is always present but becomes much more important for submerged bodies because the density of water is so much higher than the density of air. The forces and moments required for accelerating the fluid that surrounds the rigid body enters into the equations in the same way as the inertia of the rigid body. We can therefore write the kinetic energy of the surrounding fluid as $T_A = \frac{1}{2}\nu^{\mathsf{T}} M_A \nu$. The matrix M_A, however, is not necessarily in the same form as M_{RB}.

In general the inertia matrix M_A is not symmetric. For completely submerged vehicles we can assume that $M_A(\eta) > 0$, i.e., it is symmetric and positive definite (see Definition 9.3), and if we also consider an ideal fluid and assume that the vehicle is operating at low velocities with no waves or currents present, the added mass is given by a constant matrix M_A. Because $T_A = \frac{1}{2}\nu^{\mathsf{T}} M_A \nu$, it is reasonable to write the total inertia matrix of the underwater vehicle as

$$M = M_{RB} + M_A. \tag{10.2}$$

It is important to keep in mind that the inertia matrix M does not necessarily possess the same properties as M_{RB} because of the added mass. Under the conditions mentioned above, however, we can associate both symmetry and positive definiteness with the inertia matrix (10.2). If we further assume that the vehicle has three planes of symmetry, which is often the case, we can neglect the off-diagonal elements and write the inertia matrix M_A as (Fossen 2011)

$$M_A = M_A^{\mathsf{T}} = -\operatorname{diag}(X_{\dot{u}}, Y_{\dot{v}}, Z_{\dot{w}}, K_{\dot{p}}, M_{\dot{q}}, N_{\dot{r}}) \tag{10.3}$$

where $-X_{\dot{u}}\dot{u}$ is the added mass force X in the direction of the x-axis due to an acceleration $\dot{u}$ in the direction of the x-axis.

For a cylindrical rigid body with mass m, length l, and radius r submerged in water with density ρ the elements of the inertia matrix of added mass can be found as (Fossen 1994)

$$\begin{aligned} X_{\dot{u}} &= -0.1m, & K_{\dot{p}} &= 0, \\ Y_{\dot{v}} &= -\pi\rho r^2 l, & M_{\dot{q}} &= -\frac{1}{12}\pi\rho r^2 l^3, \\ Z_{\dot{w}} &= -\pi\rho r^2 l, & N_{\dot{r}} &= -\frac{1}{12}\pi\rho r^2 l^3. \end{aligned} \tag{10.4}$$

We note that rotational motion around the axis through the center of the cylinder (in our case the x-axis) does not generate any flow in the fluid and therefore $K_{\dot{p}} = 0$. As a result we can only guarantee positive semidefiniteness, $M_A \geq 0$. This is not a problem though, because we can still write $M = M_{RB} + M_A > 0$.

The surrounding fluid also adds terms to the Coriolis matrix. Similarly to C_{RB} the Coriolis matrix C_A can always be parameterized so that $C_A = -C_A^{\mathsf{T}}$ and is found in the same way as C_{RB} by replacing M_{RB} with M_A. For a submerged vehicle in an ideal fluid we can write the Coriolis matrix C_A as Fossen (2011)

$$C_A = -C_A^{\mathsf{T}} = \begin{bmatrix} 0 & 0 & 0 & 0 & -Z_{\dot{w}}w & Y_{\dot{v}}v \\ 0 & 0 & 0 & Z_{\dot{w}}w & 0 & -X_{\dot{u}}u \\ 0 & 0 & 0 & -Y_{\dot{v}}v & X_{\dot{u}}u & 0 \\ 0 & -Z_{\dot{w}}w & Y_{\dot{v}}v & 0 & -N_{\dot{r}}r & M_{\dot{q}}q \\ Z_{\dot{w}}w & 0 & -X_{\dot{u}}u & N_{\dot{r}}r & 0 & -K_{\dot{p}}p \\ -Y_{\dot{v}}v & X_{\dot{u}}u & 0 & -M_{\dot{q}}q & K_{\dot{p}}p & 0 \end{bmatrix}. \tag{10.5}$$

The explicit expression for the entries in the added mass Coriolis matrix for a cylinder can be found in (10.4).

The Coriolis matrix can now be written as

$$C = C_{RB} + C_A. \tag{10.6}$$

We see that we can therefore find the Coriolis matrix C including the added mass C_A in the same way that we did in Chap. 6 by replacing M_{RB} with M in the expressions.

Hydrodynamic Damping The third term in Eq. (10.1) represents the dissipative drag and lift forces caused when a rigid body moves in water. The damping matrix $D(\nu)$ can be divided into a linear and a non-linear part. Furthermore, for a completely submerged vehicle with three symmetry planes we can neglect the off-diagonal terms, as we did with the added mass inertia matrix M_A. The damping terms of an underwater vehicle can therefore be written in terms of diagonal matrices representing the linear and non-linear contributions as

$$\begin{aligned} D(\nu) &= D_l + D_{nl}(\nu) \\ &= -\operatorname{diag}(X_u, Y_v, Z_w, K_p, M_q, N_r) \\ &\quad - \operatorname{diag}\big(X_{u|u|}|u|, Y_{v|v|}|v|, Z_{w|w|}|w|, K_{p|p|}|p|, M_{q|q|}|q|, N_{r|r|}|r|\big). \end{aligned} \tag{10.7}$$

The damping parameters are in general quite hard to find, but restricting ourselves to diagonal matrices in this form allows us to find relatively good approximations of the damping forces that act on a completely submerged vehicle.

Gravity and Buoyancy Forces and Moments The restoring forces in Eq. (10.1) arise due to the gravitational and buoyancy forces that act on an underwater vehicle.

The gravitational forces are given by the total mass of the vehicle and act on the center of gravity. These forces are always present and act in the direction of the center of the Earth with an acceleration of $g = 9.81$ m/s^2.

A submerged vehicle will take up the space of the fluid and the volume of the fluid displaced by the vehicle generates buoyancy forces. Let ρ be the density of the fluid and ∇ be the volume displaced by the vehicle, then the buoyancy forces are given by the weight of the displaced fluid $\rho\nabla$. Both the gravitational and buoyancy forces act in the direction of the world frame z-axis and the linear forces are given by

$$F_g^0 = \begin{bmatrix} 0 \\ 0 \\ -mg \end{bmatrix}, \qquad F_b^0 = \begin{bmatrix} 0 \\ 0 \\ \rho\nabla g \end{bmatrix} \tag{10.8}$$

in the inertial frame $\mathscr{F}_0$ and by

$$\begin{aligned} F_g &= R_{0b}^{-1} F_g^0 = \begin{bmatrix} -mg\sin\theta \\ mg\cos\theta\sin\phi \\ mg\cos\theta\cos\phi \end{bmatrix}, \\ F_b &= R_{0b}^{-1} F_b^0 = \begin{bmatrix} \rho\nabla g\sin\theta \\ -\rho\nabla g\cos\theta\sin\phi \\ -\rho\nabla g\cos\theta\cos\phi \end{bmatrix} \end{aligned} \tag{10.9}$$

in the body frame $\mathscr{F}_b$.

Denote the location of the center of gravity by $r_g^b = [x_g\ y_g\ z_g]^\mathsf{T}$ and the center of buoyancy by $r_b^b = [x_b\ y_b\ z_b]^\mathsf{T}$. Then the corresponding moments are given by $r_g^b \times F_g$ and $r_b^b \times F_b$ and the gravitational and buoyancy forces can be collected in one vector $N(\eta) = N_g(\eta) + N_b(\eta)$ given by

$$N(\eta) = \begin{bmatrix} -(\rho\nabla g - mg)\sin\theta \\ (\rho\nabla g - mg)\cos\theta\sin\phi \\ (\rho\nabla g - mg)\cos\theta\cos\phi \\ (y_b\rho\nabla g - y_g mg)\cos\theta\cos\phi + (z_b\rho\nabla g - z_g mg)\cos\theta\sin\phi \\ -(z_b\rho\nabla g - z_g mg)\sin\theta + (x_b\rho\nabla g - x_g mg)\cos\theta\cos\phi \\ (x_b\rho\nabla g - x_g mg)\cos\theta\sin\phi - (y_b\rho\nabla g - y_g mg)\sin\theta \end{bmatrix}. \tag{10.10}$$

Note that for a neutrally buoyant vehicle where $\rho\nabla g = mg$ the linear forces disappear while the moments do not. The moments only disappear from the equations if $r_g^b = r_b^b$. The center of gravity and the center of buoyancy with the corresponding forces acting on the rigid body are illustrated in Fig. 10.2.

Actuation Matrix One great challenge of modeling surface and underwater vehicles, compared to for example robotic manipulators, is the complex and highly non-linear characteristics of the actuators. There are often several different types

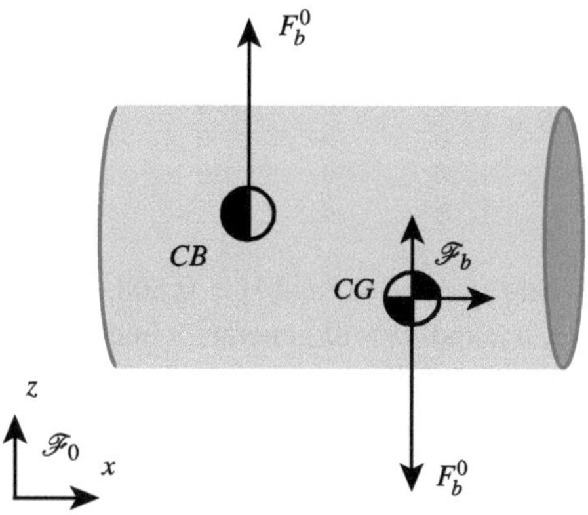

Fig. 10.2 Gravitational and buoyancy forces acting on the center of gravity (CG) and the center of buoyancy (CB), respectively

of actuators with different characteristics that can be used to control an underwater vehicle. The most commonly found actuators are (Fossen 2002)

- *Propellers*: Many underwater vehicles are equipped with propellers mounted aft. The propellers are used to generate a forward motion and is the main source of actuation for vehicles that need to locomote over large areas and with high velocities.
- *Rudders*: Rudders are normally used in conjunction with the propellers to change the steering angle of the vehicle when the vehicle is moving with high velocity.
- *Tunnel thrusters*: For low speed maneuvering, tunnel thruster are propeller units mounted in the hull of the vessel generating a force in the direction of the transverse tube.
- *Azimuth thrusters*: While the tunnel thruster generate a force in a constant direction in the body frame $\mathscr{F}_b$, azimuth thrusters can be rotated by an angle α and generate a force in a direction given by the thruster angle.

The relation between the wrench acting on the vehicle τ and the control input of the thrusters u is highly non-linear. However, it is common to approximate this with a linear relation

$$\tau = Bu \tag{10.11}$$

where $B \in \mathbb{R}^{6\times p_u}$ is a known constant matrix, u is the p_u-dimensional vector of control inputs and p_u is the number of thrusters, propellers, rudders, sterns, etc. The control allocation matrix B is shown in Antonelli (2006) for different underwater vehicles. The ODIN AUV, for example, has 8 fixed thrusters and a control allocation matrix B given by

$$B = \begin{bmatrix} \sin\frac{\pi}{4} & -\sin\frac{\pi}{4} & -\sin\frac{\pi}{4} & \sin\frac{\pi}{4} & 0 & 0 & 0 & 0 \\ \sin\frac{\pi}{4} & -\sin\frac{\pi}{4} & -\sin\frac{\pi}{4} & -\sin\frac{\pi}{4} & 0 & 0 & 0 & 0 \\ 0 & 0 & 0 & 0 & -1 & -1 & -1 & -1 \\ 0 & 0 & 0 & 0 & l_1\sin\frac{\pi}{4} & l_1\sin\frac{\pi}{4} & -l_1\sin\frac{\pi}{4} & -l_1\sin\frac{\pi}{4} \\ 0 & 0 & 0 & 0 & l_1\sin\frac{\pi}{4} & -l_1\sin\frac{\pi}{4} & -l_1\sin\frac{\pi}{4} & l_1\sin\frac{\pi}{4} \\ l_2 & -l_2 & l_2 & -l_2 & 0 & 0 & 0 & 0 \end{bmatrix} \tag{10.12}$$

where $l_1 = 0.381$, and $l_2 = 0.508$. This tells us that the first four thrusters, i.e., u_1, u_2, u_3, and u_4 will generate a linear force in the xy-plane and a rotational momentum around the z-axis of the body frame $\mathscr{F}_b$. Similarly, the last four thrusters will generate a linear force in the direction of the z-axis and momentum around the x- and y-axis, also in the body frame $\mathscr{F}_b$. Because we have eight thrusters to generate a 6-DoF motion we have an overactuated control problem. The optimal control input u can be found as a solution to an optimization problem. This is beyond the scope of this book, but the control allocation problem is discussed in detail in Fossen (2002).

On the other hand, if we have fewer actuators than degrees of freedom we have an underactuated control problem. This is the case when the control allocation matrix has more rows than columns. As an example, the Phantom S3 which is manufactured by Deep Ocean Engineering has only four thrusters and the control allocation matrix becomes (Antonelli 2006)

$$B = \begin{bmatrix} * & * & 0 & 0 \\ 0 & 0 & * & * \\ 0 & 0 & * & * \\ 0 & 0 & * & * \\ * & * & 0 & 0 \\ * & * & 0 & 0 \end{bmatrix}. \tag{10.13}$$

Environmental Disturbances Waves, wind, and ocean currents are always present in marine systems. Wind and waves only affect surface or close to surface vehicles and can be neglected in an underwater environment. The most important environmental disturbance for underwater vehicles is the ocean currents. Ocean currents can reach velocities of up to several meters per second and need to be compensated for in the control loop.

We will assume that the ocean current ν_c is expressed in the inertial frame. Then the relative velocity of the vehicle, i.e., the velocity with respect to its surrounding fluid is given in the body-fixed frame $\mathscr{F}_b$ by

$$\nu_r = \nu - R_{0b}^{-1}\nu_c. \tag{10.14}$$

Since the currents are normally slow varying we can assume that $\dot{\nu}_c = 0$. This relative motion of the fluid affects the hydrodynamics of the vehicle, but not the rigid body motion. The effects of the current are therefore included in the dynamics by using ν_r in the derivation of hydrodynamic terms and the dynamics of the underwater vehicle in the presence of currents is given by

$$\underbrace{M_{RB}\dot{\nu} + C_{RB}(\nu)\nu}_{\text{Rigid body}} + \underbrace{M_A\dot{\nu} + C_A(\nu_r)\nu_r + D(\nu_r)\nu_r}_{\text{Hydrodynamic terms}} + N(\eta) = Bu$$
$$M\dot{\nu} + C_{RB}(\nu)\nu + C_A(\nu_r)\nu_r + D(\nu_r)\nu_r + N(\eta) = Bu \tag{10.15}$$

where we have used that $\dot{\nu}_r = \dot{\nu}$.

10.2.1 Full State Space Dynamics in Vector Form

We have now found the dynamics of an underwater vehicle in terms of the position vector η and velocity vector ν in Eq. (10.1). Following the train of thought in Sect. 6.3 we can eliminate the velocity variables ν from the equations and write the dynamics in terms of position variables η and velocity variables $\dot{\eta}$. We then obtain a complete description of the vehicle dynamics, including the hydrodynamic effects and environmental disturbances with state variables η and $\dot{\eta}$.

The vector representation of the dynamics of an underwater vehicle can be written in terms of the position variables η and velocities $\dot{\eta}$ as

$$\tilde{M}(\eta)\ddot{\eta} + \tilde{C}(\eta,\dot{\eta})\dot{\eta} + \tilde{D}(\eta,\dot{\eta})\dot{\eta} + \tilde{N}(\eta) = \tilde{\tau} \tag{10.16}$$

where

$$\tilde{M}(\eta) = J_b^{-\mathsf{T}}(\eta) M J_b^{-1}(\eta), \tag{10.17}$$
$$\tilde{C}(\eta,\dot{\eta}) = J_b^{-\mathsf{T}}(\eta)\big(C(\eta,\dot{\eta}) - M J_b^{-1}(\eta)\dot{J}_b(\eta)\big)J_b^{-1}(\eta), \tag{10.18}$$
$$\tilde{D}(\eta,\dot{\eta}) = J_b^{-\mathsf{T}}(\eta) D(\eta,\dot{\eta}) J_b^{-1}(\eta), \tag{10.19}$$
$$\tilde{N}(\eta) = J_b^{-\mathsf{T}}(\eta) N(\eta), \tag{10.20}$$
$$\tilde{\tau} = J_b^{-\mathsf{T}}(\eta)\tau. \tag{10.21}$$

Note that Eqs. (10.16)–(10.21) are only valid when $J^{-1}(\eta)$ is non-singular, i.e., for $\theta \neq \pm\frac{\pi}{2}$.

The expressions for $\tilde{M}(\eta)$, $\tilde{C}(\eta,\dot{\eta})$, and $\tilde{\tau}$ were found in Eqs. (6.62)–(6.64). The matrices representing the hydrodynamic effects are found in the same way, i.e., we use $\nu = V_{0b}^B = J_b^{-1}(\eta)\dot{\eta}$ and rewrite $D(\nu)\nu$ as

$$D(\nu)\nu = D\big(J_b^{-1}(\eta)\dot{\eta}\big)J_b^{-1}(\eta)\dot{\eta}. \tag{10.22}$$

Recall that we premultiply the equations with $J_b^{-\mathsf{T}}(\eta)$ (Eq. (10.21)) so that the control signal is collocated with the new velocity variables $\dot{\eta}$. The new damping matrix is then found to be

$$\begin{aligned}\tilde{D}(\eta,\dot{\eta}) &= J_b^{-\mathsf{T}}(\eta)D\big(J_b^{-1}(\eta)\dot{\eta}\big)J_b^{-1}(\eta)\\ &= J_b^{-\mathsf{T}}(\eta)D(\eta,\dot{\eta})J_b^{-1}(\eta).\end{aligned} \tag{10.23}$$

The buoyancy and gravitational vector $\tilde{N}(\eta)$ does not depend on the velocity vector ν and is therefore found simply by $\tilde{N}(\eta) = J_b^{-\mathsf{T}}(\eta)N(\eta)$.

10.3 AUV-Manipulator Dynamics

The velocity transformation matrix of an underwater vehicle-manipulator system is given by the kinematic relations found in Chap. 5 as

$$J(\xi) = \begin{bmatrix} R_{0b}(\Theta) & 0 & 0\\ 0 & T_\Theta(\Theta) & 0\\ 0 & 0 & I\end{bmatrix}. \tag{10.24}$$

The kinematics of the vehicle-manipulator system given by $\dot{\xi} = J(\xi)\zeta$ is thus the same regardless of the environment in which it operates, as expected. In other words, *kinematically* speaking the vehicle and manipulator arm are decoupled which means that the vehicle velocity does not affect the joint velocities and vice versa.

The dynamics of an AUV-manipulator system is given by Antonelli (2006)

$$\dot{\xi} = J(\xi)\zeta, \tag{10.25}$$

$$M(q)\dot{\zeta} + C(q,\zeta)\zeta + D(q,\zeta)\zeta + N(\xi) = \tau \tag{10.26}$$

where $\xi = [\eta^{\mathsf{T}}\ q^{\mathsf{T}}]^{\mathsf{T}}$, $\zeta = [\nu^{\mathsf{T}}\ \dot{q}^{\mathsf{T}}]^{\mathsf{T}}$.

$M(q) \in \mathbb{R}^{(6+n)\times(6+n)}$ is the inertia matrix including added mass. The added mass is in the same form as the rigid body inertia matrix M_{RB} so the inertia matrix of an underwater VM system depends only on the joint positions and not on the location of the vehicle. Also, the Coriolis and centripetal matrix $C(q,\zeta) \in \mathbb{R}^{(6+n)\times(6+n)}$ is found in the same way by substituting M_{RB} with $M = M_{RB} + M_A$. Therefore, the Coriolis matrix is also independent of the vehicle pose η.

The damping matrix $D(q,\zeta) \in \mathbb{R}^{(6+n)\times(6+n)}$ represents the dissipative forces. From Eq. (10.7) is it clear that for a sumberged VM system operating at low velocities the damping matrix becomes diagonal. This is a good approximation for most underwater VM systems due to the low velocities that these systems normally operate at. Also the damping matrix is independent of the vehicle position η.

The vector $N(\xi)$ representing the gravitational and buoyancy forces, however, does depend on the vehicle pose ϕ, θ, and ψ. This is easily seen from Eq. (10.10) which depends on the Euler angles. The gravitational and buoyancy forces also depend on the joint positions of the manipulator.

Finally, τ is the vector of control forces and moments working on the mechanism and is given by

$$\tau = \begin{bmatrix} \tau_V \\ \tau_q \end{bmatrix} = \begin{bmatrix} B & 0 \\ 0 & I \end{bmatrix} u \tag{10.27}$$

where $u = [u_V^\mathsf{T}\ u_q^\mathsf{T}]^\mathsf{T}$ is the control input. For the manipulator arm this relation is linear, i.e., the control signal u_i affects only the moments τ_i of joint i. This is one of the advantages of choosing the generalized coordinates q as the joint positions in this way. As we saw in the previous section this is not the case with the vehicle and the control allocation matrix depends on the type and location of the actuators of the vehicle.

10.3.1 Operational Space Approach

Alternatively we can write the dynamics in terms of the vector $\varsigma = [\eta^\mathsf{T}\ (x_{0e}^0)^\mathsf{T}]^\mathsf{T} \in \mathbb{R}^{12}$ where x_{0e}^0 is the end-effector position/orientation vector in the inertial frame. We then first write

$$\dot{\varsigma} = J_w(\xi)\zeta \tag{10.28}$$

where

$$J_w(\xi) = \begin{bmatrix} J_b(\eta_2) & 0 \\ J_e(\eta_{0e,2})\mathrm{Ad}_{g_{eb}} & J_e(\eta_{0e,2})J_{m,g}(q) \end{bmatrix} \in \mathbb{R}^{12\times(6+n)}. \tag{10.29}$$

The dynamics in operational space is now found as in Chap. 8, but we also need to include the damping terms and the gravity and buoyancy.

The operational space dynamic equations of an underwater vehicle manipulator system is given by

$$\bar{M}(\xi)\ddot{\varsigma} + \bar{C}(\xi,\zeta)\dot{\varsigma} + \bar{D}(\xi,\zeta)\dot{\varsigma} + \bar{N}(\xi,\zeta)\dot{\varsigma} = \bar{\tau}. \tag{10.30}$$

where we define

$$\bar{M}(\xi) = J_w^{-\mathsf{T}}(\xi)M(q)J_w^{-1}(\xi)$$

$$\begin{aligned}
\bar{C}(\xi,\dot{\xi}) &= J_w^{-\mathsf{T}}(\xi)\big(J_w^{-\mathsf{T}}(\xi)C(q,\dot{\xi})J_w^{-1}(\xi) - M(q)J_w^{-1}(\xi)\dot{J}_w(\xi)J_w^{-1}(\xi)\big)\\
\bar{D}(\xi,\zeta) &= J_w^{-\mathsf{T}}(\xi)D(\xi,\dot{\xi})J_w^{-1}(\xi)\\
\bar{N}(\eta) &= J_w^{-\mathsf{T}}(\xi)N(\xi),\\
\bar{\tau}(\xi) &= J_w^{-\mathsf{T}}(\xi)\tau.
\end{aligned}$$

10.4 Configuration States

The configuration space of underwater vehicles, and also some underwater vehicle-manipulator systems, is very large both in translation and orientation. This means that the configuration space should be valid for all attitudes of the rigid body. As we have seen, this is not the case when the Euler angles are used to represent the attitude of the vehicle.

In this section we show how to derive the AUV-manipulator dynamics without the presence of singularities. The inertia matrix of the AUV is derived in two steps. First, M_{RB} is found from (8.30). Then the added mass $M_A = M_A^{\mathsf{T}} > 0$ is found from the hydrodynamic properties and we get $M = M_{RB} + M_A$. We can now use M instead of M_{RB} to derive the Coriolis and centripetal matrix which gives us $C = C_{RB} + C_A$ (Fossen 2002). As the configuration space of an AUV can be described by the matrix Lie group $SE(3)$ we can follow the approach presented in Chap. 8.3 to obtain a well-defined formulation of the dynamic equations. Following the mathematics of Eqs. (8.136)–(8.139), the Coriolis matrix is then found as (From et al. 2010a, 2010b)

$$C(Q,v) = \sum_{k=1}^{n} \frac{\partial M}{\partial q_k}\dot{q}_k - \frac{1}{2}\begin{bmatrix} 2\tilde{\mathrm{ad}}_{(M(q)v)_V} & 0 \\ \frac{\partial^{\mathsf{T}}}{\partial q}\Big([M_{VV}\ M_{qV}^{\mathsf{T}}]\big[\begin{smallmatrix} V_{0b}^B \\ \dot{q}\end{smallmatrix}\big]\Big) & \frac{\partial^{\mathsf{T}}}{\partial q}\Big([M_{qV}\ M_{qq}]\big[\begin{smallmatrix} V_{0b}^B \\ \dot{q}\end{smallmatrix}\big]\Big)\end{bmatrix}. \tag{10.31}$$

The dynamic equations can now be written as

$$M(q)\dot{v} + C(q,v)v + D(q,v)v + N(Q) = \tau. \tag{10.32}$$

Here, $v = [(V_{0b}^B)^{\mathsf{T}}\ \dot{q}^{\mathsf{T}}]^{\mathsf{T}}$ where V_{0b}^B is the velocity state of the AUV and $\dot{q}$ the velocity state of the manipulator, and $Q = \{g_{0b}, q\}$ where $g_{0b} \in SE(3)$ determines the configuration space of the AUV (non-Euclidean) and q the configuration space of the manipulator (Euclidean). We note that the singularity in (10.25) is eliminated and the state space (Q, v) is valid globally. The rigid body properties of the dynamics is thus found in the conventional way and it only remains to add the hydrodynamic effects and the gravity and buoyancy forces.

$D(q,v)$ is found in the same way as for the conventional approach. We see this because the damping matrix does not depend on the vehicle pose g_{0b}, only the joint positions q and the velocities V_{0b}^B and $\dot{q}$. The damping matrix that corresponds to the vehicle can therefore be found as in Eq. (10.7) in terms of the variables V_{0b}^B. The same is the case for the manipulator arm.

The gravitational and buoyancy forces $N(Q)$ need to be treated somewhat differently using the formulation in (10.32) compared to the previous section. In Eq. (10.10) we found the gravitational and buoyancy forces in terms of the Euler angles (ϕ, θ, ψ). In this section we will use the matrix representation R_{0b} to describe the orientation. Let the wrench associated with the gravitational force of link i with respect to coordinate frame $\mathscr{F}_i$ be given by

$$F_g^i = \begin{bmatrix} f_g \\ \hat{r}_g^i f_g \end{bmatrix} = -m_i g \begin{bmatrix} R_{0i} e_z \\ \hat{r}_g^i R_{0i} e_z \end{bmatrix} \tag{10.33}$$

where $e_z = [0\ 0\ 1]^\mathsf{T}$ and r_g^i is the center of mass of link i expressed in frame $\mathscr{F}_i$. In our case $\mathscr{F}_i$ is chosen so that r_g^i is in the origin of $\mathscr{F}_i$ so we have $r_g^i = 0$. The equivalent joint torque associated with link i is given by

$$N_g^i(Q) = J_i^\mathsf{T}(q) \mathrm{Ad}_{g_{0i}}^\mathsf{T}(Q) F_g^i(Q) \tag{10.34}$$

where J_i is the geometric Jacobian and $\mathrm{Ad}_{g_{0i}} = \mathrm{Ad}_{g_{0b}} \mathrm{Ad}_{g_{bi}}$ is the transformation from the inertial frame to frame i. We note that both R_{0i} and $\mathrm{Ad}_{g_{0i}}$ depend on the vehicle configuration with respect to the inertial frame. The total effect of the gravity from the vehicle and the links is then given by

$$N_g(Q) = \sum_{i=b}^{n} N_g^i(Q) \tag{10.35}$$

which enters (8.118) in the same way as τ.

Similarly the buoyancy forces are found by denoting the center of buoyancy of each link as $\hat{r}_b^i$. Due to the choice of the link reference frame with the origins in the center of gravity, the center of buoyancy is normally not equal to zero. The total effects of the gravitational and buoyancy forces of link i then becomes

$$F_i = F_g^i + F_b^i = -m_i g \begin{bmatrix} R_{0i} e_z \\ 0 \end{bmatrix} + \nabla \rho g \begin{bmatrix} R_{0i} e_z \\ \hat{r}_b^i R_{0i} e_z \end{bmatrix}. \tag{10.36}$$

References

Antonelli, G. (2006). *Underwater robots. Motion and force control of vehicle-manipulator systems*. Berlin: Springer.

Casalino, G., Angeletti, D., Bozzo, T., & Marani, G. (2001). Dexterous underwater object manipulation via multi-robot cooperating systems. In *Proceedings 2001 ICRA: Vol. 4. IEEE international conference on Robotics and automation* (pp. 3220–3225)

Choi, H., Hanai, A., Choi, S., & Yuh, J. (2003). Development of an underwater robot, ODIN-III. In *IEEE/RSJ international conference on proceedings of intelligent robots and systems*, Las Vegas, NV, USA (Vol. 1, pp. 836–841).

Evans, J., Redmond, P., Plakas, C., Hamilton, K., & Lane, D. (2003). Autonomous docking for intervention-auvs using sonar and video-based real-time 3d pose estimation. In *Proceedings of OCEANS* (Vol. 4, pp. 2201–2210).

Faltinsen, O. M. (1990). *Sea loads on ships and offshore structures*. Cambridge: Cambridge University Press.

Fossen, T. I. (1994). *Guidance and control of ocean vehicles*. Chichester: Wiley.

Fossen, T. I. (2002). *Marine control systems*. Trondheim: Marine Cybernetics AS. 3rd printing.

Fossen, T. I. (2011). *Handbook of marine craft hydrodynamics and motion control*. New York: Wiley.

From, P. J., Duindam, V., Pettersen, K. Y., Gravdahl, J. T., & Sastry, S. (2010a). Singularity-free dynamic equations of vehicle-manipulator systems. *Simulation Modelling Practice and Theory*, *18*(6), 712–731.

From, P. J., Pettersen, K. Y., & Gravdahl, J. T. (2010b). Singularity-free dynamic equations of AUV-manipulator systems. In *Proceeding of symposium on intelligent autonomous vehicles*, Lecce, Italy.

Lane, D., Davies, J., Casalino, G., Bartolini, G., Cannata, G., Veruggio, G., Canals, M., Smith, C., O'Brien, D., Pickett, M., Robinson, G., Jones, D., Scott, E., Ferrara, A., Angelleti, D., Coccoli, M., Bono, R., Virgili, P., Pallas, R., & Gracia, E. (1997). Amadeus: advanced manipulation for deep underwater sampling. *IEEE Robotics & Automation Magazine*, *4*(4), 34–45.

Moore, S., Jensen, V., Bohm, H., & Johnston, N. (2009). *Underwater robotics: science, design and fabrication*. MATE Center/Monterey Peninsula College.

Sanz, P. J., Ridao, P., Oliver, G., Melchiorri, C., Casalino, G., Silvestre, C., Yvan, P., & Turetta, A. (2010). TRIDENT: a framework for autonomous underwater intervention missions with dexterous manipulation capabilities intelligent autonomous vehicles. In *Proceeding of symposium on intelligent autonomous vehicles*, Lecce, Italy.

Sutton, R., & Roberts, G. (2006). *Advances in unmanned marine vehicles*. Institution of Engineering and Technology.

Wang, H. H., Rock, S. M., & Lee, M. J. (1996). OTTER: the design and development of an intelligent underwater robot. *Autonomous Robots*, *3*(2–3), 297–320.

Yuh, J., Choi, S., Ikehara, C., Kim, G., McMurty, G., Ghasemi-Nejhad, M., Sarkar, N., & Sugihara, K. (1998). Design of a semi-autonomous underwater vehicle for intervention missions (SAUVIM). In *Proceedings of the international symposium on underwater technology* (pp. 63–68).

Chapter 11
Spacecraft-Manipulator Systems

Over the last decades robotic technology has emerged as the safest and most cost efficient solution to a wide variety of tasks frequently encountered in space exploration. The main reason for this is the extreme costs involved with transporting humans to extraterrestrial locations like orbiting space stations and satellites. Furthermore, manned missions pose great dangers to astronauts and should be avoided whenever it is possible to perform such missions more safely using robotic solutions either autonomously or remotely operated from the Earth.

This chapter is concerned with robotic solutions utilized in orbiting missions, i.e., satellites and space stations that orbit the Earth and have one or more robotic manipulators mounted on them. These robots thus operate in a free-fall environment. Rovers and other kinds of wheeled robots used for missions to other planetary bodies are treated in the next chapter.

11.1 Introduction

Orbiting robotic systems consist of a free-floating base, typically a satellite or a space station, with a robotic manipulator attached to it. In this chapter we will discuss these spatial orbiting robotic systems in detail, i.e., VM systems in a free-fall environment with an unactuated free-floating or an actuated free-flying base. The free-fall environment poses several challenges to the operation and control of these robots that are not present in Earth-based systems.

The main difference between space manipulators and Earth-based robots is the lack of a fixed base on which the robot is mounted. An important property of space robots that is not present in fixed-base manipulators is therefore that the forward kinematics map does not only depend on the geometry of the mechanism, but also on the inertial properties of the spacecraft and the manipulator arm. Furthermore, the dynamic coupling between the manipulator and the supporting base is very present. Because of this dynamic coupling the reaction forces and moments that arise when the robot arm is in motion affect the position and orientation of the base. This base

P.J. From et al., *Vehicle-Manipulator Systems*, Advances in Industrial Control,
DOI 10.1007/978-1-4471-5463-1_11,

motion thus needs to be included in the mathematical model of the space robot and compensated for in the control law. In some cases we can use this dynamic coupling to control the motion of the base using the robotic manipulator.

Due to the lack of a fixed base there is no natural way to choose the inertial reference frame for space robots. A frame attached to the robot base, for example, will not be inertial because of the motion of the base caused by the manipulator motion. However, the center of gravity of the robotic system can be chosen as an inertial frame if this is constant. If the center of gravity is not constant, for example due to external disturbances or the use of thrusters, we can choose the center of gravity at time zero, i.e., assuming no external forces. Another option is to choose a point on other nearby objects, such as space stations or satellites that the robot arm is to perform work on.

A wide variety of robotic manipulators have been and are currently operating is space. This includes research satellites with robotic arms such as the Japanese ETS-VII satellite and operating arms such as the Canadarms mounted on the space shuttle and the International Space Station. Even though the number of systems operating in outer space in not very large, space robotics has become a fairly mature field with several successful missions over the last few decades.

11.1.1 Operating in Space

Operating robotic systems in zero gravity or free fall environments is very different from operating robots mounted on a fixed base on Earth. First of all, for most robotic applications the gravitational forces can be approximated to zero and can therefore be removed from the dynamic equations. We can do this if the mechanism is operating in an area with negligible gravitational forces—which we may call a zero gravity environment—or when the gravitational forces are canceled by the centrifugal forces—which is called a free fall environment and occurs in orbiting robotic systems. When operating in a free fall environment it is important to remember that the gravitational forces are not absent, but for motions with a time duration much shorter than one orbit period (most often the case for robotic operations performed on satellites and space stations), neglecting the gravitational forces is a very accurate approximation (Bryson 1994). In this chapter we will therefore write the dynamics without the presence of gravitational forces. For orbiting systems this does not mean that the gravitational forces are not present, but that the trajectory of the robotic system is such that they are canceled by the centrifugal forces, and therefore left out of the equations.

Secondly, the lack of a fixed base and reference point represent a challenge when modeling spacecraft-manipulator systems. Because the systems that we are interested in are constantly orbiting the Earth, they will be in constant acceleration and therefore also in constant motion with respect to any choice of inertial reference frame. It is, however, common to limit the "world" that we are interested in to the objects that orbit the Earth and other nearby objects. The reference frame attached to some point on the robotic system or an object that the robot is to perform work on is then chosen as the inertial reference frame. It is important to note that this is only a valid choice as an inertial reference frame if we limit our view of the world in this way, i.e., that other objects, such as the Earth, sun and moon, and other external forces do not affect the motion of our system. This is normally a very good approximation, for example when the gravitational forces from the Earth are canceled by the centrifugal forces and can be neglected for the other objects in the solar system.

A wide variety of space manipulation concepts have been proposed over the last decades. Depending on whether or not actuation is utilized to control the spacecraft position and orientation, we can divide these systems into free-flying and free-floating space manipulators. In a free-flying robotic system the position and orientation of the base, in our case the spacecraft, is actively controlled by the spacecraft's actuators. This allows us to completely control both the base configuration and the manipulator arm. If the control objective is to control the end-effector motion this results in a redundant system so that the end-effector, in addition to some secondary task can be controlled. This task is most often to maintain a fixed orientation of the spacecraft to allow for continuous communication and so on. It is important to keep in mind, however, that due to the dynamic coupling between the manipulator and the spacecraft, the motion of the manipulator arm will constantly affect the motion of the base. The main concern with these systems is therefore the excessive fuel consumption required to compensate for the dynamic coupling between the manipulator and the base while maintaining a constant base attitude.

A free-floating robotic system, on the other hand, does not use spacecraft actuation to compensate for the manipulator motion. The spacecraft motion is therefore not controlled directly, but arises as a result of the dynamic coupling between the manipulator and the base. In this case we can either choose not to control the spacecraft motion at all, or we can use the manipulator arm to obtain the required motion also for the spacecraft. In many cases it is necessary to generate a manipulator motion which guarantees that the spacecraft orientation remains almost constant so that antennas and other instruments point in the right direction. Free-floating space robots of this kind possess a global redundancy that can be utilized to obtain this secondary task, at least to some extent. Global redundancy arises due to the non-holonomic characteristics of free-floating robots and is discussed in detail in this chapter. Global redundancy is different from the local redundancy that is found in redundant fixed-base manipulators because it does not allow for internal motion. The redundancy arises because different paths to the same end-effector location will lead to different orientations of the spacecraft. This allows us to control the orientation of the spacecraft by choosing a certain path in joint space that takes us to the desired end-effector configuration and at the same time controls the spacecraft orientation.

Another property of free-floating robots is the fact that the forward kinematics map cannot be found in closed form. This is clear because the attitude of the spacecraft does not only depend on the current joint positions, but also the path taken to reach the configuration, as discussed above. Furthermore, the kinematics map depends on the inertial properties of the VM system. This is due to the dynamic coupling between the manipulator and the base, which is not present in fixed-base systems.

Also the dynamics of free-floating space robots is different from the fixed base case. When the spacecraft has a total mass close to the mass of the manipulator arm, there is a very strong dynamic coupling between the manipulator and the base. Due to this dynamic coupling the motion of the manipulator affects the motion of the base, which can be compensated for either by the spacecraft actuators or by applying the appropriate manipulator control. Likewise, the motion of the base obviously affects the motion of the manipulator arm. The dynamic coupling decreases as the mass of the spacecraft increases, and if the mass of the manipulator arm is negligible compared to the spacecraft, the manipulator arm behaves as a fixed-base manipulator.

The cost of bringing equipment to space is extremely high, and there is therefore much focus on weight. As a result of this the robotic arms that operate in space are designed to be light weight and are therefore flexible and with relatively small actuators. This complicates the control of these systems due to low frequency resonances and nonlinear actuator saturation (Dubowsky and Papadopoulos 1993). Much work has thus been done on modeling flexible structures in zero-gravity environments. This is, however, outside the scope of this book and we refer to Yoshikawa et al. (1996), Meirovich and Kwak (1989) and Nenchev et al. (1999) for details on this topic.

There are also other more technical challenges of operating in space. Lubrication under ultrahigh vacuum conditions as one example has shown to cause problems for systems that are to operate over longer periods of time (Dube et al. 2003). Synthetic lubricants are therefore used with success in several space systems. It is also worth noting that what lubricant is used affects the friction characteristics of the joints and should therefore be accounted for in the dynamic models.

Space systems also experience extreme temperature variations from high temperatures when exposed to direct sun light and far lower temperatures when in the shade. This can affect both the mechanical and dynamic properties of the system, in addition to the electrical system of the spacecraft. In most cases these effects need to be taken into account to guarantee robust and accurate operation.

Even though space exploration is challenging compared to Earth-based systems, space manipulator design also benefits from the free-fall environment in some areas. It is, for example, possible to construct robots with extremely high redundancy and with several joints. This type of robots, which may resemble snake robots (Liljebäck et al. 2013), are able to support their own weight in space due to the small gravitational forces. On Earth, however, such robots would collapse due to their own weight when the number of joints becomes too large. This allows for more redundant robots in space than on Earth, and also more fault tolerant robots because the robots can continue operation even after several joint failures. Similarly, the free-fall environment allows for effective control with very small actuators. In fact, most space manipulators are able to handle very high payloads in space, while they cannot even withstand the weight of the manipulator arm itself if they were to be placed in the Earth's gravity field.

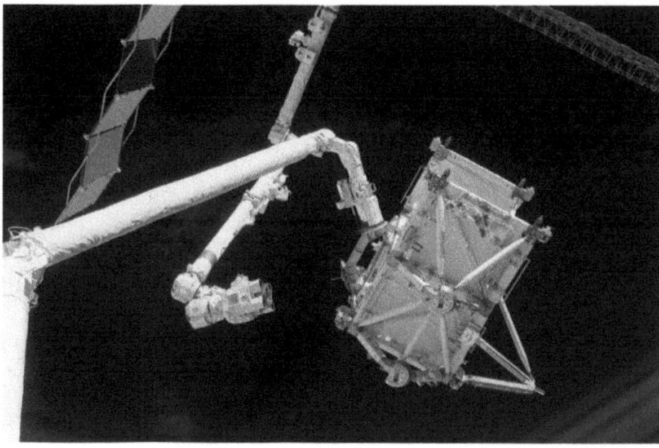

Fig. 11.1 The Shuttle Remote Manipulator System of the Space Shuttle Discovery hands over a new section to the Canadarm2 of the International Space Station. Courtesy of NASA

11.1.2 Space Exploration

The history of space exploration is short but diverse, and a wide variety of vehicles have been launched into space. The first satellite was launched by the Soviet Union in 1957. The Sputnik 1 satellite completed a total of 1.440 orbits around the Earth before it entered into the atmosphere the year after. The first space probe to successfully perform an interplanetary mission was the American Mariner 2 which passed within 35.000 kilometers of the planet Venus in 1962.

The first robotic arm to be launched into space was the Shuttle Remote Manipulator System (SRMS), also known as the Canadarm, mounted on the Space Shuttle. The arm was launched in November 1981 and has been in operation until today. The Canadarm is a 15.2 meter long arm with 6 degrees of freedom and a gripper attached to the end. The total weight of the arm is 410 kg, but it can retrieve payloads of up to 293 tons due to the free-fall environment. It is interesting to note that the actuators are not powerful enough to lift even the weight of the arm itself when on Earth. The Canadarm has been used for over 100 operations and is normally used for repairing, retrieving, and deploying satellites; assisting humans during extravehicular activities; and for remote inspection tasks. The Canadarm can be seen handling an object in Fig. 11.1.

Another early mission including a robotic arm in space was the MIR space station. This Soviet Union/Russian space station was the first modular space station and was in operation from 1986 to 2001. MIR was a microgravity research laboratory which conducted research in biology, astronomy, meteorology, and physics, to name a few. Four of the modules were equipped with the Lyappa robotic arm used for assembling the modules of the space station.

In 1997 the National Space Development Agency of Japan (NASDA) launched the Engineering Test Satellite No. 7 (ETS-VII), the first ever satellite to be equipped with a robotic arm. The ETS-VII performed several successful docking operations using the manipulator arm. The main objectives of the project were (Oda et al. 1996):

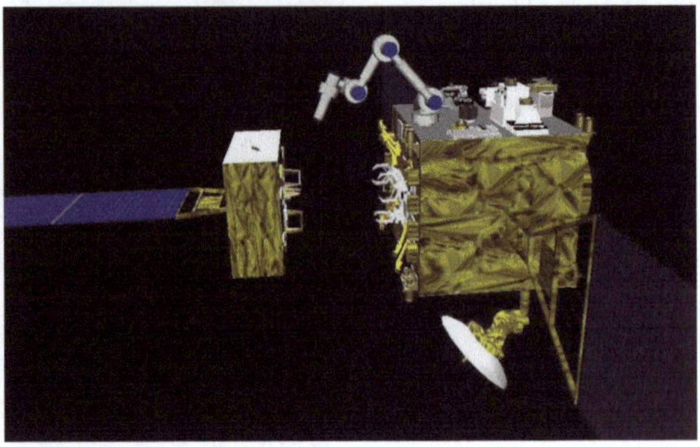

Fig. 11.2 Illustration of a docking operation performed by the ETS-VII research satellite. Courtesy of Tohoku University

- performance evaluation of a satellite-mounted robotic system;
- coordinated control of the satellite attitude and robot arm;
- teleoperation of the robot arm;
- demonstration of in-orbit satellite servicing.

The system consisted of two modules. The main module, the chaser satellite, had a mass of 2.500 kg, while the secondary module, the target satellite, had a mass of 400 kg. A 2 meter long 6-DoF robot arm with a total mass of 150 kg was mounted on the chaser satellite. The dynamic coupling between the robot and the base is therefore substantial. The project demonstrated the successful execution of experiments which gained insight into operations such as docking, fuel transfer, assembling, and berthing. Several successful docking operations were performed with the chaser and target satellite.

The ETS-VII project differs from other space project in that its main intention was to gain insight into free-floating space robots and space robotics in general. Also teleoperation tasks with a time delay of up to 7 seconds were demonstrated (Oda et al. 1996; Inaba and Oda 2000). Operation of the robot arm with almost no change in the attitude orientation was demonstrated during a point-to-point task (Nenchev et al. 1999). An illustration of the ETS-VII robotic satellite is shown in Fig. 11.2.

The Japanese also developed a robotic arm which was mounted on the Japanese Experiment Module at the International Space Station (ISS). The JEMRMS arm is a macro-micro manipulator arm where the 6-DoF macro manipulator performs the rough positioning and the 6-DoF micro manipulator performs the fine manipulation (Abiko and Yoshida 2003). The macro arm has a length of 10 meters and weighs 780 kg while the micro manipulator is 2,2 meters long and weighs 190 kg. The JEMRMS arm is shown in Fig. 11.3 together with the Canadarm2.

The first humanoid robot to operate in space is the Robonaut developed at NASA's Johnson Space Center (JSC) (Fig. 11.4). Robonaut is a dual-arm robot designed to perform dexterous manipulation and other complex tasks. The Robonaut is to operate side by side with humans, but it is also intended for extravehicular operations too dangerous for humans. The robot arrived at the International Space Station in February 2011 where it was set to operate at a fixed location inside the station.

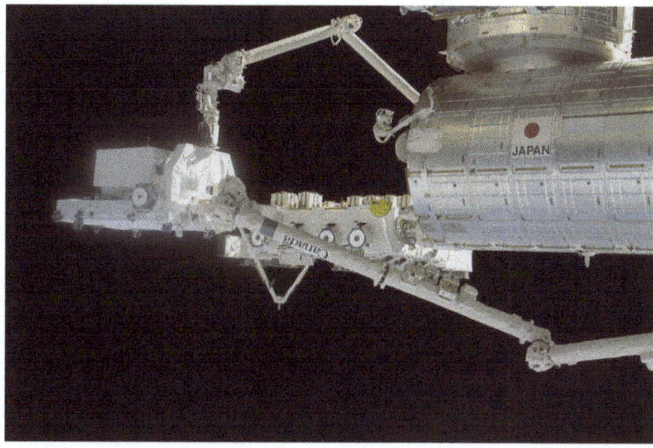

Fig. 11.3 The JEMRMS arm (*top*) and the Canadarm2 (*bottom*) working at the Japanese Experiment Module. Courtesy of NASA

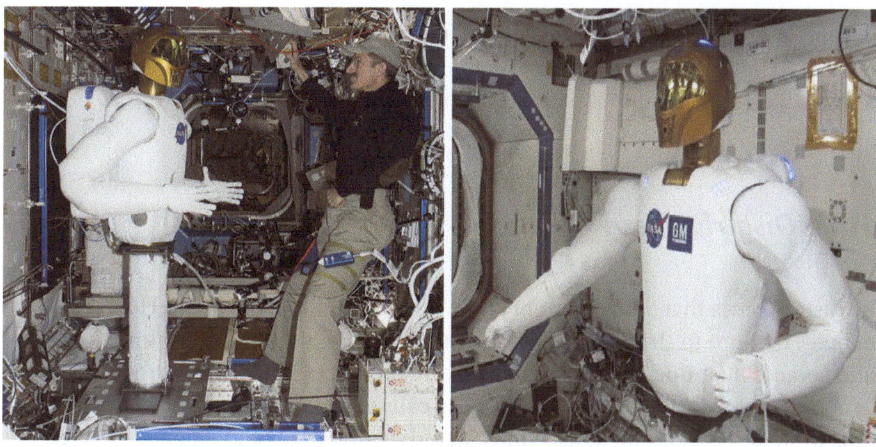

Fig. 11.4 The Robonaut humanoid robot installed at the International Space Station. Courtesy of NASA

A larger and more advanced version of the SRMS arm is the MSS arm (Mobile Servicing System), which is mounted on the International Space Station. The MMS arm consists of a mobile base, a more advanced version of the Canadarm, called the Canadarm2, and dexterous manipulator hand. The Canadarm2 is a 17.6 meters long arm with seven degrees of freedom and has a total mass of 1.800 kg. The robot is mounted on a rail which allows it to move along the platform. The Canadarm2 also has automatic collision avoidance and the force and moment sensors provide haptic feedback to the operator. The Canadarm and Canadarm2 robotic arms are shown in a joint hand-off operation in Fig. 11.1.

In March 2008 the last part of the Mobile Servicing System was added to the International Space Station. The Canada Hand, or the Special Purpose Dexterous Manipulator (SPDM), is a two-armed robot but is often referred to as a hand because it can be attached to the Canadarm2 robotic arm and taken to any location on the space station (Fig. 11.5). Each arm is 3.35 meters long and the total weight of

Fig. 11.5 The DEXTRE dual-arm robot at the International Space Station. Courtesy of NASA

the hand is 1662 kg. The first mission of the SPDM hand was performed in February 2011. The total weight of the robotic system is approximately 5.000 kg, which accounts for 1–1.5 % of the total weight of ISS.

11.1.3 Disturbances

The disturbances that are present in space are in general quite small. However, continuous exposure to these disturbances can affect the motion of orbiting objects to quite some extent due to the absence of a fixed base. Earth-based robotic systems are not extremely affected by disturbances caused by effects such as small changes in the system momentum, solar winds, aerodynamic drag, and so on. However, because there is very little resistance to motion in free-floating systems compared to systems with a fixed base on Earth, disturbances that can be neglected on Earth normally need to be taken into account in the modeling and control of spacecraft and satellites.

One example of a disturbance that can be neglected in Earth-based systems is the solar radiation pressure. Solar radiation is caused by the momentum of the photons that hit a surface. This is a rather weak force which at the Earth's distance from the sun is about 4.5×10^{-6} Newtons per square meter if the radiation is absorbed and approximately half that if it is reflected. By intelligent design of the spacecraft the effects of the solar radiation can be either decreased or exploited to control the spacecraft. Solar radiation is therefore also used as actuation of free-floating satellites and spacecrafts, see Sect. 11.1.4.

Gas leakage is another important source of disturbance in free-floating systems. Gas leakage is always present, but also in this case the design of the spacecraft can reduce the effects of the disturbance. In the same way as solar radiation can be used as control, a controlled gas leakage can be used to control the position and orientation of the spacecraft. This is discussed in more detail in the next section.

Furthermore, the Earth's gravity gradient and magnetic fields affect the motion of spacecraft in high orbital altitude. For spacecraft and satellites operating at lower altitudes, i.e., below 1000 km, aerodynamic drag also needs to be taken into account.

Finally, at spacecraft where astronauts move around, the attitude will be affected by this motion. The same is the case with other moving mechanical parts, such as a robotic arm mounted on the spacecraft. This motion is caused by the conservation of momentum, so if one part of the system moves, the remaining parts of the mechanical system will move so that the momentum remains constant. This effect is a rather big disturbance, but the principle can also be used very efficiently as actuation to control both linear and angular motion.

It is important to note that all the disturbances mentioned above can also be used for control. A thruster, discussed in the next section, is of course just a controlled gas leakage used for control purposes. Similarly, solar radiation can also be used as control torque by positioning the solar sails so that the radiation causes the desired motion. Also, the gravity gradient can be used for stabilization through the use of a gravity boom. Spacecraft actuators are discussed in more detail in the next section.

11.1.4 Actuators

Spacecraft control is mainly concerned with controlling the attitude of the spacecraft. In some special cases, however, also the translational motion needs to be controlled. Controlling the orientation and controlling the translational motion of a spacecraft are quite different in nature because the former can be performed while under conservation of linear momentum while the latter requires a change in the linear momentum of the satellite. The same concepts can be transferred to vehicle-manipulator systems in orbital motion.

Bryson (1994) divide spacecraft control into fast and slow attitude control. Fast control is applied to motion with a large control bandwidth compared to the orbit rate. This is normally the case for robotic motion which has a high frequency compared to the orbit frequency of the satellite. In this case the disturbance torques discussed in the previous section are treated as disturbances and generally canceled, as opposed to attempting to leverage the forces for control purposes.

For slow attitude control, on the other hand, the control bandwidth is comparable to that of the spacecraft orbit time. In this case the gravitational and magnetic torques are used as control torques to stabilize the spacecraft without the use of fuel-consuming thrusters. Slow attitude control of the spacecraft is applied in order to maintain the orientation of the satellite fixed during the orbit. However, for robotic motions where the bandwidth is large, thrusters generally need to be used to control the spacecraft motion.

An important source of control for attitude control in robotic operations is thrusters. Thrusters used in space are normally controlled in an on-off fashion so that they are either open or closed, and do not provide a continuous control torque like normal thrusters. Due to the small torques required for spacecraft attitude control, the opening of the thruster valve is very small which may allow for ice particles and dust to get stuck in the valve openings. This will again lead to gas leakage which causes disturbances as discussed above and also reduces the gas supply and operation time of the system.

Continuous thrusters are more sensitive to particles and dust than on-off thrusters. Because of this, on-off thrusters are the preferred choice because they reduce gas leakage and thus also the resulting disturbances. On-off thrusters are typically opened for as little as a few milliseconds, but can be fired repeatedly several times in a row. Each firing of the thruster leads to a finite change in the spacecraft angular velocity. It this therefore possible to control the spacecraft orientation to an almost constant value, but not exactly to zero.

Because the bang-bang control described above cannot control the attitude error to zero, but is typically limited to an accuracy of 0.1 to 1 degrees, attitude control is often performed with thrusters in combination with other types of actuation. Reaction wheels can be used to remove the remaining error in the attitude control and the thrusters are only fired whenever the reaction wheels reach saturation, when large motions are required, or when the wheels are de-spun. Reaction wheels are driven by proportional electromagnetic motors and the control is obtained by transferring unwanted angular momentum from the spacecraft to the reaction wheels and stabilizing the spacecraft attitude in this way.

In slow control, the gravity and magnetic fields can also be used for control. Torques arise when the electric currents on the spacecraft interact with the Earth's magnetic field. Similarly the gravity field can produce a torque by perturbing the pitch angle slightly from the equilibrium. Both these torques are rather small and can also be used in combination with reaction wheels.

Furthermore, conservation of the momentum vector can be used to keep the attitude of a spacecraft in a specified direction. When a rigid body rotates around its principal axis of inertia it will maintain this axis fixed in the inertial space. This concept has been used in satellite control from the very beginning of space exploration and was also used on the first satellite in synchronous orbit in 1963. Gimbaled momentum wheels is another passive stabilizing system which counteracts drift of the spin axis due to external disturbances.

Up until now we have only considered actuation and control of the attitude of the spacecraft. For certain operations also the translational motion of spacecraft needs to be controlled. In rendezvous tasks, for example, where two spacecraft or satellites approach each other in a docking operation, the translational motion needs to be controlled very accurately.

Translational motion means a change in the spacecraft center of gravity with respect to the inertial space. Translational motion in space is obtained by Newton's third law which states that action and reaction forces between two rigid objects are equal but of opposite signs (Newton 1687). In other words, to every action—in our case the firing of a thruster—there is an equivalent and opposed reaction—the motion of the spacecraft. By action Newton means momentum, so every particle that is fired from the thruster with a linear momentum mv generates an equivalent force on the spacecraft. Thus, due to the huge difference in mass between the particles being fired from the thruster and the spacecraft, the motion on the spacecraft is limited even when the velocity of the mass exiting the thrusters is very high. Because thrusters fire large amounts of particles at extremely high velocities, they are well suited to generate translational motion for free-floating and free-flying objects.

In Sect. 11.1.3 we mentioned that solar radiation can be used to control spacecraft and satellites. This is obtained by mounting large solar sails on the spacecraft. Because the momentum caused by the photons is rather small, these sails need to be very big to be effective. The first successful use of solar sails in space was performed by the Japan Aerospace Exploration Agency (JAXA) which succeeded in controlling the spacecraft using solar sails in the IKAROS project (Interplanetary Kite-craft Accelerated by Radiation Of the Sun). This spacecraft, which was launched in 2010, used solar sails as the main mean of propulsion for the first time in history.

11.1.5 Coordinate Frames

The inertial reference frame is a frame that is defined so that Newton's laws of motion hold. This means that the frame is not accelerating and a body represented in this frame will satisfy Newton's first law, i.e., if it initially is at rest it will remain at rest, and if it is traveling with a constant velocity, it will continue to do so as long as no forces are applied (Jekeli 2000). Furthermore, the inertial reference frame is defined so that Newton's second and third laws are also satisfied.

An inertial reference frame as the one described above is always a simplification of the real world. Any reference frame that we define will be under acceleration, but for engineering purposes we can always find a frame that is sufficiently inertial for our application. For Earth-based applications, for example, we defined the inertial frame as a frame attached to the surface of the Earth, which is clearly a simplification because the Earth is accelerating with respect to the sun and the distant stars. Because the gravitational forces of the sun do not affect the motion of object on the Earth to any extent, and nor does the rotation of the Earth around its own axis, this is a good simplification for most Earth-based robotic systems. However, there are also applications for which this choice of coordinate frame will not hold. One example is large-scale motions, such as the motion of the water in the oceans, where the gravity of the moon and the Coriolis and centripetal forces that arise when the Earth is spinning need to be taken into account to obtain an accurate mathematical model.

We see that the inertial reference frame needs to be defined depending on the application and system at hand. Similarly to the way the Coriolis and centripetal forces are added to the equations when the frame is in rotation, we need to compensate for the varying gravitational forces for extraterrestrial motion. In this chapter we will therefore assume that the frames are in fact inertial, and we will compensate for the non-inertial effects by adding these forces. For orbiting systems this allows us to choose a frame that moves with the orbiting system as inertial. Using the equivalence principle we can then modify Newton's laws to account for the non-inertial effects and we can proceed as if the frame were inertial, even though it strictly speaking is non-inertial and accelerating. It turns out that a frame with constant velocity and that is free falling in the gravitational fields of the Earth, sun, moon and the planets is a very good choice of inertial reference frame for orbiting objects for which the motion is of short duration compared to the orbiting time. All motions of interest in robotics are of short duration, i.e., seconds or minutes, compared to the orbiting time of the spacecraft or satellite around the Earth, for which the orbiting time is hours or days.

11.2 Free-Floating and Free-Flying Rigid Bodies

Most objects that orbit the Earth are single rigid bodies. This includes satellites, spacecraft, and space stations. The dynamic equations of single rigid bodies in space are obtained straight forward from Euler's equations of motion.

11.2.1 Spacecraft Kinematics

The kinematics of a single rigid body such as a satellite or spacecraft is given by the velocity transformation matrix. The time derivatives of the Euler angles $\eta_2 = [\phi\ \theta\ \psi]^{\mathsf{T}}$ then relate to the body velocities $\omega_{0b}^B = [p\ q\ r]^{\mathsf{T}}$ by

$$\omega_{0b}^B = \begin{bmatrix} 1 & 0 & -\sin\theta \\ 0 & \cos\phi & \cos\theta\sin\phi \\ 0 & -\sin\phi & \cos\theta\cos\phi \end{bmatrix} \begin{bmatrix} \dot{\phi} \\ \dot{\theta} \\ \dot{\psi} \end{bmatrix}. \tag{11.1}$$

This equation gives us the inverse of the velocity transformation matrix. The velocity transformation matrix $T_{0b}(\eta_2)$ is given by

$$J_{b,o}(\eta_2) = T_{0b}(\eta_2) = \begin{bmatrix} 1 & \frac{\sin\theta\sin\phi}{\cos\theta} & \frac{\sin\theta\cos\phi}{\cos\theta} \\ 0 & \cos\phi & -\sin\phi \\ 0 & \frac{\sin\phi}{\cos\theta} & \frac{\cos\phi}{\cos\theta} \end{bmatrix}. \tag{11.2}$$

The kinematics of single rigid bodies were studied in detail in Chap. 3, and is also treated in detail in Bryson (1994) and Tatnall et al. (2011).

11.2.2 Spacecraft Dynamics

The attitude dynamics of a single rigid body is given by Euler's equations of motion, i.e.,

$$I_x\dot{\omega}_x - (I_y - I_z)\omega_y\omega_z = \tau_1 \tag{11.3}$$

$$I_y\dot{\omega}_y - (I_z - I_x)\omega_z\omega_x = \tau_2 \tag{11.4}$$

$$I_z\dot{\omega}_z - (I_x - I_y)\omega_x\omega_y = \tau_3 \tag{11.5}$$

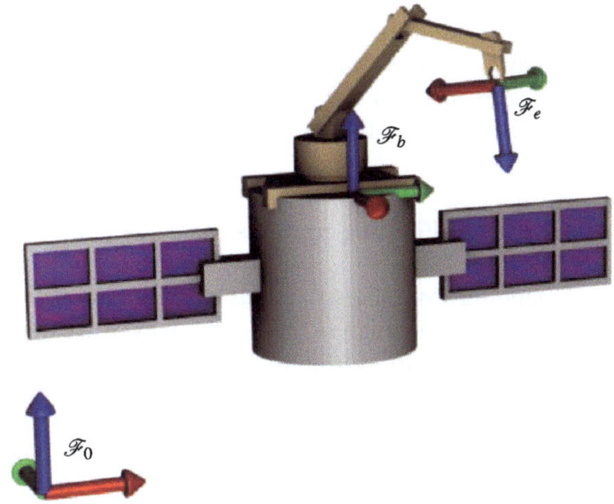

Fig. 11.6 Model setup for a robot attached to a spacecraft with coordinate frame $\mathscr{F}_b$, inertial reference frame $\mathscr{F}_0$, and end-effector frame $\mathscr{F}_e$. Courtesy of Elsevier

where $\tau = [\tau_1\ \tau_2\ \tau_3]^{\mathsf{T}}$ is the spacecraft control, for example reaction wheels. These equations are only valid if the center of gravity does not move with respect to the inertial frame. If thrusters are used to actuate the spacecraft so that it moves with translational motion with respect to the inertial frame the 6-DoF dynamic equations need to be applied. Finally the external forces discussed in Sect. 11.1.3 and the actuation discussed in Sect. 11.1.4 need to be added to the dynamic equations. The details on the dynamics of single rigid bodies are found in Chap. 6.

11.3 Modeling of Free-Floating and Free-Flying Robots

In this section we will study modeling of free-floating and free-flying spatial vehicle-manipulator systems, such as the one illustrated in Fig. 11.6. We are thus concerned with aspects such as kinematic and dynamic coupling between the spacecraft and the manipulator, local and global redundancy, non-holonomic spacecraft behavior, dynamic singularities, and other characteristics of spatial VM systems.

11.3.1 Kinematics of Spatial Vehicle-Manipulator Systems

There are many types of models available for spacecraft-manipulator systems depending on the actuation available or used to control the spacecraft. Firstly, when the spacecraft is fully actuated, we will have a large workspace and this also allows for interaction with the environment such as other spacecraft and space stations. A fully actuated spacecraft does, however, require a lot of fuel, which is a precious resource in space. Spacecraft-manipulator systems are therefore often operated without the use of thrusters to reduce fuel consumption. Secondly, we will therefore study the free-floating case with no spacecraft actuators. This does, however reduce the effective workspace of the manipulator and does not allow for interaction with the environment. We will therefore treat the two cases separately.

11.3.1.1 Free-Floating Systems

The mapping from the spacecraft and joint velocities to the end-effector velocities is given by the Jacobian in the normal way. If we denote the end-effector position in terms of the position variables and Euler angles, we can use the expression that we found in Eq. (5.42) and write (From et al. 2011a, 2010)

$$\begin{bmatrix} \dot{\eta} \\ \dot{\eta}_{0e} \end{bmatrix} = \begin{bmatrix} J_b(\eta_2) & 0 \\ J_e(\eta_{0e,2})\mathrm{Ad}_{g_{be}}^{-1}(q) & J_e(\eta_{0e,2})J_{m,g}(q) \end{bmatrix} \begin{bmatrix} V_{0b}^B \\ \dot{q} \end{bmatrix}. \tag{11.6}$$

This gives us the spacecraft and end-effector velocities from the velocity state of the system. It turns out, however, that the spacecraft and manipulator velocities are not decoupled. In fact, in the special case of a free-floating system with no external disturbances, the spacecraft velocities can be written in terms of the manipulator velocities. This relation can be derived by the conservation of momenta in the mechanism.

The linear momentum of the mechanism represented in the inertial frame is given by

$$P^0 = m_t \dot{r}_{cm} = \sum_{i=b}^{n} m_i \dot{p}_{0i}, \tag{11.7}$$

where m_i is the mass of rigid body i and $m_t = \sum_{i=b}^{n} m_i$ is the total mass of the mechanism. Following the notation in Fig. 11.7 we can write $p_{0i} = r_{cm} + \rho_i$ where r_{cm} is the vector from $\mathscr{F}_0$ to the center of mass (CM) of the whole system $\mathscr{F}_{cm}$, and ρ_i is the vector from $\mathscr{F}_{cm}$ to the center of mass of body i. External forces that act on the robot will affect the linear motion of r_{cm}. However, since the positions of the center of mass of all the links ρ_i are given with respect to the center of mass of the whole system, we can write the useful property

$$\sum_{i=b}^{n} m_i \rho_i = 0 \tag{11.8}$$

which is also the conservation of linear momentum when no external forces are present.

Similarly, the angular momentum of the mechanism is given by

$$\Pi^0 = \sum_{i=b}^{n} \left(I_i \omega_{0i}^B + m_i p_{0i} \times \dot{p}_{0i} \right) \tag{11.9}$$

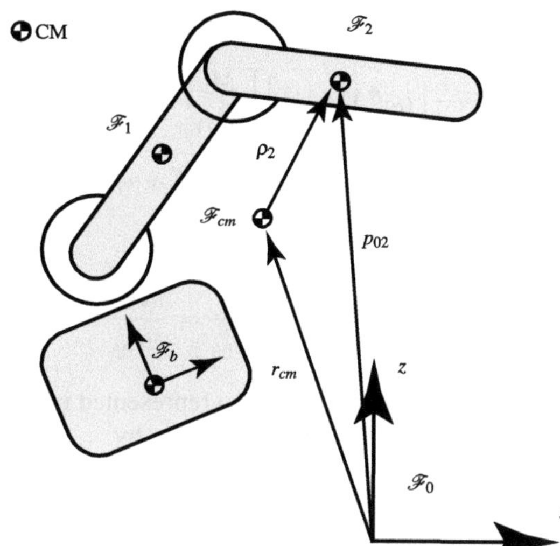

Fig. 11.7 The vector p_{0i} from the inertial frame $\mathscr{F}_0$ to the center of mass of link i can be written as $p_{0i} = r_{cm} + \rho_i$, here illustrated for the second link

which is also conserved for free-floating robotic systems with no external forces. The conservation of angular momentum makes the motion of the spacecraft's dependent variables non-holonomic, i.e., given a manipulator velocity $\dot{q}$ the spacecraft velocity is given. To find the spacecraft velocities we first write the dynamic equations in Lagrangian form (recall that the kinematic relations depend on the inertial properties of the system) as

$$\frac{\mathrm{d}}{\mathrm{d}t}\frac{\partial L}{\partial \omega_{0b}^{B}} - \frac{\partial L}{\partial \Theta} = 0 \tag{11.10}$$

$$\frac{\mathrm{d}}{\mathrm{d}t}\frac{\partial L}{\partial \dot{q}} - \frac{\partial L}{\partial q} = 0. \tag{11.11}$$

We know that the kinetic energy of the system does not depend on the pose of the system, here denoted by Θ. We can therefore write

$$\frac{\partial L}{\partial \Theta} = 0 \tag{11.12}$$

and Eq. (11.10) reduces to

$$\frac{\mathrm{d}}{\mathrm{d}t}\frac{\partial L}{\partial \omega_{0b}^{B}} = 0. \tag{11.13}$$

This means that $\frac{\partial L}{\partial \omega_{0b}^{B}}$ is a constant quantity. This quantity represents the conservation of angular momentum, and $\Pi = \frac{\partial L}{\partial \omega_{0b}^{B}}$ is in fact the angular momentum of the mechanism. If we write the kinetic energy as

$$\begin{aligned}\mathscr{K} &= \begin{bmatrix}(\omega_{0b}^B)^\mathsf{T} & \dot{q}^\mathsf{T}\end{bmatrix}\begin{bmatrix}M_V & M_{qV}^\mathsf{T}\\ M_{qV} & M_q\end{bmatrix}\begin{bmatrix}\omega_{0b}^B\\ \dot{q}\end{bmatrix}\\ &= (\omega_{0b}^B)^\mathsf{T} M_V \omega_{0b}^B + \dot{q}^\mathsf{T} M_{qV}\omega_{0b}^B + (\omega_{0b}^B)^\mathsf{T} M_{qV}^\mathsf{T}\dot{q} + \dot{q}^\mathsf{T} M_q \dot{q} \end{aligned}\tag{11.14}$$

we can therefore write

$$\Pi = \frac{\partial L}{\partial \omega_{0b}^B} = \frac{\partial \mathscr{K}}{\partial \omega_{0b}^B} = M_V \omega_{0b}^B + M_{qV}^\mathsf{T}\dot{q}\tag{11.15}$$

which is the angular momentum represented in the base frame. The angular momentum in the inertial frame is then given by

$$\Pi^S = R_{0b}\big(M_V \omega_{0b}^B + M_{qV}^\mathsf{T}\dot{q}\big).\tag{11.16}$$

Here M_V is the inertia matrix of the entire system when seen from the center of mass, shown in the upper left block in Eq. (8.31), i.e., we add the inertias of all the rigid bodies in the system and transform these to the same frame. M_{qV}^T represents the inertias that arise when the rigid bodies in the system move relative to one another, shown in the upper right block in (8.31). Furthermore, if we assume that no external forces are applied and the total momentum is zero (no thrusters or similar actuators) we can write the spacecraft velocities in terms of the manipulator velocities by rewriting Eq. (11.16) as

$$0 = M_V \omega_{0b}^B + M_{qV}^\mathsf{T}\dot{q}\tag{11.17}$$

$$\omega_{0b}^B = -M_V^{-1} M_{qV}^\mathsf{T}\dot{q}.\tag{11.18}$$

This defines the spacecraft velocity given the inertias of the spacecraft and manipulator arm that arise due to the conservation of angular momentum.

The constraint Jacobian J_C defined by

$$J_C = -M_V^{-1} M_{qV}^\mathsf{T}\tag{11.19}$$

defines mapping from the manipulator velocities to the spacecraft velocities as

$$\omega_{0b}^B = -M_V^{-1} M_{qV}^\mathsf{T}\dot{q}\tag{11.20}$$

for a free-floating space robot.

We can now write the end-effector velocities in terms of the joint velocities only and eliminate ω_{0b}^B from the equations. If we assume that the translational motion of the spacecraft is decoupled from the rotational motion (see Sect. 11.3.2) we can simplify Eq. (11.6) to

$$\begin{bmatrix} \dot{\eta}_2 \\ \dot{\eta}_{0e} \end{bmatrix} = \begin{bmatrix} J_{b,\omega} & 0 \\ J_S & J_M \end{bmatrix} \begin{bmatrix} \omega_{0b}^B \\ \dot{q} \end{bmatrix} \tag{11.21}$$

and write

$$\dot{\eta}_{0e} = J_S \omega_{0b}^B + J_M \dot{q} \tag{11.22}$$

$$= -J_S M_V^{-1} M_{qV}^{\mathsf{T}} \dot{q} + J_M \dot{q} \tag{11.23}$$

$$= \left(J_M - J_S M_V^{-1} M_{qV}^{\mathsf{T}}\right) \dot{q}. \tag{11.24}$$

In the space robotics literature, the Jacobian

$$J_G = J_M - J_S M_V^{-1} M_{qV}^{\mathsf{T}} \tag{11.25}$$

is normally denoted the Generalized Jacobian Matrix (GJM).

We see that the generalized Jacobian is a mere modification of the standard manipulator Jacobian that gives the mapping from the joint velocities to the end-effector velocities. For free-floating systems this mapping also needs to take the relative inertia of the base and the arm into account. It is interesting to note that this Jacobian matrix, which is a strictly kinematic mapping between two velocity spaces, actually depends on the inertial properties of the spacecraft and the manipulator. This is a characteristic property of free-floating space robots. We also see that if the inertia of the spacecraft becomes very large, the generalized Jacobian reduces to the manipulator Jacobian and the system will behave like a fixed-base manipulator, as expected.

One interesting consequence of the above is the non-holonomic redundancy that arises in free-floating space robots. Non-holonomic redundancy is studied in detail in Nakamura and Mukherjee (1993). To see how this redundancy arises we will first write the state space Q, the end-effector workspace g_{0e}, and the control space u of a 6-DoF manipulator on a spacecraft in vector form as

$$Q^\vee = \begin{bmatrix} \phi \\ \theta \\ \psi \\ q_1 \\ \vdots \\ q_6 \end{bmatrix} \in \mathbb{R}^9, \qquad g_{0e}^\vee = \begin{bmatrix} \phi_{0e} \\ \theta_{0e} \\ \psi_{0e} \\ x_{0e} \\ y_{0e} \\ z_{0e} \end{bmatrix} \in \mathbb{R}^6, \qquad u = \begin{bmatrix} \tau_1 \\ \vdots \\ \tau_6 \end{bmatrix} \in \mathbb{R}^6. \tag{11.26}$$

Here we have left out translational motion because we assume that the center of gravity of the VM system is constant, see Sect. 11.3.2. In free-floating systems only the joint variables are free independent variables, so the dimension of the control is equal to the number of joints, while the spacecraft variables are dependent, and given by (11.20).

Because the end-effector workspace, i.e., $g_{0e}^{\vee} = [\phi_{0e}\ \theta_{0e}\ \psi_{0e}\ x_{0e}\ y_{0e}\ z_{0e}] \in \mathbb{R}^6$ has a lower dimension than the state space $Q^{\vee} \in \mathbb{R}^9$ there is redundancy present in the system: When controlling the end effector towards a given configuration, the final values of the spacecraft and manipulator depend not only on the end-effector configuration but also on the path taken to reach this configuration. This allows us to control the final configuration of the 9 states of $Q^{\vee}$ by choosing a suitable path. This redundancy can be used to control the spacecraft attitude utilizing the manipulator joints and at the same time control the desired configuration of the end effector, even when the spacecraft has no actuation.

It is important to note, however, that this redundancy is different from the local redundancy found in redundant fixed-base manipulators. For a standard 6-DoF manipulator we see that because the input state, i.e., our control forces $u \in \mathbb{R}^6$, has the same dimension as the task space $g_{0e}^{\vee} \in \mathbb{R}^6$ that we want to control, this kind of redundancy does not allow for internal motion. This is different from local redundancy where the input space has a higher dimension ($\mathbb{R}^n, n > 6$) than the space of the end effector ($\mathbb{R}^6$). Global redundancy does therefore not unfold itself locally, but only becomes visible on a global scale: Assume a desired end-effector configuration. Then, due to the non-holonomic nature of the spacecraft-manipulator system, several different paths can be taken to reach the same end-effector configuration and each of these paths will lead to a different spacecraft attitude. We can therefore choose a path for which we obtain both the desired end-effector configuration and control the spacecraft attitude using only the 6 actuators in the manipulator joints. Because of this we will call this kind of redundancy a global redundancy, or non-holonomic redundancy.

Many control schemes have been developed to take advantage of this global redundancy. In Nakamura and Mukherjee (1993), for example, a principal Lyapunov-like functions is used to reach the desired end-effector configuration while a secondary function is used for joint limit avoidance, obstacle avoidance, and so on. The approach exploits the non-holonomic nature of free-flying space robots. In Oriolo and Nakamura (1991) the second-order constraint on the generalized coordinates that arise in robotic manipulators with passive joints (joints without actuation) are treated. The approach is similar to the case when the spacecraft has no actuation and the constraints are in general non-integrable.

We see that both local and global redundancy arise because the state space has a higher dimension than the task space of the end effector. When the input space has the same dimension as the state space ($u \in \mathbb{R}^{m+n}$) we have local redundancy with respect to the end-effector task—which is the case for an actuated spacecraft—and when the input space has the same dimension as the task space ($u \in \mathbb{R}^n$) we have global redundancy—which is the case when only the n-DoF manipulator, and not the spacecraft, is actuated.

It is important to note, however, that global redundancy does not exclude local redundancy, and the other way around. A redundant manipulator with $n > r$ joints mounted on a free-floating unactuated spacecraft will be both locally and globally redundant.

11.3.1.2 Fully Actuated Base

When the base is fully actuated reaction wheels are used to control the attitude of the spacecraft and thrusters can be used to control both the attitude and the translational motion. In this case the spacecraft actuators can compensate for the motion of the robot arm in such a way that the robot behaves like a fixed-base manipulator and also has the same workspace as fixed-base manipulators. This is often a preferred operating mode because it guarantees that the spacecraft attitude is kept constant allowing for continuous communication with the Earth.

The spacecraft thrusters can also be used to take the manipulator close to other objects in its vicinity giving it a very large workspace. This is for example necessary in berthing applications where the two spacecraft need to approach each other before the robot arm can attach to the other spacecraft.

Interaction with the environment will create linear and angular momenta in the spacecraft-manipulator system. Thus, to be able to apply continuous and accurate interaction forces, some kind of actuation at the spacecraft is needed. Spacecraft actuation is therefore used during interaction tasks to avoid undesired linear and angular momenta.

11.3.1.3 Dynamics Singularities

For fixed-base manipulators the singularities are found from the manipulator Jacobian by studying its determinant, i.e., the singularities arise whenever $\det[J_{m,g}(q)] = 0$. This is thus a purely kinematic relation and the singularities are found in joint space from the manipulator geometry. Because we can find a one-to-one mapping from the joint space to the workspace, the singularities are also fixed in the workspace variables.

For free-floating robots, however, the mapping from joint space to the end-effector velocities in the inertial frame is given by the Generalized Jacobian matrix

$$J_G = J_M - J_S M_V^{-1} M_{qV}^{\mathsf{T}}. \tag{11.27}$$

For free-floating manipulators the singularities are found by the points for which the determinant of the Generalized Jacobian matrix is zero, i.e., $\det[J_G(q)] = 0$. There are several fundamental differences between the singularities of fixed-base and free-floating manipulators when it comes to singularities:

- First of all, the Generalized Jacobian matrix J_G depends on the inertial properties of the system. Thus, also the location of the singularities in the inertial frame depend on the inertial properties. This is thus not purely dependent on the kinematic relations like for fixed-base manipulators, and denoted *dynamic singularities*.
- Like kinematic singularities found in fixed-base manipulators, the dynamic singularities are fixed in *joint space*. However, because there does not exist a one-to-one mapping from joint space to the end-effector workspace for free-floating manipulators these singularities are not fixed in the workspace of the vehicle-manipulator system.
- The spacecraft orientation depends on what path was taken to reach the position in joint space. Thus, also the location of the dynamic singularities in the workspace depend on the path taken to reach the current configuration.

Even though there are many differences between fixed-base and free-floating systems when it comes to singularities, the interpretation of dynamic singularities is the same as that of kinematic singularities: the Jacobian must be of full rank to guarantee that the end-effector can move in all directions. Thus for a given end-effector configuration and the path taken to reach this configuration the Generalized Jacobian tells us whether the end-effector can move in all directions of the workspace or not. Whether the Generalized Jacobian loses rank or not depends on the geometric and inertial properties of the manipulator and the spacecraft.

11.3.1.4 Actuated Spacecraft and Redundant Manipulators

Also space manipulators may possess local redundancy in the same way as redundant fixed-base robots. This is the case when the degree of mobility is higher than the dimension of the task space. We will denote the number of actuated joints as n. We can also allow for spacecraft actuation as long as the conservation of momentum and Eq. (11.20) still hold. If this is the case we will not distinguish the manipulator actuation from the spacecraft actuation and denote the dimension of the two combined as n. The dimension of the task space of the spacecraft is denoted r_s and the task space of the end effector is denoted r.

When local redundancy is present we can treat this in more or less the same way as with fixed-base robots (Nenchev et al. 1992; Caccavale and Siciliano 2001). Let $\eta_{1,d}(t)$ and $\eta_{2,d} = R_d^\vee(t)$ represent the desired position and orientation of the end-effector and similarly $\eta_{1,c}(t)$ and $\eta_{2,c}(t)$ the current configuration. A simple control scheme called the closed-loop inverse kinematics (CLIK) algorithm is obtained by writing the joint velocities in terms of the end-effector configuration as (Caccavale and Siciliano 2001)

$$\dot{q} = J_{m,a}^{-1}(q) \begin{bmatrix} \dot{\eta}_{1,d} + k_{Pp}(\eta_{1,d} - \eta_{1,c}) \\ \dot{\eta}_{2,d} + k_{Po}(\eta_{2,d} - \eta_{2,c}) \end{bmatrix} \tag{11.28}$$

for some controller gains k_{Pp}, k_{Po}. We can represent the orientation error in geometric quantities, for example the quaternion error, which allows us to write the equations in terms of the geometric Jacobian $J_{m,g}$ as

$$\dot{q} = J_{m,g}^{-1}(q)\big(V_{0e}^{B} + K_P \Delta e\big) \tag{11.29}$$

where Δe represents the position and orientation error collocated with V_{0e}^{B}.

When redundancy is present in the system we can utilize the null space to add additional control objectives in addition to the end-effector configuration. Firstly, if $n \geq r_s + r$ we have redundancy that can be utilized to control some secondary objective. We let our principal control objective be given by Eq. (11.29), i.e., the end-effector motion. In addition we would like to control the spacecraft attitude to a desired orientation. We then introduce the null space projection in the normal way and write this as (Caccavale and Siciliano 2001)

$$\dot{q} = \underbrace{J_{m,g}^{\dagger}(q)\big(V_{0e}^{B} + K_P \Delta e\big)}_{\text{end-effector control}} + \underbrace{\big(I - J_{m,g}^{\dagger} J_{m,g}\big)}_{\text{null space}} J_C^{\mathsf{T}} K_C e_C \tag{11.30}$$

where $(I - J_{m,g}^{\dagger} J_{m,g})$ ensures that the additional velocity contribution is projected into the null space and does not affect the end-effector velocity, e_C is the error of the spacecraft attitude and J_C is the corresponding Jacobian of the constraint task. Note that we use the pseudoinverse $J_{m,g}^{\dagger}$ instead of $J_{m,g}^{-1}$ in the end-effector control in (11.30) because it is to be calculated anyway for the null space motion. In the case that $n = r_s + r$, i.e., we have exactly the required actuator torques to control both the end-effector motion and the spacecraft attitude, then J_C can be found by Eq. (11.20) as

$$J_C = -M_V^{-1} M_{qV}^{\mathsf{T}}. \tag{11.31}$$

Equation (11.30) will then guarantee that both the end effector and the spacecraft follow the desired motion, as long as they are not conflicting. Highest priority is given to the end-effector motion.

We note that if $n > r_s + r$ we can add additional control objectives in addition to the spacecraft attitude, such as collision avoidance and joint constraint avoidance. On the other hand, if $n \leq r_s + r$ but $n > r$, then we can control the manipulator end effector completely, but only partially control the spacecraft rotational velocity using the $n - r$ redundant degrees of freedom available. In addition the *final spacecraft configuration* can be controlled using the global redundancy discussed previously.

11.3.2 Dynamics of Spatial Vehicle-Manipulator Systems

The equations of motion of a spacecraft-manipulator system can be written in the normal way as

$$M(x)\dot{v} + C(x, v)v = \tau. \tag{11.32}$$

Different choices of the state variables will give very different structures of the equations. We can for example choose the position of the center of gravity of either the spacecraft or the entire VM system to represent translational motion. If we choose the spacecraft position and orientation as the position states we will get the state space

$$x = \begin{bmatrix} p_{0b} \\ \Theta_{0b} \\ q \end{bmatrix}, \qquad v = \begin{bmatrix} v_{0b}^B \\ \omega_{0b}^B \\ \dot{q} \end{bmatrix} \tag{11.33}$$

where p_{0b} is the position of the center of mass of the spacecraft, Θ_{0b} is the Euler angles representing the spacecraft attitude, v_{0b}^B and ω_{0b}^B are the linear and angular velocity of the spacecraft, and q is the joint positions of the manipulator. The dynamic equations now take the form

$$\begin{bmatrix} M_v & M_{\omega v}^\mathsf{T} & M_{qv}^\mathsf{T} \\ M_{\omega v} & M_\omega & M_{q\omega}^\mathsf{T} \\ M_{qv} & M_{q\omega} & M_q \end{bmatrix} \begin{bmatrix} \dot{v}_{0b}^B \\ \dot{\omega}_{0b}^B \\ \ddot{q} \end{bmatrix} + \begin{bmatrix} C_v & C_{v\omega} & C_{vq} \\ C_{\omega v} & C_\omega & C_{\omega q} \\ C_{qv} & C_{q\omega} & C_q \end{bmatrix} \begin{bmatrix} v_{0b}^B \\ \omega_{0b}^B \\ \dot{q} \end{bmatrix} = \begin{bmatrix} \tau_v \\ \tau_\omega \\ \tau_q \end{bmatrix} \tag{11.34}$$

and do not have a particularly simple structure. The control torques are given by $\tau = [\tau_v^\mathsf{T} \; \tau_\omega^\mathsf{T} \; \tau_q^\mathsf{T}]^\mathsf{T}$ where τ_v is the spacecraft forces generated by thrusters, τ_ω is the spacecraft moments generated by thrusters, momentum gyros or reaction wheels, and τ_q is the manipulator torques.

It is, however, possible to use the center of mass of the whole system r_{cm} to represent the translational motion instead of the center of mass of the spacecraft. Then $v = [\dot{r}_{cm}^\mathsf{T} \; (\omega_{0b}^B)^\mathsf{T} \; \dot{q}^\mathsf{T}]^\mathsf{T}$ where $\dot{r}_{cm}$ is the linear velocity of the center of mass of the spacecraft-manipulator system. The motion of the center of mass is decoupled from the angular velocity ω_{0b}^B and the inertia matrix for a free-flying spacecraft-manipulator system can be written as (Dubowsky and Papadopoulos 1993)

$$M = \begin{bmatrix} m_t I & 0 & 0 \\ 0 & M_\omega & M_{q\omega}^\mathsf{T} \\ 0 & M_{q\omega} & M_q \end{bmatrix} \tag{11.35}$$

where m_t is the total mass of the system. The Coriolis matrix will take a similar overall structure and the final dynamics is given by

$$\begin{bmatrix} m_t I & 0 & 0 \\ 0 & M_\omega & M_{q\omega}^\mathsf{T} \\ 0 & M_{q\omega} & M_q \end{bmatrix} \begin{bmatrix} \ddot{r}_{cm} \\ \dot{\omega}_{0b}^B \\ \ddot{q} \end{bmatrix} + \begin{bmatrix} 0 \\ c_\omega(q, \omega_{0b}^B, \dot{q}) \\ c_q(q, \omega_{0b}^B, \dot{q}) \end{bmatrix} = \begin{bmatrix} \tau_v \\ \tau_\omega \\ \tau_q \end{bmatrix} \tag{11.36}$$

where we recognize the first line as the motion of the center of mass of the whole system given in Eq. (11.7).

Other models are also available depending on the actuators available to control the spacecraft. In the case of a free-floating space robot, i.e., $\tau_v, \tau_\omega = 0$, and with no external forces, the center of mass does not accelerate and the system linear momentum is constant $\dot{r}_{cm} = 0$. This can be used to simplify the equations to an n-dimensional system in terms of the joint variables only. To see this we first reduce Eq. (11.36) to

$$\begin{bmatrix} M_\omega & M_{q\omega}^\mathsf{T} \\ M_{q\omega} & M_q \end{bmatrix} \begin{bmatrix} \dot{\omega}_{0b}^B \\ \ddot{q} \end{bmatrix} + \begin{bmatrix} C_\omega & C_{\omega q} \\ C_{q\omega} & C_q \end{bmatrix} \begin{bmatrix} \omega_{0b}^B \\ \dot{q} \end{bmatrix} = \begin{bmatrix} 0 \\ \tau_q \end{bmatrix}. \tag{11.37}$$

To remove ω_{0b}^B from the dynamics we will write the kinetic energy as (Dubowsky and Papadopoulos 1993)

$$\begin{aligned} \mathscr{K} &= \begin{bmatrix} (\omega_{0b}^B)^\mathsf{T} & \dot{q}^\mathsf{T} \end{bmatrix} \begin{bmatrix} M_\omega & M_{q\omega}^\mathsf{T} \\ M_{q\omega} & M_q \end{bmatrix} \begin{bmatrix} \omega_{0b}^B \\ \dot{q} \end{bmatrix} \\ &= (\omega_{0b}^B)^\mathsf{T} M_\omega \omega_{0b}^B + \dot{q}^\mathsf{T} M_{q\omega} \omega_{0b}^B + (\omega_{0b}^B)^\mathsf{T} M_{q\omega}^\mathsf{T} \dot{q} + \dot{q}^\mathsf{T} M_q \dot{q} \\ &= \dot{q}^\mathsf{T} M_{q\omega} M_\omega^{-\mathsf{T}} M_\omega M_\omega^{-1} M_{q\omega}^\mathsf{T} \dot{q} - \dot{q}^\mathsf{T} M_{q\omega} M_\omega^{-1} M_{q\omega}^\mathsf{T} \dot{q} \\ &\quad - \dot{q}^\mathsf{T} M_{q\omega} M_\omega^{-\mathsf{T}} M_{q\omega}^\mathsf{T} \dot{q} + \dot{q}^\mathsf{T} M_q \dot{q} \\ &= \dot{q}^\mathsf{T} M_{q\omega} M_\omega^{-\mathsf{T}} M_{q\omega}^\mathsf{T} \dot{q} - \dot{q}^\mathsf{T} M_{q\omega} M_\omega^{-1} M_{q\omega}^\mathsf{T} \dot{q} - \dot{q}^\mathsf{T} M_{q\omega} M_\omega^{-\mathsf{T}} M_{q\omega}^\mathsf{T} \dot{q} + \dot{q}^\mathsf{T} M_q \dot{q} \\ &= \dot{q}^\mathsf{T} (M_q - M_{q\omega} M_\omega^{-1} M_{q\omega}^\mathsf{T}) \dot{q} \end{aligned} \tag{11.38}$$

where we have used $\omega_{0b}^B = -M_\omega^{-1} M_{q\omega}^\mathsf{T} \dot{q}$.

The dynamics of a free-floating VM system can be written as

$$M_r \ddot{q} + C_r \dot{q} = \tau_q \tag{11.39}$$

with

$$M_r = M_q - M_{q\omega} M_\omega^{-1} M_{q\omega}^\mathsf{T} \tag{11.40}$$

and where C_r is found in the normal way from M_r.

The attitude of the spacecraft is recovered from

$$\omega_{0b}^B = -M_\omega^{-1} M_{q\omega}^\mathsf{T} \dot{q}. \tag{11.41}$$

We see from Eq. (11.39) that the motion of the VM system is completely determined by the joint variables. Because Eq. (11.39) only represents an n-DoF motion we cannot control both the manipulator and the spacecraft using only the robot actuators. Rather, the spacecraft attitude is given as a function of the joint velocities given by Eq. (11.41). However, since the spacecraft motion is subject to nonholonomic constraints we can control the final configuration of the spacecraft, as we have already seen.

11.3.2.1 Canceling the Manipulator Motion

For actuated spacecraft as described in Sect. 11.3.1.2 we can use the spacecraft actuators to remove the effects of the dynamic coupling so that the spacecraft orientation remains fixed. We will assume that no external forces are present and write the spacecraft dynamics as

$$\begin{bmatrix} M_V & M_{qV}^{\mathsf{T}} \\ M_{qV} & M_q \end{bmatrix} \begin{bmatrix} \dot{\omega}_{0b}^B \\ \ddot{q} \end{bmatrix} + \begin{bmatrix} C_V & C_{Vq} \\ C_{qV} & C_q \end{bmatrix} \begin{bmatrix} \omega_{0b}^B \\ \dot{q} \end{bmatrix} = \begin{bmatrix} \tau_\omega \\ \tau_q \end{bmatrix} \tag{11.42}$$

The motion of the manipulator arm will thus create a motion in the spacecraft due to the conservation of angular momentum. This disturbance can be canceled if the dynamic model of the VM system is available. Consider the standard augmented PD control law

$$\tau_\omega = \tau_{ff} + \tau_{\mathrm{PD}} \tag{11.43}$$

where

$$\tau_{ff} = \underbrace{M_{qV}^{\mathsf{T}} \ddot{q}_d + C_{Vq} \dot{q}_d}_{\text{dynamic coupling terms}} \tag{11.44}$$

$$\tau_{\mathrm{PD}} = \underbrace{K_{P,\omega}\left(\left(\omega_{0b}^B\right)_d - \omega_{0b}^B\right) + K_{D,\omega}\left(\left(\dot{\omega}_{0b}^B\right)_d - \dot{\omega}_{0b}^B\right)}_{\text{PD-controller}}. \tag{11.45}$$

The manipulator is controlled in the normal way as

$$\tau_q = \underbrace{K_{P,q}(q_d - q) + K_{D,q}(\dot{q}_d - \dot{q})}_{\text{PD-controller}}. \tag{11.46}$$

Ideally, this control law will cancel the effects that the manipulator motion has on the spacecraft attitude and at the same time follow the desired manipulator trajectory. This control may, however, in some cases require excessive control torques, which is a precious resource in outer space.

11.4 The Dynamically Equivalent Manipulator Approach

We have seen that the dynamic coupling between the manipulator and the spacecraft complicates both the modeling and the control of spatial VM systems. Under certain conditions, however, the space manipulator can be treated as a fixed-base manipulator with modified geometric and inertial properties. Liang et al. (1997a, 1997b) showed that a free-flying space robot will behave exactly like an imaginary fixed-base robot with the base located in the center of mass of the free-flying robot.

The kinematic and dynamic properties of this fixed-base robot are found from the kinematic and dynamic properties of the free-flying robot with some quantitative modifications. In the following, we refer to the real spacecraft-manipulator system as a space manipulator and the imaginary manipulator as the dynamically equivalent manipulator (DEM).

The dynamically equivalent manipulator approach has shown to be very useful for modeling spatial robots as it allows us to treat the space robot as if it were mounted on a fixed base. The approach can also be used for control because equal control torques will produce the same trajectories. More specifically, it can be shown that a given sequence of actuator torques acting on the DEM will produce the same joint trajectories for the space manipulator. We can show that any motion in joint space for the space manipulator is equal to that of the DEM for revolute joints and that it is simply related for prismatic joints. Finally, the DEM approach is very useful for workspace analysis.

Robustness of spacecraft-manipulator systems is a major concern for space operators and robust mathematical models are important for both simulation and control of these systems. We therefore derive the singularity-free dynamic equations of the DEM first presented in From et al. (2011b). We will see that, similar to the conventional approach, the DEM obtained using this framework has the same kinematic and dynamic properties as the space manipulator. If we assume that the center of gravity of the VM system is fixed we can use the results from Sect. 11.3.2 and write the state space of the spacecraft as $SO(3)$. Then, using the framework introduced in Sect. 8.3.3.1 we can formulate the DEM dynamics without the singularities that normally arise when the Euler angles are adopted and present the singularity-free dynamic equations of the DEM.

Earlier work was presented by Vafa and Dubowsky (1987) who presented the Virtual Manipulator approach which addressed the modeling and control of a robot mounted on a free-floating base. The virtual manipulator is a fixed base manipulator where the spacecraft is modeled as a spherical joint and the kinematic properties of the virtual manipulator are the same as for the space manipulator. The virtual manipulator is an idealized massless kinematic chain describing the kinematic relations between the space manipulator and the virtual manipulator. Similarly, Liang et al. (1997a, 1998) mapped the free-floating space manipulator system into a fixed-base manipulator and it was shown that the DEM mapping preserved both the kinematic and dynamic properties of the space manipulator. The dynamically equivalent manipulator can thus be used for simulation, control, and dynamic analysis in addition to workspace analysis and trajectory planning made possible with the virtual manipulator.

The virtual manipulator and DEM concepts have proven very important when it comes to modeling and control of space manipulators. In Parlaktuna and Ozkan (2004) an adaptive control scheme of free-floating space manipulators was derived based on the dynamically equivalent manipulator concept. By using the fixed-base manipulator the dynamics could be linearly parameterized and an adaptive control law was developed to control the system in joint space.

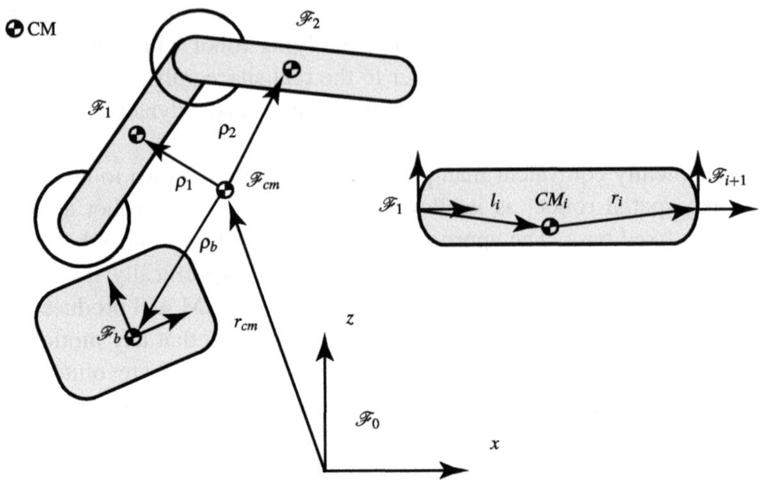

Fig. 11.8 Schematic illustration of a space manipulator

11.4.1 Mathematical Formulation

The dynamic equations of the free-floating space manipulator can be derived from Lagrange's equations. The Lagrangian of the space manipulator is then given by the kinetic energy only, i.e.,

$$\mathscr{K} = \sum_{i=b}^{n+1} \left[\frac{1}{2} \dot{\rho}_i^{\mathsf{T}} m_i \dot{\rho}_i + \frac{1}{2} \left(\omega_{0i}^B\right)^{\mathsf{T}} R_{i0}^{\mathsf{T}} D_i R_{0i} \omega_{i0}^B \right] \tag{11.47}$$

for both the spacecraft and the links, which is different from Eq. (8.29) in that the inertia matrix depends on the configuration of both the spacecraft and the joints. The spacecraft orientation is of course dependent, and can therefore be found from the manipulator joint motion. m_i is the mass and D_i the 3×3 inertia tensor of link i, and ρ_i is the distance from the center of mass of the VM system to the center of mass of link i, as shown in Fig. 11.8.

Similarly, we can define a fixed-base manipulator with a spherical first joint and kinetic energy

$$\mathscr{K}' = \sum_{i=b}^{n+1} \left[\frac{1}{2} \left(v_{0i}'\right)^{\mathsf{T}} m_i' v_{0i}' + \frac{1}{2} \left(\omega_{0i}'\right)^{\mathsf{T}} \left(R_{i0}'\right)^{\mathsf{T}} D_i' R_{0i}' \omega_{i0}' \right] \tag{11.48}$$

where v_{0i}' is the velocity of link i with respect to the base and ω_{0i}' is still the body twist, even though we have left out the superscript B.

It can be shown that the kinematic and dynamic parameters of the space manipulator can be mapped to the DEM by modifying the kinematic and dynamic

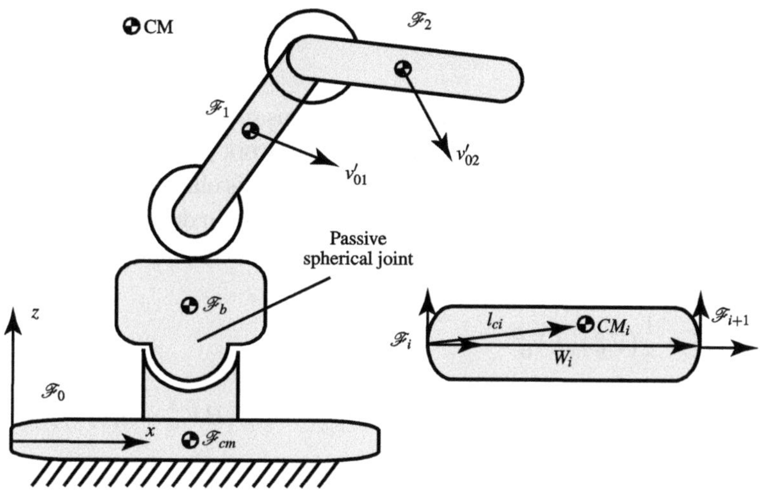

Fig. 11.9 Schematic illustration of the DEM

parameters in the following way (Liang et al. 1998; Parlaktuna and Ozkan 2004):

$$
\begin{aligned}
m'_i &= m_i \frac{(\sum_{k=1}^{n+1} m_k)^2}{\sum_{k=1}^{i-1} m_k \sum_{k=1}^{i} m_k}, \quad i = 2 \dots n+1, \\
D'_i &= D_i, \quad i = 1 \dots n+1, \\
W_1 &= \frac{r_1 m_1}{\sum_{k=1}^{n+1} m_k}, \\
W_i &= r_i \left(\frac{\sum_{k=1}^{i} m_k}{\sum_{k=1}^{n+1} m_k} \right) + l_i \left(\frac{\sum_{k=1}^{i-1} m_k}{\sum_{k=1}^{n+1} m_k} \right), \quad i = 2 \dots n+1, \\
l_{c1} &= 0, \\
l_{ci} &= l_i \left(\frac{\sum_{k=1}^{i-1} m_k}{\sum_{k=1}^{n+1} m_k} \right), \quad i = 2 \dots n+1,
\end{aligned}
\tag{11.49}
$$

where the vector W_i connecting joint i with joint $i+1$ of the DEM is given by r_i and l_i of the space manipulator where r_i is the vector connecting the center of mass of link i and joint $i+1$ and l_i is the vector connecting joint i with the center of mass of link i. l_{ci} is the vector connecting joint i and the center of mass of link i in the DEM. See Fig. 11.9. We refer to Liang et al. (1998) and Parlaktuna and Ozkan (2004) for details.

11.4.2 Configuration States

In this section we reformulate the dynamic equations of a space manipulator and its dynamically equivalent manipulator using the framework presented in Sect. 8.3.3.1. This removes the singularities in the representation, but is otherwise similar. Assume no spacecraft actuation, so the center of mass of the system does not accelerate, i.e., $\dot{r}_{cm} = 0$. Then the kinetic energy of link i of the space manipulator is given by

$$\begin{aligned}
\mathcal{K}_i &= \frac{1}{2}\left(V_{0i}^B\right)^\mathsf{T} I_i V_{0i}^B \\
&= \frac{1}{2}\left(\left(\tilde{V}_{0b}^B\right)^\mathsf{T} H^\mathsf{T} + \dot{q}^\mathsf{T} J_i(q)^\mathsf{T}\right)\mathrm{Ad}_{g_{ib}}^\mathsf{T} I_i \mathrm{Ad}_{g_{ib}}\left(H\tilde{V}_{0b}^B + J_i(q)\dot{q}\right) \\
&= \frac{1}{2}\left(\left(\omega_{0b}^S\right)^\mathsf{T} H^\mathsf{T} + \dot{q}^\mathsf{T} J_i(q)^\mathsf{T}\right)\mathrm{Ad}_{g_{ib}}^\mathsf{T} I_i \mathrm{Ad}_{g_{ib}}\left(H\omega_{0b}^S + J_i(q)\dot{q}\right) \\
&= \frac{1}{2}\begin{bmatrix}\left(\omega_{0b}^S\right)^\mathsf{T} & \dot{q}^\mathsf{T}\end{bmatrix} M_i(q)\begin{bmatrix}\omega_{0b}^S \\ \dot{q}\end{bmatrix} = \frac{1}{2}\upsilon^\mathsf{T} M_i(q)\upsilon
\end{aligned} \tag{11.50}$$

where

$$M_i(q) = \begin{bmatrix} H^\mathsf{T}\mathrm{Ad}_{g_{ib}}^\mathsf{T} I_i \mathrm{Ad}_{g_{ib}} H & H^\mathsf{T}\mathrm{Ad}_{g_{ib}}^\mathsf{T} I_i \mathrm{Ad}_{g_{ib}} J_i \\ J_i^\mathsf{T}\mathrm{Ad}_{g_{ib}}^\mathsf{T} I_i \mathrm{Ad}_{g_{ib}} H & J_i^\mathsf{T}\mathrm{Ad}_{g_{ib}}^\mathsf{T} I_i \mathrm{Ad}_{g_{ib}} J_i \end{bmatrix} \tag{11.51}$$

and the inertia matrix is given by substituting this into (8.30) with H given in the normaly way. The configuration space is then given by $Q = \{R_{0b}, q\}$.

Similarly, we can define a fixed-base manipulator with a spherical first joint, also with configuration space $SO(3)$. The corresponding inertia matrices are then given by

$$M_i'(q) = \begin{bmatrix} H^\mathsf{T}\mathrm{Ad}_{g_{ib}'}^\mathsf{T} I_i' \mathrm{Ad}_{g_{ib}'} H & H^\mathsf{T}\mathrm{Ad}_{g_{ib}'}^\mathsf{T} I_i' \mathrm{Ad}_{g_{ib}'} J_i' \\ (J_i')^\mathsf{T}\mathrm{Ad}_{g_{ib}'}^\mathsf{T} I_i' \mathrm{Ad}_{g_{ib}'} H & (J_i')^\mathsf{T}\mathrm{Ad}_{g_{ib}'}^\mathsf{T} I_i' \mathrm{Ad}_{g_{ib}'} J_i' \end{bmatrix}, \tag{11.52}$$

where I_i' and the kinematic relations used to compute R_{0i}' and J_i' are found from (11.49). Thus, we have $\tilde{V}_{0b}^B = \tilde{V}_{0b}'^B$ as required. Specifically, the inertia tensor of link i is given by $I_i' = \left[\begin{smallmatrix} m_i' I & 0 \\ 0 & D_i' \end{smallmatrix}\right]$ and the upper left part of M_b' is given by

$$M_{b,VV}' = H^\mathsf{T}\mathrm{Ad}_{g_{bb}'}^\mathsf{T} I_b' \mathrm{Ad}_{g_{bb}'} H = D_b' = \begin{bmatrix} J_x' & 0 & 0 \\ 0 & J_y' & 0 \\ 0 & 0 & J_z' \end{bmatrix} \tag{11.53}$$

which also represents the inertial properties of the spherical base link. The Coriolis matrix is then given by

$$C'(Q, v) = \sum_{k=1}^{n} \frac{\partial M'}{\partial q_k} \dot{q}_k - \frac{1}{2} \begin{bmatrix} 2(\widehat{M'(q)v})_{\tilde{V}} & 0 \\ \frac{\partial^{\mathsf{T}}}{\partial q}\left(\begin{bmatrix} M'_V & (M'_{qV})^{\mathsf{T}} \end{bmatrix} \begin{bmatrix} V_{0b}^B \\ \dot{q} \end{bmatrix}\right) & \frac{\partial^{\mathsf{T}}}{\partial q}\left(\begin{bmatrix} M'_{qV} & (M'_q)^{\mathsf{T}} \end{bmatrix} \begin{bmatrix} V_{0b}^B \\ \dot{q} \end{bmatrix}\right) \end{bmatrix} \tag{11.54}$$

where $(M'(q)v)_{\tilde{V}}$ is the vector of the first three entries of the vector $M'(q)v$ (corresponding to $\tilde{V}_{0b}^B = \omega_{0b}^S$).

The dynamic equations of the space manipulator can now be represented by the DEM dynamics

$$M'(q)\dot{v} + C'(Q, v)v = \tau, \tag{11.55}$$

with M' given as in (11.52) and C' as in (11.54). Here, $v = [(\omega_{0b}^B)^{\mathsf{T}}\ \dot{q}^{\mathsf{T}}]^{\mathsf{T}}$ where ω_{0b}^B is the velocity state of the spherical joint of the DEM (and thus also the spacecraft) and $\dot{q}$ the velocity state of the manipulator of the DEM (and the space manipulator), and $Q = \{R_{0b}, q\}$ where $R_{0b} \in SO(3)$ determines the configuration of the spherical joint/spacecraft and q the configuration of the manipulators of the DEM and space manipulator. We note that the singularity that normally arises when using the Euler angles is eliminated and the state space (Q, v) is valid globally. Similar to the conventional approach, the DEM described by (11.55) has the same kinetic and dynamic properties as the space manipulator and if the same actuator torques $\tau(t) = \tau'(t)$ are applied on both the DEM and the space manipulator, this will produce the same joint trajectory $q(t) = q'(t)$ for $\forall t \in [t_0, \infty]$ if $q(t_0) = q'(t_0)$.

References

Abiko, S., & Yoshida, K. (2003). An effective control strategy of Japanese experimental module remote manipulator system (jemrms) using coupled and un-coupled dynamics. In *Proceeding of the international symposium on artificial intelligence, robotics and automation in space: I-SAIRAS, NARA*, Japan.

Bryson, A. E. (1994). *Control of spacecraft and aircraft*. Princeton: Princeton University Press.

Caccavale, F., & Siciliano, B. (2001). Quaternion-based kinematic control of redundant spacecraft/manipulator systems. In *Proceedings of IEEE international conference on robotics and automation*, Seoul, Korea.

Dube, M., Bollea, D., Jones, W. R. Jr., Marchetti, M., & Jansen, M. J. (2003). A new synthetic hydrocarbon liquid lubricant for space applications. *Tribology Letter*, *15*(1), 3–8.

Dubowsky, S., & Papadopoulos, E. (1993). The kinematics, dynamics and control of free-flying and free-floating space robotic systems. *IEEE Transactions on Robotics and Automation*, *9*(5), 531–541.

From, P. J., Pettersen, K. Y., & Gravdahl, J. T. (2010). In *Singularity-free dynamic equations of spacecraft-manipulator systems. International Astronautical Congress*, Prague, Czech Republic.

From, P. J., Pettersen, K. Y., & Gravdahl, J. T. (2011a). Singularity-free dynamic equations of spacecraft-manipulator systems. *Acta Astronautica*, *69*(11–12), 1057–1065.

From, P. J., Pettersen, K. Y., & Gravdahl, J. T. (2011b). Singularity-free formulation of the dynamically equivalent manipulator mapping for space manipulators. In *AIAA SPACE*, Long Beach, California, USA.

Inaba, N., & Oda, M. (2000). Autonomous satellite capture by a space robot: world first on-orbit experiment on a Japanese robot satellite ETS-vii. In *Proceedings of IEEE international conference on robotics and automation* (Vol. 2, pp. 1169–1174).

Jekeli, C. (2000). *Inertial navigation systems with geodetic applications*. Walter De Gruyter Inc.

Liang, B., Xu, Y., & Bergerman, M. (1997a). Dynamically equivalent manipulator for space manipulator system. 1. In *Proceedings of IEEE international conference on robotics and automation* (Vol. 4, pp. 2765–2770).

Liang, B., Xu, Y., & Bergerman, M. (1998). Mapping a space manipulator to a dynamically equivalent manipulator. *ASME Journal of Dynamic Systems, Measurement, and Control*, *120*(1), 1–7.

Liang, B., Xu, Y., Bergerman, M., & Li, G. (1997b). Dynamically equivalent manipulator for space manipulator system. 2. In *Proceedings of IEEE/RSJ international conference on intelligent robots and systems* (Vol. 3, pp. 1493–1499).

Liljebäck, P., Pettersen, K. Y., Stavdahl, Ø., & Gravdahl, J. T. (2013). *Snake robots modelling, mechatronics, and control*. Berlin: Springer.

Meirovich, L., & Kwak, M. K. (1989). State equations for a spacecraft with maneuvering flexible appendages in terms of quasi-coordinates. *Applied Mechanics Reviews*, *42*(11), 161–170.

Nakamura, Y., & Mukherjee, R. (1993). Exploiting nonholonomic redundancy of free-flying space robots. *IEEE Transactions on Robotics and Automation*, *9*(4), 499–506.

Nenchev, D., Umetani, Y., & Yoshida, K. (1992). Analysis of a redundant free-flying spacecraft/manipulator system. *IEEE Transactions on Robotics and Automation*, *8*(1), 1–6.

Nenchev, D., Yoshida, K., Vichitkulsawat, P., & Uchiyama, M. (1999). Reaction null-space control of flexible structure mounted manipulator systems. *IEEE Transactions on Robotics and Automation*, *15*(6), 1011–1023.

Newton, S. I. (1687). *Philosophiæ Naturalis Principia Mathematica*. Londini.

Oda, M., Kibe, K., & Yamagata, F. (1996). Ets-vii, space robot in-orbit experiment satellite. In *Proceedings of IEEE international conference on robotics and automation* (Vol. 1, pp. 739–744).

Oriolo, G., & Nakamura, Y. (1991). Free-joint manipulators: motion control under second-order nonholonomic constraints. In *Proceedings of IEEE international workshop on intelligent robots and systems*, Osaka, Japan (pp. 1248–1253).

Parlaktuna, O., & Ozkan, M. (2004). Adaptive control of free-floating space manipulators using dynamically equivalent manipulator model. *Robotics and Autonomous Systems*, *46*(3), 185–193.

Tatnall, A. R. L., Farrow, J. B., Bandecchi, M., & Francis, C. R. (2011). *Spacecraft system engineering* (4th ed.). Chichester: Wiley.

Vafa, Z., & Dubowsky, S. (1987). On the dynamics of manipulators in space using the virtual manipulator approach. In *Proceedings of IEEE international conference on robotics and automation*, North Carolina (pp. 579–585).

Yoshikawa, T., Harada, K., & Matsumoto, A. (1996). Hybrid position/force control of flexible-macro/rigid-micro manipulator systems. *IEEE Transactions on Robotics and Automation*, *12*(4), 633–640.

Chapter 12
Field Robots

Field robotics is a large and diverse research area which deals with mobile robots of different types that operate in unstructured and often dynamic environments. We have already studied underwater robots and space robots in the previous chapters, so we will focus on land-based robots. Most robots that operate on land use wheels for locomotion, so in this chapter we will mainly focus on wheeled robots. Also legged robots are being developed, and are found especially in research and education. Legged robots are particularly efficient in rough and uneven terrain but have yet to reach the same level of utility as their wheeled counterparts. Snake robots are also being used in research and education and are expected to play an important role in field robotics in the future Liljebäck et al. (2013).

Field robotics is a large and diverse research area that include topics such as mechanical design, locomotion and actuation, sensor technology, mapping and navigation, perception, and the control of the robot, possibly with some level of autonomy and cognition. Most of these topics are outside the scope of this book, and we refer to other references such as Siegwart and Nourbakhsh (2004), Dudek and Jenkin (2000), and Choset et al. (2005). In this chapter we will focus on the mechanical design and the kinematic and dynamic modeling of these robots.

12.1 Introduction

Field robots are mainly used for operation of plants and fields that cover large areas, such as in agriculture and environmental mapping, or for exploration tasks in remotely located areas, such as deserts and distant planets (Lazinica 2006). Common for these robots is that they tend to have a high degree of autonomy and are often only supervised or partially controlled by an operator.

One promising application area for field robots is agricultural automation (Ceccarelli 2012). Because of the large geographical areas that these fields cover and the extremely repetitive tasks that need to be performed, vehicle-manipulator systems are well suited for agricultural applications. Agricultural robots are normally

P.J. From et al., *Vehicle-Manipulator Systems*, Advances in Industrial Control,
DOI 10.1007/978-1-4471-5463-1_12,

Fig. 12.1 The Autonomous Prime Mover can navigate autonomously in the field. Courtesy of Carnegie Mellon University and Washington State University

wheeled, car-like robots that move either autonomously or under guidance through the fields. The manipulator arm can then be used for pruning, harvesting, and sample taking, among other tasks. With increasing labor costs and more focus on the ergonomics of the working environment, agricultural robotics is becoming one of the most promising application areas for vehicle-manipulator systems.

Another important area of field robotics is exploration of remotely located areas (Stachniss 2009; Iagnemma and Dubowsky 2004). Mobile robots with manipulator arms have already been on several missions to distant planets and they have collected valuable information about these extremely remote locations. Also these robots normally use wheels for locomotion, sometimes in combination with legs to cope with rough terrain.

12.1.1 Earth-Based Systems

There are several different kinds of vehicle-manipulator systems that are being used or are under development for Earth-based applications. As already mentioned, agricultural applications are surging as one of the most promising areas for these robots. Several autonomous robots, with or without robotic arms, have been developed for pruning, harvesting, and weed control, to name a few.

One autonomous robot that is being developed for use in orchards is the Autonomous Prime Mover from Carnegie Mellon University (Bergerman 2012). This vehicle is able to autonomously drive through the rows with trees in the orchards. The vehicle is shown in Fig. 12.1 where we can see that it is equipped with a platform that can carry the workers so that they don't have to climb up and down the ladder every time they need to move.

There are many autonomous vehicles that are being used or developed for agricultural purposes, but very few that utilize robotic arms. The main reason for this is the complex manipulation required for handling fruit and vegetables and the advanced vision required for locating the fruit. There are, however, other simpler tasks such as weed control that can be performed very accurately using robotic arms or specifically designed devices. This has the potential of reducing the amount of pesticides used and also improving the overall quality of the food.

Other applications of wheeled robots on Earth include mining robots that can alleviate human workers from a hostile working environment and robots that operate in distant areas such as in the desert or in arctic areas. In both cases a manipulator arm may be useful both for interaction tasks or for sampling and observation during exploration missions.

12.1.2 Space Robots

The search for intelligent life has resulted in several missions to other planets. One of the places most likely to be inhabited by simple life is the planet Mars and several research missions to the planet Mars have been performed over the last decades. The main objective of these mission has been to collect samples in search for life and other scientific experiments. These systems have therefore been equipped with robotic arms.

The first successful soft landings on Mars were NASA's Viking 1 and 2, both launched in 1975. The Viking spacecraft were designed to take pictures of the surface of Mars from the orbiter and also to take close-up pictures and samples from the surface. The sampler arm stretched out from the vehicle and consisted of a collector head for sampling, in addition to a temperature sensor and a magnet. The Viking orbiters lead to several findings regarding water on Mars and the photos showed huge river valleys and probable signs of ancient rainfall. The lander did not successfully prove biological findings on Mars, but more recent analyses of the collected soil may suggest the presence of organic molecules in the soil. The Viking robots sent the first color pictures from Mars back to Earth. Figure 12.2 shows two pictures taken by the Viking robots, one in which we can see the manipulator arm.

Another successful mission to Mars was the Phoenix spacecraft which reached the planet in 2008. The spacecraft carried a 2.35 meter long robotic arm that was used to scoop up soil and ice samples for analysis. The arm managed to dig deep enough to reach areas with water in the form of ice, and confirmed the presence of shallow surface water on Mars.

Also several robotic missions to the Moon have been carried out with success. The Surveyor 3 mission that landed on the Moon in 1967 was the first robot with a robotic arm able to scoop up the surface soil and send pictures back to Earth (Fig. 12.3). Also the 1968 Surveyor 7 carried a robotic arm.

12.1.3 Locomotion

This chapter is concerned with mobile robots that operate in more or less remote environments either on Earth or on distant planets. This type of environments require

Fig. 12.2 The Viking 1 mission to Mars. The *top photo* shows the soil sampler mounted on the Viking 1 lander vehicle. The *bottom photo* shows the first photo ever taken of the surface of Mars, taken on July 20, 1976. Courtesy of NASA

Fig. 12.3 The Surveyor 3 with a simple robotic arm landed on the moon in 1967. Courtesy of NASA

advanced locomotion, both mechanically speaking and in the control. Biological systems have always been a source of inspiration for mechanical engineers. Most animals that live on land use legs for locomotion. Large mammals typically have four legs, insects have six legs, while there are also quite a few mammals, and also birds, that are able to balance on two legs only (Siegwart and Nourbakhsh 2004). Field robots normally have four or more legs to guarantee stability in rough terrain and when external forces act on the robot.

Two legs are not sufficient to obtain static stability of a legged robot. For a two-legged robot advanced and fast active control is required to keep the robot from tipping over. In general, three contact points with the ground are required to guarantee static stability. A three-legged robot is stable whenever the center of gravity is kept inside the ground contact points. The same is the case for four-legged robots. Four-legged robots also require active or passive control of all the legs in order to guarantee that all the legs are in contact with the ground at all times.

The main advantage of legged robots is their ability to move in rough and uneven terrain. This ability allows the robot to operate in virtually any terrain and in the presence of large obstacles. However, this mobility comes at a price: the speed and the energy consumption. Legged robots normally move a lot slower than for example wheeled robots and they also use more energy. Furthermore, the mechanical complexity due to the number of legs required and the degrees of freedom in each leg makes these robots rather expensive, they are more exposed to mechanical failure, and require advanced control of the legs and the body.

One alternative to legged robots, especially when operating in flat terrain, is wheeled robots. Wheels are the most popular choice of locomotion for robots due to their speed and low energy consumption compared to their legged counterparts. Furthermore, wheels normally only have one or two degrees of freedom and are therefore also kinematically simpler than legged robots.

Wheeled robots are the preferred choice whenever the robot is required to move over large areas or with high velocities and when the terrain is sufficiently flat. Furthermore, we will see that several different kinds of wheel configurations exist, which make wheeled robots useful in many different types of terrain, from flat ground in environments constructed by humans, to rather uneven terrain in more remote locations.

Common for all wheeled robots is the non-holonomic nature that arise due to the resistance to pure sideways motion of the robot. As a result the velocity space of wheeled robots is of a lower dimension than the configuration space. This has positive effects on the stability of the robot but complicates the modeling and control.

12.1.4 Mobility and Configuration Spaces

The first thing that we need to determine in the design of mobile robots is what space the robot is to operate in and how it is to locomote in this space. The first is denoted the configuration space of the robot and tells us what the world that the robot is to operate in looks like in terms of mathematical spaces. Underwater and airborne robots operate in the 3-dimensional Euclidean space while seaborne and land robots are normally assumed to operate in the plane, at least for planning and control purposes, as will be discussed further in Chap. 13. All robots, of course, live in the 3-dimensional Euclidean space $SE(3)$, but for wheeled mobile robots we are mainly concerned with the location of the chassis which can be described by the x- and y-coordinates for position and ψ for attitude. The heave motion (the z-coordinate) is normally not included in the model because it cannot be controlled using the actuation that is available on a wheeled robot. The same is the case with the pitch and roll motions. If the robot is equipped with an active suspension, however, also these motions can be controlled. This is not discussed further in this book, but we refer to Messuri and Klein (1985) and Freitas et al. (2009) for more details.

Mobility, on the other hand, is not concerned with what the robot's world looks like, but rather how the robot itself is able to move in its environment. The dimensionality of the mobility cannot be higher than that of the configuration space. If the mobility and configuration space have the same dimension the robot is able to move in all the directions of the configuration space, which for $SE(2)$ are motions in the directions x, y, and ψ. The dimension of the mobility can, however, be lower than the dimension of the configuration space. This is the case when an instantaneous motion in one of the directions of the configuration space cannot be realized. The classical example is a wheeled mechanism which does not allow for lateral motion due to the kinetic constraints of the wheels.

The dimensionality of the configuration space is given by the degrees of freedom (DoF) of the *attainable robot configurations*, i.e., all the independent components that describe how the robot can be placed in its environment.

As we have seen, $SE(3)$ has 6 DoF while $SE(2)$ has 3 DoF.

Mobility describes the independent components of the *admissible velocities* of the robot. The mobility describes what instantaneous motions the kinematic constraints allow for, and is denoted the differential degree of freedom (DDoF).

In some cases we can have more degrees of freedom (DoF) than differential degrees of freedom (DDoF), which is the case when non-holonomic constraints are present.

12.1.5 Non-holonomic Motion

A fundamental and important topic in mobile robotics is the concept of non-holonomic constraints. A non-holonomic constraint is a kinematic constraint imposed on the robot chassis which cannot be expressed as an explicit function of the position variable only, but requires also the velocity variables to describe the constraint. In other words, a non-holonomic constraint requires a differential relationship such as the time derivative of the position state. Furthermore, the constraints cannot be expressed in terms of the position variables only by integrating the velocity variables. Non-holonomic constraints are therefore often referred to as non-integrable constraints (Siegwart and Nourbakhsh 2004).

One common example of non-holonomic constraints is found in wheeled robots with Ackerman steering, i.e., the wheel configuration found in normal cars where the front wheels turn and the rear wheels are fixed. We know from experience that it is impossible for a car to obtain instantaneous lateral motion. This kinematic constraint is a non-holonomic constraint. This is fairly straight forward to see because this constraint needs to be expressed in terms of the lateral velocity, more explicitly it requires that the lateral velocity in the body frame is always zero. It is not possible to express this velocity using position variables only.

When non-holonomic constraints are present we will use the term non-holonomic robot, and conversely we will term it a holonomic robot when no non-holonomic constraints are present. The only way a robot can obtain a configuration space with a higher degree of freedom than the dimensionality of the mobility is through non-holonomic constraints (Siegwart and Nourbakhsh 2004). We can therefore conclude the following:

Definition 12.1 (Non-holonomic robot) A non-holonomic robot is a robot with non-holonomic constraints. Non-holonomic constraints are present if and only if DDoF < DOF.

If no non-holonomic constraints are present we can conclude that the robot is holonomic:

Definition 12.2 (Holonomic robot) A holonomic robot is a robot without non-holonomic constraints. A robot is holonomic if and only if DDoF = DOF.

We see that there is a clear disadvantage when non-holonomic constraints are present because the robot cannot instantaneously move in any direction of the workspace. Furthermore, we have less freedom to choose the path which takes us to the desired end configuration and it is not always possible to control each component independently in such a way that they reach the end configuration at the same time.

There are also some positive aspects of non-holonomic constraints. A car-like mechanism, for example, can sustain lateral motion very well, which is taken advantage of during turn. When a car is turning, centrifugal forces will try to push the car out of its trajectory. However, because of the non-holonomic constraints of car-like mechanisms, these centrifugal forces are effectively counteracted by the sliding constraints.

12.2 Modeling of Wheeled Robots

Wheeled mobile robots present us with both kinematic and dynamics considerations that differ from the robots that we have studies so far. Firstly, the kinematics of a mobile robot is mainly concerned with the position of the chassis in its environment, i.e., the location of the body frame $\mathscr{F}_b$, with respect to the inertial frame $\mathscr{F}_0$. However, the motion is generated by the wheels, which are often numerous. A large number of wheels gives us several degrees of freedom for control: one degree of freedom is available to control the velocity of each wheel, and for certain wheel types we can also control the angle of the wheel, which adds another degree of freedom to be controlled.

We have seen that the mobility of a wheeled robot is normally fairly low: robots with standard wheels allow for forward and turning motions. This difference in dimensionality between the mobility and the degrees of freedom that can be controlled, arise due to kinematic constraints between the chassis and the wheels, and between the wheels themselves. These constraints are important to understand in order to find the chassis motion from the motion of the wheels. Furthermore, this kinematic relation is required for control.

There are also some dynamic aspects that need to be considered particularly for wheeled mobile robots. The center of gravity, for example, needs be low enough so that the robot does not tip over during high-speed turns. A high center of mass does normally not affect the static stability, but when the robot is in motion, dynamic considerations must be made.

12.2.1 Chassis Kinematics

The main control objective of wheeled mobile robots is to control the velocity and position of the robot chassis. We will study the wheel kinematics in detail in the next section, for now we will assume that these are satisfied and look at how the robot chassis moves as a function of the motion of the wheels.

The position of the robot can be written either in the body or the spatial frame. The orientation of the robot is the same regardless of the frame in which it is represented, while the relation between the position variable in the body and inertial frame is given by the rotation matrix in the normal way so we have the relation

$$\begin{bmatrix} x_{0b} \\ y_{0b} \\ \psi_{0b}^S \end{bmatrix} = \begin{bmatrix} \cos\psi_{0b}^B & -\sin\psi_{0b}^B & 0 \\ \sin\psi_{0b}^B & \cos\psi_{0b}^B & 0 \\ 0 & 0 & 1 \end{bmatrix} \begin{bmatrix} x_{0b}^B \\ y_{0b}^B \\ \psi_{0b}^B \end{bmatrix} \tag{12.1}$$

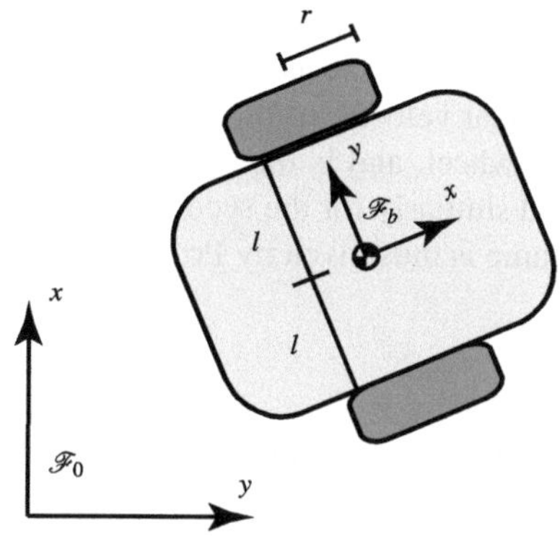

Fig. 12.4 A differential-drive robot

where the superscript B denotes that the variables are denoted in the body frame. For planar motion we can find the same relation for the velocities i.e.,

$$\begin{bmatrix} \dot{x}_{0b}^0 \\ \dot{y}_{0b}^0 \\ \dot{\psi}_{0b}^S \end{bmatrix} = \begin{bmatrix} \cos\psi_{0b}^B & -\sin\psi_{0b}^B & 0 \\ \sin\psi_{0b}^B & \cos\psi_{0b}^B & 0 \\ 0 & 0 & 1 \end{bmatrix} \begin{bmatrix} \dot{x}_{0b}^B \\ \dot{y}_{0b}^B \\ \dot{\psi}_{0b}^B \end{bmatrix} \tag{12.2}$$

where $\dot{x}_{0b}^0$ and $\dot{y}_{0b}^0$ are the linear velocities as seen from the inertial frame (and not the spatial velocities).

Now that we can define the velocities in the body frame we can find the body velocities from the wheel velocities. How the chassis moves when the wheels move depends on the geometry of the robot and what kind of wheels are used.

We will look at a simple example, the differential-drive robot which consist of a chassis and two standard and fixed wheels attached to each side of the robot on the same axis. A simple illustration of a differential-drive robot is shown in Fig. 12.4. It is interesting to note that this kind of robot can actually be made statically stable by placing the center of gravity at a point lower than the wheel axis. Alternatively it can be stabilized by a passive castor wheel. We start with the following simple but important property:

Property 12.1 For a differential-drive robot the linear and angular contributions of each wheel can be added to find the total velocity of the chassis. This means that the instantaneous contribution of each wheel can be computed as if the other wheel were fixed.

This is only possible, of course, because we have two degrees of freedom available for control (the velocity of each wheel), which is the same as the degree of freedom of the mobility of the robot. There are therefore no kinematic constraints on the velocity of the wheels and each wheel can be controlled freely.

To find the chassis velocities we will follow the train of thought in Siegwart and Nourbakhsh (2004) and Property 12.1. We are interested in the speed of the body at the center point denoted by $\bar{\mathcal{F}}_b$ (not necessarily at the center of mass). The speed of a frame attached to wheel 1 is $r\omega_1$ where r is the wheel radius and ω_1 is the angular velocity of the wheel. The speed of the center point $\bar{\mathcal{F}}_b$ is half the speed at the wheel, and is thus given by $\dot{x}^B_{0b,1} = \frac{1}{2}r\omega_1$ for a positive rotation of the wheel, and similarly for the second wheel. The total instantaneous forward motion in body frame is thus given by Property 12.1 as

$$\dot{x}^B_{0b} = \frac{1}{2}r(\omega_1 + \omega_2). \tag{12.3}$$

We assume no slippage, so none of the wheels contribute to lateral motion in the body frame (because we assume that the wheel and chassis frames are always aligned), so we have $\dot{y}^B_{0b} = 0$.

If only one wheel moves this wheel will move in a circle with a $2l$ diameter. Once again the speed at the wheel is $r\omega_1$, so the angular contribution of this wheel is

$$\dot{\psi}^B_{0b,1} = \frac{r\omega_1}{2l}. \tag{12.4}$$

The total angular velocity at the point $\bar{\mathcal{F}}_b$ (and any other point on the chassis, for that sake) is therefore

$$\dot{\psi}^B_{0b} = \frac{r(\omega_1 - \omega_2)}{2l}. \tag{12.5}$$

We can now find the body velocities of the chassis as a function of the wheel velocities as

$$\begin{bmatrix} \dot{x}^B_{0b} \\ \dot{y}^B_{0b} \\ \dot{\psi}^B_{0b} \end{bmatrix} = \begin{bmatrix} \frac{r}{2} & \frac{r}{2} \\ 0 & 0 \\ \frac{r}{2l} & -\frac{r}{2l} \end{bmatrix} \begin{bmatrix} \omega_1 \\ \omega_2 \end{bmatrix}. \tag{12.6}$$

We see that while the kinematic relation in (12.2) does not take the admissible velocities into account, we can now rewrite (12.6) as

$$\begin{bmatrix} \dot{x}^0_{0b} \\ \dot{y}^0_{0b} \\ \dot{\psi}^S_{0b} \end{bmatrix} = \begin{bmatrix} \cos\psi^B_{0b} & -\sin\psi^B_{0b} & 0 \\ \sin\psi^B_{0b} & \cos\psi^B_{0b} & 0 \\ 0 & 0 & 1 \end{bmatrix} \begin{bmatrix} \frac{r}{2} & \frac{r}{2} \\ 0 & 0 \\ \frac{r}{2l} & -\frac{r}{2l} \end{bmatrix} \begin{bmatrix} \omega_1 \\ \omega_2 \end{bmatrix} \tag{12.7}$$

which also includes the kinematic constraints on the wheels.

We can also take a more formal approach based on the framework presented earlier in the book. Let m represent the dimensionality of the configuration space (DoF) and $\bar{m}$ the dimensionality of the of the mobility (DDoF). The definition of a non-holonomic joint can then be written as Duindam and Stramigioli (2008)

Definition 12.3 (Globally parameterized non-holonomic joint) A globally parameterized non-holonomic joint is a globally parameterized holonomic joint together with a restriction on the allowed instantaneous velocities, given by an equation

$A^{\mathsf{T}}(Q)v = 0$ for some differentiable matrix $A(Q) \in \mathbb{R}^{m\times(m-\bar{m})}$ of constant rank $(m - \bar{m})$. In addition a differentiable rank $\bar{m}$ matrix $S(Q) \in \mathbb{R}^{m\times\bar{m}}$ that satisfies $A^{\mathsf{T}}(Q)\bar{S}(Q) = 0$ for all Q must exist.

We will now represent the non-holonomic constraint of the differential-drive robot that we looked at above in terms of the matrices $A(Q)$ and $\bar{S}(Q)$ above. This is fairly straight forward if we keep in mind that no lateral motion in the body frame is allowed. The kinematic constraint, which has dimension one, can then be written as

$$A^{\mathsf{T}}(Q)v = \begin{bmatrix} 0 & 1 & 0 \end{bmatrix} \begin{bmatrix} u \\ v \\ r \end{bmatrix} = v = 0 \tag{12.8}$$

which does not allow for motion in the direction of the y-axis of the body frame. Once the A-matrix is found we can find the $\bar{S}$-matrix as

$$\bar{S}(Q) = \begin{bmatrix} 1 & 0 \\ 0 & 0 \\ 0 & 1 \end{bmatrix} \tag{12.9}$$

for which $A^{\mathsf{T}}(Q)\bar{S}(Q) = 0$.

Finally we will note that we can include non-holonomic constraints quite effectively by the selection matrix described in detail in Chap. 3. For holonomic robots in $SE(3)$ the velocity state is fully determined by only three variables and we choose H so that

$$V_{0b}^{b} = H\tilde{V}_{0b}^{b} \tag{12.10}$$

with

$$H = \begin{bmatrix} 1 & 0 & 0 \\ 0 & 1 & 0 \\ 0 & 0 & 0 \\ 0 & 0 & 0 \\ 0 & 0 & 0 \\ 0 & 0 & 1 \end{bmatrix}. \tag{12.11}$$

In this case both the mobility and the configuration space have dimensionality three so no non-holonomic constraints are present. If, on the other hand, non-holonomic constraints are present, such as for wheeled mechanisms, we can incorporate these constraints by the selection matrix

$$\bar{H} = \begin{bmatrix} 1 & 0 \\ 0 & 0 \\ 0 & 0 \\ 0 & 0 \\ 0 & 0 \\ 0 & 1 \end{bmatrix} \tag{12.12}$$

and choosing the correct velocity state, which in our case reduces to $\tilde{V}_{0b}^{b} = [u\ r]^{\mathsf{T}}$. The dynamics is then found by substituting $\bar{H}$ and $\tilde{V}_{0b}^{b}$ into the formalism presented in Sect. 3. We see that this is a very straight forward way of including the non-holonomic constraints because we restrict the velocity space by reducing the dimension of the selection matrix without reducing the dimension of the configuration space of the robot.

12.2.2 Wheel Kinematics

For the differential-drive robot in the previous section the chassis velocity is found from the wheel velocities by a simple geometric analysis. For mobile robots with more wheels or wheels with more degrees of freedom and more complex geometry we need to express the wheel kinematics explicitly. This relation is then used to describe how the chassis moves when the motion of each individual wheel is known.

We will concentrate on the two most common wheels: fixed wheels—as the rear wheels on a car—and steered wheels—as found at the car's front wheels. We will follow the derivation found in Siegwart and Nourbakhsh (2004), and we also refer to this reference for more details and other wheel types.

There are two constraints that restrict the wheel motion. The first is the rolling constraint, i.e., the velocity of the wheel given by $r\omega_i$ must correspond with the chassis velocity at the point $\mathscr{F}_{w_i}$ where the wheel is attached. The second constraint is the sliding constraint which prevents lateral motion of the wheel, i.e., the component of the wheel's motion in the direction of the wheel axis is zero. We will write the kinematic constraints of the wheel in terms of the distance from the chassis center of mass l, the wheel radius r, the angle to the position of the wheel α, and the current orientation of the wheel β, corresponding to Fig. 12.5.

12.2.2.1 Rolling Constraints

The rolling constraint of a fixed wheel forces the velocity of the point on the chassis where the wheel is attached, i.e., the wheel frame $\mathscr{F}_{w_i}$, to be equal to the velocity of the wheel. The velocity of the wheel is given simply by the angular velocity and the radius of the wheel as $r\omega_i$. The velocity of the frame $\mathscr{F}_{w_i}$ is given by

$$v_{\omega_i} = \dot{x}_{0b}^{B} \sin(\alpha_i + \beta_i) - \dot{y}_{0b}^{B} \cos(\alpha_i + \beta_i) - l_i \dot{\psi}_{0b}^{B} \cos\beta_i. \tag{12.13}$$

The rolling constraint can therefore be written as

$$\begin{bmatrix} \sin(\alpha_i + \beta_i) & -\cos(\alpha_i + \beta_i) & -l_i \cos\beta_i \end{bmatrix} \begin{bmatrix} \dot{x}_{0b}^{B} \\ \dot{y}_{0b}^{B} \\ \dot{\psi}_{0b}^{B} \end{bmatrix} - r\omega_i = 0. \tag{12.14}$$

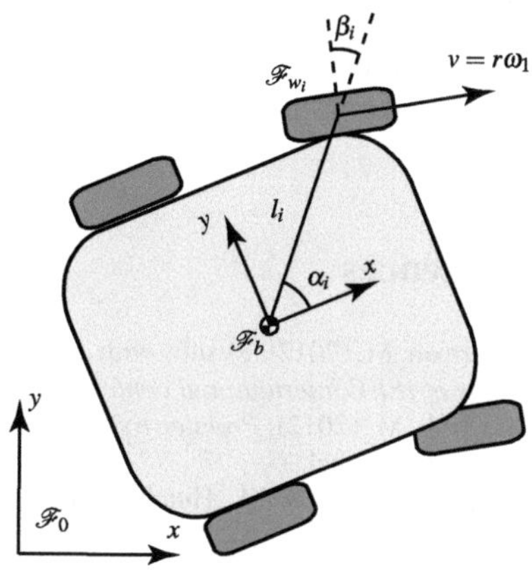

Fig. 12.5 The kinematics of fixed and steered wheels

Steered wheels are identical to fixed wheels with the additional degree of freedom which allows it to rotate the steering angle β_i. The rolling constraint of a steered wheel is therefore identical to that of the fixed wheel with the exception that the steering angle $\beta(t)$ is time varying and can be used for control if actuated. The rolling constraint of a steered wheel therefore becomes

$$\left[\sin\left(\alpha_i + \beta_i(t)\right) \; -\cos\left(\alpha_i + \beta_i(t)\right) \; -l_i \cos\beta_i(t)\right] \begin{bmatrix} \dot{x}_{0b}^B \\ \dot{y}_{0b}^B \\ \dot{\psi}_{0b}^B \end{bmatrix} - r\omega_i = 0. \tag{12.15}$$

12.2.2.2 Sliding Constraints

The sliding constraint of a fixed wheel limits the motion of the wheel frame $\mathscr{F}_w$ to be orthogonal to the wheel axis. There is therefore no motion in the direction of the wheel axis, which can be written as

$$\left[\cos(\alpha_i + \beta_i) \; \sin(\alpha_i + \beta_i) \; l_i \sin\beta_i\right] \begin{bmatrix} \dot{x}_{0b}^B \\ \dot{y}_{0b}^B \\ \dot{\psi}_{0b}^B \end{bmatrix} = 0. \tag{12.16}$$

Also the sliding constraint is the same for steering wheels as for fixed wheels, with the additional freedom in the wheel steering angles. The sliding constraint for steering wheels is therefore given by

$$\left[\cos\left(\alpha_i + \beta_i(t)\right) \; \sin\left(\alpha_i + \beta_i(t)\right) \; l_i \sin\beta_i(t)\right] \begin{bmatrix} \dot{x}_{0b}^B \\ \dot{y}_{0b}^B \\ \dot{\psi}_{0b}^B \end{bmatrix} = 0. \tag{12.17}$$

Using these kinematic constraints we can find a more complete description of the chassis motion in terms of the velocity and orientation of each wheel. The wheel constraints are also necessary to obtain the control allocation matrix similar to the approach presented in Chap. 10. We refer to Siegwart and Nourbakhsh (2004) for more details on modeling and control of mobile wheeled robots.

References

Bergerman, M. (2012). Results with autonomous vehicles operating in specialty crops. In *Proceedings of IEEE international conference on robotics and automation* (pp. 1829–1835).

Ceccarelli, M. (2012). *Premier reference source. Service robots and robotics: design and application*. IGI Global.

Choset, H., Lynch, K. M., Hutchinson, S., Kantor, G. A., Burgard, W., Kavraki, L. E., & Thrun, S. (2005). *Principles of robot motion: theory, algorithms, and implementations*. Cambridge: MIT Press.

Dudek, G., & Jenkin, M. (2000). *Computational principles of mobile robotics*. New York: Cambridge University Press.

Duindam, V., & Stramigioli, S. (2008). Singularity-free dynamic equations of open-chain mechanisms with general holonomic and nonholonomic joints. *IEEE Transactions on Robotics*, *24*(3), 517–526.

Freitas, G., Lizarralde, F., Hsu, L., & Dos Reis, N. R. S. (2009). Kinematic reconfigurability of mobile robots on irregular terrains. In *Proceedings of IEEE international conference on robotics and automation* (pp. 823–828).

Iagnemma, K., & Dubowsky, S. (2004). *Springer tracts in advanced robotics. Mobile robots in rough terrain: estimation, motion planning, and control with application to planetary rovers*. Berlin: Springer.

Lazinica, A. (2006). *Mobile robots: towards new applications*. I-Tech Education and Publishing.

Liljebäck, P., Pettersen, K. Y., Stavdahl, Ø., & Gravdahl, J. T. (2013). *Snake robots modelling, mechatronics, and control*. Berlin: Springer.

Messuri, D., & Klein, C. (1985). Automatic body regulation for maintaining stability of a legged vehicle during rough-terrain locomotion. *IEEE Journal of Robotics and Automation*, *1*(3), 132–141.

Siegwart, R., & Nourbakhsh, I. R. (2004). *Introduction to autonomous mobile robots*. Scituate: Bradford Company.

Stachniss, C. (2009). *Springer tracts in advanced robotics. Robotic mapping and exploration*. Berlin: Springer.

Chapter 13
Robotic Manipulators Mounted on a Forced Non-inertial Base

A class of vehicle-manipulator systems that differs from the ones presented earlier in this book—where both the vehicle and the manipulator motions were controlled using the manipulator and vehicle actuators—is manipulators mounted on a forced base. These are robotic systems consisting of a manipulator arm attached to a vehicle that cannot be controlled or it is not desirable to control, for example due to the excessive actuator forces required.

One example that will be studied in detail in this chapter is robots and cranes mounted on ships and other marine vessels. In this case the motion of the base will affect not only the kinematic relations of the manipulator end effector with respect to the world frame, but also the manipulator dynamics. Due to the dynamic coupling between the base and the manipulator, the non-inertial base motion will create forces in the manipulator that are not present for fixed-base robots. Forces that arise due to the non-inertial motion of the base are called inertial forces and need to be included in the mathematical models to guarantee for robust and accurate control.

13.1 Introduction

Up until now we have only discussed vehicle-manipulator systems where both the vehicle and the manipulator were either free-floating or controlled through the actuators in the system. Furthermore, the external forces have been considered rather small so that they are effectively canceled through the feedback control loop. In this chapter we will study systems where the vehicle motion arises due to external forces that act on the vehicle, as opposed to motion created by the actuators, in such a way that this motion affects the performance of the overall system to such an extent that it should be included either in the motion planning or the control of the robot. To this end, the disturbance is often obtained by estimation techniques using advanced sensor systems or by modeling the external disturbances.

High- and low-frequency disturbances affect the manipulator motion in different ways. While low-frequency motion can be handled on a kinematic level, i.e., by

P.J. From et al., *Vehicle-Manipulator Systems*, Advances in Industrial Control,
DOI 10.1007/978-1-4471-5463-1_13, © Springer-Verlag London 2014

compensating for the motion in the motion planner, the high frequency motion of the vehicle also affects the dynamic properties of the manipulator. In this case the forces that arise due to the dynamic coupling need to be compensated for in the control loop. For most *vehicles* the high-frequency motion is normally not compensated for: A ship in high sea will filter all high-frequency motions from the measurements so that the controller does not try to compensate for this motion. The motion due to waves and wind cannot be eliminated using the ship's actuators, and even if we to a certain extent could compensate for these disturbances, the wear and tear on the actuators would shorten their operational life drastically and would also lead to an excessive fuel consumption. The high-frequency motions are therefore filtered before they enter the control loop. Similarly, a robotic rover as discussed in the previous chapter will not be able to compensate for uneven terrain for the same reasons as the ship. Thus, the control of the robot will in most cases only require the rover to follow the desired trajectory in the plane and not try to compensate for high-frequency motion and errors in heave z, roll ϕ, and pitch θ. These motion components are either not taken into account in the control of the system or compensated for using the robotic arm.

Several systems with manipulators mounted on seaborne platform have been studied in literature (From et al. 2009). Love et al. (2004) study the impact of wave generated disturbances on the tracking control of a manipulator mounted on a ship. The use of cable robots for loading and unloading cargo between two ships has also been addressed by Kitarovic et al. (2005) and Oh et al. (2005). In the Ampelmann project (Salzmann 2007), a Stewart platform is mounted on a ship and is used to compensate for the motion of the ship by keeping the platform still with respect to the world frame. Kosuge et al. (1992) and Kajita and Kosuge (1997) address the control of robots floating on the water utilizing vehicle restoring forces.

13.1.1 Seaborne Platforms

Several marine vessels have one or more robotic arms or cranes installed for cargo handling, load and reload operations offshore, anchoring, the deployment of underwater AUVs, and even offshore cabling. Common for all these systems is that they will have to sustain very large inertial forces due to the possibly extreme conditions offshore. Offshore manipulators are therefore powerful arms and are often built with hydraulic actuation and very high stiffness. As a result, these arms are normally very heavy mechanisms which leads to high inertias when the vessel is in motion (Lebans et al. 1997; Kajita and Kosuge 1997; Kosuge et al. 1992).

13.1.2 Active Heave Compensation

One of the success stories of offshore robotic systems is the implementation of active heave compensation (From et al. 2009). When transferring cargo from a ship to a

(a) The Hydralift AHC from National Oilwell Varco is an active heave compensation system designed for subsea load handling from marine vessels and rigs

(b) The Helia crane from Scantrol is mounted at the ASL shipyard in Singapore and facilitates loading to and off-loading from moving vessels. Courtesy of Scantrol AS.

Fig. 13.1 Active heave compensation systems avoid that the motion of the vessel is transferred to the load and allows for safer load handling. The cranes can either be mounted on a moving vessel, load cargo to/from a moving vessel from a fixed location, or transfer cargo between two moving vessels

fixed installation, it is desirable to eliminate the ship motion and let the operator focus only on the relative distance between the cargo and the site where the cargo is to be off-loaded. When off-loading cargo from a ship in motion to an inertially fixed location, this means that the arm keeps a constant distance between the cargo and the seabed as long as no input is given from the operator. The operator can then focus only on the relative motion between the cargo and the onshore location which makes transfer of the load both safer and simpler. Figure 13.1 shows two active heave compensating systems, one which is mounted on a moving base (left) and one which is mounted onshore but operates on moving vessels (right).

A similar system is the Ampelmann self stabilizing platform (Salzmann 2007) which allows for safe transfer of personnel from a moving ship to a fixed installation such as a platform. Inertial sensors measure the ship motion which is compensated for using a Stewart platform so that the far end of the bridge is kept constant at the entry point at the fixed installation.

Another example of a crane system that is used on offshore vessels is the MDH arm mounted on a 2-DoF gantry crane shown in Fig. 13.2. Robotic manipulators and cranes are being used more frequently on ships and other marine vessels during operation at high sea. Several tasks, such as handling anchors, cables, and other equipment are performed more safely using robotic arms than humans. As these systems need to be operable at all times and under all conditions, i.e., also when the robotic systems is under the influence of extreme inertial forces, advanced control and motion planning algorithms need to be implemented.

13.1.3 Land Vehicles

Also for land vehicles with robotic arms the inertial forces can be substantial. Robotic arms mounted on land vehicles are in general equipped with smaller actu-

Fig. 13.2 The TRIPLEX MDH system is a crane mounted on a rail with an additional 2 degrees of freedom. The system can be used for handling loads on the ship deck (*top photo*) and for cabling (*bottom illustration*). Courtesy of Triplex

ators than their marine counterparts, which requires advanced motion planning and control to guarantee good performance. In general the environmental disturbances on a land vehicle that moves in rough terrain are of a higher frequency than marine systems and they are also more difficult to predict due to the non-deterministic motion of the vehicle. Because of this, any compensation of the inertial forces for ground vehicles relies heavily on measuring the state of the robot at all times.

The use of robotic manipulators on land vehicles with large non-inertial motion components is still fairly limited. Most of the field robots that we studied in the previous chapter are not subject to any large base velocities or accelerations and therefore almost no inertial forces enter into the manipulator dynamics. There are, however, a few applications where robotic arms are mounted on robots that move

(a) The Seekur Jr used for indoor applications where little or no inertial forces act on the manipulator arm

(b) In outdoor applications in rough terrain the robot motion can be large and the inertial forces that act on the manipualtor arm are substantial

Fig. 13.3 The Seekur Jr Mobile Manipulator is a wheeled mobile base with a 5-DoF manipulator arm that can be used for both indoor and outdoor applications. Courtesy of Adept MobileRobots

with high velocity or in rough terrain. One example is in agricultural applications where a robotic rover or vehicle locomotes in rough and uneven terrain. The robotic arm is thus influenced by the inertial forces, and this needs to be compensated for to be able to perform accurate manipulation tasks in the field. Cost is one of the most important factors in agricultural robotics so these arms are often light-weight and with relatively small actuators. The manipulation tasks include harvesting and picking fruit and vegetables, pruning, and so on—all of which are extreme precision tasks which cannot be performed without compensating for the inertial forces.

Some experimental platforms consisting of a rover with a robotic arm are also available. These are in general smaller platforms that can move with relatively high velocities. If the manipulator arm is to perform some kind of work during this high-velocity motion—such as holding a camera, monitoring, or surveillance of a plant or area—also these motions need to be compensated for. One example of such a robotic platform is the Seekur Jr shown in Fig. 13.3.

13.2 Dynamics of Manipulators on a Forced Base

In this chapter we consider systems in the form shown in Fig. 13.4. For a vehicle-manipulator system where the base motion is completely or partially governed by disturbances that are not compensated for in the control loop, it is reasonable to write the dynamics in a block-diagonal form separating the part of the system with the forced motion, normally the vehicle, from the part that we can control and for which the control objective is defined, normally the manipulator. We will therefore write the dynamics as

$$\begin{bmatrix} M_V & M_{qV}^{\mathrm{T}} \\ M_{qV} & M_q \end{bmatrix} \begin{bmatrix} \dot{V}_{0b}^{B} \\ \ddot{q} \end{bmatrix} + \begin{bmatrix} C_V & C_{Vq} \\ C_{qV} & C_q \end{bmatrix} \begin{bmatrix} V_{0b}^{B} \\ \dot{q} \end{bmatrix} = \begin{bmatrix} \tau_V \\ \tau_q \end{bmatrix} \tag{13.1}$$

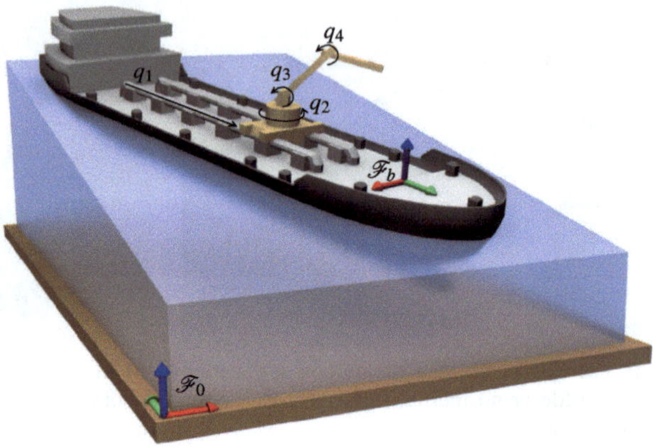

Fig. 13.4 Model setup for a four-link robot attached to a non-inertial base with coordinate frame $\mathcal{F}_b$. Frame $\mathcal{F}_0$ denotes the inertial reference frame. Courtesy of IEEE

with τ_V a wrench of control and external forces acting on the vehicle expressed in coordinates $\mathcal{F}_b$ (such that it is collocated with V_{0b}^B) and τ_q is the manipulator control torque.

We will assume that the motion of the base is fully determined by external forces that are neither known nor of interest. We further assume that the relative pose g_{0b}, velocity V_{0b}^B, and acceleration $\dot{V}_{0b}^B$ of the base relative to the inertial frame are known, or at least estimated, from measurements. We thus assume that the torques applied to the internal robot joints do not influence the motion of the platform, which is a reasonable assumption in applications with a relatively small robot attached to a large moving base.

Since the vehicle motion is not controlled we regard this as a disturbance and write the manipulator dynamics in the following way:

$$M_q \ddot{q} + C_q \dot{q} + \underbrace{M_{qV} \dot{V}_{0b}^B + C_{qV} V_{0b}^B}_{\text{inertial forces}} = \tau. \tag{13.2}$$

This equation partially separates the usual robot dynamics (first two terms) from the inertial forces (last two terms), although the matrix C_q generally still depends on V_{0b}^B. For a static base frame ($V_{0b}^B = 0$), the equations reduce to the regular dynamics of an n-link robotic mechanism on a fixed base. Note that for constant V_{0b}^B, the terms due to the motion of the base generally do not cancel out, since a constant twist can also contain (non-inertial) angular components (From et al. 2010, 2011).

The terms C_q and C_{qV} can be written more explicitly as

$$C_q = \sum_{k=1}^{n} \frac{\partial M_q}{\partial q^k} \dot{q}^k - \frac{1}{2} \frac{\partial^T}{\partial q} \left(\begin{bmatrix} M_{qV} & M_q^T \end{bmatrix} \begin{bmatrix} V_{0b}^B \\ \dot{q} \end{bmatrix} \right),$$

$$C_{qV} = \sum_{k=1}^{n} \frac{\partial M_{qV}}{\partial q^k} \dot{q}^k - \frac{1}{2} \frac{\partial^T}{\partial q} \left(\begin{bmatrix} M_V & M_{qV}^T \end{bmatrix} \begin{bmatrix} V_{0b}^B \\ \dot{q} \end{bmatrix} \right).$$

Finally we include the gravitational forces in the normal way as

$$\tau_g^i = J_i(q)\mathrm{Ad}_{g_{0i}}^{\mathrm{T}}(Q)F_g^i(Q) \tag{13.3}$$

where J_i is the geometric Jacobian and $\mathrm{Ad}_{g_{0i}} = \mathrm{Ad}_{g_{0b}}\mathrm{Ad}_{g_{bi}}$ is the transformation from the inertial frame to frame $\mathscr{F}_i$. We note that both R_{0i} and $\mathrm{Ad}_{g_{0i}}$ depend on the base configuration with respect to the inertial frame. The total effect of the gravity from all the links is then given by $N(Q) = \sum_{i=1}^{n} \tau_g^i$ which enters Eq. (13.2) the same way as the control torque. The dynamics is then written as

$$M_q\ddot{q} + C_q\dot{q} + \underbrace{M_{qV}\dot{V}_{0b}^B + C_{qV}V_{0b}^B}_{\text{inertial forces}} + \underbrace{J_i(q)\mathrm{Ad}_{g_{0i}}^{\mathrm{T}}(Q)F_g^i(Q)}_{\text{kinematic disturbances}} = \tau. \tag{13.4}$$

13.3 Motion Planning and Control

In this section we will look briefly at how we can use the models that we derived earlier in this chapter to compensate for the effects of the moving base. In general there are two ways to deal with the inertial forces: we can either try to compensate for the effects in the controller by simply canceling the forces, or alternatively we can deal with the forces in the motion planning algorithm to either reduce or leverage the forces. What approach to choose depends on the system at hand and to what extent the base motion can be predicted.

13.3.1 Canceling the Inertial Forces

If we look at the inertial forces as a disturbance that can be measured for example by an IMU on the vehicle we can simply choose to cancel the effects of the moving base in the feed-forward term in the controller. Consider the dynamics in Eq. (13.1) where τ_q is the control and τ_V is represent external forces that we cannot control. Furthermore consider the control law

$$\tau_q = \tau_{ff} + \tau_{\mathrm{PD}} \tag{13.5}$$

where

$$\tau_{ff} = \underbrace{M_q\ddot{q}_d + C_q\dot{q}_d}_{\text{tracking terms}} + \underbrace{M_{qV}\dot{V}_{0b}^B + C_{qV}V_{0b}^B}_{\text{compensation for inertial forces}} - \underbrace{\sum_{n=1}^{n}(J_i\mathrm{Ad}_{g_{0i}}^{\mathrm{T}}F_g^i)}_{\text{gravity compensation}}, \tag{13.6}$$

$$\tau_{\mathrm{PD}} = \underbrace{K_P(q_d - q) + K_D(\dot{q}_d - \dot{q})}_{\text{PD-controller}}. \tag{13.7}$$

This is the standard augmented PD control law which in our case also compensates for the inertial forces (From et al. 2010, 2011).

τ_{PD} is the standard proportional and derivative terms that guarantee that the manipulator follows the desired trajectory given by q_d, $\dot{q}_d$. The term τ_{ff} consists of the feed-forward terms. Firstly, the tracking terms and the gravity compensation together with τ_{PD} constitute the standard augmented PD control law. These terms compensate for the manipulator dynamics in the normal way. It is important to note, however, that $C_q(V_{0b}^B, q, \dot{q})$ depends on the velocity of the vehicle and the gravity compensation depends on the vehicle attitude with respect to the inertial frame.

The final term (the second term in (13.6)) is an additional term that arises when the manipulator is mounted on a moving base. We see that both the (possibly constant) velocities and the accelerations of the base enter into the manipulator dynamics. This term thus tries to eliminate the effects of the moving base on the manipulator arm that arises due to the dynamic coupling. Equation (13.5) compensates for the dynamic coupling and changes in the direction of the gravitational forces in the body frame. If the desired trajectory is defined in the world frame, also the kinematic disturbances (as illustrated in Eq. (13.4)) due to the motion of the base frame with respect to the world frame needs to be included in the motion planner.

13.3.2 Leveraging the Inertial Forces

When large inertial forces are present, canceling these terms as we did in the previous section may be very energy demanding. Continuous compensation in this way will also result in undesirable wear and tear on the manipulator due to the continuous actuation of the robot joints. Therefore, instead of regarding these terms as disturbances, we can in some cases find a trajectory for which the non-inertial and gravitational terms coincide with the tracking terms. In this way the inertial forces will contribute to the desired motion instead of working against it. This will reduce the wear and tear on the manipulator, requires less actuator torques, and allows for more accurate manipulation. Whenever we have sufficient freedom in the motion planning to plan the trajectory in this way—for example in a point-to-point task without hard time constraints—and also have accurate predictions of the base motion available, we should try to take advantage of the inertial forces in this way. This topic was studied in detail in From et al. (2010, 2011).

We will now look at a simple point-to-point task for a robotic manipulator on a moving base. Given the dynamic equations, the initial position q_0, and desired end position q_{des} in joint coordinates, we want to find the optimal trajectory given by the minimum of the cost function P, i.e.

$$P_{min} = \min_{q(t)} \int_{t=t_0}^{t_1} P\big(\tau_q(t)\big)\, dt \tag{13.8}$$

where $P(\tau_q(t))$ is some cost function representing for example the torque required for the motion,

$$\begin{aligned} q(t_0) &= q_0 \\ q(t_1) &= q_{des} \end{aligned} \tag{13.9}$$

are the vectors describing the initial and end positions of all the joints and

$$M_q\ddot{q} + C_q\dot{q} + M_{qV}\dot{V}_{0b}^B + C_{qV}V_{0b}^B - N(Q) = \tau \tag{13.10}$$

determines the dynamics of the system.

We see that if we know or are able to estimate the future motion of the base, we can use this information to find the optimal path in joint space that minimizes the cost function $P(\tau)$. It is important to note, however, that the global solution to this problem is very complex. We will not in general be able to find an optimal solution to this problem in real time, but we will in some cases be able to find approximate solutions that can greatly enhance the overall performance. This was illustrated theoretically and experimentally in From et al. (2010, 2011).

References

From, P. J., Duindam, V., Gravdahl, J. T., & Sastry, S. (2009). Modeling and motion planning for mechanisms on a non-inertial base. In *Proceedings of international conference of robotics and automation*, Kobe, Japan (pp. 3320–3326).

From, P. J., Gravdahl, J. T., & Abbeel, P. (2010). On the influence of ship motion prediction accuracy on motion planning and control of robotic manipulators on seaborne platforms. In *Proceedings of international conference of robotics and automation*, Anchorage, Alaska, USA.

From, P. J., Gravdahl, J. T., Lillehagen, T., & Abbeel, P. (2011). Motion planning and control of robotic manipulators on seaborne platforms. *Control Engineering Practice*, *19*(8), 809–819.

Kajita, H., & Kosuge, K. (1997). Force control of robot floating on the water utilizing vehicle restoring force. In *Proceedings of IEEE/RSJ international conference on intelligent robot and systems*, Grenoble, France (Vol. 1, pp. 162–167).

Kitarovic, J., Tomas, V., & Cisic, D. (2005). The electronic and informatics age—a new stage in developing highly effective ships. In *Proceedings of international ELMAR symposium*, Zadar, Croatia (pp. 385–388).

Kosuge, K., Okuda, M., & Fukuda, T. (1992). Motion control of manipulator/vehicle system floating on water. In *Proceedings of IEEE international workshop on advanced motion control*, Nagoya, Japan (pp. 506–511).

Lebans, G., Wilkie, K., Dubay, R., Crabtree, D., & Edmonds, T. (1997). Telerobotic shipboard handling system. In *MTS/IEEE OCEANS*, Halifax, Nova Scotia, Canada (pp. 1237–1241).

Love, L. J., Jansen, J. F., & Pin, F. G. (2004). On the modeling of robots operating on ships. In *Proceedings of IEEE international conference on robotics and automation*, New Orleans, LA, USA (pp. 2436–2443).

Oh, S.-R., Mankala, K., Agrawal, S., & Albus, J. (2005). Dynamic modeling and robust controller design of a two-stage parallel cable robot. *Multibody System Dynamics*, *13*(4), 385–399.

Salzmann, D. C. (2007). Ampelmann prototype—developing a motion compensating platform for offshore access. In *Proceedings of European wind energy conference*, Milano, Italy.

Appendix
Implementation and Proofs

A.1 Computing the Partial Derivatives of the Inertia Matrix

To compute the Coriolis matrix we need the partial derivatives of the terms in the inertia matrix which are found by the partial derivatives of the Adjoint and Jacobian matrices. The partial derivatives of the inertia matrix $M(q_1, \dots, q_n)$ with respect to $q_1, \dots, q_n$ are computed by

$$
\begin{aligned}
&\frac{\partial M(q_1, \dots, q_n)}{\partial q_k} \\
&\quad = \sum_{i=k}^{n} \left(\begin{bmatrix} H^{\mathsf{T}} \\ J_i^{\mathsf{T}} \end{bmatrix} \begin{bmatrix} \frac{\partial^{\mathsf{T}} \mathrm{Ad}_{g_{ib}}}{\partial q_k} I_i \,\mathrm{Ad}_{g_{ib}} + \mathrm{Ad}_{g_{ib}}^{\mathsf{T}} I_i \frac{\partial \,\mathrm{Ad}_{g_{ib}}}{\partial q_k} \end{bmatrix} \begin{bmatrix} H & J_i \end{bmatrix} \right) \\
&\qquad + \sum_{i=k+1}^{n} \begin{bmatrix} 0 & H^{\mathsf{T}} \mathrm{Ad}_{g_{ib}}^{\mathsf{T}} I_i \,\mathrm{Ad}_{g_{ib}} \frac{\partial J_i}{\partial q_k} \\ \frac{\partial^{\mathsf{T}} J_i}{\partial q_k} \mathrm{Ad}_{g_{ib}}^{\mathsf{T}} I_i \,\mathrm{Ad}_{g_{ib}} H & \frac{\partial^{\mathsf{T}} J_i}{\partial q_k} \mathrm{Ad}_{g_{ib}}^{\mathsf{T}} I_i \,\mathrm{Ad}_{g_{ib}} J_i + J_i^{\mathsf{T}} \mathrm{Ad}_{g_{ib}}^{\mathsf{T}} I_i \,\mathrm{Ad}_{g_{ib}} \frac{\partial J_i}{\partial q_k} \end{bmatrix}.
\end{aligned} \tag{A.1}
$$

The Coriolis matrix is thus found by the partial derivatives of the Adjoint and Jacobian matrices with respect to each joint variable q_k.

A.1.1 Computing the Partial Derivatives of $\mathbf{Ad}_{g_{ij}}$

The main computational burden when computing the Coriolis matrix is on the computation of the partial derivatives of M with respect to q for which we need the partial derivatives of the Adjoint matrices, also with respect to q. To compute these one can use a relatively simple relation. If we express the velocity of joint k as $V_{(k-1)k}^{(k-1)} = X_k^k \dot{q}_k$ for constant X_k^k, then the following holds (From et al. 2010):

P.J. From et al., *Vehicle-Manipulator Systems*, Advances in Industrial Control,
DOI 10.1007/978-1-4471-5463-1, © Springer-Verlag London 2014

Proposition A.1 *The partial derivatives of the Adjoint matrix is given by*

$$\frac{\partial \operatorname{Ad}_{g_{ij}}}{\partial q_k} = \begin{cases} \operatorname{Ad}_{g_{i(k-1)}} \operatorname{ad}_{X_k^k} \operatorname{Ad}_{g_{(k-1)j}} & \textit{for } i < k \le j \\ -\operatorname{Ad}_{g_{i(k-1)}} \operatorname{ad}_{X_k^k} \operatorname{Ad}_{g_{(k-1)j}} & \textit{for } j < k \le i \\ 0 & \textit{otherwise} \end{cases}$$

Proof (By computing the spatial velocity) To prove this, we start by writing out the spatial velocity of frame $\mathscr{F}_k$ with respect to $\mathscr{F}_{(k-1)}$ when $i < k \le j$:

$$\hat{X}_k \dot{q}_k = \hat{V}^{(k-1)}_{(k-1)k} = \dot{g}_{(k-1)k} g^{-1}_{(k-1)k} = \frac{\partial g_{(k-1)k}}{\partial q_k} g_{k(k-1)} \dot{q}_k$$

where $\hat{X} = \left[\begin{smallmatrix} \hat{X}_\omega & X_v \\ 0 & 0 \end{smallmatrix}\right]$. If we compare the first and the last terms, we get

$$\frac{\partial R_{(k-1)k}}{\partial q_k} = \hat{X}_\omega R_{(k-1)k}, \tag{A.2}$$

$$\frac{\partial p_{(k-1)k}}{\partial q_k} = \hat{X}_\omega p_{(k-1)k} + X_v. \tag{A.3}$$

We can use this relation in the expression for the partial derivative of $\operatorname{Ad}_{g_{(k-1)k}}$:

$$\begin{aligned}
\frac{\partial \operatorname{Ad}_{g_{(k-1)k}}}{\partial q_k} &= \begin{bmatrix} \frac{\partial R_{(k-1)k}}{\partial q_k} & \frac{\hat{p}_{(k-1)k}}{\partial q_k} R_{(k-1)k} + \hat{p}_{(k-1)k} \frac{\partial R_{(k-1)k}}{\partial q_k} \\ 0 & \frac{\partial R_{(k-1)k}}{\partial q_k} \end{bmatrix} \\
&= \begin{bmatrix} \hat{X}_\omega R_{(k-1)k} & \widehat{\hat{X}_\omega p_{(k-1)k}} R_{(k-1)k} + \hat{X}_v R_{(k-1)k} + \hat{p}_{(k-1)k} \hat{X}_\omega R_{(k-1)k} \\ 0 & \hat{X}_\omega R_{(k-1)k} \end{bmatrix} \\
&= \begin{bmatrix} \hat{X}_\omega R_{(k-1)k} & \hat{X}_\omega \hat{p}_{(k-1)k} R_{(k-1)k} + \hat{X}_v R_{(k-1)k} \\ 0 & \hat{X}_\omega R_{(k-1)k} \end{bmatrix} \\
&= \begin{bmatrix} \hat{X}_\omega & \hat{X}_v \\ 0 & \hat{X}_\omega \end{bmatrix} \begin{bmatrix} R_{(k-1)k} & \hat{p}_{(k-1)k} R_{(k-1)k} \\ 0 & R_{(k-1)k} \end{bmatrix} \\
&= \operatorname{ad}_{X_k^k} \operatorname{Ad}_{g_{(k-1)k}}
\end{aligned} \tag{A.4}$$

where we have used that

$$\hat{X} \hat{p} = \widehat{(\hat{X} p)} + \hat{p} \hat{X}. \tag{A.5}$$

It is now straight forward to show that

$$\begin{aligned}
\frac{\partial \operatorname{Ad}_{g_{ij}}}{\partial q_k} &= \operatorname{Ad}_{g_{i(k-1)}} \frac{\partial \operatorname{Ad}_{g_{(k-1)k}}}{\partial q_k} \operatorname{Ad}_{g_{kj}} \\
&= \operatorname{Ad}_{g_{i(k-1)}} \operatorname{ad}_{X_k^k} \operatorname{Ad}_{g_{(k-1)k}} \operatorname{Ad}_{g_{kj}} \\
&= \operatorname{Ad}_{g_{i(k-1)}} \operatorname{ad}_{X_k^k} \operatorname{Ad}_{g_{(k-1)j}} .
\end{aligned} \tag{A.6}$$

The proof is similar for $j < k \leq i$. □

Proof (By direct computation) We can also prove Proposition A.1 by direct computation. For $i < k \leq j$ the proof is shown by

$$
\begin{aligned}
&\frac{\partial \operatorname{Ad}_{g_{ij}}}{\partial q_k} \\
&= \operatorname{Ad}_{g_{i(k-1)}} \frac{\partial \operatorname{Ad}_{g_{(k-1)k}}}{\partial q_k} \operatorname{Ad}_{g_{kj}} \\
&= \begin{bmatrix} R_{i(k-1)} & \hat{p}_{i(k-1)} R_{i(k-1)} \\ 0 & R_{i(k-1)} \end{bmatrix} \begin{bmatrix} \frac{\partial R_{(k-1)k}}{\partial q_k} & \frac{\hat{p}_{(k-1)k}}{\partial q_k} R_{(k-1)k} + \hat{p}_{(k-1)k} \frac{\partial R_{(k-1)k}}{\partial q_k} \\ 0 & \frac{\partial R_{(k-1)k}}{\partial q_k} \end{bmatrix} \\
&\times \begin{bmatrix} R_{kj} & \hat{p}_{kj} R_{kj} \\ 0 & R_{kj} \end{bmatrix} \\
&= \begin{bmatrix} R_{i(k-1)} \frac{\partial R_{(k-1)k}}{\partial q_k} R_{kj} & \begin{bmatrix} R_{i(k-1)} \frac{\partial R_{(k-1)k}}{\partial q_k} \hat{p}_{kj} R_{kj} \\ + R_{i(k-1)} \frac{\hat{p}_{(k-1)k}}{\partial q_k} R_{(k-1)j} \\ + R_{i(k-1)} \hat{p}_{(k-1)k} \frac{\partial R_{(k-1)k}}{\partial q_k} R_{kj} \\ + \hat{p}_{i(k-1)} R_{i(k-1)} \frac{\partial R_{(k-1)k}}{\partial q_k} R_{kj} \end{bmatrix} \\ 0 & R_{i(k-1)} \frac{\partial R_{(k-1)k}}{\partial q_k} R_{kj} \end{bmatrix} \\
&= \begin{bmatrix} R_{i(k-1)} \hat{X}_\omega R_{(k-1)k} R_{kj} & \begin{bmatrix} R_{i(k-1)} \hat{X}_\omega R_{(k-1)k} \widehat{\hat{p}_{kj} R_{kj}} \\ + R_{i(k-1)} ((\hat{X}_\omega p_{(k-1)k}) + \hat{X}_v) R_{(k-1)j} \\ + R_{i(k-1)} \hat{p}_{(k-1)k} \hat{X}_\omega R_{(k-1)k} R_{kj} \\ + \hat{p}_{i(k-1)} R_{i(k-1)} \hat{X}_\omega R_{(k-1)k} R_{kj} \end{bmatrix} \\ 0 & R_{i(k-1)} \hat{X}_\omega R_{(k-1)k} R_{kj} \end{bmatrix} \\
&= \begin{bmatrix} R_{i(k-1)} \hat{X}_\omega R_{(k-1)j} & \begin{bmatrix} R_{i(k-1)} \hat{X}_v R_{(k-1)j} + R_{i(k-1)} \hat{X}_\omega \hat{p}_{(k-1)j} R_{(k-1)j} \\ + \hat{p}_{i(k-1)} R_{i(k-1)} \hat{X}_\omega R_{(k-1)j} \end{bmatrix} \\ 0 & R_{i(k-1)} \hat{X}_\omega R_{(k-1)j} \end{bmatrix} \\
&= \begin{bmatrix} R_{i(k-1)} \hat{X}_\omega & R_{i(k-1)} \hat{X}_v + \hat{p}_{i(k-1)} R_{i(k-1)} \hat{X}_\omega \\ 0 & R_{i(k-1)} \hat{X}_\omega \end{bmatrix} \begin{bmatrix} R_{(k-1)j} & \hat{p}_{(k-1)j} R_{(k-1)j} \\ 0 & R_{(k-1)j} \end{bmatrix} \\
&= \begin{bmatrix} R_{i(k-1)} & \hat{p}_{i(k-1)} R_{i(k-1)} \\ 0 & R_{i(k-1)} \end{bmatrix} \begin{bmatrix} \hat{X}_\omega & \hat{X}_v \\ 0 & \hat{X}_\omega \end{bmatrix} \begin{bmatrix} R_{(k-1)j} & \hat{p}_{(k-1)j} R_{(k-1)j} \\ 0 & R_{(k-1)j} \end{bmatrix} \\
&= \operatorname{Ad}_{g_{i(k-1)}} \operatorname{ad}_{X_k^k} \operatorname{Ad}_{g_{(k-1)j}}
\end{aligned} \tag{A.7}
$$

where we have used Eq. (A.5) and

$$
\hat{p}_{(k-1)j} = (\widehat{R_{(k-1)k} p_{kj}}) + \hat{p}_{(k-1)k}. \tag{A.8}
$$

□

A.1.2 Computing the Jacobian and Its Partial Derivatives

The Jacobian $\bar{J}_i$ of link i is given by

$$J_i(q) = \begin{bmatrix} \mathrm{Ad}_{g_{0\bar{1}}} X_1^1 & \mathrm{Ad}_{g_{0\bar{2}}} X_2^2 & \mathrm{Ad}_{g_{0\bar{3}}} X_3^3 & \cdots & \mathrm{Ad}_{g_{0\bar{i}}} X_i^i & 0_{(n-i)\times 6} \end{bmatrix}. \quad \text{(A.9)}$$

When the partial derivatives of the Adjoint map are found as in the previous section we can also use these to find the partial derivatives of the Jacobian, i.e.,

$$\frac{\partial J_i}{\partial q_k} = \begin{bmatrix} 0_{(k-1)\times 6} & \frac{\partial \, \mathrm{Ad}_{g_{b\bar{k}}}}{\partial q_k} X_k^k & \frac{\partial \, \mathrm{Ad}_{g_{b\overline{(k+1)}}}}{\partial q_k} X_{k+1}^{k+1} & \cdots & \frac{\partial \, \mathrm{Ad}_{g_{b\bar{i}}}}{\partial q_k} X_i^i & 0_{(n-i)\times 6} \end{bmatrix}. \quad \text{(A.10)}$$

A.1.3 Implementation

In this section we will show how to compute the Coriolis matrix for implementation purposes. We will show this for $SO(3)$ for which the Coriolis matrix is given by

$$C(Q, v) = \sum_{k=1}^{n} \frac{\partial M}{\partial q_k} \dot{q}_k - \frac{1}{2} \begin{bmatrix} \widehat{(M(q)v)_{\tilde{V}}} & 0 \\ \frac{\partial^{\mathsf{T}}}{\partial q} \left(\begin{bmatrix} M_{VV} & M_{qV}^{\mathsf{T}} \end{bmatrix} \begin{bmatrix} \tilde{V}_{0b}^{B} \\ \dot{q} \end{bmatrix} \right) & \frac{\partial^{\mathsf{T}}}{\partial q} \left(\begin{bmatrix} M_{qV} & M_{qq}^{\mathsf{T}} \end{bmatrix} \begin{bmatrix} \tilde{V}_{0b}^{B} \\ \dot{q} \end{bmatrix} \right) \end{bmatrix} \quad \text{(A.11)}$$

where $(M(q)v)_{\tilde{V}}$ is the vector of the first three entries of the vector $M(q)v$ (corresponding to $\tilde{V}_{0b}^{B} = \omega_{0b}^{B}$).

We first define the vector

$$\left(M(q)v\right)_{\tilde{V}} = \begin{bmatrix} (M(q)v)_1 \\ (M(q)v)_2 \\ (M(q)v)_3 \end{bmatrix} = \begin{bmatrix} M_{VV} & M_{qV}^{\mathsf{T}} \end{bmatrix} \begin{bmatrix} V_{0b}^{B} \\ \dot{q} \end{bmatrix} \quad \text{(A.12)}$$

which gives

$$\mathrm{ad}_{(M(q)v)_{\tilde{V}}} = \begin{bmatrix} 0 & -(M(q)v)_3 & (M(q)v)_2 \\ (M(q)v)_3 & 0 & -(M(q)v)_1 \\ -(M(q)v)_2 & (M(q)v)_1 & 0 \end{bmatrix}. \quad \text{(A.13)}$$

The lower part of the second part of (A.11) is calculated as

$$\frac{\partial^{\mathsf{T}}}{\partial q}\left(\begin{bmatrix} M_{VV} & M_{qV}^{\mathsf{T}} \end{bmatrix}\begin{bmatrix} V_{0b}^{B} \\ \dot{q} \end{bmatrix}\right)$$

$$= \begin{bmatrix} \frac{\partial (Mv)_1}{\partial q_1} & \frac{\partial (Mv)_2}{\partial q_1} & \frac{\partial (Mv)_3}{\partial q_1} \\ \frac{\partial (Mv)_1}{\partial q_2} & \frac{\partial (Mv)_2}{\partial q_2} & \frac{\partial (Mv)_3}{\partial q_2} \\ \vdots & & \vdots \\ \frac{\partial (Mv)_1}{\partial q_n} & \frac{\partial (Mv)_2}{\partial q_n} & \frac{\partial (Mv)_3}{\partial q_n} \end{bmatrix}$$

$$= \begin{bmatrix} \sum_{i=1}^{3+n} \frac{\partial M_{1i}}{\partial q_1} v_i & \sum_{i=1}^{3+n} \frac{\partial M_{2i}}{\partial q_1} v_i & \sum_{i=1}^{3+n} \frac{\partial M_{3i}}{\partial q_1} v_i \\ \sum_{i=1}^{3+n} \frac{\partial M_{1i}}{\partial q_2} v_i & \sum_{i=1}^{3+n} \frac{\partial M_{2i}}{\partial q_2} v_i & \sum_{i=1}^{3+n} \frac{\partial M_{3i}}{\partial q_2} v_i \\ \vdots & & \vdots \\ \sum_{i=1}^{3+n} \frac{\partial M_{1i}}{\partial q_n} v_i & \sum_{i=1}^{3+n} \frac{\partial M_{2i}}{\partial q_n} v_i & \sum_{i=1}^{3+n} \frac{\partial M_{3i}}{\partial q_n} v_i \end{bmatrix} \tag{A.14}$$

$$\frac{\partial^{\mathsf{T}}}{\partial q}\left(\begin{bmatrix} M_{qV} & M_{qq}^{\mathsf{T}} \end{bmatrix}\begin{bmatrix} V_{0b}^{B} \\ \dot{q} \end{bmatrix}\right) = \begin{bmatrix} \sum_{i=1}^{3+n} \frac{\partial M_{(m+1)i}}{\partial q_1} v_i & \cdots & \sum_{i=1}^{3+n} \frac{\partial M_{(m+n)i}}{\partial q_1} v_i \\ \sum_{i=1}^{3+n} \frac{\partial M_{(m+1)i}}{\partial q_2} v_i & \cdots & \sum_{i=1}^{3+n} \frac{\partial M_{(m+n)i}}{\partial q_2} v_i \\ \vdots & \ddots & \vdots \\ \sum_{i=1}^{3+n} \frac{\partial M_{(m+1)i}}{\partial q_n} v_i & \cdots & \sum_{i=1}^{3+n} \frac{\partial M_{(m+n)i}}{\partial q_n} v_i \end{bmatrix} \tag{A.15}$$

and is thus also given by the partial derivative of the elements of the inertia matrix. We thus only need to compute the partial derivative $\frac{\partial M(q)}{\partial q_i}$ once and use the result in the both in the first and second part of (A.11). This will simplify the computation of the dynamic equations for an arbitrary n-link mechanism mounted on a vehicle.

References

From, P. J., Duindam, V., Pettersen, K. Y., Gravdahl, J. T., and Sastry, S. (2010). Singularity-free dynamic equations of vehicle-manipulator systems. *Simulation Modelling Practice and Theory*, 18(6), 712–731.

Index

P.J. From et al., *Vehicle-Manipulator Systems*, Advances in Industrial Control,
DOI 10.1007/978-1-4471-5463-1, © Springer-Verlag London 2014

MIX
Papier aus verantwortungsvollen Quellen
Paper from responsible sources
FSC® C105338

If you have any concerns about our products,
you can contact us on
ProductSafety@springernature.com

In case Publisher is established outside the EU,
the EU authorized representative is:
Springer Nature Customer Service Center GmbH
Europaplatz 3, 69115 Heidelberg, Germany

Printed by Libri Plureos GmbH
in Hamburg, Germany